Microwave Radio Links

WILEY SERIES IN TELECOMMUNICATIONS
AND SIGNAL PROCESSING

John G. Proakis, Editor
Northeastern University

A complete list of the titles in this series appears at the end of this volume.

Microwave Radio Links

From Theory to Design

Carlos Salema

A JOHN WILEY & SONS, INC., PUBLICATION

Translated from *Feixes Hertzianos,* by Carlos Salema. Published by IST Press, Lisbon, Portugal, 1998.

Published by John Wiley & Sons, Inc., Hoboken, New Jersey.
Published simultaneously in Canada.

Library of Congress Cataloging-in-Publication Data is available.

ISBN 0-471-42026-3

Printed in the United States of America.

10 9 8 7 6 5 4 3 2 1

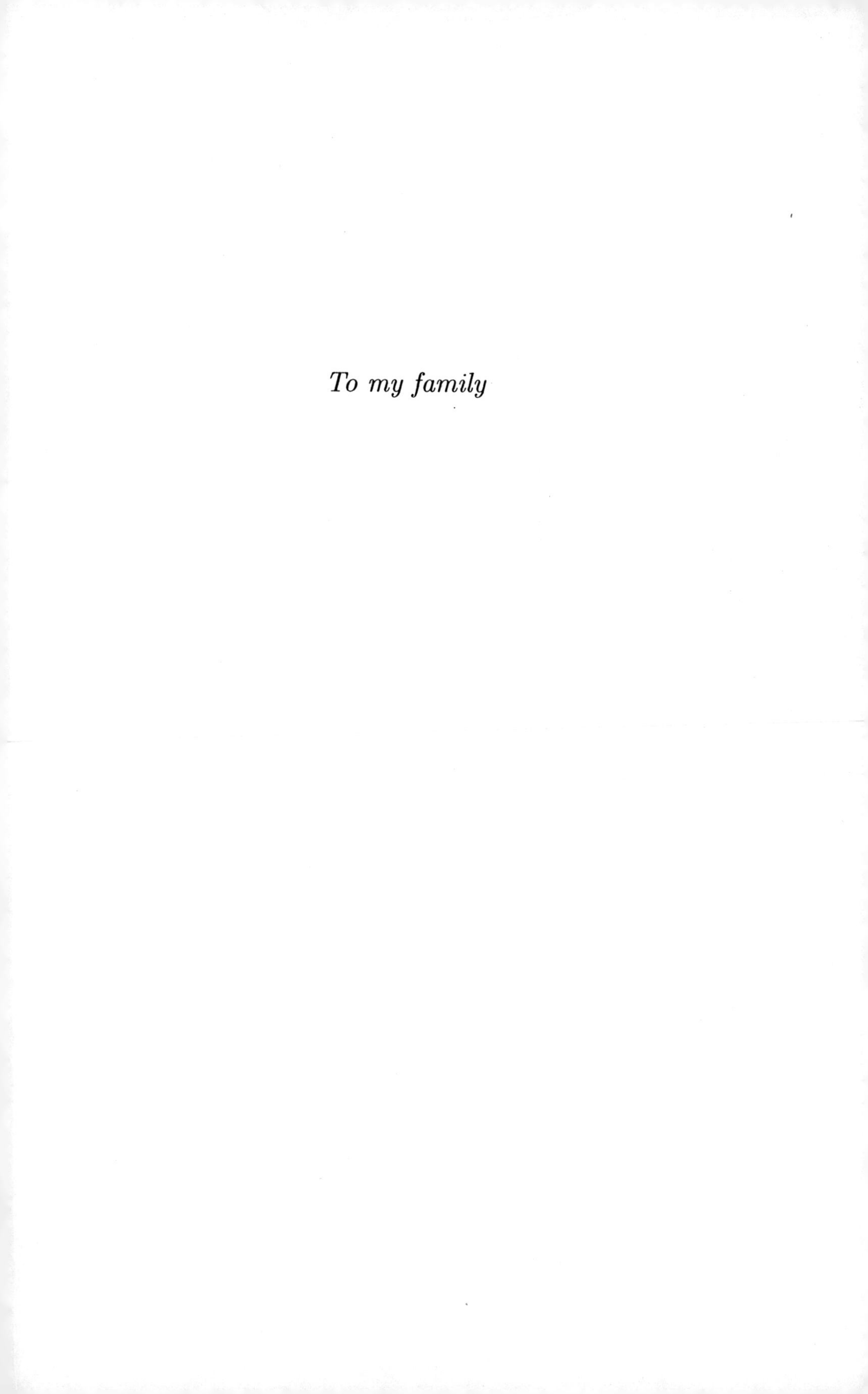

To my family

Contents

List of Figures

List of Tables

Preface

This book incorporates 28 years of teaching experience in telecommunication systems to final year undergraduate students in electrical engineering.

The systems approach to telecommunications appears to be (almost) absent from the curriculum of many engineering degrees, even when the basics are appropriately covered and dealt with.

Thus it does not come as a surprise that many students have difficulties in integrating and inter-relating knowledge acquired in widely different subjects and to fill in unavoidable gaps. In addition undergraduate students are rarely exposed to international standards and recommendations and have little familiarity with the issues of quality in telecommunication services.

All the above shortcomings may be overcome with on-the-job training, but experience shows the systems approach is highly motivating and tends to be appreciated by prospective employers, especially for the first job.

Project work is a necessary part of the systems approach. As is usually the case in engineering, project work makes use of a number of models of reality from the simplest to the most complex. Easy access to sophisticated computer based models often drives the student away from the basic phenomena, and what may be worse, severely limits judgement on the validity of the model and the quality of the results.

In this book we take the subject of microwave radio links, which is rapidly regaining interest particularly in relation to cellular radio and broadband fixed wireless access, and try to provide a solution to both issues: integration of different subjects to solve a real-life problem, emphasis on the basics and on model limitations.

The book offers comprehensive coverage of the subject, from antennas and radio propagation to modulation and noise, from basic economics to reliability and quality standards. In each topic we start from the basic models, where a nomogram or a simple calculator suffices, and proceed to more sophisticated models where software packages are required.

It is obviously impossible to include every subject, but at least, an attempt has been made to cover most topics in an integrated way.

Acknowledgments

I would like to express my deepest thanks to all who, directly or indirectly, contributed to this book, namely:

- My students, for whom this book was written, and to whom I am indebted for many suggestions;
- To João Paulo Cunha, José Costa Leal, Luis Rosa Santos, Eduardo Morgado, José Neto de Aguiar and many others, whose names I can no longer remember, for valuable information and remarks;
- To my colleagues, namely Profs. António Rodrigues, Carlos Fernandes, Carlos Fernandes, Fernando Pereira, Francisco Cercas, João Pires, João Luís Sobrinho, Mário Figueiredo and Paula Queluz and Jaime Afonso and Paulo Correia, for their help in revising the text;
- To my colleagues Profs. Joaquim Moura Ramos and Eduardo Borges Pires, who help to obtain the required copyright permissions;
- To Prof. Martin Tomlinson, from Universidade de Plymouth, for the encouragement and support in the writing Chapter 6;
- To Mr. Paulo Tribolet, who prepared most of the figures;
- To Prof. Francisco Miguel Dionísio for sorting out quite a few software installation problems;

- To Professor Donald Knuth and Leslie Lamport, the authors of the TEX and LATEX , the software used to prepare the current (English) version of this book and the original version (in Portuguese) which, for the last 20 years, survived more than 10 different computers and operating systems, without requiring any modification;
- To the International Telecommunications Union (ITU), for the copyright permission CPY/LAU/97/19, for the figures referred to in the text, whose selection is the sole responsibility of the author [1];
- To Andrew Inc., for the permission to reproduce, from catalog 37, the figures mentioned in the text;
- To Valmont/Microflet, 3575 25th Street S.E., Salem, Oregon 97302, USA, for the permission to reproduce the figures referred to in the text;
- To Instituto Superior Técnico and to Instituto de Telecomunicações, for the logistics, the environment and the working conditions provided;
- To my wife, Helen, for her immense love and forbearance.

[1]The Recommendations mentioned in the text may be acquired at the at the International Telecommunication Union, General Secretariat - Sales and Marketing Services, Place des Nations - CH - 1211 Genève 20, Suisse

1

Introduction

1.1 MICROWAVE RADIO RELAY LINKS

Above about 30 MHz the ionosphere is no longer the dominant factor in setting up radio links. As the frequency increases, and the wavelength decreases, practical antennas get more and more directive and most of the transmitted power is confined to a progressively narrowing beam.

Propagation is normally along line-of-sight, that is, between antennas that "see" each other. When the link length exceeds about 50 kilometers, or the terrain imposes it, intermediate relay stations, which operate as repeaters, become necessary.

Very high capacity (over 1000 telephone channels) links may be established at frequencies above about 1 GHz given the large bandwidth available. The absence of a physical connection makes these links attractive, when compared with fiber optic cables, for long distances (over some tens of kilometers), particularly when there are obstacles, such as rivers and mountain ridges, on the path.

This book introduces the reader to the design of microwave radio links. It is conceived as a textbook for a final year graduate course (or an M.Sc. subject) in Telecommunications Engineering, and assumes that the reader has mastered the basics of propagation and radiation of electromagnetic waves and of modulation and noise.

After a brief historic review, this chapter describes a typical microwave radio link, introduces the techniques and the hardware, and presents the principles upon which frequency plans are based.

Chapter 2 deals with propagation in the atmosphere in stable conditions, that is, in conditions which are valid, on average, for at least 10 s, including the effects of the presence of the Earth, obstacles, atmospheric gases and rain. Although emphasis is on line-of-sight links and links with propagation by diffraction, troposcatter links are referred to briefly.

Varying propagation conditions, leading to fading, and diversity techniques to counteract fading are described in Chapter 3.

Chapters 3 and 5 are devoted to link performance, as a function of the received power, for analog and digital links, respectively, in accordance to the applicable recommendations for international links.

Chapter 6 deals with error control codes which are becoming increasingly popular in digital links as a measure to continuously monitor link performance and, in some cases, to correct transmission errors.

Finally, Chapter 7, the principles and theory presented in the previous chapters are applied and the reader is introduced to the design of microwave radio links.

In the appendices we include:

- detailed frequency plans for the main microwave bands;
- the derivation of the output signal-to-noise ratio for frequency and phase demodulators as a function of the input carrier to noise ratio and the comparison, under various criteria, of these two types of modulation;
- typical costs of microwave radio links components;
- link calculations for a typical microwave radio link.

Recommendations of the International Telecommunications Union – Radiocommunications, referred to by the acronym ITU-R, that replaced the International Commission for Radiocommunications (CCIR)[1] are often quoted, sometimes extensively. Less frequently, recommendations from the International Telecommunications Union – Telecommunications, ITU-T, previously CCITT, related to transmission systems in general, are also referred to. These texts were previously revised, once every four years, albeit not necessarily in full. At present, between meetings of the General Assembly, new recommendations may be approved and existing ones may be revised by written procedure. As far as possible the latest available edition, on CD-ROM, dated September 2000, is used although references to earlier versions are made when necessary. In many cases reasoning behind ITU-R recommendations is discussed in detail.

Although focusing on point-to-point links the basic principles dealt with in this book remain all important for radio communication systems in general, for satellite communications and for mobile communications. Space (and time)

[1]From the French *Commission Consultive Internacional pour les Radiocommunications.*

have imposed some selection of the topics included. In spite of its importance, in drawing up frequency plans, the subject of interferences is not discussed in detail.

Formerly almost restricted to point-to-point long-haul links, microwave radio links have, recently, been losing part of their former importance in the network backbone, mainly due to the progress in optical communications systems, but are increasingly being used for applications such as the wireless local loop (WLL), fixed wireless access (FWA) and broadband wireless access.

1.2 HOW IT ALL STARTED[2]

The first microwave radio link was possibly one set up in 1931 across the English Channel, between Calais (in France) and St. Margarets Bay (in England). With capacity for one telephone or one telegraph channel, it was operated at 1700 MHz, used 3 m parabolic antennas and 1 W amplitude modulation transmitters. Two years later, in 1933, the first permanent link was installed between the airports of Lympne (England) and St. Inglevert (in France).

Difficulties in achieving adequate multi-channel performance with amplitude modulation led to the first experiments with frequency modulation, at the time a relatively unknown technique.

In 1936, the British General Post Office opened the first multi-channel link 65 kilometers long, over the sea, between Scotland and Northern Ireland. It was operated at 65 MHz, used amplitude modulation and carried 9 voice channels.

In Germany the first single-channel systems, operating at 500 MHz, were produced by Lorenz and Telefunken, in 1935. In 1939, frequency modulation was introduced and the Stuttgart system deserves a special reference. It boasted a ten-voice channel capacity, operated at 1300 MHz and was initially equipped with a magnetron, later (1942) substituted by a traveling wave tube.

In the United States, Bell Laboratories started experimenting with multi-channel systems in 1941 and tested a 12-voice channel, amplitude modulation system, between Cape Charles and Norfolk, both in Virginea.

The end of the war, in 1945, enabled the commercial use of microwave links. Required capacities were however well beyond the 6- to 12-voice channels typical of military equipment.

In 1945, the American Telegraph and Telephone Corporation (AT&T), had in operation a multi-channel frequency modulation system, with a number of voice channels and a music channel, between New York and Philadelphia. A new 100-telephone channel link, at 4 GHz , between New York and Boston, followed in 1947. This equipment was the basis for an improved one, known

[2]This section is based on Helmut Carl [2].

as TD-2, which was used to set up, in 1949, the first permanent television link between New York and Philadelphia. One year later the total length of the network of television links had reached 12 000 kilometers. In 1955, AT&T started the development of an 1800-channel system at 6 GHz.

Meanwhile, in Europe, analog systems with increasing capacity were being developed: at first 60-voice channels, then 240, 300, 480 and 600 channels, operating at 2, 4 and 6 GHz.

Time division multiplexing was making its debut in France in 1932. The first patents on pulse modulation and pulse coded modulation, due to Deloraine and Reeves, date from 1938. From 1945 onwards, International Telegraph and Telephone Corporation (IT&T) developed 24 telephonic channels digital systems.

The first interconnection of European microwave radio links for television took place in 1953, for the coronation of Queen Elizabeth II.

The sixth CCIR Plenary Assembly, gathered in Geneva in 1951, approved Recommendation 40 (335, since 1963) which considered microwave radio links (above 30 MHz) equivalent to metallic circuits in order to achieve CCITT recommended performance for international telephone circuits. When this recommendation was approved it was by no means sure that it would be met in the foreseeable future, as it was the case five years later.

In the eighth CCIR Plenary Assembly (Warsaw, 1956) it was possible to set up the basic recommendations for microwave radio links which were revised and improved in the Los Angeles (1959) and Geneva (1963) meetings.

Although the CCIR Plenary Assemblies after 1966 introduced few changes, the 1986 Plenary Assembly decide to redraw some of the more important recommendations related to digital links.

After 1960 the network of microwave radio links, at the time mostly analog, increased steadily and in many regions become the backbone of interurban telecommunication networks.

Television coverage implies a dense network of microwave radio links between the production and control centers and the main transmitters. In addition, mobile microwave radio links are required to link mobile studios used to transmit live programs and the production centers.

As stated before, digital links were initially limited to the lower capacities, known as T1 and E1, respectively, for 24 and 30 voice channels, and binary rates of 1.5 and 2 Mbit/s[3]. This was mainly due to the difficulty of producing reliable and economic electronic circuits, with a large number of active components, based on vacuum tubes. Later, the introduction of solid-state devices and, above all, large and very large scale integrated circuits, able to concentrate in a small volume, with low power consumption and at a mod-

[3]According to the ITU-T Recommendation G.701, Mbit/s is the abbreviation of "Million Binary Digits per second".

erate cost, a very large number of active devices made high-capacity digital microwave radio links very attractive.

From the mid-1770s onwards, digital microwave radio links of increasing capacity came on the market. A few years later digital microwave radio links up to the first synchronous hierarchy (STM-1), for up to 2349 voice channels and 155.52 Mbit/s) were available.

The use of digital microwave radio links for television started on the early 1990s. This is due to the high binary rate (216 Mbit/s) required to transmit, without compression, a color television signal, according to the ITU-R[4] Recommendation BT.601-5, with contribution quality (links between production centers). Reducing this very high binary rate without significantly impairing the quality is still an active research subject. Available codecs[4] achieve distribution quality (slightly inferior to contribution quality) with binary rates between 5 and 10 Mbit/s. In the beginning these codecs were bulky and very expensive devices but, once the signal processing algorithms were standardized, it became economically feasible to develop special purpose integrated circuits enabling a significant reduction in bulk and cost. With the progressive digitalization of television studios and given the advantages of digital microwave radio links it is most likely that they will soon replace existing analog links in television chains.

1.3 MICROWAVE RADIO LINK DESCRIPTION

For the reader to grasp the main features of a microwave radio link, we will start by a short description of a typical interurban, point-to-point, high capacity link, with a total length of 300 km, operating in the 6 GHz frequency band.

The end stations are located in hill-tops or mountain-tops to achieve line-of-sight paths. These places must be fairly close to the sources and destinations of the traffic and, in as far as possible, should have easy access. In order to minimize the number of repeater stations, line-of-sight paths as long as possible are sought. Easy access and availability of a reliable electric power supply is highly desirable but difficult to come by in many cases.

The end stations are linked to the traffic source and destination centers by multiplex circuits, either frequency division or time division, on coaxial, or more frequently now, on optical fiber cables.

Figure 1.1 shows a much simplified block diagram of the link where, for simplicity, only one repeater is represented.

[4]According to ITU-T Recommendation G.701 [5] a codec is a set of a coding and a decoding device, that is a device that transforms the original analog signal into a digital one and another that rebuilds (decodes) the original signal from the digital signal. the transformation of the analog signal includes sampling, quantization and coding of samples.

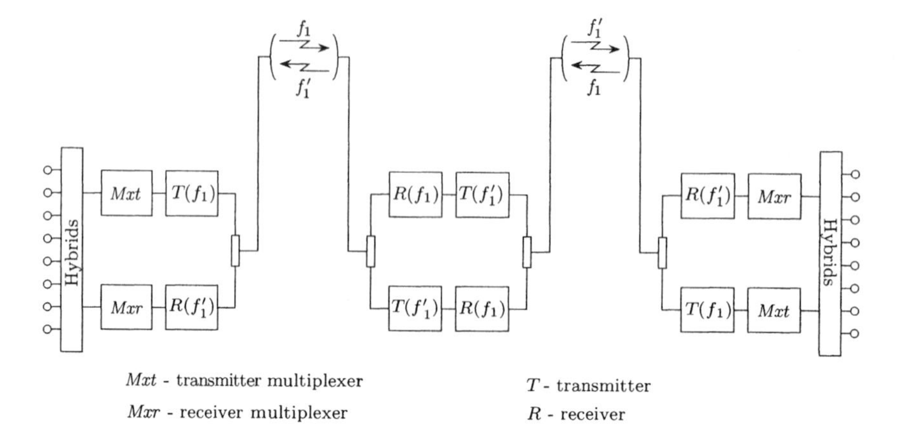

Fig. 1.1 A much simplified block diagram of a microwave radio link with one repeater.

In most cases link capacity is not concentrated in a single transmitter-receiver, as shown in Figure 1.1 but divided into m identical transmitter-receiver sets, with an additional *hot* [5] standby set, capable to take over any of the other sets, automatically, in case of malfunction, with no service disruption. This configurations is usually known as an $m+1$, or in general an $m+r$, where m stands for the number of working systems and r for the number of reserve systems. Reserve systems are customary in microwave radio links included in the public network even when there is a single operating system ($1+1$ configuration).

In a given station all transmitters and receivers involved in the link with the next station use the same antenna, mounted on a tower or a mast[6]. According to the station importance and the antenna height above ground level, towers may be (see Figure 1.2):

- simple, self-supporting, metallic structures, for heights up to about 6 m;
- simple, guyed, metallic structures for heights up to about 100 m;
- moderately complex, self-standing metallic structures for heights up to about 100 m;
- complex, metallic, concrete or metallic and concrete, self-standing structures for heights between about 30 and 300 m.

[5]Opposite to hot standby we may define cold standby as a standby system that requires a definite amount of time to take over a working system.

[6]Guyed towers are usually known as masts.

Low-height metallic self-supporting tower

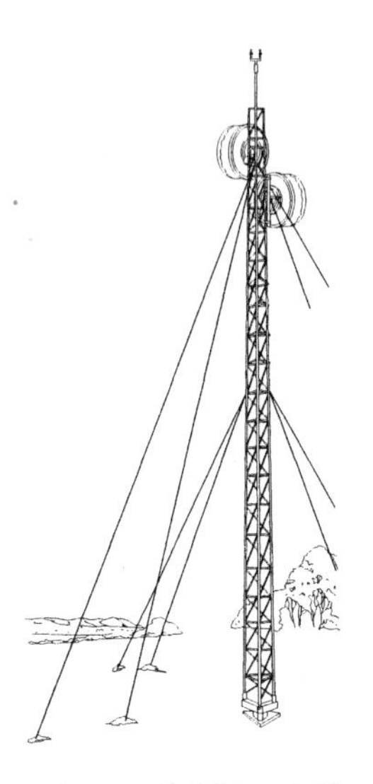

Average-height metallic guyed tower (mast)

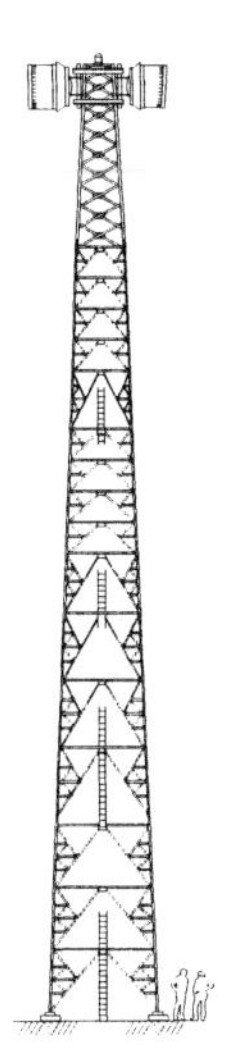

High metallic self-supporting tower

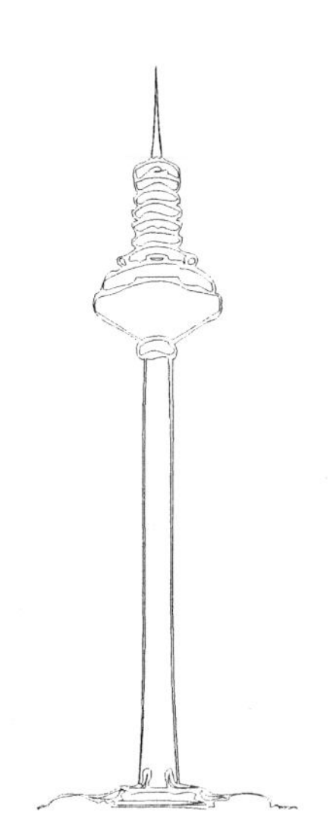

High concrete tower

Fig. 1.2 Antenna towers for microwave radio links (adapted from [1]).

When the tower is a simple structure, transmitters and receivers are usually installed in a building or simply a container at the tower base. In large stations, with tower heights above about 30 m, it often pays to locate the radio equipment close to the antennas near the tower top.

Transmitters and receivers are connected to the antenna by coaxial cable or, for frequencies above about 2 GHz, by metallic waveguides. Filters and, for frequencies above 2 GHz, associations of filters and circulators, as shown in Figure 1.3, enable sharing of a single antenna by up to four transmitters and/or receivers.

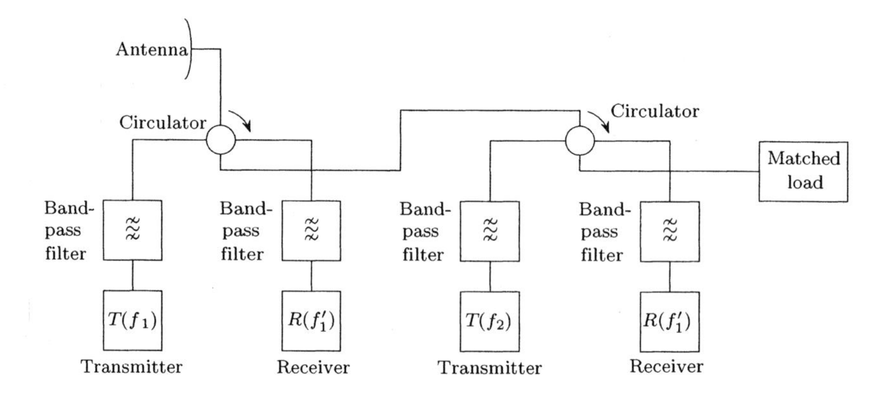

Fig. 1.3 Circulators and filters enable two transmitters and two receivers to share the same antenna.

For frequencies above about 1 GHz most antennas are reflectors fed at the focus by a gooseneck open-ended waveguide (Figure 1.4).

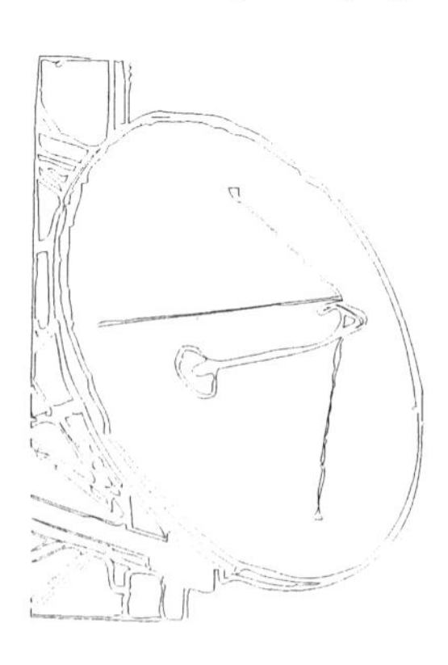

Fig. 1.4 Parabolic antenna fed at the focus (adapted from [1]).

The reflector is a spun parabola with a diameter between 1 and 4 m. This type of antenna, made popular by satellite home receivers, is usually known as a parabolic reflector.

In special cases hog-horn antennas may be used. As shown in Figure 1.5 the hog-horn antenna can be thought of as an association of a pyramidal horn with a parabolic reflector.

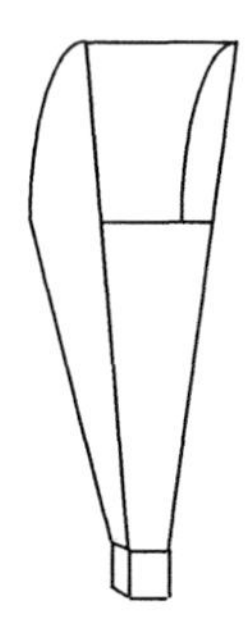

Fig. 1.5 Hog-horn antenna.

For frequencies below about 1 GHz and according to the required polarization, bandwidth or gain, Yagi-Uda antennas and helix antennas, either single or in arrays, with or without reflector planes, are the most common solutions.

Transmitters and receivers are full solid state, unless output powers in excess of a few tens of watts are required, in which case transmitter power stages may use traveling wave tubes (twt).

Receivers are mostly double conversion heterodyne, with intermediate frequencies at about 1 GHz and 70 MHz. For simplicity and low cost, radio frequency amplifier stages are often omitted, unless very low (below about 5 dB) noise factors are required.

In some cases transmitters (and receivers) may consiste of two separate units: one handling signals from base band up to and including the intermediate frequency (or IF) and the other for the radio frequency stages, installed close to the antenna. Interconnection of the two units at the intermediate frequency (often at 70 MHz or 140 MHz) using coaxial cable minimizes feeder losses.

Since each radio section uses a pair of frequencies (transmit/receive), both receivers and transmitters are crystal controlled and factory tuned. Nowadays synthesizers have become more and more common enabling on-site adjustment of the operating frequency.

As referred to in Section 1.2 microwave radio links may transmit either analog or digital signals. The most important analog signals, in terms of bandwidth, are television and frequency multiplexed telephony, where the baseband may reach 10 MHz. Digital signals used to be mostly time multiplexed telephony, but nowadays include from multiplexed telephony, to high-speed (broadband) data and video.

Station siting for microwave radio links being mostly determined by propagation conditions, it rarely coincides with traffic origin or destination. Since multiplexing equipment (multiplexers), both analog and digital, tends to be placed near traffic origin and destination, a coaxial or fiber optic link between both is often required.

In relay stations transmitters and receivers may be interconnected in a number of ways, offering increased complexity and flexibility:

- intermediate frequency;
- baseband;
- (telephonic) channel or group of channels.

Intermediate frequency interconnection is the simplest and most economic of all forms, but the transmitted signal is not available at the relay stations. For analog systems it has the advantage of not requiring modulators and demodulators and thus it minimizes intermodulation noise. For digital systems it avoids pulse recovering and thus leads to a considerably lower performance.

Baseband interconnection implies the use of modulators and demodulators in relay stations thus making the baseband signal available. It is often used in analog television radio links, because it simplifies the connection of a mobile link to the fixed network, and in digital links whenever digital signals are to be recovered.

Interconnection at the level of channel or group channels, requires, in addition to modulators and demodulators, multiplex equipment. It is only used when the relay station doubles as an end station for local traffic.

Synchronous digital hierarchies (also known as SDH) have an important advantage over the older plesiochronous digital hierarchies (or PDH) in that they enables the use of simple add-and-drop multiplexer (between any hierarchy levels) rather than the full set of multiplexer-demultiplexer units previously required.

Besides point-to-point links, as the one previously described, microwave radio links may also be used in a point-to-multipoint configuration, where a number of remote stations is served by a single base station. Point-to-multipoint radio systems are not symmetric as point-to-point and thus two directions must be considered: downstream, that is, from the base to the remote stations, and upstream.

In point-to-multipoint systems, the base station antenna is either omnidirectional (in the horizontal plane), defining a cell area of coverage, or slightly more directional covering a sector of the cell. In the downstream direction the same signal is broadcast to all remote stations in the entire cell (or sector), whereas in the upstream direction the available spectrum is shared between active remote stations, by a multiple access scheme.

The main types of multiple access schemes are:

- Frequency division multiple access (FDMA), where the available bandwidth is divided into channels, each to be dedicated exclusively for transmission to one active remote station;

- Time division multiple access (TDMA), where the same bandwidth is shared by all remote stations, which are allows to transmit only during their allocated time slot;

- Code division multiple access (CDMA)). where all remote stations may transmit simultaneously in the same bandwidth and are separated by coding[7].

1.4 FREQUENCY PLANS

Each transmitter-receiver pair of a unidirectional link, plus its antennas, waveguides and the propagation medium between the antennas, is known as a radioelectric section. In a radioelectric section the carrier, modulated by analog or digital signals, occupies one radio channel (or simply one channel). In a bidirectional link one channel is known as the go-channel and the other as the return-channel. The distribution of channels in the available frequency spectrum follows the ITU-R recommendations which strive for a compromise between waste of a valuable and scarce resource (*the radio spectrum*) and equipment costs.

The radio spectrum is divided in frequency bands which are allocated on an international basis for such uses as terrestrial fixed and mobile, maritime and satellite communications, navigational aids, broadcast, radio astronomy and amateur radio. The band structure is adjusted every four years at the World Administrative Radio Council (or WARC).

ITU-R divides the world into three regions. Region 1 includes Europe, Africa and part of Asia (ex-USSR). Region 2 covers North, Central and South America, and Region 3 India, Southeast Asia, Japan and Australia. Within each region national (and regional) agencies are responsible for detailed spectrum management.

Two bodies are influential in setting regional plans: the Conference of European Post and Telecommunications (CEPT) in Europe and the Federal Communications Commission (FCC) in the United tates of America.

ITU-R frequency plans are based on the following principles:

[7]CDMA may be thought of as a room where a number of widely different, and unrelated, languages are simultaneously being used by different people to communicate. If each subset of persons can only speak and understand one language, communication is possible even if the same frequency band is used simultaneously by all participants.

- Radio sections corresponding to each of the directions of a bidirectional link should use different channels; although it is theoretically possible to share the same channel this sharing is difficult to implement.

- Adjacent radio sections, belonging to the same link, cannot use the same go-channel, due to the risk of feedback between the transmitter and the receiver at the relay station, as antennas are normally not capable of providing sufficient front-to-back isolation (see Figure 1.6).

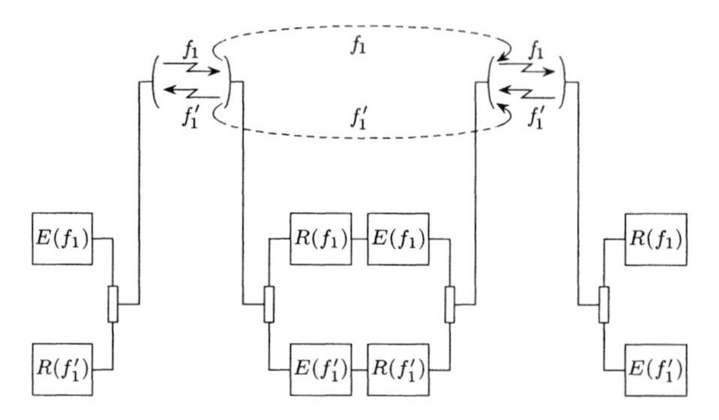

Fig. 1.6 Adjacent radio sections belonging to the same link cannot use the same go channel.

- Adjacent radio sections, belonging to the same link, may use the same channels, provided the go channel in one radio section become the return channel in the other section and vice-versa (see Figure 1.1).

- Two frequency plans, in which adjacent radio sections alternately make use of the same go and return channel may suffer from interference between channels in non adjacent sections, in abnormally good propagation conditions, if the directions of propagation are very similar (see Figure 1.7).

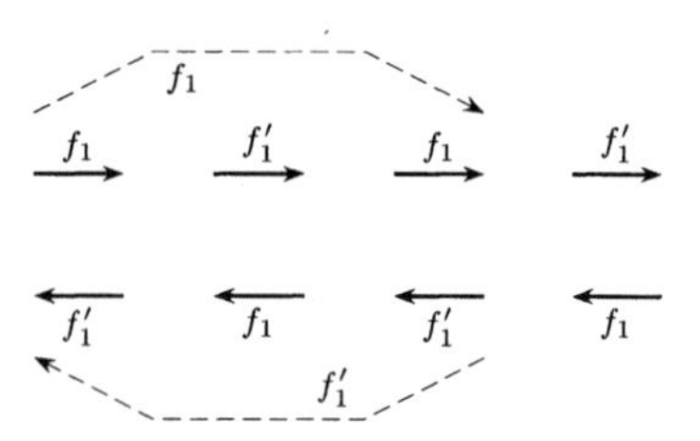

Fig. 1.7 Risk of interference in two-frequency plans.

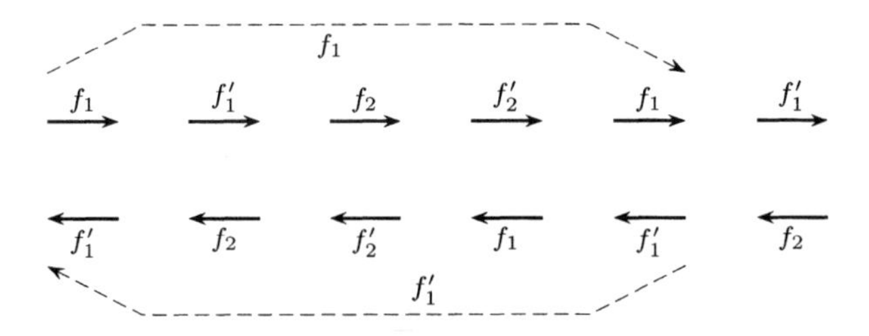

Fig. 1.8 A possible four-frequency plan.

- In the rare cases where a two-frequency plan is not suitable, it is possible to implement a plan using more frequencies, for instance four, increasing the distance between sections using the same channel in the same direction and thus decreasing the risk of interference (see Figure 1.8).

Link frequency plans, particularly for international links, are organized so as to achieve the best possible use of the radio spectrum and to ease system interconnections. These plans are based on the ITU-R Recommendation F.746-4 [4]. The analysis of those frequency plans enables us to identify a few common principles:

- The available bandwidth, in each frequency band (usually of a few hundreds of MHz) is divided in two approximately equal sub-bands, within which radio frequency channels, or simply channels are defined. In each station, and for each direction, all transmitter channels are located in the same sub-band and all receiver channels in the other sub-band.
- Each radio frequency channel is identified by an integer, n for the go channels and n' for the return channels.
- The central frequency for channel n is denoted by f_n, for channels in the lower sub-band, and f_n' for channels in the upper sub-band.
- Spacing — absolute value of the difference between the center frequencies of adjacent channels — is constant in each sub-band. Spacing between the upper channel in the lower sub-band and the lower channel in the upper sub-band is, generally, higher than the spacing within the sub-bands.
- Some frequency plans foresee the installation of auxiliary links, in the same band as the main links, occupying the upper and lower edges of the available sub-bands and (or) the central gap.
- In order to decrease channel spacing without pushing filter channel specifications too hard, some frequency plans make use of orthogonal polarizations, which in terrestrial microwave radio links, are restricted to horizontal and vertical polarizations.

- When the same antenna is shared by various (3 or 4) channels, these are usually distributed within the whole band, as shown in Table 1.4.

Number of go channels	Channels in the same antenna
6	1,3,5 2,4,6
8	1,3,5,7 2,4,6,8
12	1,5,9 2,6,10 3,7,11 4,8,12
16	1,5,9,11 2,6,10,14 3,7,11,15 4,8,12,16
20	1,8,15 2,9,16 ... 6,13,20

Table 1.1 Preferred channel groupings in the same antenna, as a function of the number of go channels defined in the band.

According to the use of polarizations and channel spacing, frequency plans may be classified as:

- simple, when a single polarization is used;
- alternated or interleaved band re-use, when adjacent channels have a non zero spacing and use orthogonal polarizations;
- co-channel band re-use, if channels are grouped in pairs, sharing the same central frequency with polarization discrimination.

Alternated or interleaved band re-use frequency plans may be:

- simply alternated or simply interleaved, when the corresponding go and return channels use the same polarization, thus enabling the same polarization to be used in each hop;
- doubly alternated or doubly interleaved, when corresponding go and return channels use orthogonal polarizations.

The layout of radio channels in simple, simply interleaved, doubly interleaved and co-channel band re-use frequency plans is shown in Figure 1.9.

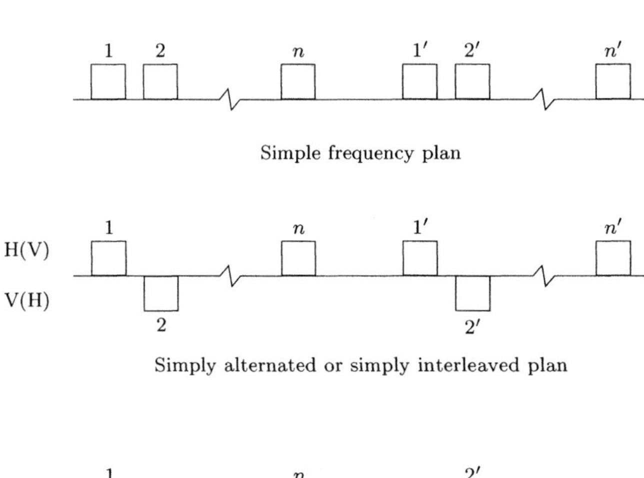

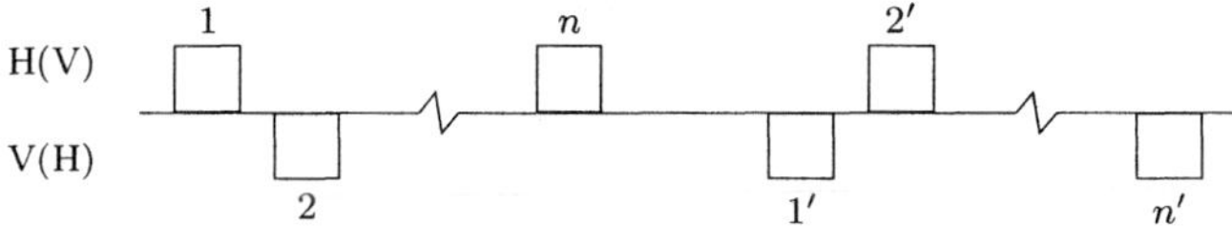

Doubly alternated or doubly interleaved plan

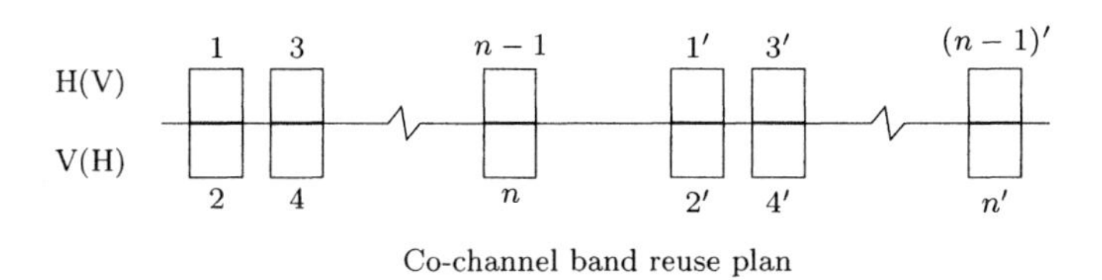

Co-channel band reuse plan

Fig. 1.9 Layout of radio channels in simple, simply interleaved, doubly interleaved and co-channel band re-use frequency plans.

Whenever the type of interleaved frequency band re-use plan is omitted we will assume that both are acceptable.

For very low capacity microwave radio links national authorities may specify frequency bands below 1 GHz. For all the other cases, line-of-sight or near line-of-sight microwave radio links use the 1.4, 2, 4, 5, 6 (lower and upper), 7, 8, 10, 11, 12, 13, 14, 15, 18, 23, 27, 31, 38 and 55 GHz frequency bands, where radio channels are organized as detailed in Annex A. It is worth mentioning that recommendations referred to are applicable only to international links but they are very often applied to national links too.

Trans-horizon radio links with propagation by tropospheric scattering make use of frequency bands reserved for fixed services, but frequency allocation is performed on a case-by-case basis by the concerned administrations, taking into account ITU-R Recommendation F.698-2 [4].

REFERENCES

1. Andrew Corporation, *Andrew Catalog 37*, Andrew Co., Orland Park, IL, 1997.
2. Carl, H., *Radio Relay Systems*, Macdonald and Co., London, 1966.
3. Huurdeman, A., *Radio-Relay Systems*, Artech House, Boston, 1995.
4. ITU, *ITU-R Recommendations on CD-ROM*, UIT, Geneva, 2000.
5. ITU, *ITU-T Recommendations on CD-ROM*, UIT, Geneva, 1997.

2

Propagation

2.1 FREE SPACE PROPAGATION

In point-to-point radio link design the first problem is the derivation of an expression linking the transmitter output power p_T and receiver input power p_R.

To solve this problem let us consider a very simple model shown in Figure 2.1, which consists of two antennas, immersed in free space, in vacuum. The transmitter is connected to one of the antennas, through a filter, using a waveguide. The receiver is connected to the other antenna, also through a filter and a waveguide.

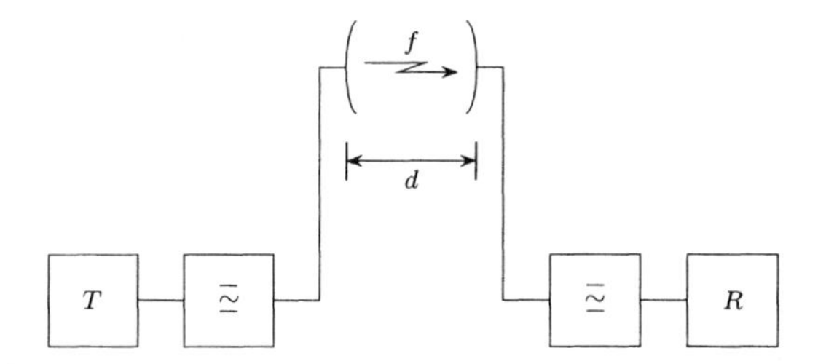

Fig. 2.1 Simplified scheme of a unidirectional point-to point radio link in free space.

Let:

- d – distance between antennas;

- f – operating frequency;
- a_T,a_R – transmitter and receiver filter plus waveguide gains[1];
- g_T – transmitter antenna gain in the direction of the receiver antenna, relative to the isotropic loss less antenna;
- g_R – receiver antenna gain in the direction of the transmitter antenna, relative to the isotropic loss less antenna.

If the two antennas are far away enough from each other, the power density Σ_R created by the transmitter antenna at the receiver antenna is given by

$$\Sigma_R = p_T \, a_T \, \frac{g_T}{4\pi d^2}. \tag{2.1}$$

The product $p_T a_T g_T$ is sometimes known as the equivalent isotropic radiated power (in short *eirp*) of the transmitter. The *eirp* concept is mostly used in satellite communications specifications.

The available power at the receiver input may be derived from the power density at the receiver antenna as

$$p_R = \Sigma_R \, a_R \, a_{ef_R}, \tag{2.2}$$

where a_{ef_R} stands for the effective receiver antenna aperture in the direction of the transmitter antenna.

The effective aperture a_{ef_R} and the gain g of the receiver antenna are related by

$$g = \frac{4\pi}{\lambda^2} \, a_{ef_R}, \tag{2.3}$$

where λ is the free space wavelength corresponding to the operating frequency f.

Substituting the value of Σ_R given in (2.1) and the value of a_{ef_R} from (2.3) in (2.2) we get a relation between the receiver input power and the transmitter output power, also known as the Friis formula [25]

$$p_R = p_T \, a_T \, a_R \, g_T \, g_R \left(\frac{\lambda}{4\pi d} \right)^2. \tag{2.4}$$

We will denote the free space attenuation, or free space basic transmission loss, as a_0

$$a_0 = \left(\frac{\lambda}{4\pi d} \right)^2. \tag{2.5}$$

Recalling that

$$\lambda = \frac{c}{f}, \tag{2.6}$$

[1] In practice, these gains are smaller than unity and thus correspond to losses.

where c is the free space propagation velocity of electromagnetic waves in vacuum, the free space attenuation becomes

$$a_0 = \left(\frac{c}{4\pi f d}\right)^2. \tag{2.7}$$

For convenience, microwave radio link design quantities are usually expressed in logarithmic units (dB). Choosing an arbitrary (but often 1 W or 1 mW) value of power p_o we define

$$\begin{aligned}
P_R &= 10\ \log_{10}\left(\frac{p_R}{p_o}\right),\\
P_T &= 10\ \log_{10}\left(\frac{p_T}{p_o}\right),\\
G_T &= 10\ \log_{10}(g_T),\\
G_R &= 10\ \log_{10}(g_R),\\
A_T &= -10\ \log_{10}(a_T),\\
A_R &= -10\ \log_{10}(a_R),\\
EIRP &= P_T - A_T + G_T.
\end{aligned}$$

To avoid misunderstandings, in the following we will generally use lower-case letters to represent variables expressed in linear units and upper-case letters for variables expressed in logarithmic units. The few exceptions to this rule will be duly marked.

Taking decimal logarithms in (2.5) and making use of variables expressed logarithmic units we have

$$P_R = P_T - A_T - A_R + G_T + G_R - A_0, \tag{2.8}$$

where

$$A_0 = -20 \log_{10}\left(\frac{\lambda}{4\pi d}\right), \tag{2.9}$$

or

$$A_0 = -20 \log_{10}\left(\frac{c}{4\pi f d}\right). \tag{2.10}$$

Substituting c by 2.998×10^8 m/s, expressing the frequency f in GHz and the distance d in kilometers we have[2], from (2.10)

$$A_0 = 92.4 + 20\ \log_{10}(d_{[\mathrm{km}]}) + 20\ \log_{10}(f_{[\mathrm{GHz}]}). \tag{2.11}$$

The nomogram in 2.2 may be used to estimate free space attenuation for usual values of distance and frequency.

[2]With the frequency in MHz we have

$$A_0 = 32.448 + 20\ \log_{10}(d_{[\mathrm{km}]}) + 20\ \log_{10}(f_{[\mathrm{GHz}]}).$$

ITU-R[19] Recommendation P.525-2 rounds 32.448 to 32.4. Previous versions of the same Recommendation adopted 32.5.

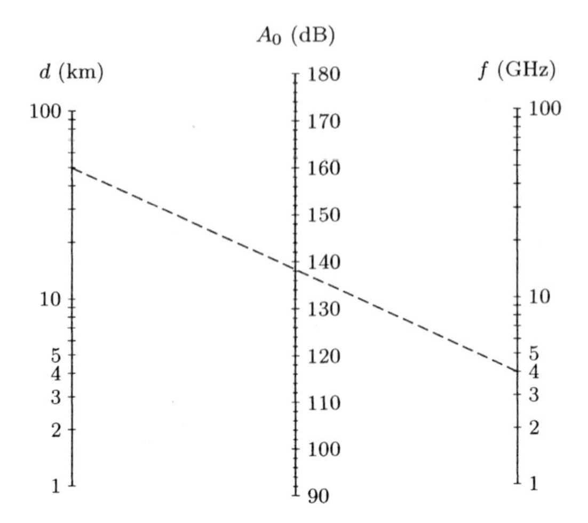

Fig. 2.2 Nomogram showing free space attenuation A_0 as a function of the distance d (in km) and frequency f (in GHz). In the example $d = 50$ km, $f = 4$ GHz and $A_0 \approx 138$ dB.

As it was already stated, the previous expressions assume that antennas are sufficiently apart from each other. It is usual to take as the minimum distance d_{min} the far-field distance of the largest antenna

$$d_{min} = \frac{2\, d_a^2}{\lambda}, \tag{2.12}$$

where d_a is the largest aperture dimension of either the transmitter or the receiver antenna. The nomogram in Figure 2.3 may be used to calculate d_{min}, as a function of frequency f, for different values of d_a.

For parabolic reflector antennas, commonly used in microwave radio links, gain G relative to a loss less isotropic antenna and expressed in base 10 logarithmic units (dBi), diameter d_a (in m) and wavelength λ (in m) or frequency f (in GHz), may be derived from (2.3), substituting a_{ef_R} by its value in terms of the diameter d_a

$$G = 20\,\log_{10}\left(\frac{\pi d_a}{\lambda}\right) + 10\,\log_{10}\eta, \tag{2.13}$$

or

$$G = 20\,\log_{10}\left(\frac{\pi\, d_a\, f}{0.3}\right) + 10\,\log_{10}\eta, \tag{2.14}$$

where η is the antenna aperture efficiency.

The nomogram in 2.4 may be used to estimate the gain of a parabolic reflector antenna as a function of frequency, for some common antenna diameters and a typical aperture efficiency of 0.5 (or 50 percent).

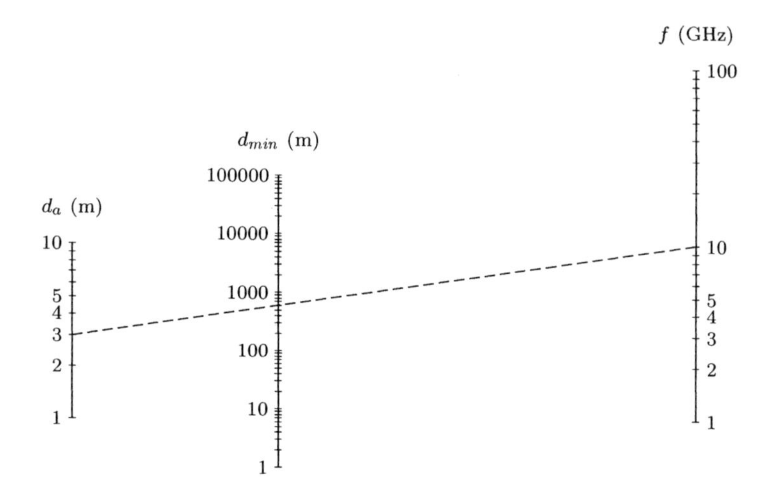

Fig. 2.3 Minimum distance (d_{min}) to ensure the validity of the Friis formula as a function of frequency (f) and antenna dimension (d_a). In the example $d_a = 3$ m, $f = 10$ GHz and $d_{min} \approx 600$ m.

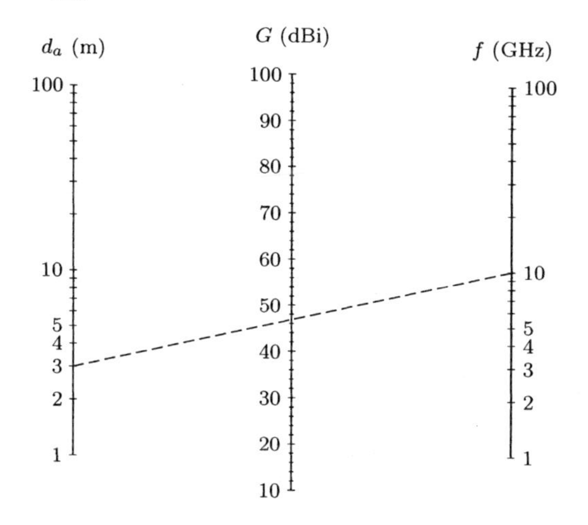

Fig. 2.4 Gain of a parabolic antenna G (in dBi) as a function of the frequency f (in GHz) and the diameter d_a (in m) for an aperture efficiency $\eta = 0.5$. In the example, $d_a = 3$ m, $f = 10$ GHz and $G \approx 47$ dBi.

ITU-R Recommendation F.699-5 [19] suggests, for a circular symmetric antenna, with diameter d_a/λ the following gain:

$$G = 20\ \log_{10}\left(\frac{d_a}{\lambda}\right) + 7.7. \tag{2.15}$$

It is easy to show that (2.13) and (2.15) lead to the same results if $\eta = 0.6$, a value deemed optimistic for link design but which should be adopted for interference calculations, as intended in the ITU-R Recommendation F.699-5.

It is appropriate now to define a few concepts following ITU-R Recommendation P.341-5 [19]. Take Figure 2.5 and let d be the distance between transmitter and receiver antennas, with gains G_T and G_R respectively, and λ the free space wavelength.

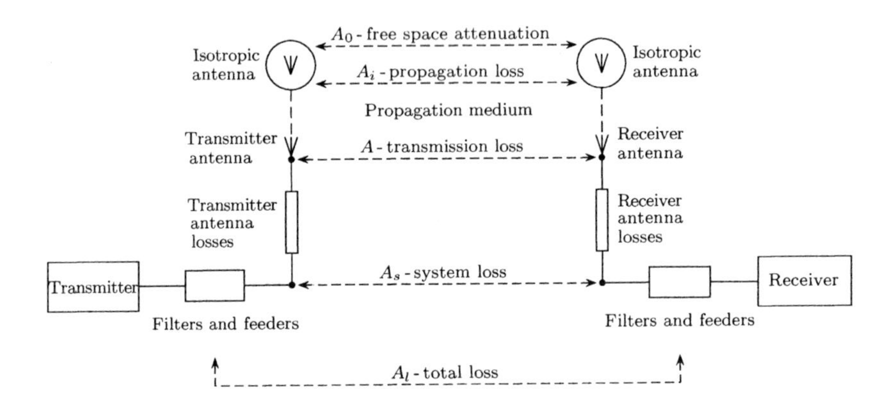

Fig. 2.5 Attenuation definitions according to ITU-R Recommendation P-341-5 [19].

- Total loss (A_l) – ratio between the transmitter output power and the receiver input power in actual operation. Measuring points must always be defined.

- System loss A_s – ratio between the power supplied to the transmitter antenna and the power available at the receiver antenna. We recall that the available power from a given source is the power supplied to a load whose impedance is equal to the conjugate of the source impedance. System loss does not include losses in the transmission lines but includes all antenna losses, ground losses, and losses in possible impedance matching circuits or terminating resistors.

- Transmission loss (A) – ratio between the power radiated by the transmitter antenna and the power available at the receiver antenna, if there were no losses in the radio-frequency circuits. Transmission loss is equal to system loss minus the attenuation due to circuits associated to the antennas.

- Basic transmission loss A_i – transmission loss if the antennas are replaced by lossless isotropic antennas, with the same polarization as the actual antennas, while keeping the same propagation conditions but neglecting the effects of the ground near the antennas[3]

$$A_i = A + G_R + G_T. \tag{2.16}$$

- Free space basic transmission loss or free space attenuation (A_0) – transmission loss that would occur if the antennas were replaced by isotropic lossless antennas and the medium between the antennas by free space vacuum

$$A_0 = 20 \log_{10} \left(\frac{4\pi d}{\lambda} \right). \tag{2.17}$$

- Ray path transmission loss (A_t) – difference between the basic transmission loss and the sum of the antenna gains (expressed in dB) for a given propagation direction

$$A_t = A_i - (G_R + G_T). \tag{2.18}$$

- Loss relative to free space (A_m) – difference between the basic transmission attenuation and the free space attenuation expressed in dB

$$A_m = A_i - A_0. \tag{2.19}$$

Loss relative to free space may be due to:

- absorption by atmospheric gases and hydrometeores (mainly rain drops);
- equivalent reflection attenuation, that is, the difference between reflection in a perfect plane reflector and reflection in non plane or finite surfaces;
- attenuation caused by polarization mismatch;
- antenna gain losses due to mismatch between the antenna and its environment, due to modifications in the equiphase surface which may cease to be plane near the receiver antenna;
- interferences between the direct ray and ground reflected rays.

In the next sections we will successively deal with the presence of the Earth and the atmosphere, the conditions in which a link may be designed as if in free space and the methods to calculate the various components of the attenuation referred to free space, for frequencies between about 50 MHz and about 50 GHz.

[3]Ground losses near the antennas are accounted for in the antenna gains rather than in the basic transmission loss.

2.2 FRESNEL ELLIPSOIDS

The propagation model defined in the previous section – antennas in free space – cannot be used in real links where the antennas are close to the Earth. Fresnel ellipsoids enable us to define the conditions under which propagation between two antennas may be taken as if in free space.

Consider a unidirectional radio link, operating at a frequency f (free space wavelength λ), with point source antennas, at a distance d, such $d \gg \lambda$. We will denote with T and R the points where the transmitter and the receiver antennas are located, respectively.

Define a cylindrical coordinate system with origin in T and axis $\overline{OZ}$ defined by points T and R (see Figure 2.6).

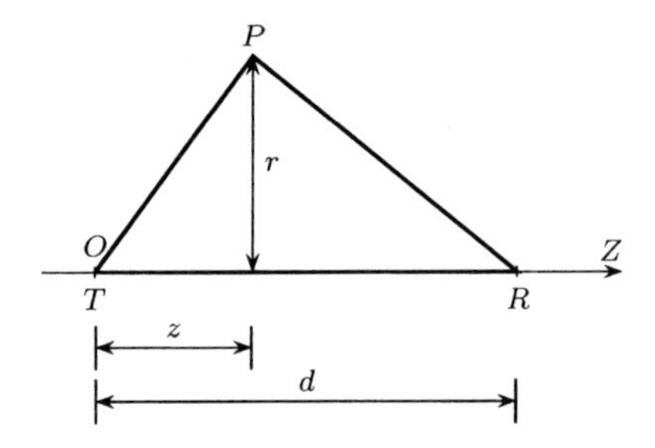

Fig. 2.6 Coordinate system used to define Fresnel ellipsoids.

Take point P with coordinates r,θ,z. By definition, point P belongs to the nth Fresnel ellipsoid if

$$\overline{TP} + \overline{PR} - d = n\,\frac{\lambda}{2}, \tag{2.20}$$

where:

- $\overline{TP}$ is the distance between T and P;
- $\overline{PR}$ is the distance between P and R.

Equation (2.20) defines an ellipsoid of revolution with axis coinciding with the direction defined by the antennas (axis $\overline{OZ}$) and foci (T and R) on the antennas. This equation may be rewritten in terms of the coordinates of point P and the distance d between antennas as

$$\sqrt{z^2 + r^2} + \sqrt{(d-z)^2 + r^2} - d = n\,\frac{\lambda}{2}. \tag{2.21}$$

Assuming that $r \ll |z|$ and that $r \ll |d - z|$ and approximating the square roots in (2.21) by the first two terms of their series expansion we have:

$$\frac{r^2}{2}\left(\frac{1}{z} + \frac{1}{d-z}\right) = n\,\frac{\lambda}{2}, \tag{2.22}$$

or

$$r = \pm\sqrt{n\,\frac{z(d-z)}{d}\lambda}\,. \tag{2.23}$$

Before using equation (2.23) it is important to evaluate the consequences of approximating the square roots by the first two terms of their series expansion. From (2.23) we get, for $z = 0$ and $z = d$, $r = 0$, that is the foci of the Fresnel ellipsoids coincide with the major axis. This result, in contradiction with the initial equation (2.20), is incorrect. It shows that equation (2.23) should not be applyed without prior estimation of the errors involved.

Let ε be the distance between the focus T and the nearby edge of the major axis. From the definition of the ellipsoid (2.20) we have

$$\varepsilon + (\varepsilon + d) - d = n\,\frac{\lambda}{2} \tag{2.24}$$

and thus

$$\varepsilon = n\,\frac{\lambda}{4}. \tag{2.25}$$

Since distance d is, in general, of the order of tens of kilometers and λ of the order of a few centimeters, we may conclude that the error derived from the use of equation (2.23) may be neglected in the graphic representation of the first Fresnel ellipsoids.

The importance of Fresnel ellipsoids stems from the fact that the attenuation between two antennas, even in the presence of (non reflecting) obstacles, is practically identical to the free space attenuation, provided obstacles do not come inside the first Fresnel ellipsoid. Under these conditions we say that points T and R are in line-of-sight, or that the link between T and R is a line-of-sight or unobstructed link.

The largest radius of the first Fresnel ellipsoid r_{1m} may be obtained from equation (2.23), taking $z = d/2$

$$r_{1m} = \sqrt{\frac{\lambda d}{4}}. \tag{2.26}$$

Equation (2.26) presumes that all variables are expressed in the same units. In practice this is often not the case and serious errors may be easily committed. It its thus preferable to rewrite (2.26) with r_{1m} in meters, f (rather than λ) in GHz and d in kilometers

$$r_{1m} = \frac{1}{2}\sqrt{\frac{0.3d}{f}}. \tag{2.27}$$

In Figure 2.7 we plot the value of the largest radius of the first Fresnel ellipsoid r_{1m} as a function of the frequency f and the distance d between antennas.

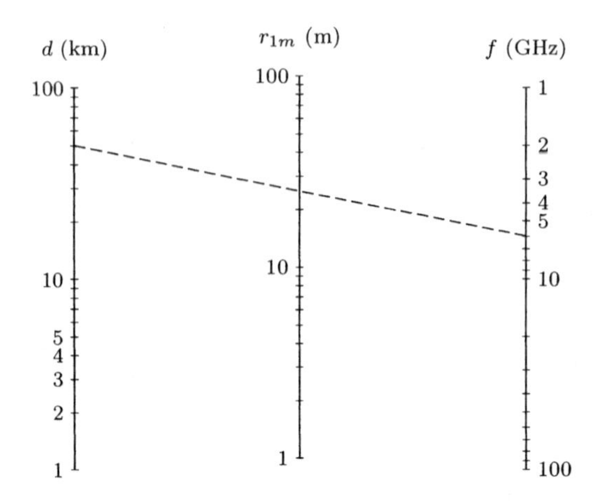

Fig. 2.7 Largest radius of the first Fresnel ellipsoid r_{1m} as a function of the frequency f and the distance d between antennas. In the example $d = 50$ km, $f = 6$ GHz and $r_{1m} \approx 25$ m.

2.3 THE EARTH AS A PLANE SURFACE

We will now proceed, step by step, improving the radio link model shown in Figure 2.1, so that it may be used for the design of a real link. We will start by considering the presence of the Earth that, at first, will be taken as a perfectly smooth, plane, homogeneous surface with dielectric constant ε_s, conductivity σ_s and permeability $\mu_s = \mu_o$.

According to geometric optics, the field at the receiver antenna (and thus the voltage at the antenna terminals) may be obtained by the vector sum of the field due to the direct ray and the field due to the ground reflected ray. If the first Fresnel ellipsoids for the direct ray and the reflected ray are unobstructed, the corresponding field values may be calculated using Friis formula (2.4).

To represent the reflected ray and the associated first Fresnel ellipsoid we recall that the effect of a perfectly reflecting Earth is equivalent to that of the images of the transmitter and receiver antennas (in relation to the Earth surface). Thus, for the reflected ray, it is as if the transmitter antenna sees the image of the receiver antenna, and the receiver antenna sees the image of the transmitter antenna. The real reflected ray path is obtained taking the fractions of the ray paths above the Earth surface (Figure 2.8). The point, on the Earth surface where the incident ray originates the reflected ray is known as the specular point.

Once the reflected ray has been defined, the associated first Fresnel ellipsoid may be plotted just like the direct ray.

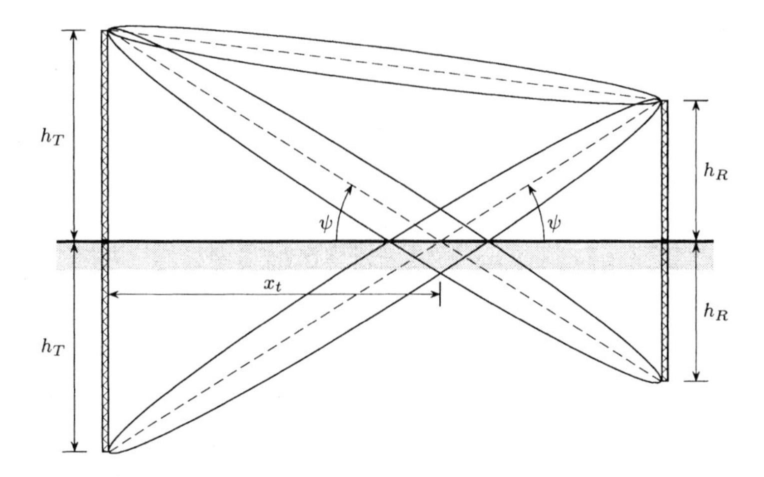

Fig. 2.8 Model for a unidirectional microwave radio link over the Earth, assumed to be a smooth, uniform, plane surface.

Further to considering that both the direct and the reflected rays are unobstructed we will assume that the receiver antenna is perfectly matched to the waveguide and the latter to the receiver, whose input impedance is taken as Z (assumed to be real). Under these conditions the voltage u_d due to the direct ray at the antenna input is

$$u_d = \sqrt{Z\, p_R}. \tag{2.28}$$

Substituting the receiver power due to the direct ray p_R by its value given in equation (2.4) we get

$$u_d = \frac{\lambda}{4\pi} \frac{\sqrt{Z\, p_T\, a_T\, a_R\, g_T\, g_R}}{\sqrt{d^2 + (h_T - h_R)^2}}, \tag{2.29}$$

where h_T and h_R are the transmitter and receiver antenna heights above the Earth surface, respectively.

The receiver input voltage u_r due to the reflected ray is given by

$$u_r = \frac{\lambda}{4\pi} \frac{\sqrt{Z\, p_T\, a_T\, a_R\, g_T^e\, g_R^e}}{\sqrt{d^2 + (h_T + h_R)^2}}\, e^{-j\Delta}\, R\, e^{j\varphi}, \tag{2.30}$$

where:

- g_T^e and g_R^e are the transmitter and receiver antenna gains in the direction of the specular point;
- $R\, e^{j\varphi}$ is the Fresnel reflection coefficient.

The delay angle Δ between voltages u_r and u_d, is given by

$$\Delta = \frac{2\pi}{\lambda}\left[\sqrt{d^2 + (h_T + h_R)^2} - \sqrt{d^2 + (h_T - h_R)^2}\right]. \tag{2.31}$$

The Fresnel reflection coefficient for horizontal polarization is

$$R_h\, e^{j\varphi_h} = \frac{\sin\psi - \sqrt{n_s^2 - \cos^2\psi}}{\sin\psi + \sqrt{n_s^2 - \cos^2\psi}} \tag{2.32}$$

and, for vertical polarization, is

$$R_v\, e^{j\varphi_v} = \frac{n_s^2\sin\psi - \sqrt{n_s^2 - \cos^2\psi}}{n_s^2\sin\psi + \sqrt{n_s^2 - \cos^2\psi}}, \tag{2.33}$$

where:

- n_s is the complex refraction index

$$\begin{aligned} n_s^2 &= \frac{\varepsilon_s}{\varepsilon_o}\left(1 - j\frac{\sigma_s}{\omega\varepsilon_s}\right) \\ &= \frac{\varepsilon_s}{\varepsilon_o} - j60\sigma\lambda; \end{aligned} \tag{2.34}$$

- ψ is the ground arriving angle:

$$\psi = \arctan\left(\frac{h_T + h_R}{d}\right). \tag{2.35}$$

In Figures 2.9 to 2.12 we plot the values of R and φ for horizontal and vertical polarization for two different surfaces:

- sea water ($\varepsilon_s = 81\,\varepsilon_o$, $\sigma_s = 5\,\Omega^{-1}\mathrm{m}^{-1}$);
- average ground ($\varepsilon_s = 15\,\varepsilon_o$, $\sigma_s = 0.005\,\Omega^{-1}\mathrm{m}^{-1}$).

Inspection of Figures 2.9 to 2.12 shows that, for microwave radio links, where $\tan\psi \ll 1$, in most cases $R_h \approx 1$, $R_v \approx 1$, $\varphi_h \approx \pi$ and $\varphi_v \approx \pi$. A notable exception is for vertical polarization over sea water.

When the distance d between antennas is much larger than the antenna height above ground, we may derive, from equations (2.29), (2.30) and (2.31), a simplified expression for the voltages due to the direct and the reflected rays

$$\frac{u_r}{u_d} = \sqrt{\frac{g_T^e\, g_R^e}{g_T\, g_R}}\; R_{h,v}\, e^{j\varphi_{h,v}}\, e^{-j\Delta}, \tag{2.36}$$

where:

$$\Delta = \frac{4\pi}{\lambda}\,\frac{h_T h_R}{d}. \tag{2.37}$$

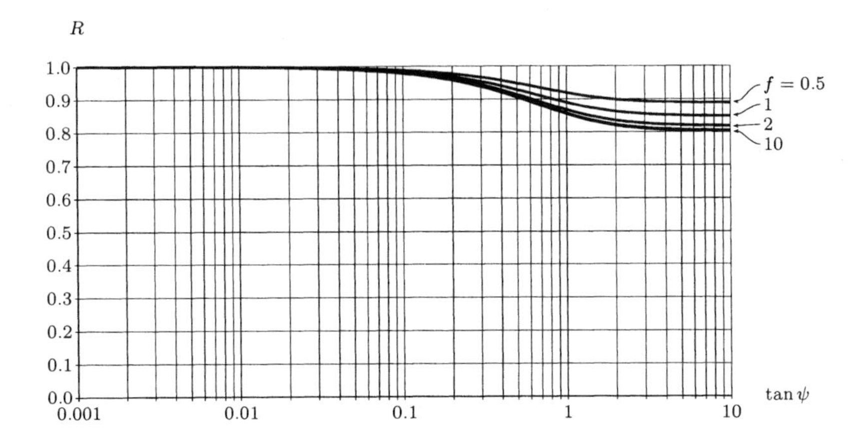

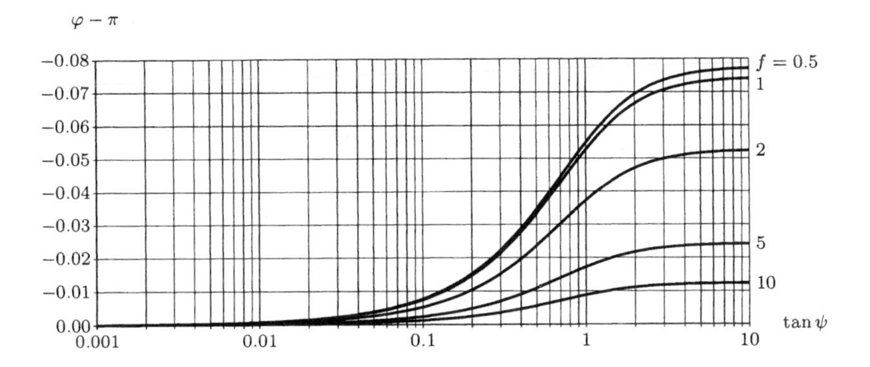

Fig. 2.9 Modulus and argument of the reflection coefficient $R_h\, e^{j\varphi_h}$ over sea water for horizontal polarization taking the frequency f in GHz as a parameter.

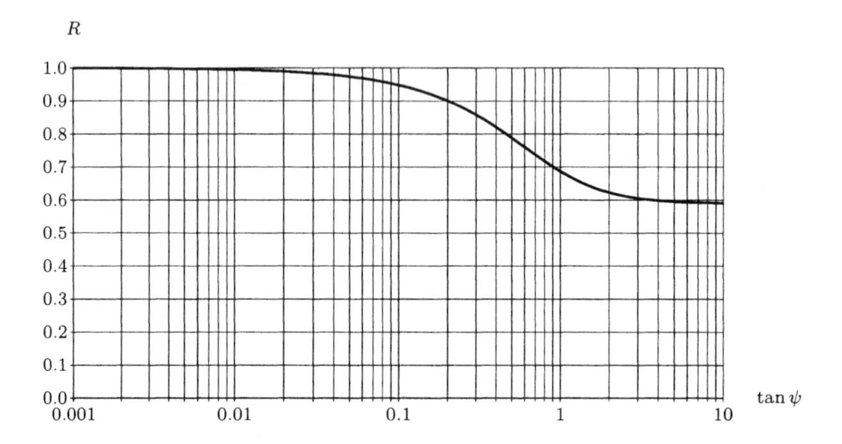

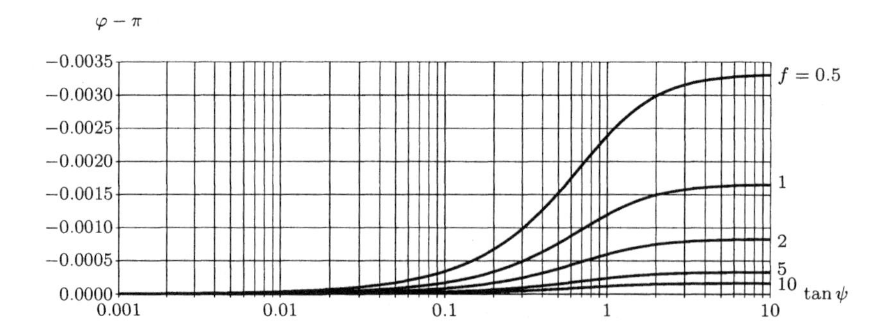

Fig. 2.10 Modulus and argument of the reflection coefficient $R_h\, e^{j\varphi_h}$ average ground for horizontal polarization taking the frequency f in GHz as a parameter.

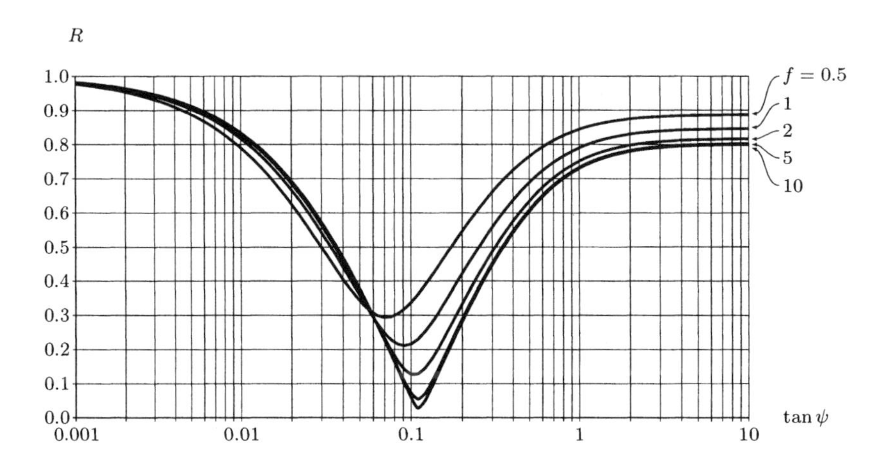

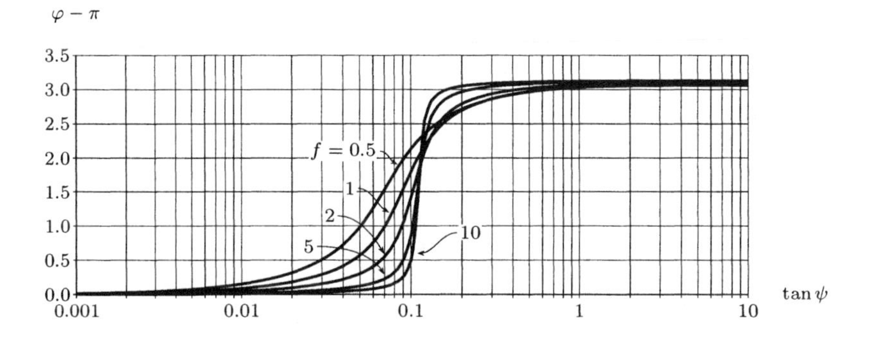

Fig. 2.11 Modulus and argument of the reflection coefficient $R_v\, e^{j\varphi_h}$ over sea water for vertical polarization taking the frequency f in GHz as a parameter.

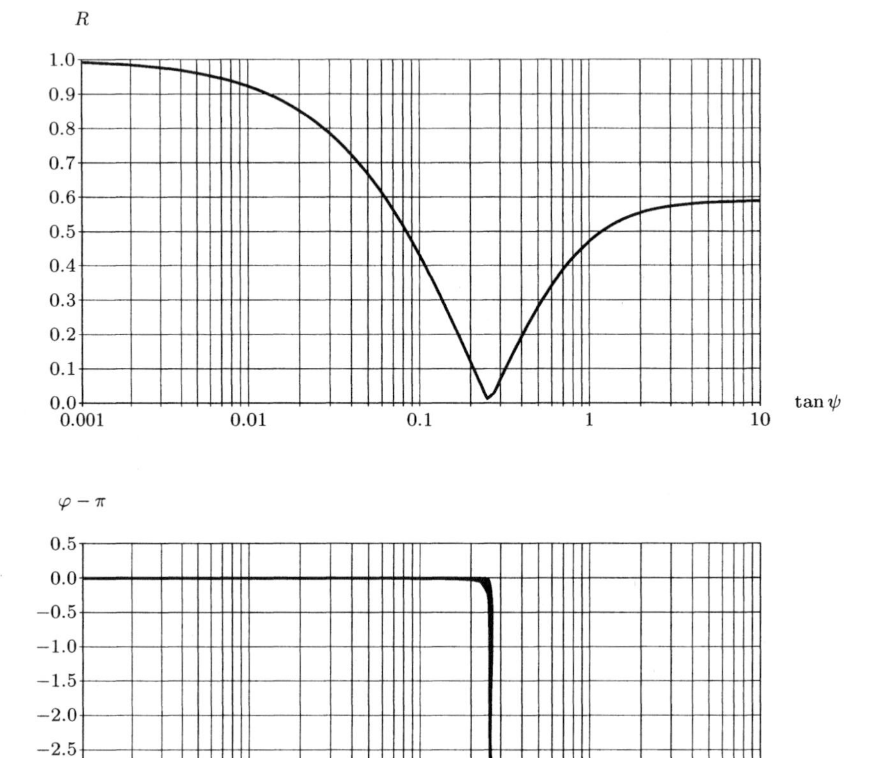

Fig. 2.12 Modulus and argument of the reflection coefficient $R_v\, e^{j\varphi_h}$ over average ground for vertical polarization taking the frequency f in GHz as a parameter.

The presence of the reflected ray becomes noticeable when the antennas are not able to provide sufficient discrimination between the direct and the reflected ray. The total received power p_{Rt} may be derived from the voltages due to the direct and reflected ray, u_d e u_r, respectively:

$$\begin{aligned} p_{Rt} &= \frac{|u_d + u_r|^2}{Z} \\ &= \frac{u_d^2}{Z}\left|1 + \frac{u_r}{u_d}\right|^2. \end{aligned} \tag{2.38}$$

Substituting (2.36) and (2.37) in (2.38), taking $R_{h,v} \approx 1$ e $\varphi_{h,v} = \pi$, usually a reasonable approximation both for horizontal and vertical polarization, and applying logarithms, we have

$$\begin{aligned} P_{Rt} &= P_R + 10\log_{10}\left[2\left(1 - \cos\Delta\right)\right] \\ &= P_R + 20\log_{10}\left[2\sin\left(\frac{\Delta}{2}\right)\right], \end{aligned} \tag{2.39}$$

where

$$\begin{aligned} P_{Rt} &= 10\log_{10}\frac{p_{Rt}}{p_0}, \\ P_R &= 10\log_{10}\frac{p_R}{p_0}, \\ \Delta &= \frac{4\pi}{\lambda}\frac{h_T h_R}{d}, \end{aligned}$$

and:

- p_0 is an arbitrary power;
- P_R is the power received through the direct ray and calculated using (2.8).

In Figure 2.13 we plot the received power as a function of the receiver antenna height above ground h_R, for a typical case, where $d = 50$ km, $h_T = 50$ m and $\lambda = 0.05$ m.

2.4 THE EARTH AS A SPHERICAL SURFACE

Modeling the Earth as a plane, perfectly reflecting surface is only acceptable in some cases. In general, we must consider not only the orography but also the Earth (almost) spherical shape.

If we assume that the Earth is a sphere with radius $r_0 = 6370$ km, the intersection of the Earth surface with plan of the link (defined by the center of the Earth and the antennas, assumed to be point sources), is a circumference.

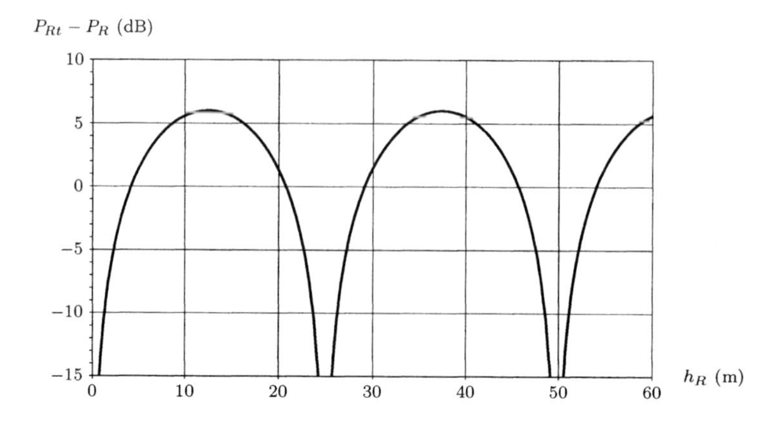

Fig. 2.13 Received power as a function of the receiver antenna height above ground, for a perfect reflector plane Earth surface ($d = 50$ km, $h_T = 50$ m, $f = 6$ GHz).

Since transversal dimensions of the first Fresnel ellipsoid and usual heights of the antennas over ground are of the order of tens of meters and link lengths of the order of tens of kilometers, it is not possible to represent graphically the vertical detail of the link, to verify the clearance of the first Fresnel ellipsoid, unless we use different units for the horizontal and for the vertical direction. If we do so, the circumference representing the Earth surface becomes an ellipsis that may be approximated by a parabola.

To plot the link profile consider (Figure 2.14) the plan of the link and a rectangular coordinate system u, v with its origin on the center of the Earth (point C). In this coordinate system the equation of a circumference with radius r_0 is given by

$$u^2 + v^2 = r_0^2, \tag{2.40}$$

or

$$v = \pm\, r_0 \sqrt{1 - \frac{u^2}{r_0^2}}. \tag{2.41}$$

If we restrict the analysis to the region close to A, where the positive part of the vertical axis crosses the circumference, we may expand the square root in (2.41) and take the first two terms of the series

$$v = r_0 \left(1 - \frac{u^2}{2\, r_0^2}\right). \tag{2.42}$$

Let us now make use of another coordinate system x, y with origin on O, near point A and such that this point has, in the new system, the following

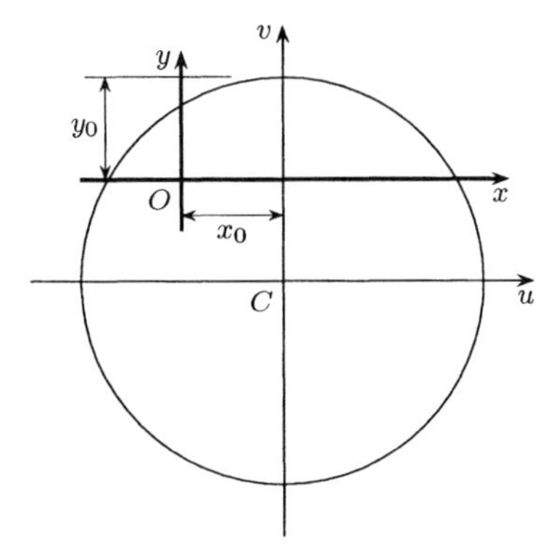

Fig. 2.14 Coordinate systems used to plot the link profile.

coordinates: x_0 e y_0. Equation (2.42) yields

$$y = y_0 - \frac{(x - x_0)^2}{2\, r_0}. \tag{2.43}$$

While x_0 and y_0 may assume any convenient values, experience shows that two choices are often made. In the first, O coincides with the transmitter or the receiver site, that is $x_0 = 0$. In the second O is on the middle of the path, that is, $x_0 = d/2$.

The value of y_0 is usually 0, $d^2/(8r_0)$ or the closest integer multiple of the vertical scale to $d^2/(8r_0)$.

Figures 2.15 and 2.16 show an example of each choice for a 50-km-long link, with 100-m high antennas, at 6 GHz.

Ploting of geometric Figures, such as the Fresnel ellipsoid, in coordinate systems with different units for each axis has a few peculiarities which should be mentioned.

Take e_x and e_y as the graphic scales for the $\overline{OX}$ and the $\overline{OY}$ axis, respectively. The real distance d between two points P_1 and P_2, with graphic coordinates (x_1, y_1) e (x_2, y_2), is

$$d = \sqrt{[e_x(x_2 - x_1)]^2 + [e_y(y_2 - y_1)]^2}. \tag{2.44}$$

For microwave radio link profiles we always have $e_x \gg e_y$. Thus, in most cases, that is, when $x_2 - x_1 \neq 0$, (2.44) may be simplified to

$$d \approx e_x(x_2 - x_1). \tag{2.45}$$

Similarly the angle $\widehat{P_1 O P_2}$ between two lines $\overline{OP_1}$ and $\overline{OP_2}$ is given by

$$\widehat{P_1 O P_2} = \arccos \frac{x_1\, x_2\, e_x^2 + y_1\, y_2\, e_y^2}{\sqrt{x_1^2\, e_x^2 + y_1^2\, e_y^2}\, \sqrt{x_2^2\, e_x^2 + y_2^2\, e_y^2}}. \tag{2.46}$$

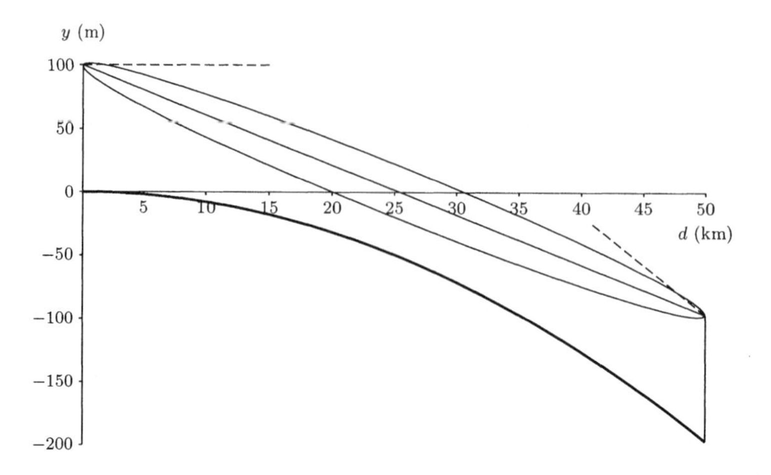

Fig. 2.15 Plot of a link profile on a spherical Earth ($d = 50$ km, $h_T = h_R = 100$ m, $f = 6$ GHz) using a coordinate system with origin at the transmitter site.

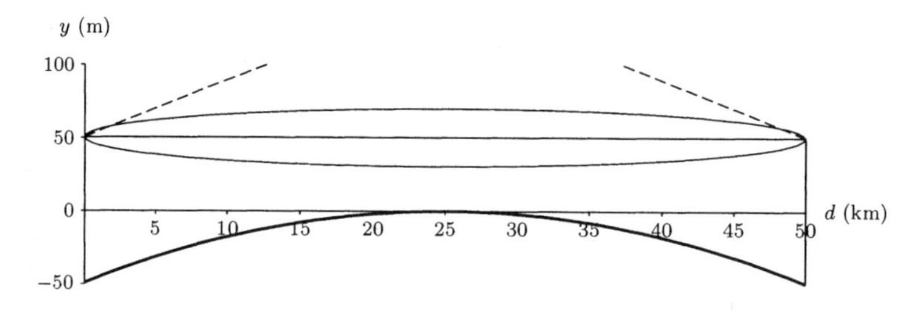

Fig. 2.16 Plot of a link profile on a spherical Earth ($d = 50$ km, $h_T = h_R = 100$ m, $f = 6$ GHz) using a coordinate system with origin at the middle of the path.

Since $e_x \gg e_y$, then:

- if $x_1 \neq 0$ and $x_2 \neq 0$

$$\widehat{P_1 O P_2} \approx 0; \tag{2.47}$$

- if $x_1 = 0$ or $x_2 = 0$

$$\widehat{P_1 O P_2} \approx \frac{\pi}{2}. \tag{2.48}$$

These approximate results mean that:

- If we exclude verticals, the angles between any straight lines are always close to zero;

- The angle between any straight line and a vertical line is always close to a right angle.

In practice, equations (2.47) and (2.48) tell us that in the plot of Fresnel ellipsoids the radius is always marked vertically above and below the direct ray as shown in Figure 2.17.

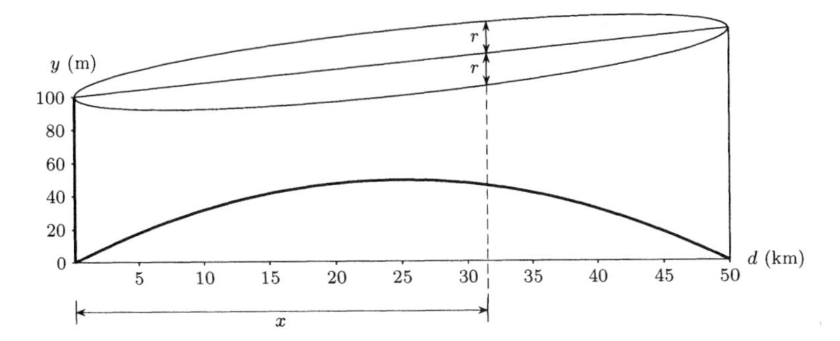

Fig. 2.17 Plot of the first Fresnel ellipsoid.

A better approximation to the angle $\widehat{P_1 O P_2}$, when $x_1 \neq 0$ and $x_2 \neq 0$ may be derived from the following exact expression:

$$\widehat{P_1 O P_2} \approx \arcsin\left[\frac{(x_1\, y_2 - x_2\, y_1)\, e_x\, e_y}{\sqrt{x_1^2\, e_x^2 + y_1^2\, e_y^2}\ \sqrt{x_2^2\, e_x^2 + y_2^2\, e_y^2}}\right], \tag{2.49}$$

which, with $e_x \gg e_y$, becomes

$$\widehat{P_1 O P_2} = \frac{e_y}{e_x}\left(\frac{y_2}{x_2} - \frac{y_1}{x_1}\right). \tag{2.50}$$

Equation (2.50) may be used to compute the antenna elevation angle from profile plot data.

To set up antennas on site we require:

- The elevation angle, defined as the angle between the direct ray and the horizontal;
- The azimuth, defined as the angle, measured on the horizontal plane between the direct ray and (true) north.

To calculate the elevation angle on the plan of the link we need to plot on this plane, the direct ray and the horizontal (tangent to the surface of the Earth) at the antenna. Since drawing the tangent, when it is not parallel to the $\overline{OX}$ axis, can lead to non-negligible errors, it is preferably to draw the tangent from the value of trigonometric tangent of the angle between the tangent and the $\overline{OX}$ axis. On the general point, with coordinates x, y, the required value is obtained from equation (2.43) as

$$\tan\varphi = -\frac{x - x_0}{r_0}. \qquad (2.51)$$

Figures 2.15 and 2.16 show the horizontal line for each antenna. As expected, although this may not be apparent at first, the elevation angle is the same for both antennas and is independent of the representation adopted for the Earth surface.

Changing the Earth shape from a plane to a sphere creates a limit, which did not exist before, to the distance at which we may achieve free space propagation between two antennas at given heights.

According to ITU-R Recommendation P.310-9[19] we define the radio horizon d_{rh} of an antenna at a height h above the Earth, with radius r, the distance measured on the Earth surface between the antenna base and the point at which the rays from the antenna are tangent to the Earth surface. It may be easily shown that

$$d_{rh} \approx \sqrt{2rh}. \qquad (2.52)$$

The maximum distance d_m, at which there are still free space propagation conditions for two antennas at heights h_T and h_R above the Earth, taken as a sphere with radius $r_0 = 6370$ km, may be estimated from the nomograms in Figures 2.18 and 2.7, as follows:

- The first estimate $d_m = \sqrt{2rh_T} + \sqrt{2rh_R}$ is obtained from the nomogram in Figure 2.18;

- With the first estimate and the operating frequency we calculate the maximum radius of the first Fresnel ellipsoid r_{1m} using the nomogram in Figure 2.7;

- A second and better estimate of d_m, sufficient for most applications, may be obtained by applying again the nomogram in Figure 2.18 with $h_T - r_{1m}$ and $h_R - r_{1m}$.

As will be dealt with in detail later (Section 2.5), the presence of the atmosphere, whose refraction index varies with altitude, modifies the ray path. For the standard atmosphere this change is equivalent to consider that the radius of the Earth is no longer its physical radius r_0, but an equivalent radius $r = 8\,500$ km. Figure 2.19 show another nomogram, similar to the one in Figure 2.18 for an equivalent Earth radius of 8 500 km.

The presence of a spherical Earth, besides creating an upper limit to free space propagation between two antennas, introduces reflections with consequences similar to those already referred to before.

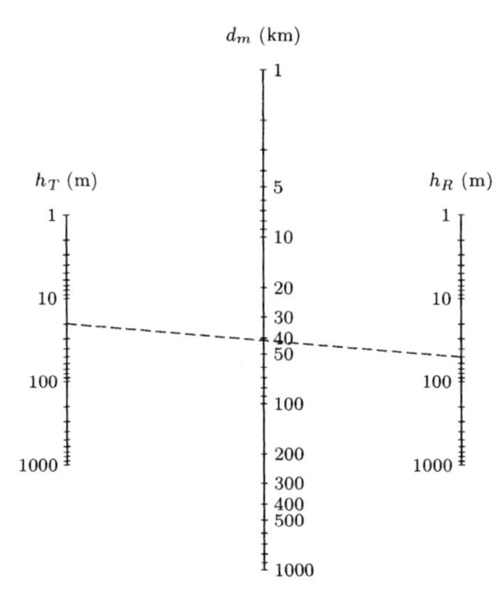

Fig. 2.18 Nomogram to estimate the distance at which the direct ray, from a transmitter antenna at a height h_T to a receiver antenna at a height h_R, is tangent to a spherical Earth with radius $r = 6\ 370$ km. In the example $h_T = 20$ m, $h_R = 50$ m and $d_m \approx 40$ km.

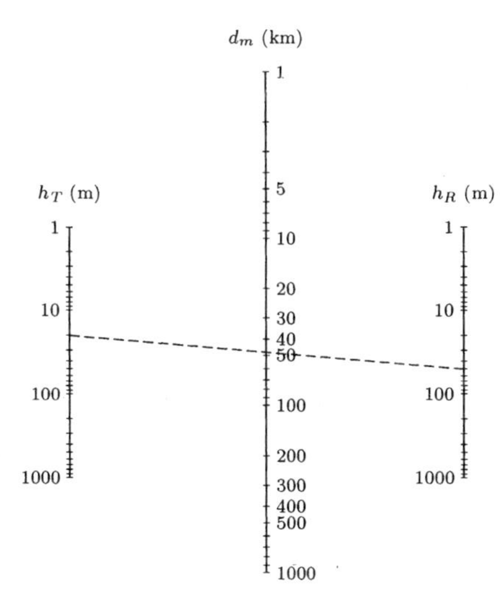

Fig. 2.19 Nomogram to estimate the distance at which the direct ray, from a transmitter antenna at a height h_T to a receiver antenna at a height h_R, is tangent to a spherical Earth with radius $r = 8\ 500$ km. In the example $h_T = 20$ m, $h_R = 50$ m and $d_m \approx 48$ km.

We will begin by determining the specular point for the general case, with two antennas at heights h_T and h_R and at a distance d. Consider the representation of the link profile in Figure 2.20 and choose $x_0 = x_t$, where x_t is the distance from the specular point to the antenna at height h_T.

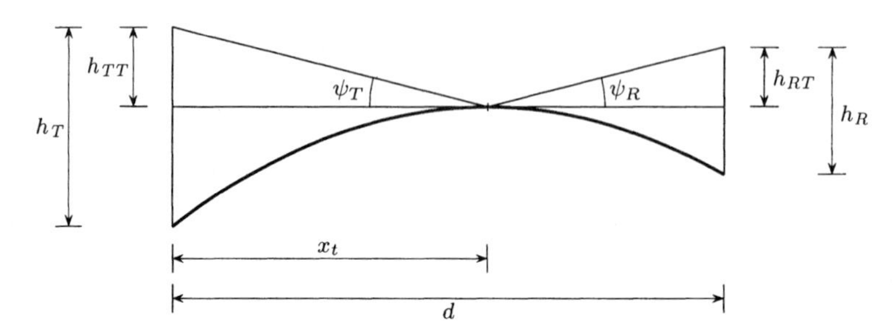

Fig. 2.20 Coordinate system used to derived the position of the specular point for a spherical Earth.

By definition of specular point, the angle of arrival of the reflected ray coming from the transmitter ψ_T is equal to the angle of arrival coming from the receiver ψ_R, and thus

$$\frac{h_{TT}}{x_t} = \frac{h_{RT}}{d - x_t}. \tag{2.53}$$

Recalling (2.43), the previous equation becomes

$$\frac{h_T - \dfrac{x_t^2}{2r_0}}{x_t} = \frac{h_R - \dfrac{(d - x_t)^2}{2r_0}}{d - x_t}, \tag{2.54}$$

hence

$$x_t^3 - 1.5\, d\, x_t^2 + [0.5\, d^2 - r_0(h_T + h_R)]x_t + r_0\, h_T\, d = 0. \tag{2.55}$$

Introducing the following normalized variables:

$$u = \frac{x_t}{d}, \tag{2.56}$$

$$r = \frac{h_R}{h_T}, \tag{2.57}$$

$$t = \frac{r_0 h_T}{d^2}, \tag{2.58}$$

equation (2.55) becomes

$$u^3 - 1.5\, u^2 + [0.5 - t(r + 1)]u + t = 0. \tag{2.59}$$

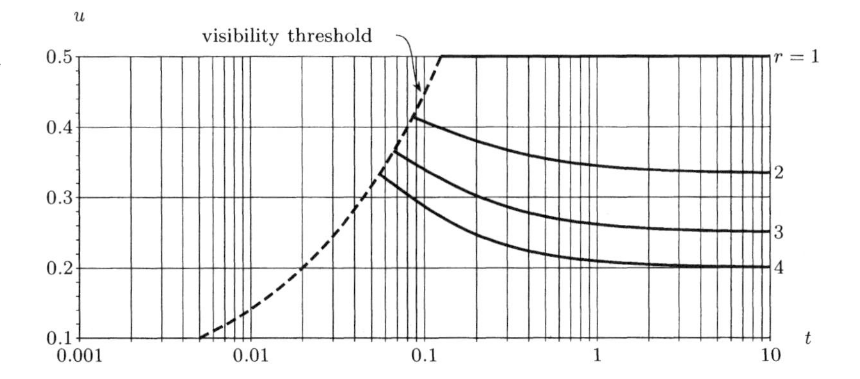

Fig. 2.21 Nomogram to calculate the position of the specular point on a spherical Earth. The dashed curve represents the radio horizon.

Equation (2.59) may be solved numerically or graphically, using the nomogram of Figure 2.21.

Since the angle of arrival ψ_T must be positive for the reflected ray to exist, that is,

$$h_T - \frac{x_t^2}{2r_0} \geq 0$$

or, writing x_t, h_T e r_0 in terms of u, r e t

$$\frac{u^2}{2\,t} \leq 1. \tag{2.60}$$

In Figure 2.21 we also represent the radio horizon defined by (2.60).

The change in the Earth shape, from plane to spherical, modifies the reflected field at the receiver antenna, both because the reflection is no longer on a plane but also because antenna heights are modified as shown in Figure 2.20. Using the same symbols as in (2.29) and (2.30) we get

$$u_d = \frac{\lambda}{4\pi}\sqrt{\frac{Z\,p_T\,a_T\,a_R\,g_T\,g_R}{[d^2 + (h_{TT} - h_{RT})^2]}}, \tag{2.61}$$

$$u_r = \frac{\lambda}{4\pi}\sqrt{\frac{Z\,p_T\,a_T\,a_R\,g_T\,g_R}{[d^2 + (h_{TT} + h_{RT})^2]}}\; e^{-j\Delta}\, R_{h,v}\, e^{j\varphi_{h,v}}\, D_v, \tag{2.62}$$

where:

$$h_{TT} = h_T - \frac{x_t^2}{2\,r_0}, \tag{2.63}$$

$$h_{RT} = h_R - \frac{(d - x_t)^2}{2\,r_0}, \tag{2.64}$$

$$\Delta = \frac{2\pi}{\lambda}\left[\sqrt{x_t^2 + h_{TT}^2} + \sqrt{(d-x_t)^2 + h_{RT}^2} - \sqrt{d^2 + (h_{RT} - h_{TT})^2}\right], \quad (2.65)$$

and the divergence factor D_v is given in Kerr [21] as

$$D_v = \frac{1}{\sqrt{1 + \frac{2x_t(d-x_t)}{r_0 d \sin\psi}}}, \quad (2.66)$$

where

$$\sin\psi = \frac{h_{TT}}{x_t}, \quad (2.67)$$

$$\sin\psi = \frac{h_{RT}}{d - x_t}. \quad (2.68)$$

In microwave radio links, where usually the distance between antennas is much larger than the antenna heights, we may derive a much simpler expression for the ratio u_r/u_d. From (2.61) –(2.65) we get

$$\frac{u_r}{u_d} = \sqrt{\frac{g_T^e\, g_R^e}{g_T\, g_R}}\, D_v\, R_{h,v}\, e^{-j\varphi_{h,v}}\, e^{-j\Delta}, \quad (2.69)$$

where

$$\Delta = \frac{2\pi}{\lambda}\left[\frac{h_{TT}^2}{2x_t} + \frac{h_{RT}^2}{2(d-x_t)} - \frac{(h_{TT} - h_{RT})^2}{2d}\right]. \quad (2.70)$$

The divergence factor may be calculated from (2.66) or as a function of the normalized variables u and t

$$D_v = \frac{1}{\sqrt{1 + \frac{2u^2(1-u)}{t\left(1 - \frac{u^2}{2t}\right)}}}. \quad (2.71)$$

The nomogram in Figure 2.22 provides a simple way to estimate D_v as a function of u and t.

The influence of the reflected ray on the received power may be derived from (2.38), where Δ is given by (2.70). If $D_v \approx 1$ equation (2.39) may be used.

2.5 ATMOSPHERE

2.5.1 Introduction

The gaseous layer surrounding the Earth is commonly known as the atmosphere [25]. It is often considered to have three regions which are roughly organized in layers, that is, stratified:

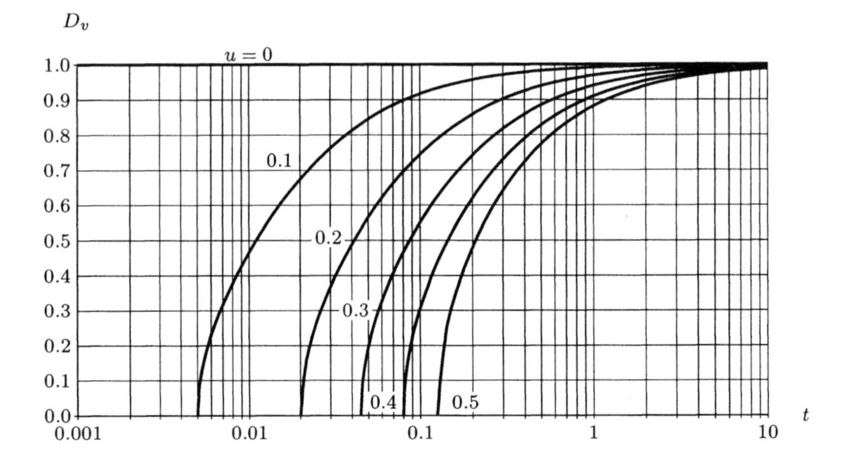

Fig. 2.22 Divergence factor D_v as a function of u and t.

1. the inner layer – troposphere – where temperature decreases with altitude (except in few well-defined sub-layers) and that extends up to 9 km over the poles and up to 17 km over the equator;
2. the middle layer – stratosphere – located between about 11 and 50 km;
3. the outer layer – ionosphere – with a high concentration of ionized particles and extending upwards from the stratosphere up to about 500 km.

The presence of the troposphere in the link model will bring five main effects:

1. Excess attenuation, depending on link length and slope, due to atmospheric gases (mainly oxygen and water vapor) and hydrometeors (rain, fog, hail, snow);
2. Modification in the ray path shape which ceases to be a straight line and becomes curved, in response to the changes in the refraction index along the path;
3. Creation of privileged directions for wave propagation – ducts – which enable signals to reach distances much larger away than would possible without atmosphere;
4. Considerable fluctuations in the amplitude of the received signal due to the existence of various signal paths, each with its own time delay, which interfere with each other; this phenomena is known as multipath fading in microwave radio links and scintillation in satellite links;

5. Scattering due to irregularities in the higher layers of the troposphere which enable signals to reach very large distances (a few hundreds of kilometers) and may be used to provide troposcatter links.

2.5.2 Excess attenuation due to gases

Atmospheric gases, particularly oxygen, water vapor and carbon dioxide, fog and hydrometeors (mainly rain) cause excess attenuation, known as atmospheric attenuation in paths that cross the atmosphere.

For microwave radio links, atmospheric attenuation is minimal in a window that extends from about 1 to about 10 GHz. In the lower part of this window, up to 2 GHz, atmospheric attenuation may be safely neglected except for very long paths. Between 2 and about 13 GHz it is usually sufficient to consider the (intense) rain attenuation which, at the higher frequencies within the range, may well become the decisive factor in link design.

The useful frequency band may extend up to 50 GHz for short links (less than 10 km) particularly if occasional interruptions due to downpours are acceptable. Another interesting frequency window for even shorter (up to about 2 km) links, with similar conditions, extends from the infrared to the near ultraviolet (300 to 1000 THz).

Usually microwave radio links use frequencies between about 1 and about 50 GHz, where the lower frequencies are better suited to the longer paths.

For a path with length d the excess loss A_a due to the atmosphere, expressed in dB, is given by

$$A_a = \int_0^d [\gamma_o(x) + \gamma_w(x)] \; dx, \tag{2.72}$$

where γ_o and γ_w are the attenuation coefficients, per unit length, due to the oxygen and to the water vapor, respectively. These coefficients are defined for the average atmospheric pressure at sea level (1 atm = 1013 hPa), at 15°C and a humidity of 7.5 grams of water vapor per cubic meter of dry air (7.5 g/m^3).

For terrestrial links, where there are no significant changes in the attenuation coefficients along the path, (2.72) may be simplified to

$$A_a = (\gamma_{o0} + \gamma_{w0})\, d. \tag{2.73}$$

Specific attenuations up to 1000 GHz, due to dry air and water vapor, may be accurately evaluated by summing up the individual resonance lines from oxygen and water vapor, together with small additional factors for the non-resonant Debye spectrum of oxygen below 10 GHz, pressure-induced nitrogen attenuation above 100 GHz and a wet continuum to account for the excess water vapor absorption found experimentally.

For altitudes up to 5 km, the attenuation coefficients, of dry air γ_o and water vapor γ_w, expressed in dB/km, may be estimated with an accuracy of ±15 percent, except for the frequency range 57 – 63 GHz, given the frequency f, in GHz, the atmospheric pressure p, in hPa, the temperature T, in centigrade, and the water vapor content ρ, in g/m^3, using the following expressions taken from the ITU-R Recommendation P.676-4 [19]):

- For frequencies $f \leq 57$ GHz:

$$\gamma_o = \left[\frac{7.34 r_p^2 r_t^3}{f^2 + 0.36 r_p^2 r_t^2} + \frac{0.3429 b \gamma_o'(54)}{(54-f)^a + b} \right] f^2 \times 10^{-3}; \tag{2.74}$$

- For $54 < f < 66$, GHz, the approximation is rather crude due to the large number of oxygen absorption lines in this frequency band

$$\gamma_o = \exp(\xi f^N); \tag{2.75}$$

where

$$\begin{aligned}
\xi = & \left[\frac{54^{-N} \log_e(\gamma_o(54)(f-57)(f-60)(f-63)(f-66)}{1944} \right. \\
& - \frac{57^{-N} \log_e(\gamma_o(57)(f-54)(f-60)(f-63)(f-66)}{486} \\
& + \frac{60^{-N} \log_e(\gamma_o(60)(f-54)(f-60)(f-63)(f-66)}{324} \\
& - \frac{63^{-N} \log_e(\gamma_o(63)(f-54)(f-60)(f-63)(f-66)}{486} \\
& \left. + \frac{66^{-N} \log_e(\gamma_o(66)(f-54)(f-60)(f-63)(f-66)}{1944} \right].
\end{aligned} \tag{2.76}$$

- For $66 \leq f < 120$ GHz

$$\gamma_o = \left[\frac{0.2296 d \gamma_o'(66)}{(f-66)^c + d} + \frac{0.286 r_p^2 r_t^{3.8}}{(f-118.75)^2 + 2.97 r_p^2 r_t^{1.6}} \right] f^2 \times 10^{-3}; \tag{2.77}$$

- For $120 \leq f \leq 350$ GHz

$$\begin{aligned}
\gamma_o = & \left[3.02 \times 10^{-4} r_p^2 r_t^{3.5} + \frac{1.5827 r_p^2 \eta^3}{(f-66)^2} \right. \\
& \left. \frac{0.286 r_p^2 r_t^{3.8}}{(f-118.75)^2 + 2.97 r_p^2 r_t^{1.6}} \right] f^2 \times 10^{-3},
\end{aligned} \tag{2.78}$$

where

$$\begin{aligned}
\gamma_o'(54) &= 2.128 r_p^{1.4954} r_t^{-1.6032} e^{-2.5280(1-r_t)}, \\
\gamma_o(54) &= 2.136 r_p^{1.4975} r_t^{-1.5852} e^{-2.5196(1-r_t)}, \\
\gamma_o(57) &= 9.984 r_p^{0.9313} r_t^{2.6732} e^{0.9563(1-r_t)},
\end{aligned}$$

$$
\begin{aligned}
\gamma_o(60) &= 15.42 r_p^{0.8595} r_t^{3.6178} e^{1.1521(1-r_t)},\\
\gamma_o(63) &= 10.63 r_p^{0.9298} r_t^{2.3284} e^{0.6287(1-r_t)},\\
\gamma_o(66) &= 1.944 r_p^{1.6673} r_t^{-3.3583} e^{-4.1612(1-r_t)},\\
\gamma_o'(66) &= 1.935 r_p^{1.6657} r_t^{-3.3714} e^{-4.1643(1-r_t)},\\
a &= \frac{\log_e\left(\frac{\eta_2}{\eta_1}\right)}{\log_e(3.5)},\\
b &= \frac{4^a}{\eta_1},\\
\eta_1 &= 6.7665 r_p^{-0.5050} r_t^{0.5106} e^{1.5663(1-r_t)} - 1,\\
\eta_2 &= 27.8843 r_p^{-0.4908} r_t^{-0.8491} e^{0.5496(1-r_t)} - 1,\\
c &= \frac{\log_e\left(\frac{\chi_2}{\chi_1}\right)}{\log_e(3.5)},\\
d &= \frac{4^c}{\chi_1},\\
\chi_1 &= 6.9575 r_p^{-0.3461} r_t^{0.2535} e^{1.3766(1-r_t)} - 1,\\
\chi_2 &= 42.1309 r_p^{-0.3068} r_t^{1.2023} e^{2.5147(1-r_t)} - 1,\\
N &= 0 \quad \text{for} \quad f \le 60\text{GHz},\\
&= -15 \quad \text{for} \quad f > 60\text{GHz},\\
r_p &= \frac{p}{1013},\\
r_t &= \frac{288}{273+T};
\end{aligned}
$$

$$
\begin{aligned}
\gamma_w = & \Bigg\{ 3.13 \times 10^{-2} r_p r_t + 1.76 \times 10^{-3} \rho r_t^{8.5} \\
& + r_t^{2.5} \left[\frac{3.84 \chi_{w1} g_{22} e^{2.23(1-r_t)}}{(f-22.235)^2 + 9.42\chi_{w1}^2} \right. \\
& + \frac{10.48 \chi_{w2} e^{0.7(1-r_t)}}{(f-183.31)^2 + 9.48\chi_{w2}^2} \\
& + \frac{0.078 \chi_{w3} e^{6.4385(1-r_t)}}{(f-321.226)^2 + 6.29\chi_{w3}^2} \\
& + \frac{3.76 \chi_{w4} e^{1.6(1-r_t)}}{(f-325.153)^2 + 9.22\chi_{w4}^2} \\
& + \frac{26.36 \chi_{w5} e^{1.09(1-r_t)}}{(f-380)^2}
\end{aligned}
$$

$$
+ \frac{17.87\chi_{w5}e^{1.46(1-r_t)}}{(f-448)^2}
$$

$$
+ frac883.7\chi_{w5}g_{557}e^{0.17(1-r_t)}(f-557)^2
$$

$$
\left. \left. + \frac{302.6\chi_{w5}g_{752}e^{0.41(1-r_t)}}{(f-752)^2} \right] \right\} f^2\rho \times 10^{-4}, \tag{2.79}
$$

where:

$$
\begin{aligned}
\chi_{w1} &= 0.9544 r_p r_t^{0.69} + 0.0061\rho, \\
\chi_{w2} &= 0.95 r_p r_t^{0.64} + 0.0067\rho, \\
\chi_{w3} &= 0.9561 r_p r_t^{0.67} + 0.0059\rho, \\
\chi_{w4} &= 0.9543 r_p r_t^{0.68} + 0.0061\rho, \\
\chi_{w5} &= 0.955 r_p r_t^{0.68} + 0.006\rho, \\
g_{22} &= 1 + \frac{(f-22.235)^2}{(f+22.235)^2}, \\
g_{557} &= 1 + \frac{(f-557)^2}{(f+557)^2}, \\
g_{752} &= 1 + \frac{(f-752)^2}{(f+752)^2}.
\end{aligned}
$$

In (2.79) the water vapor concentration ρ cannot be higher than the value corresponding to saturation for the chosen temperature. For 15°C this value is approximately 12 g/m^3.

Plots of γ_o and γ_w with frequency f, for a temperature of 15 °C, a water vapor concentration of 7.5 g/m^3 and an atmospheric pressure of 1013 hPa are shown in Figure 2.23.

Often water vapor content ρ is calculated from the relative humidity H defined as the ratio, expressed in percent, between water vapor pressure e in humid air and saturation pressure e_s at the same temperature and pressure

$$
H = 100\,\frac{e}{e_s}. \tag{2.80}
$$

Saturation pressure e_s, as a function of temperature T, in centigrade, is given in ITU-R Recommendation P.453-7 [19]

$$
e_s = 6.1121 \exp\left(\frac{17.502\,T}{240.97+T}\right) \quad \text{(hPa)} \tag{2.81}
$$

and plotted in Figure 2.24.

Expression (2.81) is valid in the range −20oC – +50°C, with an accuracy of±0.20%.

Alternatively the saturation pressure e_s, as a function of temperature T in °C may be obtained from Figure 2.24.

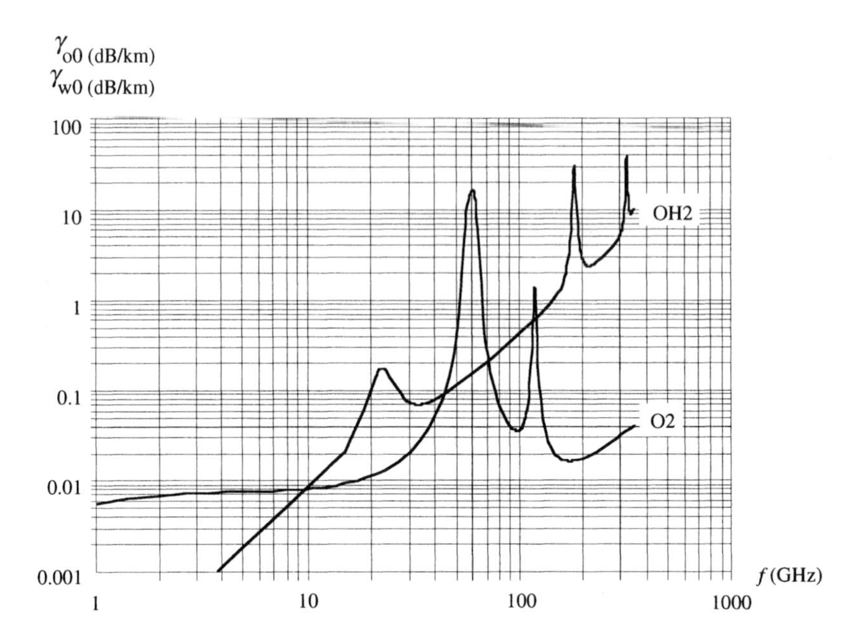

Fig. 2.23 Attenuation coefficients due to oxygen and water vapor, for an atmospheric pressure of 1013 hPa, a temperature of 15°C and a water vapor content of 7.5 g/m^3.

The relation between water vapor pressure e (in hPa) and water vapor density ρ (in g/m^3) at temperature T (in centigrade) is also given in ITU-R Recommendation P.453-6

$$\rho = 216.7 \frac{e}{T + 273.3}. \tag{2.82}$$

As an example assume typical temperate Summer (25°C and 50 percent humidity) and Winter (10°C and 85 percent humidity) conditions. From Figure 2.24 we get $e_s = 31$ hPa for $T = 25$°C and thus for 50 percent humidity $e = 15.5$ hPa. Substituting in (2.82) we get $\rho = 11.3$ g/m^3. Similarly for $T = 10$°C and 85 percent humidity we have $\rho = 7.3$ g/m^3.

It may come as a surprise to some readers that in Summer, when apparently the weather appears to be drier, water vapor content is actually higher than in winter in humid weather. This is due to the fact that human beings perceive relative humidity rather than absolute water content. This also explains why in cold climates, house heating must go together with air humidifying for comfort, and in warm climates, air conditioning must be linked with dehumidification.

ITU-R Recommendation P.836-1 [19] includes world charts of the average ground level values of ρ for February (Figure 2.25) and August (Figure 2.26).

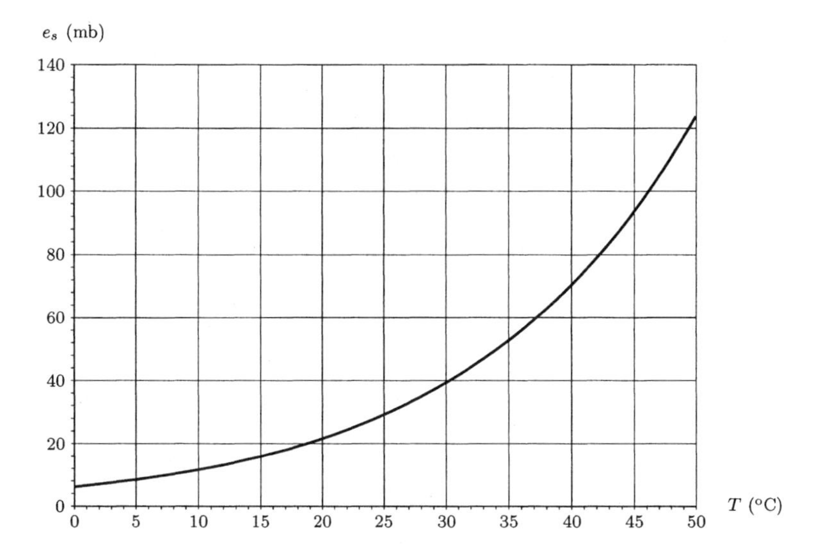

Fig. 2.24 Water vapor saturation pressure as a function of temperature.

Atmospheric attenuation, calculated from (2.73) may be neglected below 1 GHz. Between 1 and 20 GHz atmospheric attenuation does not normally exceed about 1 dB. On the contrary, for frequencies above about 10 GHz (with rain) or 20 GHz (without rain), atmospheric attenuation due to the presence of gases and hydrometeors (mainly rain) places an upper limit to the use of higher frequencies in microwave radio links.

For steep-sloped links, typical of satellite links, (2.73) is no longer valid. Then we have to revert to the method put forward by Rice [26] which corresponds to the integration of (2.72) in a standard atmosphere, or to use other approximate methods such as those in ITU-R Recommendation P.676-4 [19].

According to Rice [26], the integration of (2.72) is equivalent to the following:

$$A_a = \gamma_{o0}\, d_{eo} + \gamma_{w0}\, d_{ew} \;\; (\text{dB/km}), \tag{2.83}$$

where d_{eo} and d_{ew} are the equivalent path lengths for the attenuation by oxygen and water vapor, respectively. The values of d_{eo} and d_{ew} are functions of path slope.

According again to ITU-R Recommendation P.676-4 [19] the atmospheric attenuation for a slant path may be estimated from the zenith attenuation, that is, the attenuation obtained multiplying the specific attenuations for the oxygen and water vapor by an equivalent height. The resulting zenith attenuations are accurate within ±10 % up to about 2 km altitude, except within 0.5 GHz of the resonant lines where the accurate procedure described in Annex I of this recommendation should be used.

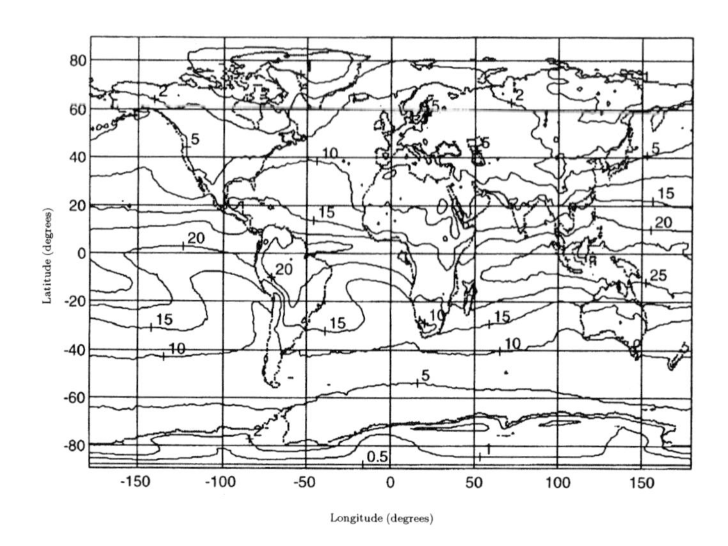

Fig. 2.25 World chart of the average ground level values of ρ for the month of February (ITU-R Recommendation P.836-1 [19]). Reproduced with ITU permission.

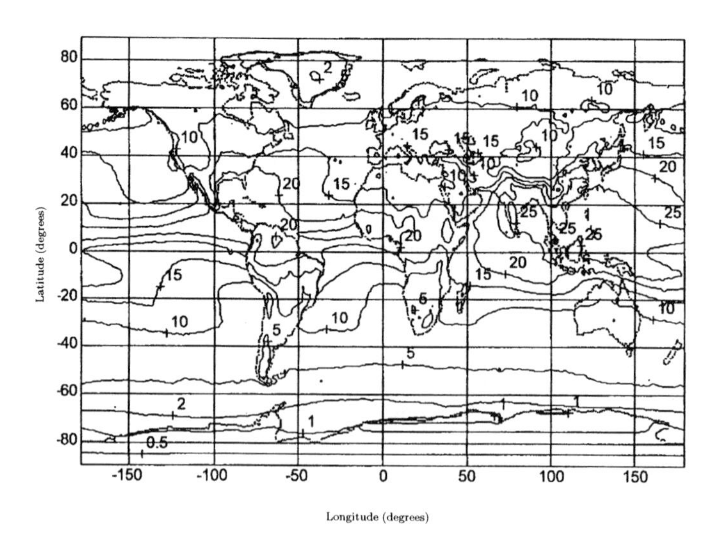

Fig. 2.26 World chart of the average ground level values of ρ for the month of August (ITU-R Recommendation P.836-1 [19]). Reproduced with ITU permission.

For dry air the equivalent height h_o is given by:

- For $1 \leq f \leq 56.7$ GHz:

$$\begin{aligned} h_o &= 5.386 - 3.32734 \times 10^{-2} f + 1.87185 \times 10^{-3} f^2 \\ &\quad - 3.52087 \times 10^{-5} f^3 + \frac{83.26}{(f-60)^2 + 1.2} \quad \text{(km)}; \end{aligned} \tag{2.84}$$

- For $56.7 < f < 63.3$ GHz:

$$h_0 = 10 \quad \text{(km)}; \tag{2.85}$$

- For $63.5 < f < 98.5$ GHz:

$$\begin{aligned} h_o &= f \left\{ \frac{0.039581 - 1.19751 \times 10^{-3} f + 9.14810 \times 10^{-6} f^2}{1 - 0.028687 f + 2.07858 \times 10^{-4} f^2} \right. \\ &\quad \left. + \frac{90.6}{(f-60)^2} \right\} \quad \text{(km)}; \end{aligned} \tag{2.86}$$

- For $98.5 \leq f \leq 350$ GHz:

$$\begin{aligned} h_o &= 5.542 - 1.76414 \times 10^{-3} f + 3.05354 \times 10^{-6} f^2 \\ &\quad + \frac{6.815}{(f - 118.75)^2 + 0.321} \quad \text{(km)}. \end{aligned} \tag{2.87}$$

The equivalent atmospheric height for water vapor h_w, for $f \leq 350$ GHz is

$$\begin{aligned} h_w &= 1.65 \left\{ 1 + \frac{1.61}{(f - 22.23)^2 + 2.91} + \frac{3.33}{(f - 183.3)^2 + 4.58} \right. \\ &\quad \left. + \frac{1.90}{(f - 325.1)^2 + 3.34} \right\} \quad \text{(km)}. \end{aligned} \tag{2.88}$$

For an elevation angle θ between 10° and 90° the equivalent path length may be taken as the slant length in a uniform atmosphere with height h_o (for the attenuation due to the oxygen) and h_w (for water vapor):

$$d_{eo} = \frac{h_o}{\sin\theta}, \tag{2.89}$$

$$d_{ew} = \frac{h_w}{\sin\theta}. \tag{2.90}$$

For temperatures other than 15°C equivalent heights must be corrected, increasing 0.1 and 1 percent, percentigrade, in fair weather and in rain, respectively.

For a very steep terrestrial link ($\theta < 10\,^{\circ}$) between two stations at altitudes h_1 and h_2 the values of h_o and h_w in (2.89) and (2.90) should be replaced by h'_o and h'_w

$$h'_o = h_o \left(e^{-h_1/h_o} - e^{-h_2/h_o}\right); \tag{2.91}$$

$$h'_w = h_w \left(e^{-h_1/h_w} - e^{-h_2/h_w}\right); \tag{2.92}$$

where the value of ρ, used to compute γ_w ,is calculated from ρ_1 for the terminal at altitude h_1 by

$$\rho = \rho_1\, e^{\frac{h_1}{2}} \tag{2.93}$$

and $h_w = 2$ km.

According to ITU-R Recommendation P.676-4 [19], for paths with slopes θ in the range 0 –10° the integration of (2.72) leads to

$$A_a = \frac{\sqrt{r}}{\cos\theta}\left[\gamma_o\sqrt{h_o}F\left(\tan\theta\sqrt{\frac{r}{h_o}}\right) + \gamma_w\sqrt{h_w}F\left(\tan\theta\sqrt{\frac{r}{h_w}}\right)\right], \tag{2.94}$$

where r is the equivalent Earth radius (defined in Section 2.5.3), which for most cases may be taken as 8 500 km, and F is a function defined as

$$F(x) = \frac{1}{0.661x + 0.339\sqrt{x^2 + 5.51}}. \tag{2.95}$$

The previous equation is applicable to the path between an Earth station at sea level and a satellite. For paths between two terrestrial stations at altitudes h_1 and h_2 (with $h_2 > h_1$), the following expression should be used

$$\begin{aligned} A_a &= \gamma_0\sqrt{h_o}\left[\frac{\sqrt{r+h_1}F(\chi_1)e^{-h_1/h_o}}{\cos(\theta_1)} - \frac{\sqrt{r+h_2}F(\chi_2)e^{-h_2/h_o}}{\cos(\theta_2)}\right] \\ &+ \gamma_w\sqrt{h_w}\left[\frac{\sqrt{r+h_1}F(\chi'_1)e^{-h_1/h_o}}{\cos(\theta_1)} - \frac{\sqrt{r+h_2}F(\chi'_2)e^{-h_2/h_o}}{\cos(\theta_2)}\right] \end{aligned} \tag{2.96}$$

where θ_1 is the path slope angle at the station at altitude h_1 and

$$\theta_2 = \arccos\left[\frac{r+h_1}{r+h_2}\cos(\theta_1)\right], \tag{2.97}$$

$$\chi_i = \tan(\theta_1)\sqrt{\frac{r+h_i}{h_o}}, \tag{2.98}$$

$$\chi'_i = \tan(\theta_1)\sqrt{\frac{r+h_i}{h_w}}, \tag{2.99}$$

with $i = 1, 2$.

As before, the value of ρ used to compute γ_w corresponds to the sea level value, obtained from ρ_1 for the terminal at the altitude h_1 by (2.93) and $h_w = 2$ km.

In Figure 2.27 we plot the equivalent path length for oxygen attenuation d_{eo} as a function of path length d, computed from (2.96), (2.98), (2.99) and (2.95), at 10 GHz, for the following two cases:

1. plane Earth ($r = \infty$) for $h_1 = 0$ and $h_2 = 200$;

2. spherical Earth ($r = 8\,500$ km) for $h_1 = 0$ e $h_2 = 200$ m.

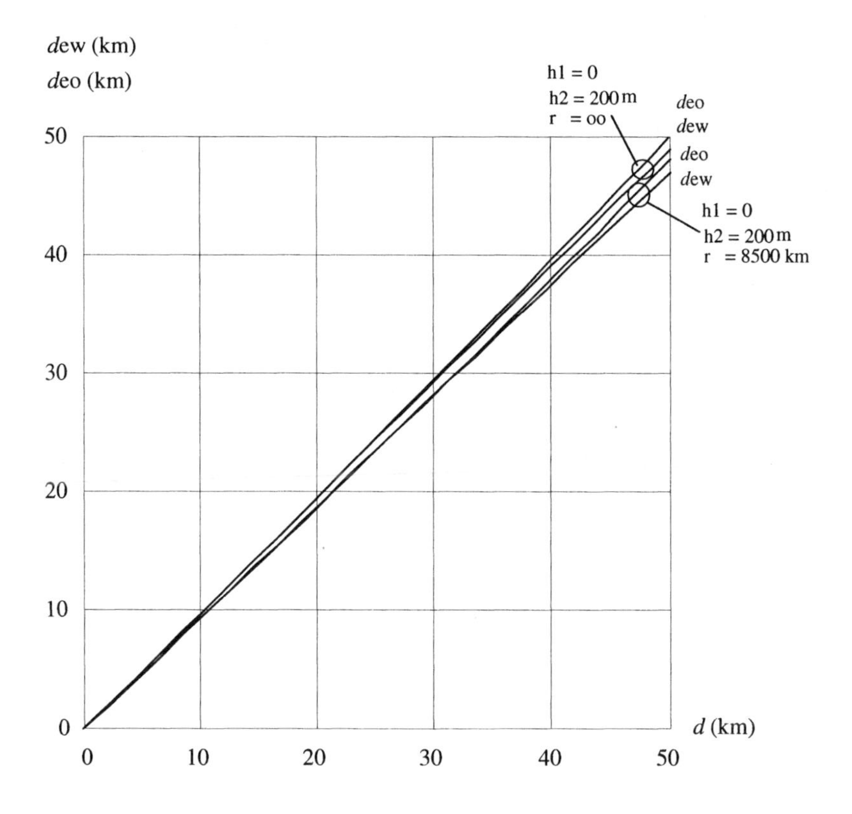

Fig. 2.27 Equivalent path length for oxygen attenuation d_{eo} as a function of path length d, at 10 GHz, for plane and spherical Earth.

As shown in Figure 2.27, the approximation used in Section 2.5.2, from which (2.73) was derived, where $d_{eo} \approx d$ e $d_{ew} \approx d$, is justified in microwave radio link design, even if it may be very slightly pessimistic.

Besides oxygen and water vapor attenuation we must consider the effects of rain and other particles in the atmosphere.

2.5.3 Excess attenuation due to hydrometeors

Hydrometeors (mainly rain) are responsible for absorption, scattering and polarizations changes in radio waves. These effects only matter for frequencies above a few GHz and for high rain intensities.

Rain attenuation may be derived from the classic theory of scattering due to Mie [21]. Assuming spherical rain drops, the attenuation per unit length (γ_R) may be related to rain intensity Ri, expressed in millimeters per hour (mm/h) by Olsen's [24] expression:

$$\gamma_r = k\, Ri^{\alpha}, \tag{2.100}$$

where k and α are functions of frequency, temperature and shape of rain drops and statistical distribution of its sizes. Since rain drop shape is not spherical, rain attenuation per unit length is different (higher) for vertical and for horizontal polarization.

Using the rain drop distribution put forward by Laws and Parsons [22] for rain intensities lower than 50 mm/h, assuming a rain drop temperature of 20 °C and a prolate ellipsoid drop shape such that its volume is the same as for a spherical drop, one may derive the values of k and α, for horizontal and vertical polarization given in Table 2.1.

For intermediate values of the operating frequency f, k and α may be interpolated from the values given in Table 2.1, using a logarithmic scale for k and f and a linear scale for α.

For linear polarization at an angle τ from the horizontal and a elevation angle θ the values of k and α are approximately given as a function of k_H, k_V, α_H and α_V as

$$k = \frac{k_H + k_V + (k_H - k_V)\cos^2\theta\cos 2\tau}{2}, \tag{2.101}$$

$$\alpha = \frac{k_H\alpha_H + k_V\alpha_V + (k_H\alpha_H - k_V\alpha_V)\cos^2(\theta)\cos(2\tau)}{2k}. \tag{2.102}$$

For circular polarization the previous expressions may be used taking $\tau = 45^\circ$.

Some experimental data suggest that for frequencies higher than 40 GHz values of k and α calculated from Table 2.1 may be underestimated and overestimated, respectively.

In Figure 2.28 we plot the rain attenuation per unit length γ_{r_H}, as given by (2.100), for horizontal polarization, with k and α as per Table 2.1, as a function of frequency f, taking as a parameter the rain intensity Ri, expressed in mm/h.

Since rain is a local phenomenon, its effect depends not only on the rain characteristics but also on its spatial distribution. According to the latter, rain may be classified in four types (see Annex I to CCIR Report 563-4 [8]):

Frequency (GHz)	k_H	k_V	α_H	α_V
1	0.0000387	0.0000352	0.912	0.880
2	0.000154	0.000138	0.963	0.923
4	0.000650	0.000591	1.121	1.075
6	0.00175	0.00155	1.308	1.265
7	0.00301	0.00265	1.332	1.312
8	0.00454	0.00395	1.327	1.310
10	0.0101	0.00887	1.276	1.264
12	0.0188	0.0168	1.217	1.200
15	0.0367	0.0335	1.154	1.128
20	0.0751	0.0691	1.099	1.065
25	0.124	0.113	1.061	1.030
30	0.187	0.167	1.021	1.000
35	0.263	0.233	0.979	0.963
40	0.350	0.310	0.939	0.929
45	0.442	0.393	0.903	0.897
50	0.536	0.479	0.873	0.868
60	0.707	0.642	0.826	0.824
70	0.851	0.784	0.793	0.793
80	0.975	0.906	0.769	0.769
90	1.06	0.999	0.753	0.754
100	1.12	1.06	0.743	0.744
120	1.18	1.13	0.731	0.732
150	1.31	1.27	0.710	0.711
200	1.45	1.42	0.689	0.690
300	1.36	1.35	0.688	0.689
400	1.32	1.31	0.683	0.684

Table 2.1 Values of k and α for horizontal and vertical polarizations as a function of frequency (ITU-R Recommendation P.838-1 [19]).

1. Stratified rain: Widespread areas with low rain rates and small embedded showers with rain rates up to 25 mm/h; rainfall is stratified horizontally.

2. Convective rain: Localized areas with relatively high rain rates, with strong ascending and descending wind currents that extend deeply in the troposphere; these areas have columnar shapes and sometimes extend up to the tropopause; very high rain rates with horizontal regions of a few kilometers are possible, but they do not last very long (tens of minutes at most); the more intense downpours in temperate climates are usually of this type.

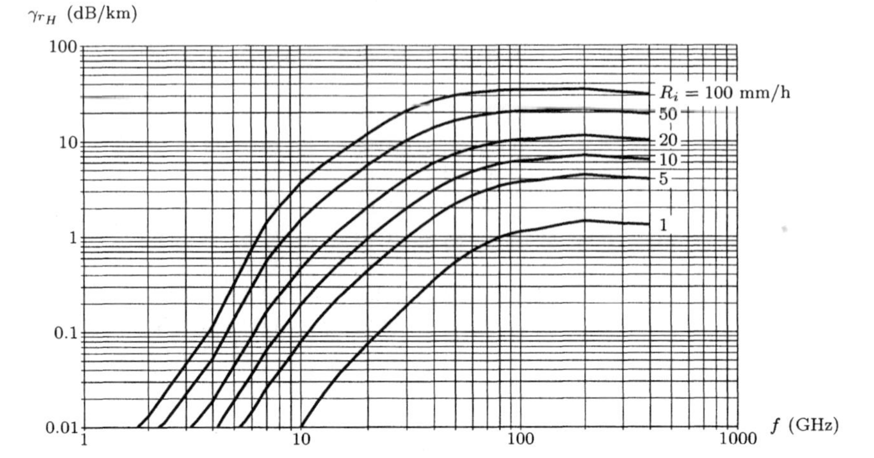

Fig. 2.28 Rain attenuation per unit length γ_{rH} as a function of frequency f, for horizontal polarization, taking rain intensity Ri in mm/h as a parameter (ITU-R Recommendation P.838-1 [19]).

3. Monsoon rain: A sequence of intense convection rain bands followed by periods of stratified rain; rain bands have typical widths of 50 km and are hundreds of kilometers long; rain rates are high and may last for several hours daily.

4. Tropical storm rain: Large organized regions of rain which may extend over hundreds of kilometers; tropical storms exhibit a number of spiral bands that end in high rainfall areas near the central region (the eye) of the storm; spiral bands also have regions with intense convective rain.

For microwave radio link design it is best to get local cumulative rainfall data. Unfortunately, for many places, there are no data on rain intensities for short periods (of the order of minutes), and we are forced to use global values according to the climatic zone of the link.

ITU-R Recommendation P.837-1 [18] divides the Earth in climatic zones and provides for each zone a cumulative distribution of rain intensity (Table 2.2).

The reader is referred to ITU-R Recommendation P.837-2 for a more accurate method to derive the cumulative rain intensity.

In a microwave radio link rain loss $A_r(p)$ exceeded during a given percentage p of the year is calculated from

$$A_r(p) = \int_0^d \gamma_r(p)\, dx, \tag{2.103}$$

Rain intensity (mm/h)		Percentage of time, per year when the rain intensity is exceeded
Zone H	Zone K	
2	1.5	1
4	4.2	0.3
10	12	0.1
18	23	0.03
32	42	0.01
55	70	0.003
83	100	0.001

Table 2.2 Cumulative rain intensity (in mm/h) according to ITU-R Recommendation P.837-1. [19]

or

$$A_r(p) = \int_0^d k(f)\, Ri(p)^{\beta(f)}\, dx. \tag{2.104}$$

ITU-R Recommendation P.530-8 [19] suggests the following method to compute rain loss, not exceeded for more than p percent of year, in a microwave radio link d kilometers long, operating at frequency f with polarization ζ:

1. Find the rain intensity $Ri_{0.01}$ only exceeded for 0.01 percent of the time (with a 1 minute integration time) preferably from local meteorologic data or, alternatively, from ITU-R Recommendation P.837-1 [19].

2. Compute the rain attenuation per unit length γ_r, for rain intensity $Ri_{0.01}$, frequency f and polarization ζ, using data from Table 2.1.

3. Compute the effective path length d_{ef} as

$$d_{ef} = \frac{d}{1 + \frac{d}{d_0}}, \tag{2.105}$$

where d_0 is given by:

$$d_0 = 35e^{-0.015 Ri_{0.01}} \quad \text{for} Ri_{0.01} \leq 100 \text{ mm/h} \tag{2.106}$$
$$= 100 \quad \text{for} Ri_{0.01} > 100 \text{ mm/h}. \tag{2.107}$$

4. Compute the rain loss not exceeded for more than 0.01 percent of time

$$A_r^{(0.01)} = \gamma_r\, d_{ef}. \tag{2.108}$$

5. Rain loss not exceeded for more than p percent of time $(0.001 < p < 1)$ is calculated from $A_r^{(0.01)}$ using

$$A_r^{(p)} = A_r^{(0.01)}\, 0.12\, p^{-(0.546 + 0.043 \log_{10} p)}. \tag{2.109}$$

6. For paths in latitudes below 30° (North or South) rain loss not exceeded for more than p percent of time ($0.001 < p < 1$) (2.109) should be substituted by

$$A_r^{(p)} = A_r^{(0.01)}\, 0.07\, p^{-(0.855+0.139\log_{10} p)}. \tag{2.110}$$

This method is valid at least for frequencies up to 40 GHz and distances up to 60 km. In temperate climates, using measured rain intensities, the average error between computed and measured loss values is less than 10 percent and the standard deviation varies between 20 and 30 percent, according to the percentage of time.

If required, the rain loss for a percentage of the time p_m relative to the "most unfavorable month" may be computed according to ITU-R Recommendation P.841-1 [19] valid for $1.9 \times 10^{-4} < p_m < 7.8$

$$p = 0.3\, p_m^{1.15}. \tag{2.111}$$

The previous equation, albeit slightly modified, may be applied to other phenomena. ITU-R Recommendation P.841-1 [19] suggests that

$$p_m = Qp, \tag{2.112}$$

where Q is given (with p in percent)

$$Q = \begin{cases} 12 & \text{for} \quad p < (\frac{Q_1}{12})^{1/\beta}, \\ Q_1 \cdot p^{-\beta} & \text{for} \quad (\frac{Q_1}{12})^{1/\beta} < p < 3, \\ Q_1 \cdot 3^{-\beta} & \text{for} \quad 3 < p < 30, \\ Q_1 \left(\frac{p}{30}\right)^{\frac{\log(Q_1 \cdot 3^{-\beta})}{\log(0.3)}} & \text{for} \quad p > 30. \end{cases} \tag{2.113}$$

Values of Q_1 and β are region and effect dependant. For planning purposes we use $Q_1 = 2.85$ and $\beta = 0.13$, which lead to (2.111). In Europe, for multipath, we may take $Q_1 = 4.0$ and $\beta = 0.13$, while for troposcatter links over land we should take $Q_1 = 3.3$ and $\beta = 0.18$ and over sea $Q_1 = 5.0$ and $\beta = 0.11$.

When a microwave radio link has more than one hop, the probability of link interruption, due to intense rain, is equal to the sum of the probabilities for each hop, provided these are long (over about 40 km). If this is not the case, the probability of link interruption is significantly less than the sum of the probabilities for each hop, and the lesser the shorter the hops and higher the number of hops.

In microwave radio link design we normally do not consider:

- attenuation due to fog, which, for frequencies below 100 GHz, does not exceed the attenuation due to low intensity rain (1 mm/h);

- attenuation due to hail, whose probability of occurrence is usually less than 0.001 percent of the time, per year.

2.5.4 Excess attenuation due to sand and dust

According to various authors referred to in CCIR Report 721-3 [8] we may state that the excess attenuation per unit length due to the presence of sand and dust in the atmosphere:

- is directly proportional to the frequency and inversely proportional to the optical visibility and depends markedly on the relative particle humidity;
- is lower than 0.1 dB/km and 0.4 dB/km, at 10 GHz, for particle concentration, respectively, of sand and clay, lower than 10^{-5}g/cm^3;
- only interferes significantly in microwave radio link operation when the optical visibility is reduced to 10 to 20 m or when particle humidity is very high.

2.5.5 Refractive effects

Besides excess losses, due to the presence of gases and hydrometeors (of which rain is the most important) the atmosphere is responsible for modifications in the direction of propagation, due to changes in the refractive index along the ray path.

The atmospheric refractive index n is a function of atmospheric pressure p (in hPa), water vapor pressure e (in hPa) and absolute temperature T (in K). According to ITU-R Recommendation P.453-6 [19] for the usual frequency range, the atmospheric refractive index may be written as

$$n = 1 + N \times 10^{-6}, \tag{2.114}$$

where N, the refractivity, is given by

$$N = \frac{77.6}{T}\left(p + \frac{4810\,e}{T}\right). \tag{2.115}$$

Water vapor pressure e may be computed for each temperature:

- from the relative humidity and the saturation pressure of water vapor e_s, given as a function of the temperature T by (2.81) and in Figure 2.24;
- from the water vapor density ρ (in g/m^3) and the temperature T, using 2.82).

In average conditions, with:

$$\begin{aligned} p &= 1017 \text{ hPa}, \\ e &= 10 \text{ hPa (50\% relative humidity)}, \\ T &= 291.3K = 18^\circ C, \end{aligned}$$

we get

$$N = 314.9,$$
$$n = 1.0003149.$$

The refractive index varies mainly with the altitude h. According to ITU-R Recommendation P.453-7 [19]

$$n(h) = 1 + a\, e^{-\frac{h}{h_0}}, \tag{2.116}$$

where a and h_0 are climate-dependent constants. For a typical atmosphere, with h in kilometers we have

$$\begin{aligned} a &= 0.000315, \\ h_0 &= 7.35. \end{aligned}$$

If (2.116) is approximated by a linear relation

$$n(h) = n_0 - \Delta n \times h, \tag{2.117}$$

a valid approximation for the lower troposphere where microwave radio link propagation mostly takes place, it is possible to demonstrate that the ray path curvature is equivalent to a straight line ray path over a spherical Earth with an equivalent radius r

$$r = k_e\, r_0, \tag{2.118}$$

where r_0 is the Earth radius ($r_0 \approx 6\,370$ km) and k_e is given by

$$k_e = \frac{1}{1 - \frac{r_0}{n_0}\,\Delta n}. \tag{2.119}$$

The value of Δn is normally calculated as the ratio of the difference of the refraction indexes at the Earth surface and at the altitude of 1 km (or 0.1 km) and difference of altitudes. The minus sign in (2.116) is not universally adopted; here we use it to simplify plotting of Δn.

For a typical atmosphere we have

$$\begin{aligned} n_0 &= 1.000315, \\ \Delta n &= \left(-\frac{\partial n}{h}\right)_{h=0} \\ &= 43 \times 10^{-6}\ \mathrm{km}^{-1}. \end{aligned}$$

Often the value of Δn is multiplied by 10^6 and the result is expressed in N/km. Thus $\Delta n = 40 \cdot 10^{-6}\ \mathrm{km}^{-1}$ becomes $\Delta N = 40$ N/km.

Experimentation shows that average values of Δn for the first 100 m lie between 70 N/km, exceeded less than 0.05 per cent of the time, and -200 N/km, exceeded for more than 99.9 per cent of the time.

For $k_e > 1$ ($\Delta n > 0$) the equivalent Earth radius increases with k_e and for ($\Delta N \approx 157$ N/km) we reach $k_e = \infty$, that is the Earth becomes plane.

For $\Delta n > 157$ N/km the value of k_e becomes negative, which means that the Earth curvature changes and the Earth surface becomes concave. For $\Delta n < 0$ the value of k_e decreases, thus decreasing the equivalent Earth radius and the radio horizon. In Figure 2.29 we represent the effect in the ray path of varying k_e.

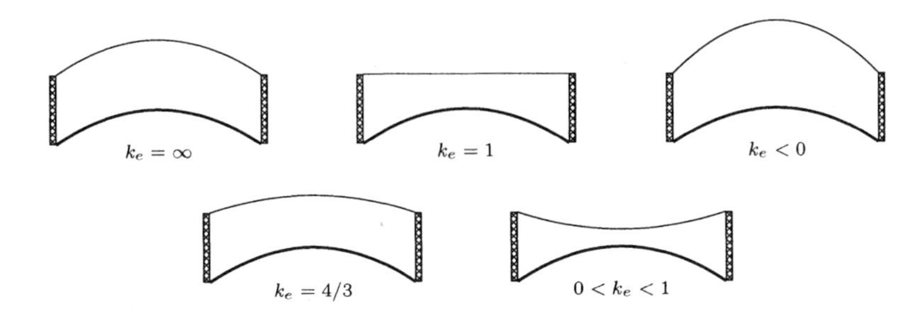

Fig. 2.29 Effect on the ray path of varying k_e, for a constant Earth radius and a varying ray path (physical model).

In microwave radio link design, instead of computing the ray path, it is much simpler to take the ray path as a straight line and to modify the Earth curvature by changing the real radius r_0 into the equivalent radius r_e. The process is shown in Figure 2.30.

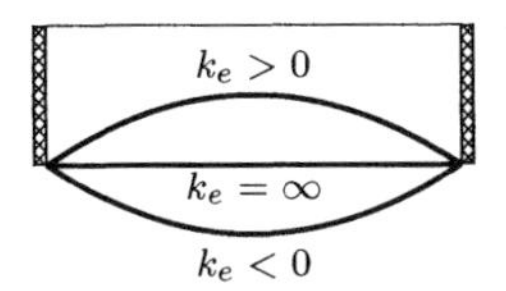

Fig. 2.30 Effect of varying k_e on the ray path, considering a constant path and changing the Earth equivalent radius (practical model).

World charts with the average values of refractivity N_0 and Δn at sea level, for February and August are given in Figures 2.31 to 2.34.

Since k_e varies with time and distance along the ray path, we must define criteria to choose the design value. Basically k_e depends on the desired quality of service and path location.

To ensure an unobstructed ray path in temperate climates it is usual to clear the first Fresnel ellipsoid for $k_e = 4/3$. Based on experimental evidence that this criterion is insufficient for long paths (over about 100 km) some designers also demand an unobstructed direct ray for $k_e = 1$.

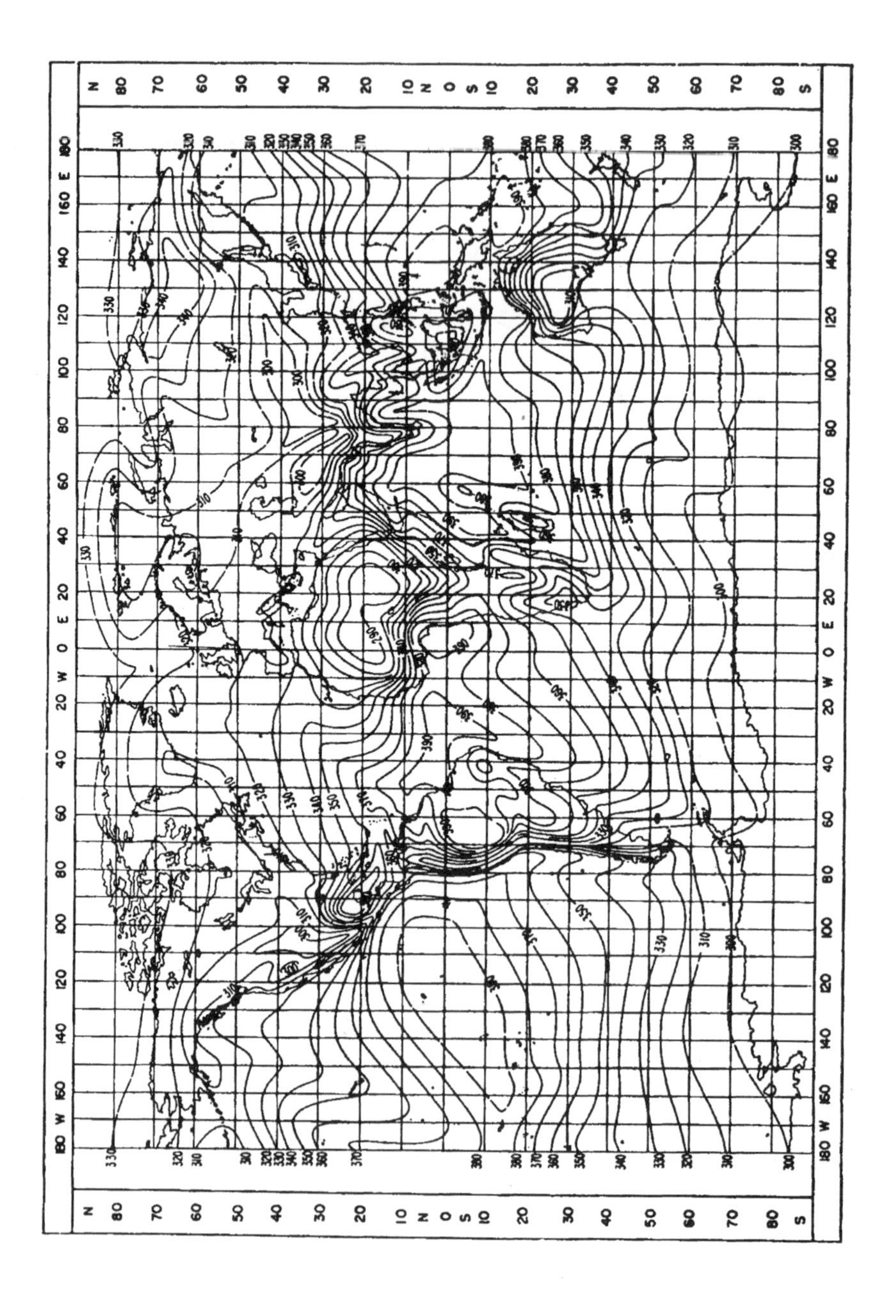

Fig. 2.31 Average sea level value of N_0 in February (ITU-R Recommendation P.453-6 [19]). Reproduced with ITU permission.

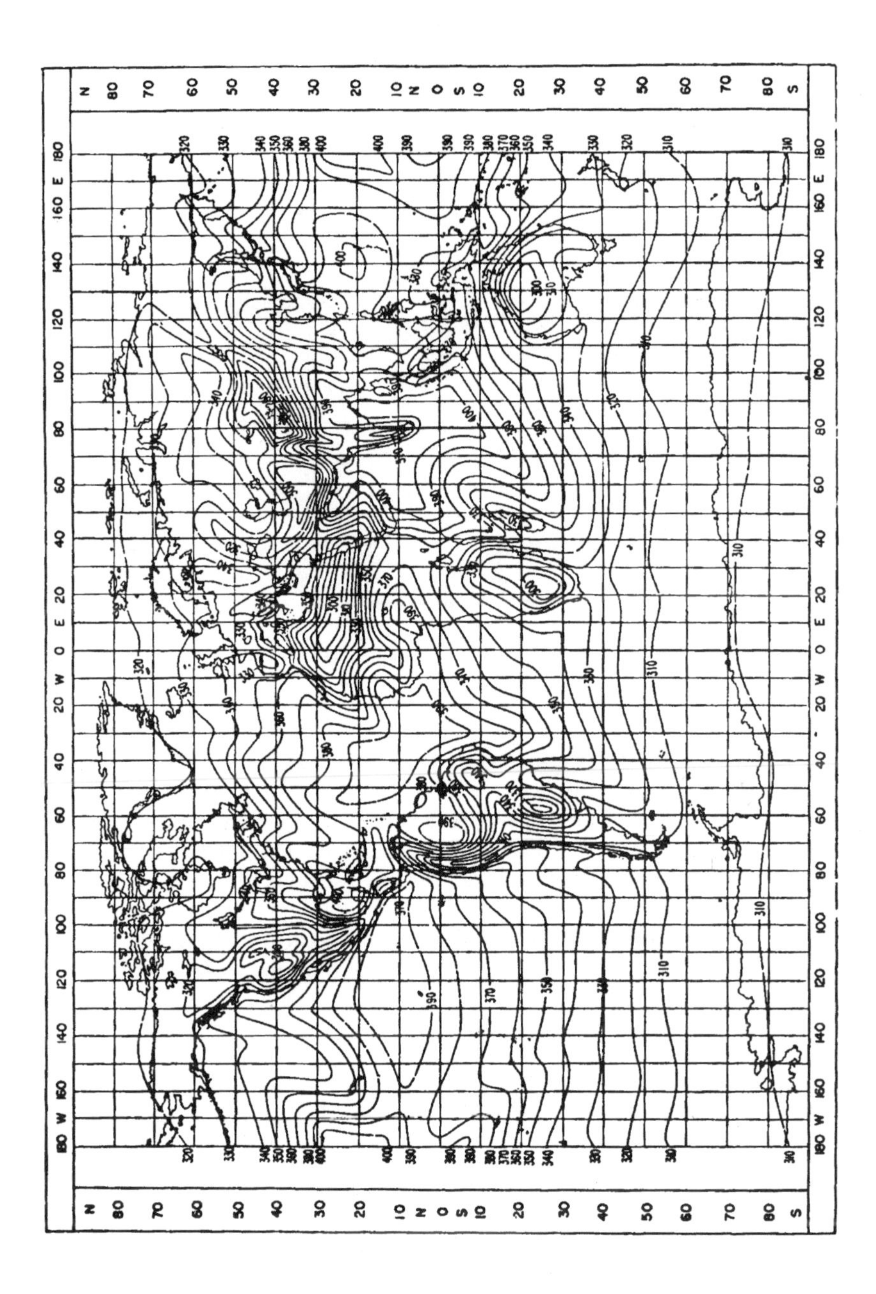

Fig. 2.32 Average sea level value of N_0 in August (ITU-R Recommendation P.453-6 [19]). Reproduced with ITU permission.

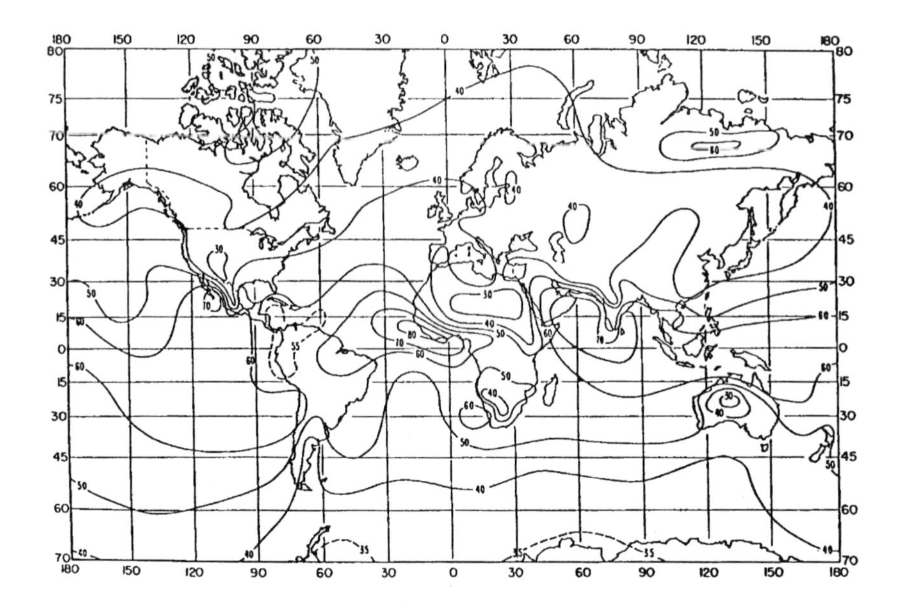

Fig. 2.33 Average sea level value of ΔN in February (ITU-R Recommendation P.453-6[19]). Reproduced with ITU permission.

Another, usually stiffer criterion, requires a clearance of at least 60 percent of the first Fresnel ellipsoid for $k_e = 0.66$. Analysis of a large number of links suggests that this criterion is likely to be excessive (and thus uneconomical).

For frequencies above 2 GHz, ITU-R Recommendation P.530-8 [19] suggests yet another criterion, where k_e and the unobstructed fraction of the first Fresnel ellipsoid are a function of path length d and propagation conditions.

Let $k_{e_{min}}(d)$, given in Figure 2.35 (due to Boithias and Battesti [5]), be the minimum value of k_e (exceeded in 99.9 % of the time) as a function of the path length d, for temperate continental climate, and r_{1e} the radius of the first Fresnel ellipsoid at the obstacle. In the design of microwave radio links we should ensure the most severe of the following criteria:

- clearance of r_{1e} for the appropriate value of k_e (in the absence of data take $k_e = 4/3$);
- clearance of $0.6r_{1e}$ for $k_{e_{min}}(d)$, for $d > 30$ km, in tropical climate;
- clearance of the direct ray, for $k_{e_{min}}(d)$, in temperate climate when there is a single obstacle to the direct ray on the path;
- clearance of $0.3r_{1e}$ for $k_{e_{min}}(d)$, in temperate climate, if obstacles in the path are extended.

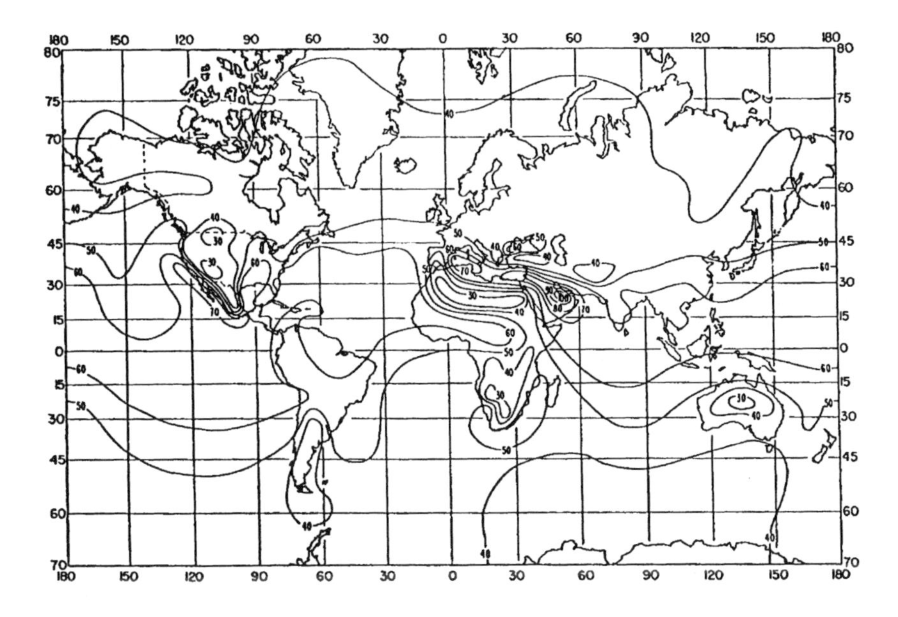

Fig. 2.34 Average sea level value of ΔN in August (ITU-R Recommendation P.453-6[19]). Reproduced with ITU permission.

For frequencies below 2 GHz, lower clearances are sometimes adopted to avoid very high masts. On the other hand, for frequencies above about 10 GHz higher clearances may be required to reduce the risk of diffraction in sub-refractive conditions.

When we use two antennas on the same mast, for space diversity (see Section 3.7) the lower antenna should ensure a clearance of $0.6r_{1e}$ to $0.3r_{1e}$ for the appropriate value of k_e (in the absence of data take $k_e = 4/3$) for an extended obstacle or $0.3r_{1e}$ to $0.0r_{1e}$ if there are only one or two isolated obstacles. The lower limits should only be applied to frequencies up to 2 GHz and when increasing masts height is impractical.

Alternatively the lower antenna may be placed so that, in normal propagation conditions, it experiences a diffraction loss of 6 dB which, for a single knife edge obstacle, corresponds to the direct ray grazing the obstacle.

We should recall that the purpose of all clearance criteria is to enable to consider the path as unobstructed, that is, with no obstacle losses. This does not mean that paths that do not obey those criteria should not be considered but rather that, in those cases, we must include the obstacle losses in the propagation loss.

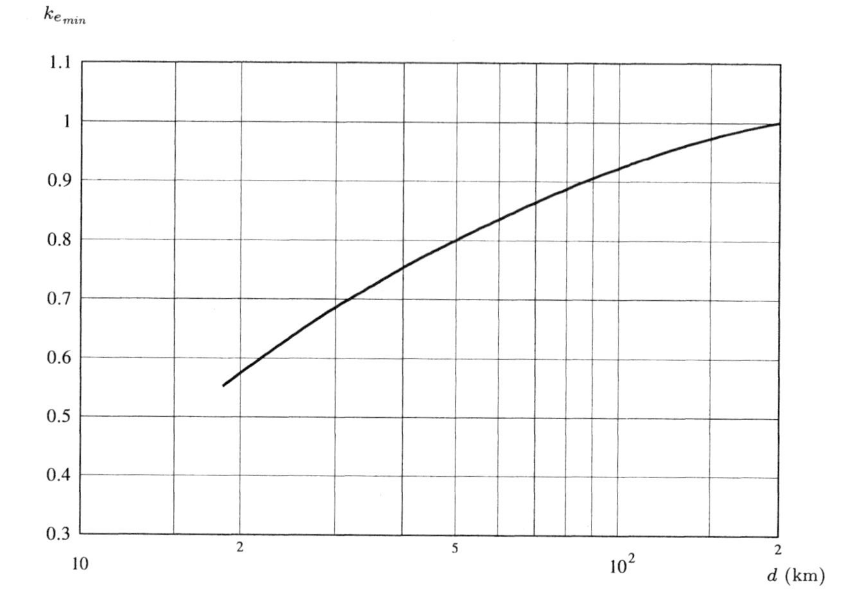

Fig. 2.35 Minimum value of k_e (exceeded during 99.9 % of time) $-k_{e_{min}}-$ as a function of path length in continental temperate climate (ITU-R Recommendation P.530-8 [19]). Reproduced with ITU permission.

2.6 TROPOSPHERIC ANOMALIES

The troposphere only behaves normally – refraction index decreasing exponentially with altitude as per (2.116) – when atmospheric turbulence is enough to achieve an adequate mix. These circumstances are almost always met over rough ground or mountainous areas. Over flat land, or in protected valleys, particularly during the night or the first hours after sunrise, tropospheric anomalies may seriously affect microwave radio link performance.

To analyze such perturbations we will look at the variation of the modified index of refraction M defined as

$$M(h) = N + 10^6 \log_e \left(1 + \frac{h}{r_0}\right), \tag{2.120}$$

where N is the refractivity given in (2.115) and r_0 is the real radius of the Earth ($r_0 = 6370$ km). Since $h \ll r_0$, $M(h)$, expressing h in kilometers, M may be approximated as

$$M = N + 157\,h. \tag{2.121}$$

From (2.117) we have

$$N = N_0 - \Delta N\, h, \tag{2.122}$$

and thus

$$M = N_0 + (157 - \Delta N)\, h. \tag{2.123}$$

For a typical atmosphere, with $N_0 = 315$ and $\Delta N = 43$

$$M = 315 + 114\, h. \tag{2.124}$$

In Figure 2.36 we plot $M(h)$ for various types of atmospheres, including the typical one, with M along the $\overline{OX}$ axis as traditionally.

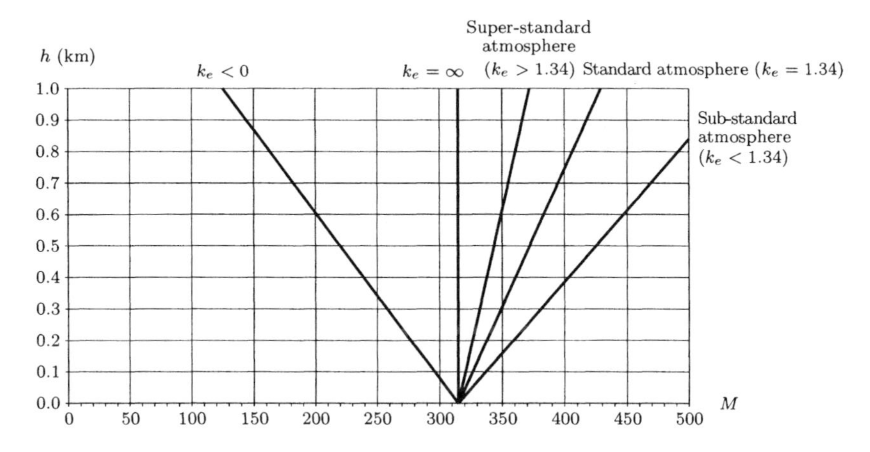

Fig. 2.36 Modified index of refraction M, for different atmospheres.

For sub-standard atmospheres, with $\Delta N < 43$, $M(h)$ remains a straight line but its slope differs from the typical atmosphere and decreases with decreasing ΔN. On the contrary, for super-standard atmospheres, with $\Delta N > 43$ the slope increases and reaches the vertical for $\Delta N = 157$, which corresponds to $k_e = \infty$. For $\Delta N > 157$ the modified refraction index decreases with altitude, which corresponds to $k_e < 0$, that is, a concave Earth surface rather than a convex one.

In certain places, due to orographic and meteorological reasons, sometimes there may be an abnormal variation of the modified refraction index. Such anomalies include surface layers with properties which are markedly different from the remaining atmosphere (both sub-standard and super-standard) and ducts. In Figure 2.37 we plot $M(h)$ for these anomalies.

Some surface layers, such as those caused by a layer of humid air over certain types of irrigated crops, or by morning fogs, give rise to strong reflections. On the other hand, ducts allow signals, with the appropriate frequencies (which depend on their vertical size), to be trapped and thus to propagate inside them with attenuations far below those of free space. Ducts may be responsible for abnormally high ranges, and may cause undesirable interferences with far away links (operating in the same frequency).

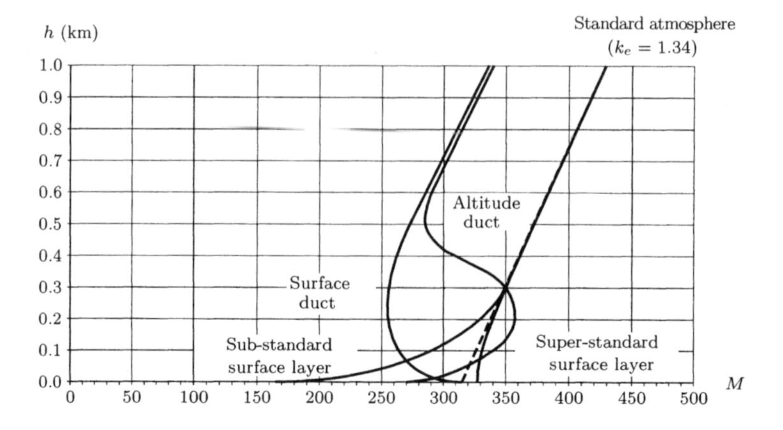

Fig. 2.37 Modified refraction index as a function of altitude for tropospheric anomalies.

Conditions that may originate ducts include temperature increase with altitude (also known as temperature inversion) and humidity decrease with altitude. These conditions arise when higher atmospheric layers are abnormally warm and dry compared with the lower ones.

Over land ducts tend to occur on clear windless Summer nights, particularly when the ground is wet. The land becomes colder and its temperature lowers but there is hardly any change in the upper atmospheric layers. During daytime, when the convective currents and the wind agitate the troposphere, ducts disappear.

Over the sea ducts may be due to warm air blown from overland which gets on top of lower temperature atmospheric layers causing a temperature inversion. These conditions are more frequent to the leeward of land masses, may arise both in daytime and during the night and last for quite some time. They are more likely at the end of the afternoon and at night, when the warm land wind blows over the sea.

Ducts over land and over sea are markedly different, since the temperature changes are far more abrupt in the former than in the latter. Over land, ducts tend to have smaller sizes and last for shorter periods than over the sea, while exhibiting significant changes during daytime.

Ducts may also form during thunderstorms: the lowering cold air may form a temperature inversion, particularly when, at the same time, there is a favorable humidity gradient. Thunderstorm ducts are neither as frequent nor as important as those previously referred to. They deserve a mention because, due to the associated radar range increase, they may be used to detect thunderstorms.

Finally we should like to recall that ducts are rather thin (from a few meters up to a few tens of meters) and thus only higher frequencies (above 1 GHz) are affected by them.

2.7 PATH PROFILE

For the design of microwave radio links we must include the actual orographic data in the path profile. Since the radius of the first Fresnel ellipsoid, that should be cleared, is of the order of tens of meters, the orographic data needs only to be analyzed along an arc of great circle or geodesic between terminals.

For distances up to about 70 km we may approximate the geodesic by a rhumb line between the terminals. This line intersects all meridians at the same angle and, on usual survey charts, is represented as a straight line. For distances larger than about 70 km and when the available maps do not represent the great circle as a straight line it is safer to compute the coordinates of a few intermediate points (about 7.5 minutes apart) along the path and approximate the great circle by an open polygonal line connecting these points.

Let t_T, g_T and t_R, g_R be, respectively, the latitude and the longitude of the stations. The longitudes are taken as positive to the east of the Greenwich meridian and negative to the west. Let x and y be the angles in each station, of the geodesic with the local meridian (true north). Assume R to be the point with the highest absolute value of the latitude. Taking

$$c = |g_R - g_T|, \tag{2.125}$$

we get[15]

$$\tan\left(\frac{y-x}{2}\right) = \cot\left(\frac{c}{2}\right) \frac{\sin\left(\dfrac{t_R - t_T}{2}\right)}{\cos\left(\dfrac{t_R + t_T}{2}\right)}, \tag{2.126}$$

$$\tan\left(\frac{y+x}{2}\right) = \cot\left(\frac{c}{2}\right) \frac{\cos\left(\dfrac{t_R - t_T}{2}\right)}{\sin\left(\dfrac{t_R + t_T}{2}\right)}. \tag{2.127}$$

Path length d is calculated as

$$d = r_0\, \theta, \tag{2.128}$$

with $r_0 = 6370$ km and θ in radians.

The value of θ is

$$\tan\left(\frac{\theta}{2}\right) = \tan\left(\frac{t_R - t_T}{2}\right) \frac{\sin\left(\dfrac{y+x}{2}\right)}{\sin\left(\dfrac{y-x}{2}\right)}. \tag{2.129}$$

According to Rice [26] to compute the coordinates t_P and g_P of intermediate points along the great circle path:

- For predominantly east-west paths and for a given difference of longitudes c_P relative to the T terminal, the latitude t_P is given by:

$$\cos(t_P) = \frac{\sin(x)\ \cos(t_T)}{\sin(y_P)}, \tag{2.130}$$

  where y_P is:

$$\cos(y_P) = \sin(x)\ \sin(c_P)\ \sin(t_T) - \cos(x)\ \cos(c_P); \tag{2.131}$$

- For predominantly north-south paths and for a latitude t_P, the longitude difference c_P in relation to terminal T is

$$\cot\left(\frac{c_P}{2}\right) = \tan\left(\frac{y_P - x}{2}\right) \frac{\cos\left(\dfrac{t_P + t_T}{2}\right)}{\sin\left(\dfrac{t_P - t_T}{2}\right)}, \tag{2.132}$$

  with

$$\sin(y_P) = \frac{\sin(x)\ \cos(t_T)}{\cos(t_P)}; \tag{2.133}$$

- For paths with azimuths close to 45°, any of the above procedures may be used.

Once the great circle path has been defined and plotted, we must plot the ground profile, using maps with level curves at 10-meter (or at most 20-meter) altitude intervals, which correspond to 1/25 000 horizontal scale. For a single path quite a few topographic sheets are usually required.

In the path profile we should include not only the ground but also the top of likely obstacles such as trees and buildings.

After plotting the path profile we determine the antenna height aiming at an unobstructed direct ray but placing antennas as low as possible. Antenna height above ground is usually limited to about 30 – 40 meters for economic reasons.

Next we plot the first Fresnel ellipsoid and adjusting antenna height as requiredcheck if the clearance criteria are met. If this is the case the path is unobstructed, otherwise it is obstructed.

Unobstructed paths behave essentially like free space paths, except for atmospheric losses and ground scattering (dealt with in the next section). Obstructed paths may be classified, according to the dominant propagation mechanism, in diffraction paths and troposcatter paths. Methods to estimate path losses for both cases are presented later on in this chapter.

2.8 GROUND SCATTERING

2.8.1 Introduction

For line-of-sight paths the received power is partly associated with the direct ray and the first Fresnel ellipsoid and partly with ground reflections or scattering.

When the ground can be assumed to be smooth and homogenous, ground reflected power may be calculated as described in Section 2.3 (plane Earth) and 2.4 (spherical Earth). In both cases the power associated with the reflected and the direct ray are of the same order of magnitude and the interference of both gives rise to very large fluctuations of the received signal power. These are dependent on the phase difference between the two paths, which in turn is a function of the path and the continuously varying atmospheric conditions.

To minimize fluctuations in the received power, also known as multipath fading, one or more of the following methods may be used:

- Avoid paths through extended flat regions (sea, lakes or swamps);
- Employ high-directivity antennas able to discriminate between the directions of the direct and the reflected rays;
- Tilt antennas upwards to improve the discrimination between direct and reflected rays, at a cost of a small (1 to 2 dB) gain loss, for the direct ray [14];
- Locate one antenna much higher than the other, so that the specular point moves closer to the lower antenna, and thus facilitates, for the latter, the discrimination between direct and reflected rays;
- Choose antenna placement and height so that the ground obstructs the reflected ray.

The effect of ground scattering on line-of-sight links may be estimated given the terrain statistical properties [27], using the method described in the following subsections.

2.8.2 Power scattered by an elementary ground area

Assume two antennas with gains g_T and g_R (in the directions of one to the other) at distance d, in free space. Let p_T be the power, at frequency f, applied to the transmitter antenna with gain g_T. The power p_R at the receiver antenna, assumed to be in the far field zone, is given by (2.4).

Assume further that the two antennas are placed, respectively, at heights h_T and h_R, above a plane, infinite, isotropic rough ground in such a way that the first Fresnel ellipsoid is cleared. Under these conditions the receiving antenna besides power p_d, associated with the direct ray, will also receives a fraction of the transmitter power p_S reflected or scattered by the ground.

Take a rectangular coordinate system centered on the transmitter antenna as shown in Figure 2.38. Assuming the validity of the far field expressions, the power dp_S scattered by the element of ground with area $dxdy$, centered on point P, with coordinates $(x, y, 0)$ is

$$dp_S = p_T \frac{\lambda^2}{4\pi} \frac{g_T^P}{4\pi(x^2 + y^2 + h_T^2)} \frac{g_R^P}{4\pi[(d-x)^2 + y^2 + h_R^2]} \sigma \, dxdy, \qquad (2.134)$$

where σ is the effective scattering section in the direction of the receiver antenna, per unit area, and g_T^P and g_R^P are the gains of the transmitter and the receiver antenna, respectively, in the direction of P .

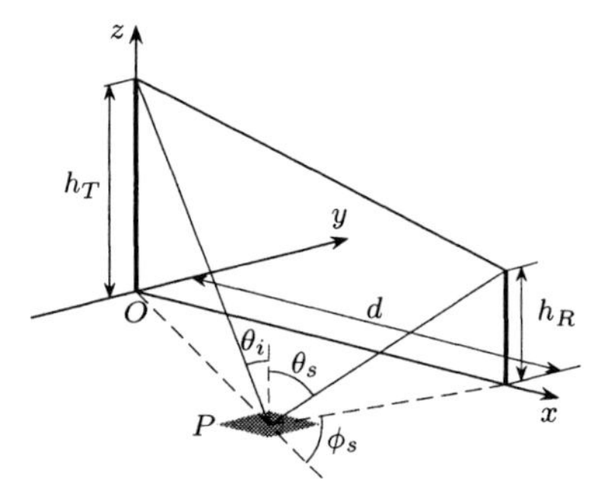

Fig. 2.38 Coordinate system for ground scattering calculations.

If:

- the wavelength is much smaller than the correlation length of the rough surface, and thus geometric optics approximations are valid,
- multiple scattering may be neglected,
- surface roughness and correlation length are Gaussian,

Barrick [2] showed that

$$\sigma = \frac{(1 + \tan^2 \gamma)^2}{s^2} e^{-\frac{\tan^2 \gamma}{s^2}} |R_{\zeta\eta}(l)|, \qquad (2.135)$$

where

$$\tan^2 \gamma = \frac{\sin^2 \theta_i - 2 \sin \theta_i \sin \theta_s \cos \phi_s + \sin^2 \theta_s}{(\cos \theta_i + \cos \theta_s)^2}, \qquad (2.136)$$

and:

- $R_{\zeta\eta}(l)$ is the Fresnel reflection coefficient on the elementary reflecting facets for polarization η of the incident ray and ζ for the reflected ray;

- s is a parameter related to the slope of the surface roughness, defined as the ratio between the standard deviation and the correlation length of the surface.

Inspection of Figure 2.38 shows that:

$$\cos\theta_i = \frac{h_T}{\sqrt{x^2+y^2+h_T^2}}, \tag{2.137}$$

$$\cos\theta_s = \frac{h_R}{\sqrt{(d-x)^2+y^2+h_R^2}}, \tag{2.138}$$

$$\cos\phi_s = \frac{x(d-x)-y^2}{\sqrt{x^2+y^2}\sqrt{(d-x)^2+y^2}}. \tag{2.139}$$

Power received through ground scattering p_S will be:

$$p_S = \int_A dp_S, \tag{2.140}$$

where the integration area A, theoretically extends to the entire surface, but practically is limited to an area which we will refer to as the active scattering area.

The chosen geometry and the nature of the equations derived do not help to visualize the results. It is thus useful to make a few restrictions and simplifying assumptions.

In the following we will assume, as is usual in the design of microwave radio links, that;

- antenna height is much smaller the distance between antennas, and thus, excluding the region close to the antennas we may take ground incidence as grazing and, for all polarizations, $|R_{\zeta\eta}| = 1$;

- s is small, that is, $s^2 \ll 1$.

Under these assumptions σ as defined in (2.135), only differs significantly from zero when $\tan\gamma$ is of the same order of magnitude as $\tan^2\ \gamma \ll 1$). Hence

$$\sigma = \frac{1}{s^2}\, e^{-\frac{\tan^2\gamma}{s^2}}. \tag{2.141}$$

Excluding the region close to the antennas and restricting y to the range $y \ll x$ and $y \ll d-x$, we obtain:

$$\cos\theta_i = \frac{h_T}{x}, \tag{2.142}$$

$$\sin\theta_i = 1 - \frac{h_T^2}{2(x^2+y^2)}, \quad (2.143)$$

$$\cos\theta_s = \frac{h_R}{d-x}, \quad (2.144)$$

$$\sin\theta_s = 1 - \frac{h_R^2}{2[(d-x)^2+y^2]}, \quad (2.145)$$

$$\cos\phi_s = 1 - \frac{y^2\, d^2}{2\, x^2 (d-x)^2}, \quad (2.146)$$

and finally:

$$\tan^2\gamma = \frac{1}{4}\left(\frac{h_R}{d-x} - \frac{h_T}{x}\right)^2 + \frac{y^2\, d^2}{[h_T d + x(h_R - h_T)]^2}. \quad (2.147)$$

2.8.3 Effective scattering area

We will begin the analysis of the scattering area along the $\overline{OX}$ axis, that is, the line joining the projections of the antennas on the scattering surface. Following we will repeat the analysis on a line normal to the previous one.

From (2.141) and (2.147) we get, for $y = 0$

$$\sigma = \frac{1}{s^2}\,\exp\left[-\frac{1}{4s^2}\left(\frac{h_R}{d-x} - \frac{h_T}{x}\right)^2\right]. \quad (2.148)$$

The maximum value of σ coincides with the specular point where

$$x_e = \frac{d}{1+\frac{h_R}{h_T}} \quad (2.149)$$

and its value is:

$$\sigma_{max} = \frac{1}{s^2}. \quad (2.150)$$

Taking the following normalized parameters

$$t = \frac{2h_T}{sd}, \quad (2.151)$$

$$u = \frac{x}{d}, \quad (2.152)$$

$$r = \frac{h_R}{h_T}, \quad (2.153)$$

equation (2.148) may be written as

$$\frac{\sigma}{\sigma_{max}} = \exp\left[-\frac{t^2}{16}\left(\frac{r}{1-u} - \frac{1}{u}\right)^2\right]. \quad (2.154)$$

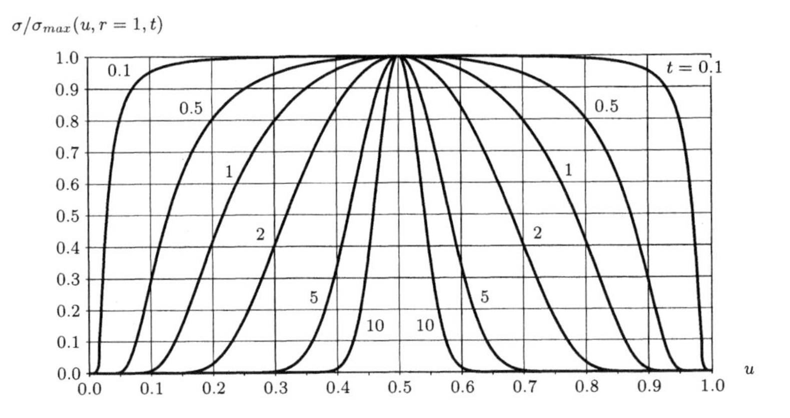

Fig. 2.39 Plot of σ/σ_{max} as a function of u for antennas at the same height ($r = 1$) with parameter t.

In Figure 2.39 we plot σ/σ_{max} as a function of u for $r = 1$ (antennas at equal height) taking t as a parameter. As shown, for $t < 1$, we get $\sigma/\sigma_{max} \approx 1$ for all u, except for the regions close to the antennas. In other words, the surface tends to act as a perfect scatterer. As the value of t increases σ/σ_{max} only differs from zero in a shrinking region, close to the specular point. The surface approaches a perfect reflector.

Note that, for antennas at the same height ($r = 1$), t represents the ratio between the tangent of the grazing angle at the specular point and the average slope of the surface roughness.

We plot on Figures 2.40 and 2.41 σ/σ_{max} as a function of u for two values of t, one corresponding to a "reflective" and the other to a "scattering" ground. It is obvious from these Figures, particularly from the former, the displacement of the maximum of σ/σ_{max} toward the specular point. In any case we get a significant increase of σ/σ_{max} in the regions near the lower antenna.

Let us define the longitudinal length ρ_a of the effective scattering area as the length, along the $\overline{OX}$ axis, where

$$\frac{\sigma}{\sigma_{max}} \geq e^{-\tau^2}, \tag{2.155}$$

τ being an arbitrary positive value larger than 1. From, (2.154) and (2.155) we get

$$\rho_a = d\,|u_1 - u_2|, \tag{2.156}$$

where u_1 e u_2 are solutions of the equation:

$$\left|\left(\frac{r}{1-u} - \frac{1}{u}\right)\right| = \frac{4\tau}{t}. \tag{2.157}$$

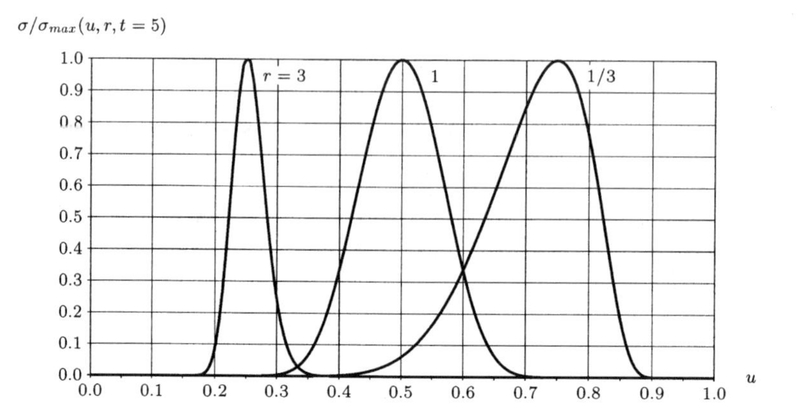

Fig. 2.40 Plot of σ/σ_{max} as a function of u, taking r as a parameter for $t = 5$ ("reflective" ground).

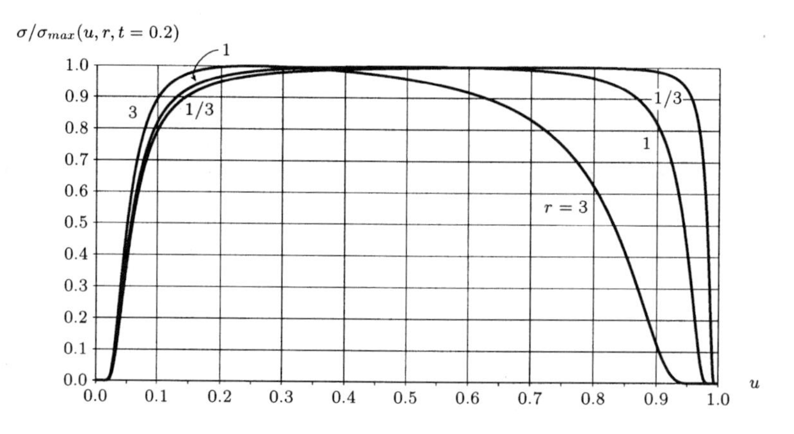

Fig. 2.41 Plot of σ/σ_{max} as a function of u, taking r as a parameter for $t = 0.2$ ("scattering" ground).

For large values of t, that is, when the ground acts as a "reflector" the function

$$f(u) = \frac{r}{1-u} - \frac{1}{u} \tag{2.158}$$

may be expanded in series around $u = 1/(r+1)$. Taking the first two terms in this expansion we get approximate values for u_1 e u_2 and hence

$$\rho_a \approx \frac{8\,\tau\,d}{t} \cdot \frac{r}{(r+1)^3}$$

$$\approx \frac{4\,\tau\, s\, d^2}{h_T} \cdot \frac{r}{(r+1)^3}. \tag{2.159}$$

For small values of t, that is, for a "scatterer" ground, neglecting the contributions of the first term of $f(u)$ in (2.158) near $u = 0$ and of the second term near $u = 1$, we obtain the following approximate value for ρ_a:

$$\rho_a \approx d - \frac{d\,t}{4\,\tau}(r+1). \tag{2.160}$$

We proceed by considering the variation of σ/σ_{max} along the $\overline{OY}$ axis, which is normal to the $\overline{OX}$ axis with the origin on the specular point. From (2.141), (2.147) and (2.150), we get

$$\frac{\sigma}{\sigma_{max}} = \exp\left[-\frac{y^2}{h_T^2\, s^2 \left(1 + \frac{r-1}{r+1}\right)^2}\right]. \tag{2.161}$$

Defining a new normalized variable

$$v = \frac{y}{h_T\, s}, \tag{2.162}$$

equation (2.161) becomes

$$\frac{\sigma}{\sigma_{max}} = \exp\left[-\frac{v^2}{(1 + \frac{r-1}{r+1})^2}\right]. \tag{2.163}$$

In Figure 2.42 we plot $\sigma/\sigma_{max}\,(v,r)$ for the same values of r used in Figures 2.40 and 2.41. As expected, the value of $\sigma/\sigma_{max}\,(v,r)$ reaches the maximum, equal to 1, for $v = 0$, that is, on the specular point, regardless of r. For antennas at the same height ($r = 1$), $\sigma/\sigma_{max}\,(v,r)$ only differs substantially from zero for $v < 3$, that is, for $y < 3\,s\,h_T$. For antennas at different heights ($r \neq 1$), the same generic conclusion holds. The maximum distance from the specular point at which σ/σ_{max} is significantly different from zero, increases, in relation to the case of antennas at the same height for $r > 1$ and decreases for $r < 1$.

For typical values of s ($s < 1$) we have thus fully justified the initial restriction placed on y, that is, $y \ll x$ and $y \ll d - x$.

We must stress that, for the most interesting cases ($s < 0.1$), σ/σ_{max} is always practically zero for values of y less than the antenna height, thus validating the usual practice of restricting the analysis of ground effects to a profile between the two antennas.

As as for the $\overline{OX}$ axis let us now define the transversal length ρ_t of the effective scattering area as the length measured along a straight line parallel to the $\overline{OY}$ axis, passing through the specular point, along which the condition in (2.155), is met, with an arbitrary value of $\tau > 1$. We have

$$\rho_t = |y_1 - y_2|,$$

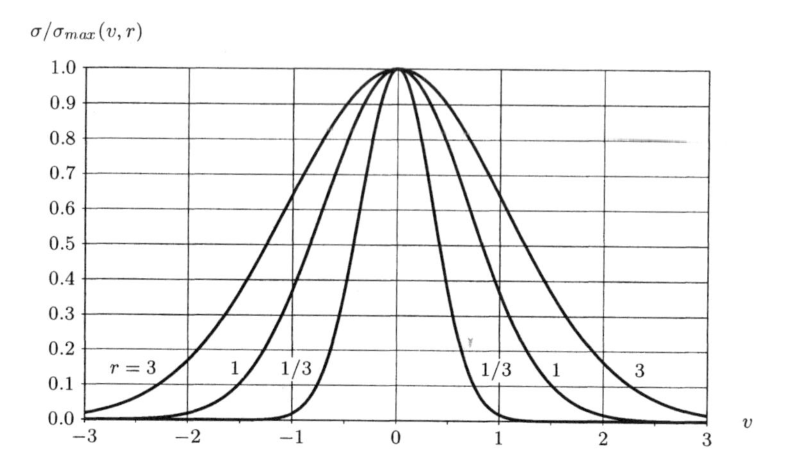

Fig. 2.42 Plot of $\sigma/\sigma_{max}(v)$ taking r as a parameter.

where y_1 and y_2 are solutions of

$$\frac{y}{h_T\, s}\left(1+\frac{r-1}{r+1}\right) = \pm\,\tau. \tag{2.164}$$

From the previous equations we may get

$$\rho_t = 2\,\tau\, s\, h_T\left(1+\frac{r-1}{r+1}\right). \tag{2.165}$$

For good "reflectors" the effective scattering area A_d, which may be approximated by the product of ρ_a and ρ_t, is independent of h_T

$$\begin{aligned} A_d &= \rho_a\,\rho_t \\ &= 16\,d^2\,\tau^2\,\frac{s^2r^2}{(r+1)^4}. \end{aligned} \tag{2.166}$$

2.8.4 Power scattered in the direction of the receiver antenna

The importance of the element of area, with dimensions $dxdy$, centered on point P, with coordinates x, y, is not only dependent on the value of the effective scattering area, but also on the position of point P in relation to the antennas. This position influences the value of the effective scattering area, the propagation losses between the antennas and point P and the antenna gains.

Assuming that:

- g_T^P and g_R^P are the transmitter and the receiver antenna gains in the direction of P, respectively,

- far field expressions are valid,
- the immediate vicinity of both antennas is excluded,
- the value of σ only differs from zero on a region close to the straight line that joins the projections of the antennas on the reference surface,
- antenna heights are both much smaller than distance between them,

we may write the power dp_S scattered by the element of area in the direction of the receiver antenna as a function of the transmitter power p_T, using (2.134)

$$dp_S = p_T\, \sigma_{max}\, g_T^P\, g_R^P\, f(x,y)\, \frac{\lambda^2}{4\pi}\, dxdy, \tag{2.167}$$

where

$$f(x,y) = \frac{\sigma(x,y)}{\sigma_{max}} \cdot \frac{1}{16\,\pi^2\, x^2 (d-x)^2} \tag{2.168}$$

and $\sigma(x,y)/\sigma_{max}$ is given from (2.141), (2.147) and (2.150), by

$$\frac{\sigma}{\sigma_{max}} = \exp\left[-\frac{t^2}{16}\left(\frac{r}{1-\frac{x}{d}} - \frac{1}{\frac{x}{d}}\right)^2\right] \exp\left\{-\frac{y^2}{s^2\, h_T^2\left[1+\frac{x}{d}(r-1)^2\right]^2}\right\}, \tag{2.169}$$

where parameters t and r are defined in (2.151) and (2.153).

The ground scattered power p_S in the direction of the receiver antenna is

$$\begin{aligned} p_S &= \int_A dp_S \\ &= p_T\, \sigma_{max}\, \frac{\lambda^2}{4\pi} \int_A g_T^P\, g_R^P\, f(x,y)\, dxdy, \end{aligned} \tag{2.170}$$

where the integration area A extends to the whole scattering surface.

Since:

$$\int_{-\infty}^{\infty} e^{-x^2/a^2}\, dx = \sqrt{\pi}\, a \tag{2.171}$$

and $f(x,y)$ only differs appreciably from zero inside the active scattering area, which is very narrow, enabling antenna gains to be taken as constant with y

$$p_S = p_T\, \sigma_{max}\, \frac{\lambda^2}{4\pi} \int_{x_1}^{x_2} g_T^P\, g_R^P\, f(x)\, dx, \tag{2.172}$$

where

$$f(x) = \frac{\sqrt{\pi}\, s\, h_T\left[1+\frac{x}{d}(r-1)\right]}{16\,\pi^2\, x^2\,(d-x)^2} \exp\left[-\frac{t^2}{16}\left(\frac{r}{1-\frac{x}{d}} - \frac{1}{\frac{x}{d}}\right)^2\right]. \tag{2.173}$$

We may rewrite (2.170) and (2.172) in terms of the free space received power p_D and variable $u = x/d$

$$\frac{p_S}{p_D} = \int_{u_1}^{u_2} gr_T \, gr_R \, g(u) \, du, \tag{2.174}$$

with

$$g(u) = \frac{t}{8\sqrt{\pi}} \frac{1 + u(r-1)}{u^2(1-u)^2} \exp\left[-\frac{t^2}{16}\left(\frac{r}{1-u} - \frac{1}{u}\right)^2\right], \tag{2.175}$$

where:

- gr_T and gr_R are the transmitter and receiver gain in the direction of point P divided by their gains in the direction of each other, respectively:

$$gr_T = \frac{g_T^P}{g_T}; \tag{2.176}$$

$$gr_R = \frac{g_R^P}{g_R}; \tag{2.177}$$

- the integration limits u_1 and u_2 are given by

$$u_1 = \frac{x_1}{d}; \tag{2.178}$$

$$u_2 = \frac{x_2}{d}. \tag{2.179}$$

It is instructive to analyze the ratio p_S/p_D defined in (2.174) for a number of simple cases.

Let us start by taking the case of isotropic transmitter and receiver antennas, that is, $gr_T = gr_R = 1$. It follows from (2.175) that

$$\frac{p_S}{p_D} = 1. \tag{2.180}$$

It is well known from geometric optics that for antennas at heights much smaller than the distance between them, over a perfect reflector plane, powers received through the direct and the reflected rays are equal. Equation (2.180) generalizes the previous result to rough planes, obeying Barrick's conditions [2]).

Another interesting case, closer to reality, is of antennas with hemispheric radiation patterns. In this case we get

$$\frac{p_S}{p_D} = \int_0^1 \frac{t}{8\sqrt{\pi}} \frac{1 + u(r-1)}{u^2(1-u)^2} \exp\left[-\frac{t^2}{16}\left(\frac{r}{1-u} - \frac{1}{u}\right)^2\right] du. \tag{2.181}$$

The ratio p_S/p_D is now a function of of t and r plotted in Figure 2.43.

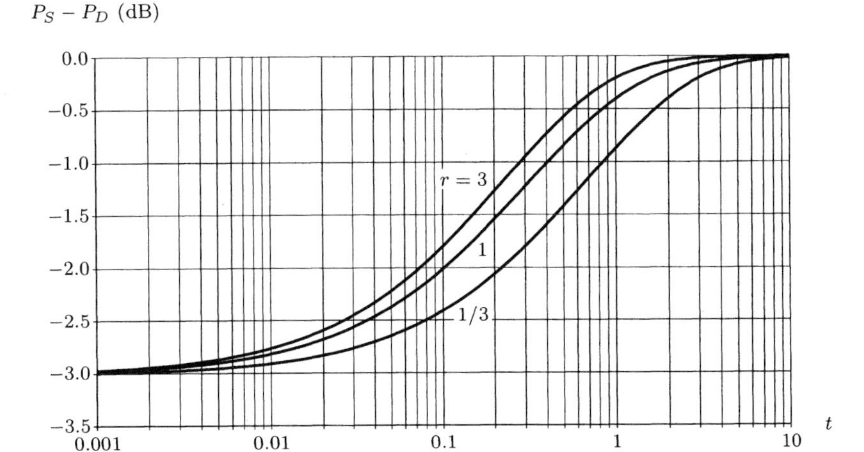

Fig. 2.43 Plot of p_S/p_D as a function of t taking r as a parameter (antennas with hemispheric radiation patterns).

From Figure 2.43 we find out that:

- For $t > 1$, we get $p_S/p_D \approx 1$, meaning that all scattered power comes from the region between the antennas;
- For lower values of t ($t < 10^{-2}$), we get $p_S/p_D \simeq 0.5$.

Finally, we will consider the case of a perfect reflector ground, that is, the limit when $t \to \infty$. From (2.173) we realize that the integrand will only be significantly different from zero in the vicinity of the specular point, for which

$$u_e = \frac{1}{1+r}. \tag{2.182}$$

In this vicinity we have

$$\begin{aligned} \frac{r}{1-u} - \frac{1}{u} &= \frac{\left(\frac{1}{1+r} + \eta\right)(1+r) - 1}{\left(\frac{1}{1+r}\right)\left(1 - \frac{1}{1+r}\right)} \\ &= \frac{\eta (r+1)^3}{r}, \end{aligned} \tag{2.183}$$

where η is a variable which tends to zero as t tends to infinity.

Thus:

$$\frac{p_S}{p_D} = \lim_{t\to\infty} \left\{ \frac{t}{8\sqrt{\pi}} \frac{1 + \frac{r-1}{r+1}}{\frac{1}{(r+1)^2}\left(1 - \frac{1}{r+1}\right)} \, gr_T \, gr_R \int_{-\infty}^{\infty} \exp\left[-\frac{t^2}{16}\frac{(r+1)^6}{r^2}\eta^2\right] d\eta \right\}, \tag{2.184}$$

or

$$\frac{p_S}{p_D} = g r_T \, g r_R, \tag{2.185}$$

a well-known result from geometric optics often used in the design of microwave radio links.

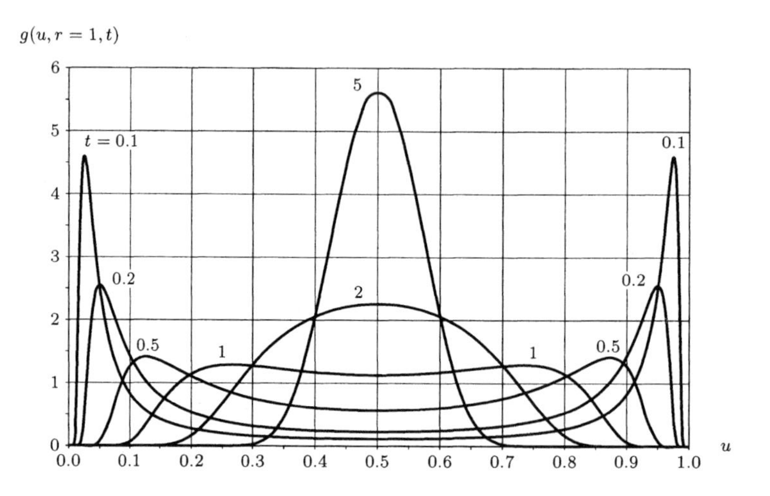

Fig. 2.44 Plot of $g(u, r, t)$ for $r = 1$.

To evaluate the importance of antenna radiation patterns we must examine, in more detail, the plot of $g(t, r, u)$ as given in (2.175) and shown in Figure 2.44 for $r = 1$. For large values of t, that is, for $t > 1$, the area contributing to the scattered power is rather limited. On the contrary, as the value of t decreases, not only this area increases, but also the major contribution for the integral comes from regions progressively closer to the antennas, irrespective of the position of the specular point. This behavior is particularly obvious for $t \leq 0.1$. Under these conditions directive antennas may strongly decrease the value of p_S/p_D. Since in real microwave radio links, highly directive antennas are almost always used, we may safely conclude that rough terrain hardly ever causes difficulties as far as ground reflections are concerned.

In Figure 2.45 we plot $g(u)$ for $t = 5$ ("reflecting" ground) taking r as a parameter. Now the ground scattered power comes only from the immediate vicinity of the specular point and the more so the larger the value of r.

In Figure 2.46 we plot$g(u)$ for $t = 0.2$ ("scattering" ground) taking r as a parameter. It should be clear that in this case the most important contribution for the ground scattered power comes from the region closer to the antennas. For antennas at different heights ($r \neq 1$) the scattered power is mainly dependent from the region closer to the lower antenna.

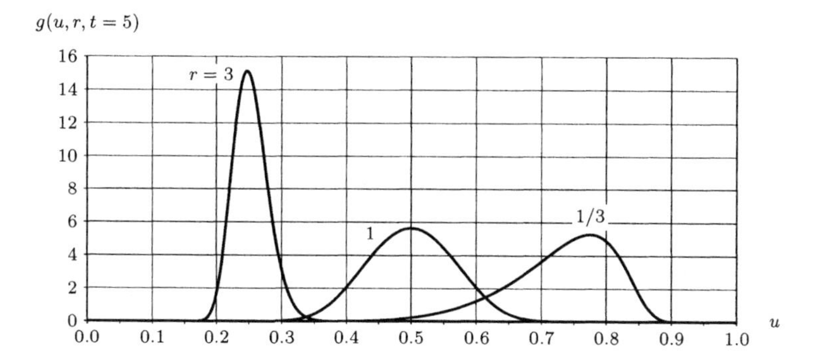

Fig. 2.45 Plot of $g(u, r, t)$ for $t = 5$.

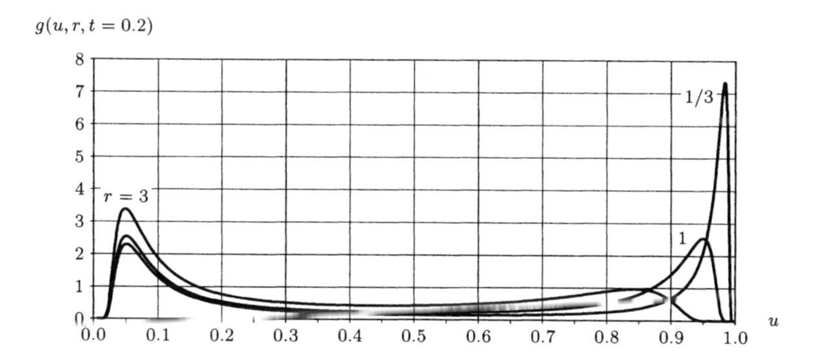

Fig. 2.46 Plot of $g(u, r, t)$ for $t = 0.2$.

2.8.5 Estimating ground scattered power

To apply the method described in the previous subsections let us start by defining the value of k_e. Usually we take the most common value in the region (often $k_e = 1.33$). We then plot the path profile, locate the antennas and establish their height (Figure 2.47). We will assume that the path is unobstructed.

We proceed by identifying the path sections which are likely to contribute to the ground scattered power. Following the previous assumptions we will exclude all profile sections which are not seen simultaneously by both antennas.

Assume that the path contains a single section likely to contribute to the ground scattered power, beginning at distance x_1 from the antenna on the left and ending at distance n_2 from the same antenna. Assume further that this section of the profile (represented in Figure 2.47 by a thick trace open

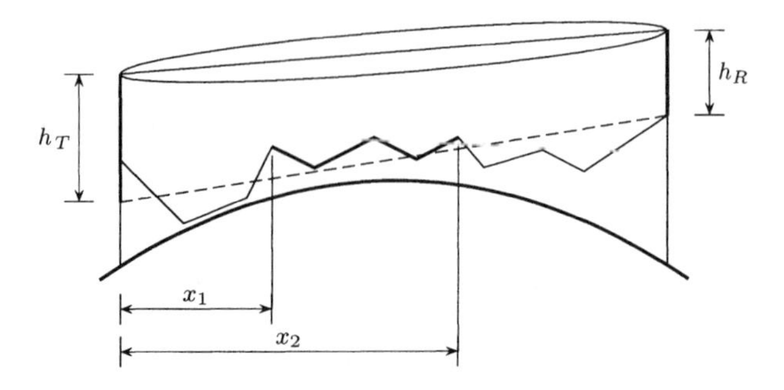

Fig. 2.47 Plot of the path profile used to estimate ground scattered power.

polygon) may be substituted by a plane rough surface (a broken line in the same Figure). Let h_E and h_R be the antenna heights above this plane surface. Assume a surface roughness s. The ratio between the scattered power p_S and the direct ray power p_D is given by (2.174).

Long path sections ($x_2 - x_1 > 10$ km) must be divided in smaller subsections and, for each one, we must define the values of x_1, x_2, h_T and h_R. In this way Earth curvature is accounted for.

In the general case, where along the path there are a number of sections likely to provide sizable contributions to the scattered power, we must take each section in turn, and add the values of the ratios p_S/p_D for each section, a procedure which often requires the use of a computer program. Alternatively, if a rough estimate is acceptable, we may use the following approximate method.

From the main path characteristics we calculate the value of t and classify the path section as "reflecting" ($t \geq 1$), "scattering" ($t \leq 0.1$) or "intermediate" ($0.1 < t < 1$). The limits, for each type of terrain, must not be taken rigidly but as orders of magnitude.

For "reflecting" terrain, we start by locating the specular point using (2.149). We then compute the length of the active scattering area (2.159) using $\tau \approx 2$.

If most of the active scattering area is located on the chosen path section

$$\frac{p_S}{p_D} \approx gr_T \, gr_R, \tag{2.186}$$

provided gr_T and gr_R do not vary along the active scattering area.

If the active scattering area does not include the path section then

$$\frac{p_S}{p_D} \approx 0. \tag{2.187}$$

For a "scattering" path section, comparing (2.174), (2.175) and (2.154) we may write

$$\frac{p_S}{p_D} = \frac{t}{8\sqrt{\pi}} \int_{u_1}^{u_2} gr_T\, gr_R \frac{1+u(r-1)}{u^2(1-u)^2} \frac{\sigma(u)}{\sigma_{max}}\, du, \tag{2.188}$$

for all u, except in the vicinity of the antennas.

All that is required now is the variation of gr_T and gr_R with u. Assume the following:

$$gr_T(\varphi)\, gr_R(\varphi) = \begin{cases} 1 & \text{if} \quad \varphi < \varphi_0 \\ 0 & \text{if} \quad \varphi > \varphi_0 \end{cases},$$

where φ is the angle with the direct ray and φ_0 is the -3 dB half beamwidth.

When φ_0 is not known (from the antenna radiation patterns) it may be derived from the following approximate expression:

$$\varphi_0 = \frac{1}{2}\sqrt{\frac{4\,\pi}{g}} \quad \text{(em radians)}. \tag{2.189}$$

The simplified assumptions about antenna radiation patterns may lead to a change in u_1 and u_2, in (2.174), since the function in the integral vanishes for u such that $\varphi > \varphi_0$ for both antennas. Excluding from the path under analysis the region which is not "seen" by both antennas, we obtain new limits u_{1m} e u_{2m}. Equation (2.181) may be rewritten as

$$\frac{p_S}{p_D} = \frac{t}{8\sqrt{\pi}} \int_{u_{1m}}^{u_{2m}} \frac{1+u(r-1)}{u^2(1-u)^2}\, du, \tag{2.190}$$

or

$$\frac{p_S}{p_D} = t\left\{[F(u_{2m}) - F(u_{1m})] + (r-1)[G(u_{2m}) - G(u_{1m})]\right\}, \tag{2.191}$$

where

$$F(u) = \frac{1}{8\sqrt{\pi}}\left[\frac{1}{1-u} - \frac{1}{u} + 2\log_e\left(\frac{u}{1-u}\right)\right], \tag{2.192}$$

$$G(u) = \frac{1}{8\sqrt{\pi}}\left[\log_e(u[1-u]) + \frac{1}{1-u}\right]. \tag{2.193}$$

To simplify calculations, we plot $F(u)$ and $G(u)$, given in (2.192) and (2.193), in Figures 2.48 and 2.49, respectively.

2.9 PROPAGATION BY DIFFRACTION

2.9.1 Introduction

In an obstructed path we have propagation by diffraction when diffraction is the dominant mechanism ensuring propagation over the obstacles. This

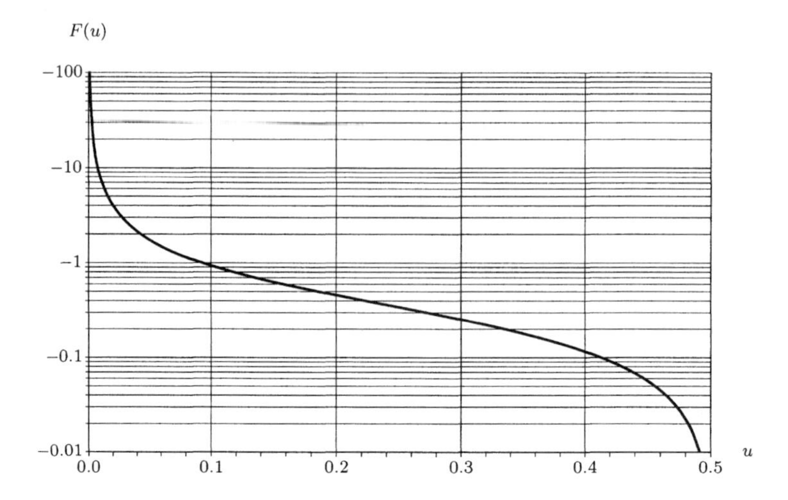

Fig. 2.48 Plot of $F(u)$ (we note that $F(1-u) = -F(u)$).

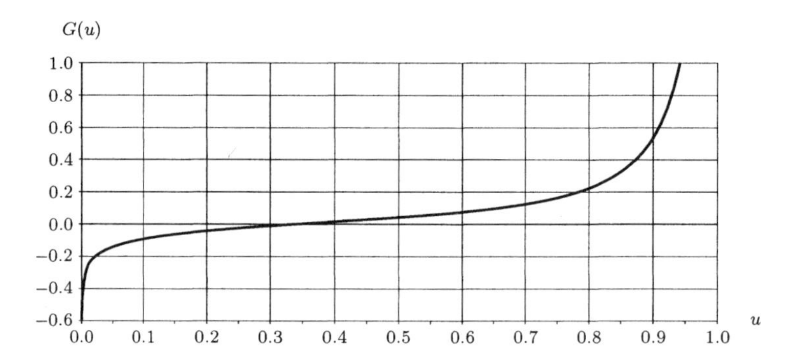

Fig. 2.49 Plot of $G(u)$.

is usually the case in paths with a common horizon, that is, paths where the radio horizon is the same from both terminals, or when radio horizons are but a few kilometers apart. When the distance between radio horizons exceed about a few tens of kilometers the dominant propagation mechanism is tropospheric scattering.

To calculate the attenuation loss caused by real obstacles these are modelled by ideal obstacles, usually with infinite length (normal to the propagation direction) and neglecting width (knife edge), or finite width and rounded top, with constant curvature (cylindrical obstacles). It is also possible [13],[7], though less common, to make use of ideal finite width obstacles whose profile, normal to the propagation direction is a rectangle or a triangle. In any case the results must be used with caution since their validity is strongly dependent on how well the ideal obstacle approximates the real one. The methods described in the following always assume that the obstacle size is much larger than the wavelength.

The existence of isolated obstacles, knife edge or rounded, leads to obstacle losses which are much lower than those which would be found if, on the same path, the obstacle did not exist and the propagation was to be by diffraction over spherical Earth. Thus the obstacle may be seen as bringing in an "obstacle gain" [11] which may reach some tens of dB. We must not mistake this "obstacle gain" with the obstacle loss (relative to free space) which, in certain cases, may be negative and thus become an "obstacle gain".

In this section we describe the ITU-R Recommendation P.526-6 [19] method to calculate the attenuation by diffraction in obstructed paths.

2.9.2 Diffraction over spherical Earth

Let us start by considering that the obstacle is the Earth surface (see Figure 2.50), assumed to be a sphere with effective radius r (in the usual conditions $r = 8500$ km).

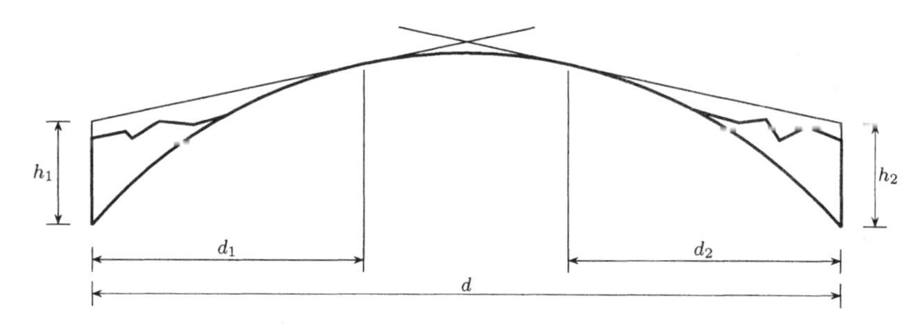

Fig. 2.50 Geometry for the calculation the diffraction losses over a spherical Earth.

Let d be the path length, h_1 and h_2 the antenna heights, and f the operating frequency (wavelength λ). The attenuation between terminals is equal to the sum of the free space attenuation plus the atmospheric attenuation plus the attenuation by diffraction A_d given by ITU-R [19] Recommendation P.526-6

$$A_{d_{[dB]}} = -[F(X) + G(Y_1) + G(Y_2)], \tag{2.194}$$

where X is the normalized path length and Y_1 and Y_2 the normalized antenna heights, given in coherent units

$$X = \beta d \left(\frac{\pi}{\lambda r^2}\right)^{1/3}, \tag{2.195}$$

$$Y_1 = 2\beta h_1 \left(\frac{\pi^2}{\lambda^2 r}\right)^{1/3}, \quad (2.196)$$

$$Y_2 = 2\beta h_2 \left(\frac{\pi^2}{\lambda^2 r}\right)^{1/3}, \quad (2.197)$$

$$(2.198)$$

where β is a parameter that depends on the nature of the ground and the wave polarization and is related with the generalized surface admittance $\mathcal{K}$ by

$$\beta = \frac{1 + 1.6\mathcal{K}^2 + 0.75\mathcal{K}^4}{1 + 4.5\mathcal{K}^2 + 1.35\mathcal{K}^4}. \quad (2.199)$$

For horizontal polarization at all frequencies, and for vertical polarization over land with $f > 20$ MHz or over sea with $f \geq 300$ MHz, we may take $\beta \approx 1$.

Functions $F(X)$ and $G(Y)$ are given by

$$F(X) = 11 + 10\log_{10}(X) - 17.6X, \quad (2.200)$$

$$G(Y) = 17.6\sqrt{Y - 1.1} - 5\log_{10}(Y - 1.1) - 8 \text{ para } Y \geq 2. \quad (2.201)$$

For $Y < 2$, $G(Y)$ is dependent on generalized surface admitance $\mathcal{K}$:

$$G(Y) \approx \begin{cases} 20\log_{10}(Y + 0.1Y^3) & \text{for } \mathcal{K} < Y < 2, \\ 2 + 20\log_{10}\mathcal{K} + 9\log_{10}(\frac{Y}{\mathcal{K}})\left[\log_{10}(\frac{Y}{\mathcal{K}}) + 1\right] & \text{for } \mathcal{K}/10 < Y < 10\mathcal{K}, \\ 2 + 20\log_{10}\mathcal{K} & \text{for } Y < \mathcal{K}/10. \end{cases} \quad (2.202)$$

$\mathcal{K}$ is a function of the wave polarization and takes the following values, in coherent units:

$$\mathcal{K}_H = \left(\frac{2\pi r}{\lambda}\right)^{-1/3} \left[(\varepsilon - 1)^2 + (60\lambda\sigma)^2\right]^{-1/4}, \quad (2.203)$$

$$\mathcal{K}_V = \mathcal{K}_H\sqrt{\varepsilon^2 + (60\lambda\sigma)^2}, \quad (2.204)$$

where

- ε is the surface relative dielectric permittivity given by the ratio of the dielectric permittivities for the surface ε_s and for vacuum ε_o

$$\varepsilon = \frac{\varepsilon_s}{\varepsilon_o};$$

- σ is the surface electric conductivity;
- λ is the wavelength.

In Figure 2.51 we plot the variation of the generalized surface admittance $\mathcal{K}$ with frequency f, for sea water ($\varepsilon = 81$ and $\sigma = 5\ \Omega^{-1}\text{m}^{-1}$) and average

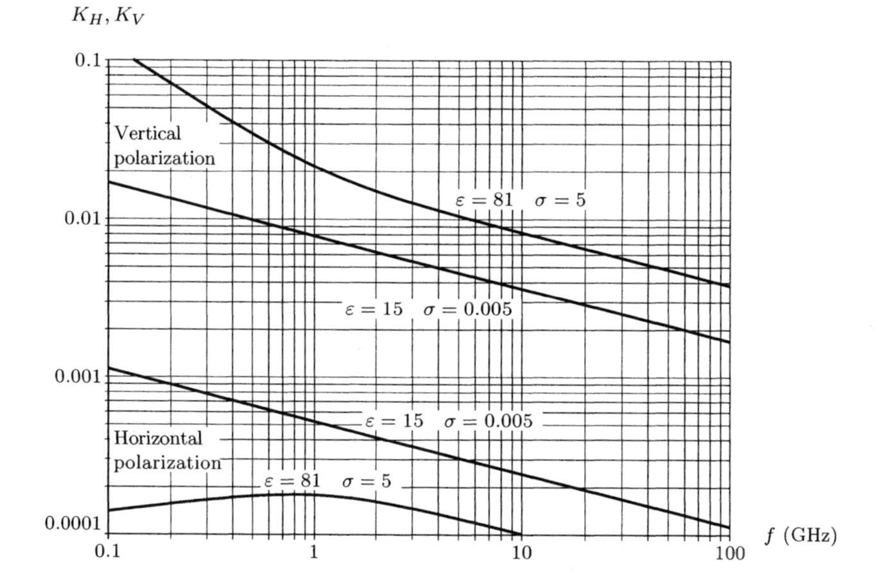

Fig. 2.51 Plot of the generalized surface admittance $\mathcal{K}$ with frequency f, for sea water ($\varepsilon = 81$ and $\sigma = 5\ \Omega^{-1}\mathrm{m}^{-1}$) and average ground ($\varepsilon = 15$ and $\sigma = 0.005\ \Omega^{-1}\mathrm{m}^{-1}$) for horizontal and vertical polarization.

ground ($\varepsilon = 15$ and $\sigma = 0.005\ \Omega^{-1}\mathrm{m}^{-1}$) for horizontal and vertical polarization.

Attenuation by diffraction A_d over the spherical Earth increases sharply with path length between radio horizons d_{de} defined as

$$d_{de} = d - \sqrt{2rh_1} - \sqrt{2rh_2}, \tag{2.205}$$

and, at most after a few tens of kilometers, the dominant propagation mechanism ceases to be diffraction and becomes tropospheric scattering.

2.9.3 Diffraction over an isolated knife edge

Assume a knife edge obstacle made up by a perfect conductor half plane, normal to the direction of propagation. In this ideal case the obstacle may be defined by a single non-dimensional parameter v given by

$$v = \pm\, h \sqrt{\frac{2\,d}{\lambda\, d_1\, d_2}}, \tag{2.206}$$

where:

- h is the obstacle height above (plus sign) or below (minus sign) the direct ray between the transmitter and receiver antennas;
- d is the distance between antennas;
- d_1 and d_2 are the distances between the obstacle and each of the antennas;
- λ is the wavelength corresponding to the operating frequency f.

Parameter v may also be expressed as a function of the radius of the first Fresnel ellipsoid r_{1e} at the obstacle. From (2.23) and (2.206) we get

$$v = \pm\, h\, \frac{\sqrt{2}}{r_{1e}}. \tag{2.207}$$

Path loss is calculated as the free space path loss plus the atmospheric loss (for distance d and frequency f) plus an additional obstacle loss A_{ol}, given in dB [13] by

$$A_{ol} = 10 \log_{10} \left\{ \frac{\left[\frac{1}{2} + Ci(-v)\right]^2}{2} + \frac{\left[\frac{1}{2} + Si(-v)\right]^2}{2} \right\}, \tag{2.208}$$

where Ci and Si are the cosine integral and sine integral functions, respectively. Function $A_{ol}(v)$ is plotted in Figure 2.52.

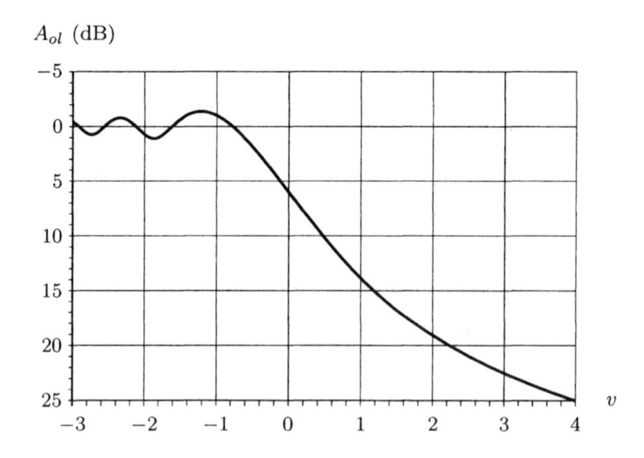

Fig. 2.52 Knife edge obstacle loss as a function of v.

From Figure 2.52, we get $A_{ol}(-0.85) \approx 0$, that is, the obstacle loss is approximately zero for $h \leq -0.6 r_{1e}$. This is the reason behind the criteria to define a line-of-sight path when 60 percent of the first Fresnel ellipsoid is unobstructed (see Section 2.5).

Close inspection of Figure 2.52 shows that for $v < 0$ the presence of the obstacle leads to a negative loss, that is, a gain. In microwave radio link design this gain is usually neglected.

For $v > -0.7$, obstacle loss (in dB) may be approximated by

$$A_{ol}(v) = 6.9 + 20\log_{10}\left(\sqrt{(v-0.1)^2+1}+v-0.1\right). \qquad (2.209)$$

Sometimes the value of v is derived from the angles (α_1 and α_2) between the top of the obstacle, as seen from the antennas, and the horizontal, or alternatively from the diffraction angle θ defined in Figure 2.53. In the following all angles are in radians:

$$v = \pm\sqrt{\frac{2}{\lambda\left(\frac{1}{d_1}+\frac{1}{d_2}\right)}}\,\theta, \qquad (2.210)$$

$$v = \pm\sqrt{\frac{2d}{\lambda}\,\alpha_1\,\alpha_2}, \qquad (2.211)$$

$$v = \pm\sqrt{\frac{2h\theta}{\lambda}}. \qquad (2.212)$$

In (2.210), (2.211) and (2.212) v is positive when the obstacle is above the direct ray between terminal antennas, and negative otherwise. The equivalence between these expressions and (2.207) for $\theta \ll 1$ is straightforward.

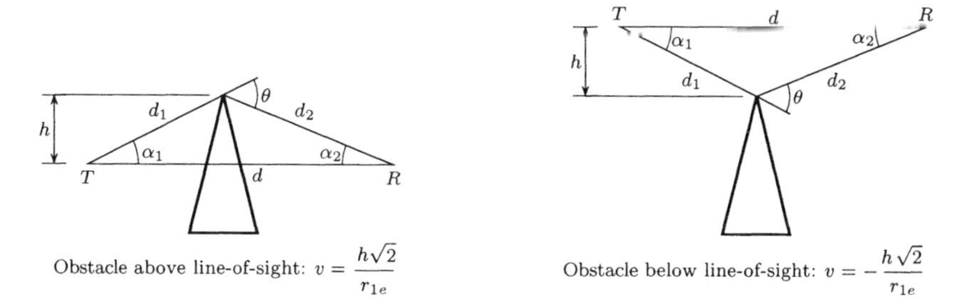

Fig. 2.53 Definition of v for obstacles above and below the line-of-sight.

2.9.4 Diffraction over a finite knife edge obstacle

When the knife edge has a finite length, normal to the direction of propagation, the obstacle loss may be derived as resulting from three knife edge obstacles, one on the top and one on each side of the original obstacle. The following approximate method may be used to estimate of the loss:

- Compute v for each of the knife edges (top and sides);

- Compute losses associated with each knife edge, A_i, with $i = 1, 2, 3$, using (2.209);

- Minimum obstacle loss A_{min} is

$$A_{min} = -20 \log_{10} \left[\frac{1}{10^{A_1/20}} + \frac{1}{10^{A_2/20}} + \frac{1}{10^{A_3/20}} \right]; \qquad (2.213)$$

- Average obstacle loss A_{med} is

$$A_{med} = -10 \log_{10} \left[\frac{1}{10^{A_1/10}} + \frac{1}{10^{A_2/10}} + \frac{1}{10^{A_3/10}} \right]. \qquad (2.214)$$

2.9.5 Diffraction over a rounded obstacle

When the knife edge approximation of a real obstacle is unsuitable it is, sometimes, possible to model it by a rounded obstacle made up of an infinite cylinder, with axis normal to the direction of propagation and radius r_{obs}, as shown in Figure 2.54.

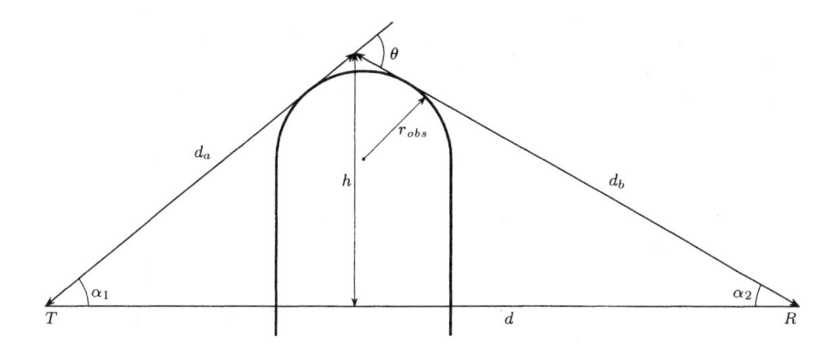

Fig. 2.54 Geometric definition of the rounded obstacle.

For $\theta > 0$ the obstacle loss A_{or} may be calculated, according to ITU-R Recommendation P.526-6 [19], as a sum of the losses due to diffraction $A_{ol}(v)$ and to the obstacle curvature $T(m, n)$

$$A_{or} = A_{ol}(v) + T(m, n). \qquad (2.215)$$

The loss due to diffraction A_{ol} is calculated, as for the knife edge obstacle, from v, given in (2.207), (2.210), (2.211) or (2.212).

The loss due to the obstacle curvature $T(m, n)$ is given, in dB, as a function of mn, by

$$T(m, n) = (8.2 + 12.0n)m^{0.73+0.27(1-e^{-1.43n})}, \qquad (2.216)$$

where

$$m = \frac{r_{obs}\left(\frac{d_a+d_b}{d_a d_b}\right)}{\left(\frac{\pi r_{obs}}{\lambda}\right)^{1/3}}, \tag{2.217}$$

$$n = \frac{h\left(\frac{\pi r_{obs}}{\lambda}\right)^{2/3}}{r_{obs}}. \tag{2.218}$$

In Figure 2.55 we plot $T(m, n)$ as a function of m and n.

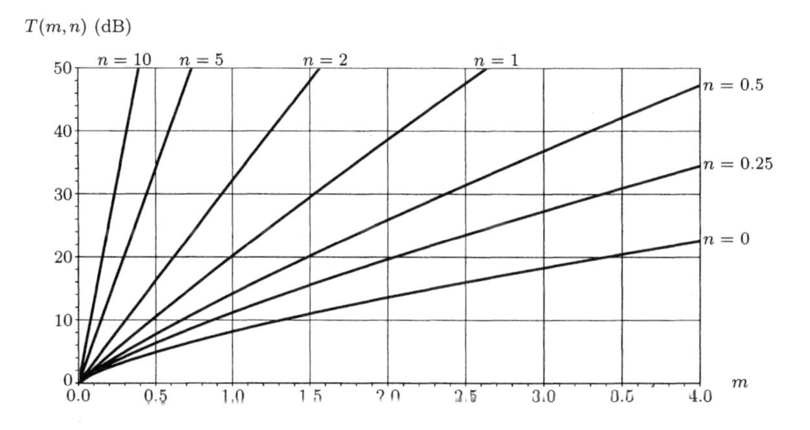

Fig. 2.55 Losses $T(m, n)$ (in dB) due to the obstacle curvature.

For $r_{obs} = 0$, we get $T(m, n) = 0$ and, thus, $A_{or} = A_{ol}$.

This method may lead to substantial errors when the obstacle radius approaches Earth radius. Thus it should not be used for paths over flat ground, lakes or the sea.

The nomogram in Figure 2.56 may be used to estimate m from the operating frequency f, in GHz, the radius of the first Fresnel ellipsoid r_{1e}, in meters, and the obstacle radius r_{obs}, in kilometers.

Lack of topographic data required to accurately define the rounded obstacle radius r_{obs} may present a problem. There are at least two methods to derive the value of r_{obs}, one due to Rice [26] and the other to Crysdale [9].

In the first method r_{obs} is defined as the geometric mean of the radii of curvature from of the radio horizons at both terminals. Let d be the path length and d_{rh_T} and d_{rh_R} the distances from each terminal to its radio horizon. Defining d_s as

$$d_s = d - d_{rh_T} - d_{rh_R} \tag{2.219}$$

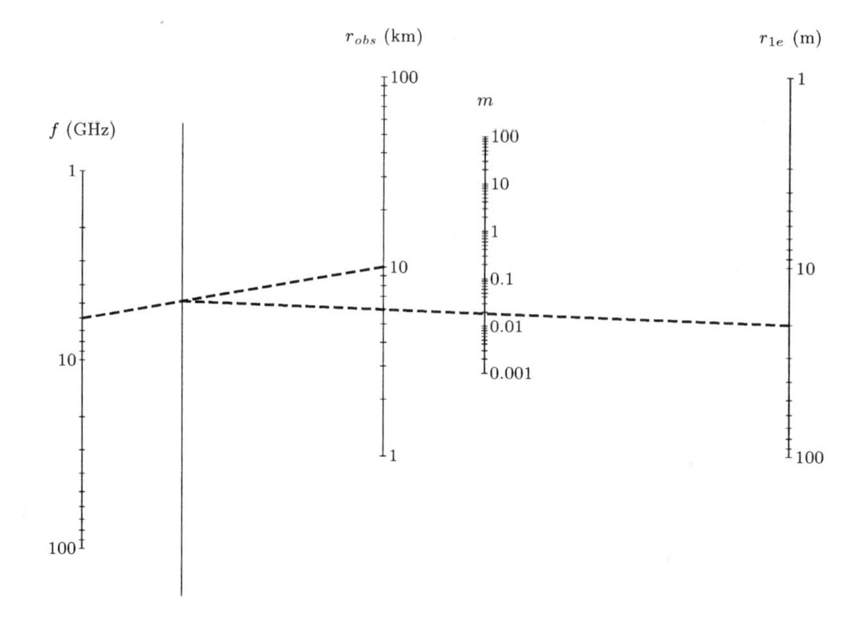

Fig. 2.56 Nomogram to estimate m from the operating frequency f, in GHz, the radius of the first Fresnel ellipsoid r_{1e}, in meters, and the obstacle radius r_{obs}, in kilometers. In the example, $f = 6$ GHz, $r_{obs} = 10$ km and $r_{1e} = 20$ m, leading to $m \approx 0.015$.

and denoting by θ (in radians) the angle between the straight lines that define the radio horizon from each terminal, we have

$$r_{obs} = \frac{2\,d_s\,d_{rh_T}\,d_{rh_R}}{\theta(d_{rh_T}^2 + d_{rh_R}^2)}. \tag{2.220}$$

In the second method we fit a parabola to the obstacle profile and define the obstacle radius as the radius of curvature of the parabola at its apex. Let the parabola be

$$y = a_0 + a_1 x + a_2 x^2. \tag{2.221}$$

The coordinates x_v and y_v of the apex are

$$x_v = -\frac{a_1}{2\,a_2}; \tag{2.222}$$

$$y_v = a_0 - \frac{a_1^2}{4\,a_2}; \tag{2.223}$$

and the radius of curvature, at the apex is

$$r_{obs} = -\frac{1}{2\,a_2}. \tag{2.224}$$

Expressions (2.221) to (2.224) require constants a_0, a_1 and a_2 to be computed from the coordinates of points near the obstacle top expressed in the same units.

2.9.6 Diffraction over multiple knife edge obstacles

In certain cases, where there is no single common horizon obstacle but rather a number of knife edge obstacles, it is possible to assimilate these obstacles to a single one. There is an exact method due to Millington [23] to compute the loss due to multiple knife edges. However due to its complexity this method is difficult to apply for more than two obstacles. For two knife edges the total loss may be computed from any of the three methods sketched in Figure 2.57:

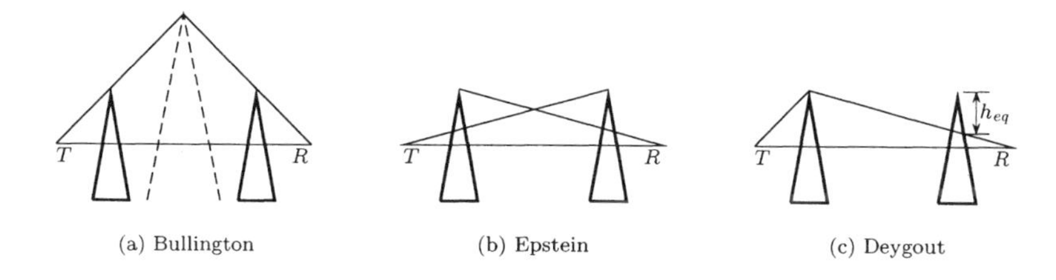

Fig. 2.57 Approximate methods to compute the total loss of two knife edge obstacles.

- Loss of a single equivalent obstacle [6], as shown in Figure 2.57 (a);
- Sum of the losses of two obstacles [12] redefined from the original ones as per Figure 2.57 (b);
- Sum of the losses due to the main obstacle and to a new obstacle redefined as the obstacle between the main obstacle and the nearest terminal antenna [10], as in Figure 2.57 (c).

The three methods have been compared for ten paths by Deygout [10] who, based on the results, derived the following conclusions:

- The single equivalent obstacle and the redefined obstacles lead to losses which are always lower than the measured values, sometimes by more than 10 dB;
- The main obstacle leads to losses within ±3dB of the measured values in 90 percent of the cases; for the remaining cases the computed loss is always higher than the measured one, with a maximum error of 6 dB.

Due to its accuracy, inherent safety and easy expansion to more than two obstacles the method of the main obstacle is strongly recommend. In the following we will describe in some detail this method for a path whose profile, for an Earth radius $k_e r_0$, shows a few obstacles inside the first Fresnel ellipsoid between terminal antennas:

1. Assimilate each obstacle, at a distance x_i from the terminal antenna on the left, to a knife edge obstacle, with height h_i relative to the direct ray between terminal antennas; as before h_i will be positive when the obstacle is above the direct ray and negative otherwise.

2. Compute the radius of the first Fresnel ellipsoid r_{1i} at x_i and the ratio h_i/r_{1i}, for each obstacle.

3. Define the main obstacle as the obstacle with the highest value of h_i/r_{1i}. When there are two or more obstacles with the same value of h_i/r_{1i}, any of them may be defined as the main obstacle.

4. Using the main obstacle, divide the original path into two, define a new direct ray line as the open polygon linking the terminal antennas and the top of the main obstacle and plot the first Fresnel ellipsoid for each of the two stretches of the new direct ray.

5. Retain from the original obstacles only those that still interfere with the two new Fresnel ellipsoids. This assumes that the original main obstacle obstructs the direct ray between terminals. If this is not the case, the resulting loss may be approximated simply by the main obstacle loss.

6. As a result of the previous step, between the main obstacle and each terminal antenna, we may either:

   - eliminate all obstacles;
   - retain a single obstacle;
   - retain two or more obstacles.

7. When all obstacles have been eliminated, the main obstacle may be taken as an isolated knife edge and the path loss calculated as the sum of the free space loss plus the atmospheric loss between terminals plus the excess loss due to the single knife edge.

8. When only one obstacle is retained we redefine its height, relative to the new direct ray, and its parameter v using the ratio of the new height and the radius of the Fresnel ellipsoid between the adjacent terminal antenna and the main obstacle, at the obstacle. The path loss is now the sum of the free space loss plus the atmospheric loss between terminals plus the excess losses due to the main obstacle and the redefined obstacle (both as knife edges).

9. When two or more obstacles are retained, we start by redefining for each of them a new height and a new value of v as previously described. If

all obstacles are below the new direct ray their contribution to the path loss is taken as the sum of the knife edge losses for each. If one or more obstacle is above the new direct ray we choose a new main obstacle as if the path was between the adjacent terminal antenna and the original main obstacle and we repeat the procedure until we have computed dealt with all obstacles.

Figure 2.58, adapted from Deygout's paper [10], provides an example for a path with five obstacles.

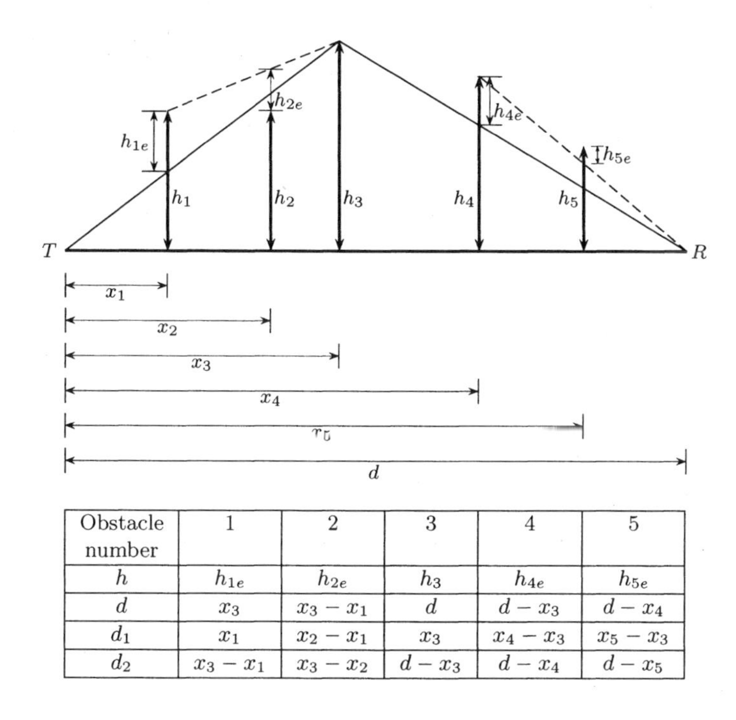

Obstacle number	1	2	3	4	5
h	h_{1e}	h_{2e}	h_3	h_{4e}	h_{5e}
d	x_3	$x_3 - x_1$	d	$d - x_3$	$d - x_4$
d_1	x_1	$x_2 - x_1$	x_3	$x_4 - x_3$	$x_5 - x_3$
d_2	$x_3 - x_1$	$x_3 - x_2$	$d - x_3$	$d - x_4$	$d - x_5$

Fig. 2.58 Use of the main obstacle method for multiple knife edges.

2.9.7 Diffraction over multiple rounded obstacles

For multiple rounded obstacles we may use an approximate method due to Assis [1], which is, in fact, a generalized versions of the main obstacle method, where the excess loss due to each obstacle is calculated using the method described for the isolated rounded obstacle.

2.9.8 Diffraction over an irregular shaped obstacle

For an irregularly shaped obstacle ITU-R Recommendation P.526-6 [19] suggests a general method which may also be applied to multiple obstacles.

Assume that the ground path between antenna terminals may be represented by a set of n points, each defined by the horizontal distance to the transmitter antenna d_i and the altitude h_i. Let d_{ij} be the horizontal distance between points i and j.

We start be identifying the main obstacle as in Deygout's method. Let v_p defined in (2.206) be the main obstacle parameter.

If $v_p < -0.78$ the path may be taken as unobstructed and the diffraction loss as zero. Otherwise Deygout's method is applied twice: one between the transmitter and the main obstacle and the other between the main obstacle and the receiver antenna. Let v_l and v_r, be the obstacles parameters, respectively. The path diffraction loss is calculated as the diffraction loss due to the main obstacle $A_{ol}(v_p)$ plus the sum of the diffraction losses due to the new main obstacles to each side of the original main obstacle, plus a correction term, all multiplied by a factor depending on the original main obstacle loss

$$A_{ir} = A_{ol}(v_p) + T\left[A_{ol}(v_l) + A_{ol}(v_r) + C\right], \qquad (2.225)$$

where

$$C = 10 + 0.04 d_{[km]}, \qquad (2.226)$$

$$T = 1.0 - \exp\left[-\frac{A_{ol}(v_p)}{6.0}\right]. \qquad (2.227)$$

Obstacle losses in (2.225) are calculated using (2.209).

In case there is a lack of detailed topographic data for the obstacle we may get a rough estimate of the obstacle loss A_{ir}, in dB, (larger than 15 dB) using the following expression, from the ITU-R Recommendation P.530-8 [19]

$$A_{ir} = \frac{20h}{r_{1e}} + 10, \qquad (2.228)$$

where, as before, h is the most significant height of the obstacle above the direct ray between terminal antennas and r_{1e} is the radius of the first Fresnel ellipsoid at that point. This equation leads to a value of the diffraction loss which falls between that of a knife edge and of a rounded obstacle.

2.10 ATTENUATION DUE TO TREES

Attenuation due to obstacles, such as terrain or buildings, may usually be estimated using one of the previously described methods. However link design sometimes faces another type of obstacle – trees – for which these methods are hardly applicable.

Attenuation due to vegetation is dealt with in ITU-R[20] Recommendation P.833-3. Two cases are considered. In the first one the transmitter (or the receiver) antenna is in free space while the receiver (or the transmitter) antenna is inside woodland. In the second one both terminal antennas are outside the vegetation and the obstruction is due to the canopy of a single tree.

In the first case the excess attenuation due to vegetation A_{tr} may be calculated as

$$A_{tr} = A_m \left[1 - \exp\left(-\frac{\gamma_{tr} d_{tr}}{A_m} \right) \right], \tag{2.229}$$

where:

- d_{tr} is the path length (in m) inside woodland;
- γ_{tr} is the excess attenuation for very short vegetative paths (in dB/m) given, as a function of the frequency f, in GHz, in the range $[1, 40]$ by

  $$\gamma_{tr} = 0.2f; \tag{2.230}$$

- A_m is the maximum attenuation for one terminal within a specific tpe and depth of vegetation (in dB).

The value of γ_{tr} varies widely due to the irregular nature of vegetation and the wide range of species, densities and water content that may be found. So the value given in (2.230) should be regarded only as typical.

Measurements in the frequency range 0.900 – 1.8 GHz, carried out in a park with tropical yield the following value for A_m (in dB).

$$A_m = 32.5 f^{0.752}, \tag{2.231}$$

where f is the frequency in GHz.

The excess attenuation A_{tr} is defined in excess to all other propagation mechanisms and not just free space loss. Using equations (2.230) and (2.231) we find out that at 1 GHz, the excess attenuation introduced by vegetation is already non-negligible (2 dB) for $d_{tr} = 10$ m and reaches a value which is likely to be decisive in link design (32.4 dB) for $d_{tr} = 1000$ m.

When both terminal antennas are outside the vegetation and the obstruction due to vegetation is short, such as paths where the obstruction is due to a canopy of a single tree, equation (2.229) no longer applies and the excess attenuation A_{tr} may be estimated as

$$A_{tr} = d_{tr} \gamma_{tr}, \tag{2.232}$$

for $f \leq 3$ GHz.

If A_{tr} is sufficiently high a lower loss path will likely exist around the vegetation. A limit to the attenuation due to these lower path losses may be calculated by modelling the tree canopy as a finite width screen as in Section 2.9.4.

For frequencies above 5 GHz the attenuation for a vegetation depth d_{tr} (in m) should be calculated as

$$A_{tr} = R_\infty d_{tr} + k\left[1 - \exp\left(-\frac{R_0 - R_\infty}{k} d_{tr}\right)\right], \qquad (2.233)$$

where f is the frequency, in GHz, and:

$$R_0 = af, \qquad (2.234)$$

$$R_\infty = \frac{b}{f^c}, \qquad (2.235)$$

$$k = k_0 - 10\log_{10}\left\{A_0\left[1 - \exp\left(-\frac{A_{min}}{A_0}\right)\right][1 - \exp(-R_f f)]\right\} \qquad (2.236)$$

Constants a, b, c, k_0, R_f and A_0 are given in Table 2.3. A_{min} (in square meters) is the minimum illumination area, defined as the product of the minimum width $\min(w_1, w_2, w_{tr})$ and the minimum height $\min(h_1, h_2, h_{tr})$ of the illuminated vegetation, which corresponds to the smaller of the two terminal antennas spot areas on the front and rear faces of the vegetation. The heights h_1 and h_2, and the widths, w_1 and w_2, are calculated using the -3 dB terminal antennas vertical and horizontal beamwidths. The height h_{tr} and width w_{tr} correspond to the vegetation assumed to be a rectangular block (see Figure 2.59).

Parameter	In leaf	Out of leaf
a	0.2	0.16
b	1.27	2.59
c	0.63	0.85
k_0	6.57	12.6
R_f	0.0002	2.1
A_0	10	10

Table 2.3 Parameters to calculate the excess attenuation due to vegetation, as given in ITU-R [20] Recommendation P.833-3.

It follows from equation (2.233) that a single tree canopy, in leaf, with $\min(w_1, w_2, w_{tr}) = 3$ m, $\min(h_1, h_2, h_{tr}) = 10$ m, leading to $A_{min} = 30$ m^2, and $d_{tr} = 3$ m, gives rise to considerable excess attenuation (from 3 dB at 5 GHz to 8.7 dB at 18 GHz), showing that trees inside the first Fresnel ellipsoid should indeed be considered as non-negligible obstacles.

2.11 TROPOSPHERIC SCATTERING

When the distance between terminal radio horizons increases, obstacle loss also increases, at first according to the laws of diffraction. When we reach a

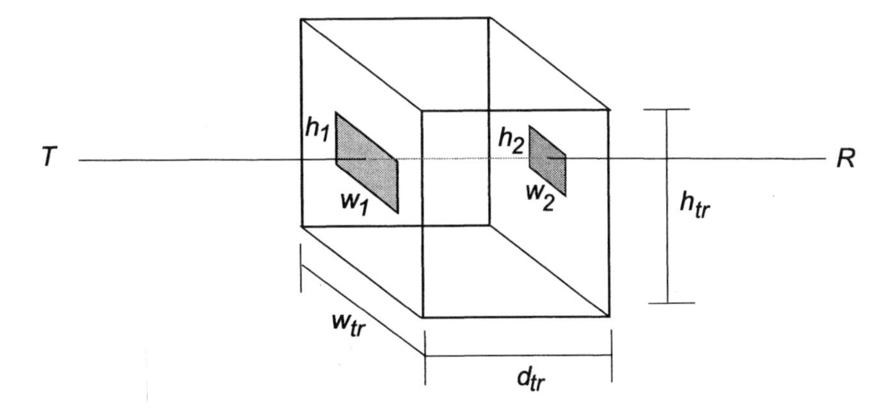

Fig. 2.59 Geometry to calculate the excess attenuation due to vegetation at frequencies above 5 GHz.

few tens of decibels, the rate of increase with distance decreases to about 0.1 dB/km and useful ranges between 300 and 1000 kilometers are possible.

The received signals, although highly variable with time, may be used in high-quality point-to-point links and should not be confused with randomly occurring signals, with much higher amplitude due to ducts or, at lower frequencies, to reflections in abnormal ionospheric layers. The presence of such signals is due to wave scattering in tropospheric heterogeneities and hence this propagation mode is known as tropospheric scattering or simply troposcattering.

Troposcatter signals have fast and slow amplitude changes. The former are due to global changes in the atmospheric refraction and do not depend much on frequency. The median of the hourly averaged signal amplitude, below the long time median, shows a log-normal distribution with standard deviation, which is a function of the climate and the distance, of a few decibels. The frequency of fast amplitude changes decreases with the operating frequency, from a few tenths of a hertz up to a few hertz, while the operating frequency decreases from a few GHz to about 300 MHz.

The loss between terminals for troposcatter links may be calculated using empiric or semi-empiric methods. The most commonly used are from the National Bureau of Standards [26], the ITU-R [17], the International Telephone and Telegraph Corporation [15] and those due to Boithias and Battesti [15] and Yeh[29] Panther [25] provides a comparison of the different methods. Here we will restrict ourselves to the method described in ITU-R Recommendation P.617-1 [19].

The median of the hourly average attenuation, between isotropic antennas, due to tropospheric scattering, $A(50)$ in dB, is given by

$$A(50) = 90 + 30\log_{10}(f) + 10\log_{10}(d) + 30\log_{10}(\theta) + M + N(H,h) + G_p, \quad (2.237)$$

where:

- f is the operating frequency, in GHz;
- d is the distance between antennas, in km;
- θ is the angle in milliradians, between two straight lines, one at each terminal, defined by the terminal antenna and its radio horizon (in median atmospheric conditions), obtained from a path profile along a great circle between terminals);
- M is a meteorological parameter, according to the climate, given in Table 2.4;
- $N(H,h)$ is a function of the altitude of the basis of the common volume, given by

$$N(H,h) = 20\log_{10}(5 + \gamma H) + 4.34\gamma h, \quad (2.238)$$

$$H = \frac{\theta d}{4 \cdot 10^3}, \quad (2.239)$$

$$h = \frac{\theta^2 r}{8 \cdot 10^6}; \quad (2.240)$$

- r is the effective Earth radius, in kilometers;
- γ is a parameter, function of the atmospheric structure for the different climates, given in Table 2.4;
- $G(p)$ is the gain of the transmitter and receiver antennas, in decibels, which for tropospheric scattering is less than the sum of the gains of these antennas

$$G(p) = G_T + G_R - L_c, \quad (2.241)$$

$$L_c = 0.07\, e^{0.055(G_T+G_R)}. \quad (2.242)$$

For the calculation of M and γ Annex I to ITU-R Recommendation P.617-1 [19] defines the following climates (in brackets we suggest typical regions):

- Type 1 — equatorial, regions between latitudes 10°N and 10°S (Congo, Ivory Coast);
- Type 2 — continental subtropical, regions between latitudes 10° and 20° (Sudan);

- Type 3 — maritime subtropical, also from regions between latitudes 10° and 20° near the see (West African Coast);
- Type 4 — desert (Sahara)
- Type 5 — mediterranean;
- Type 6 — temperate continental (France, Germany)
- Type 7a — temperate maritime, over land (England)
- Type 7b — temperate maritime, over the sea (England)
- Type 8 — polar.

Climate	1	2	3	4	6	7a	7b
M (dB)	39.60	29.73	19.30	38.50	29.73	33.20	26.60
$\gamma(km^{-1})$	0.33	0.27	0.32	0.27	0.27	0.27	0.27

Table 2.4 Meteorologic parameters for different climates, as given in ITU-R Recommendation P.617-1 [19]. Reproduced with ITU-R permission.

The hourly median as a function of the probability of it being exceeded, either annually or in the worst month, may be calculated according to the methods described in Chapter 3, Section 3.5.

Superimposed on the slow attenuation changes there are fast changes (with periods of the order of up to a few seconds) described in Chapter 3.

The tropospheric discontinuities that enable troposcatter create various paths, each with its own time delay, which give rise to signal distortion and thus limit usable bandwidth.

2.12 PASSIVE REPEATERS

In the design of microwave radio links we generally aim at achieving line-of-sight paths since, for the same distance and operating frequency, these paths provide the lowest path loss and the least expensive link equipment[4] When there are no convenient locations for the repeater stations or when the existing ones have poor or no access, we may consider using obstructed paths.

Previous sections provide methods to compute path loss for obstructed paths. In this section we describe a way to substitute an obstructed path by

[4]In most cases, we do not aim for minimum equipment cost but rather for minimum path cost, which includes equipment, infrastructure and operation costs.

one (or more) line-of-sight paths, using passive repeaters. These may provide quite an attractive alternative, particularly for the shorter distances ($d < 50$ km) and the higher frequencies ($f > 6$ GHz), in hilly regions.

Consider the common horizon path profile given in Figure 2.60. Assume that the obstacle may not be used as a terminal due to poor access and lack of infrastructures. If we install on the obstacle, a passive repeater, made up by two interconnected (by a short length of waveguide or coaxial cable) back to back antennas, the obstructed path is changed into two line-of-sight paths.

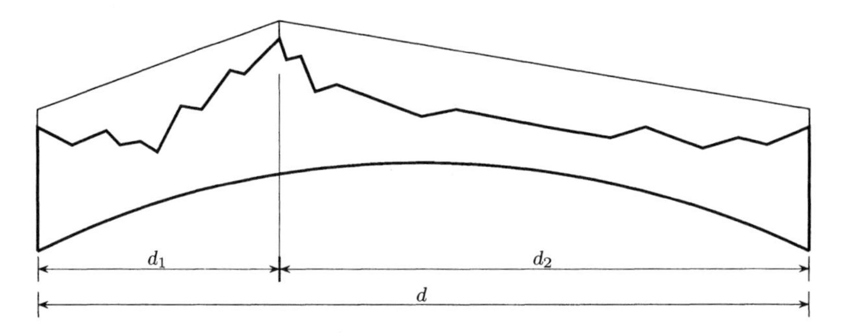

Fig. 2.60 Obstructed path with a passive repeater.

Let G_{RR} and G_{RT} be the gains (in dB), relative to the isotropic lossless antenna, of the antennas that make up the passive repeater and that are pointed to terminal T and R, respectively. If propagation between transmitter and repeater is assumed to be in free space, the received power at the passive repeater is given by

$$P_{RR} = P_T + G_T + G_{RR} - A_T - \gamma_a \, d_1 - A_{0_1}, \qquad (2.243)$$

where γ_a (in dB/km) is the atmospheric attenuation per unit length and A_{0_1} (in dB) is the free space loss

$$A_{0_1} = 92.4 + 20 \log_{10}(d_1) + 20 \log_{10}(f),$$

with d_1 in kilometers and f in GHz. Symbols here are the same as in Section 2.1.

If A_{RP} (in dB) represents the losses between antennas at the repeater, the transmitted power at the repeater is

$$P_{TR} = P_{RR} - A_{RP}, \qquad (2.244)$$

and hence the received power at R is

$$P_R = P_{TR} + G_{RT} + G_R - A_R - \gamma_a \, d_2 - A_{0_2}, \qquad (2.245)$$

where:

$$A_{0_2} = 92.4 + 20\log_{10}(d_2) + 20\log_{10}(f), \quad (2.246)$$

with d_2 in kilometers and f in GHz.

Substituting (2.244) and (2.243) in (2.245), we get:

$$P_R = P_T + G_T - A_T + G_R - A_R + G_{RR} + G_{RT} - A_{RP} - A_{01} - A_{02} - \gamma_a d. \quad (2.247)$$

If we did not make use of the passive repeater

$$P_R' = P_T - A_T - A_R + G_T + G_R - A_0 - \gamma_a d - A_{obs}, \quad (2.248)$$

where A_{obs} is the obstacle loss, in dB, and A_0 is given by

$$A_0 = 92.4 + 20\log_{10}(d) + 20\log_{10}(f),$$

with d in kilometers e f in GHz.

Comparing (2.247) and (2.248), we find out the conditions to opt for the passive repeater. For $P_R > P_R'$, we must have

$$G_{RR} + G_{RE} - A_{RP} > 92.4 + 20\log_{10}\left(\frac{d_1 d_2}{d}\right) + 20\log_{10}(f) - A_{obs}. \quad (2.249)$$

If the repeater antennas are parabolic reflectors with diameter d_a (in m), and an aperture efficiency 0.5, from (2.14), with f in GHz, we get

$$\begin{aligned} G_{RR} &= G_{RE}, \\ &= 17.4 + 20\log_{10}(d_a) + 20\log_{10}(f), \end{aligned} \quad (2.250)$$

and thus from (2.249)

$$40\log_{10}(d_a) > 57.6 - 20\log_{10}(f) + 20\log_{10}\left(\frac{d_1 d_2}{d}\right) + A_{RP} - A_{obs}. \quad (2.251)$$

Expression (2.251) shows that the passive repeater is a more attractive solution the higher the operating frequency f, the closer to one of the terminals and the higher the obstacle attenuation.

It is worth mentioning that the in previous expressions we assumed that the far field conditions were met, a necessary requirement for the calculated free space loss to be correct.

By modifying the relative position of the repeater antennas we may place the repeater in any favorable location, even outside the great circle path between link terminals. When the repeater is placed so that the repeater transfer angle — angle between the incident and emerging rays — is greater than about 50 or 60 degrees, it is often more economical to use passive repeaters made out of flat reflectors (Figure 2.61), also known as mirrors.

As in the previous case, the received power is

$$P_R = P_T + G_T - A_T + G_R - A_R + G_{RP} + A_{RP} - A_{01} - A_{02} - \gamma_a d, \quad (2.252)$$

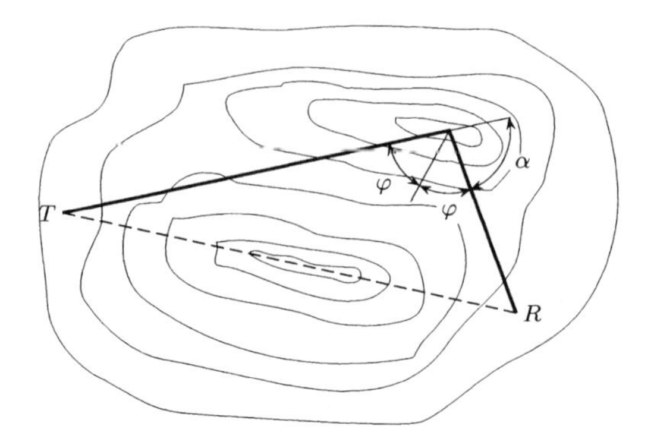

Fig. 2.61 Using a flat reflector (mirror) as a passive repeater (reflector dimensions not to scale).

where

$$G_{RP} = 20 \log_{10} \left[\frac{4\pi}{\lambda^2} \, a_{RP}, \cos(\varphi) \right], \tag{2.253}$$

a_{RP} being the physical area of the flat reflector and φ the angle of incidence on it (Figure 2.61).

The value of A_{RP} is given by

$$A_{RP} = 10 \log_{10}(\eta_{RP}), \tag{2.254}$$

where η_{RP} is the repeater "efficiency", compared with an equal sized loss-less flat reflector. For the usual sizes, the reflector "efficiency" is very close to unity (typically about 0.95).

As for the a passive repeater built with two back-to-back antennas, the use of a mirror is only worthwhile if it increases the received power.

The far field distance for a mirror may by calculated from (2.12) or estimated using the nomogram in Figure 2.3. In the usual situation where one of the link terminals is inside the near field region of the passive repeater, the repeater gain must be corrected (decreased) by a factor given in Figure 2.62, as a function of $1/K$ and L defined by

$$\frac{1}{K} = \frac{\pi \lambda d_1}{4 \, a_{ef}}, \tag{2.255}$$

$$L = d_a \sqrt{\frac{\pi}{4 a_{ef}}}, \tag{2.256}$$

where:

- a_{ef} is the effective passive area;
- d_1 is the distance between the repeater and the nearest link antenna;

- d_a is the diameter of the link antennas (assumed to be parabolic reflectors).

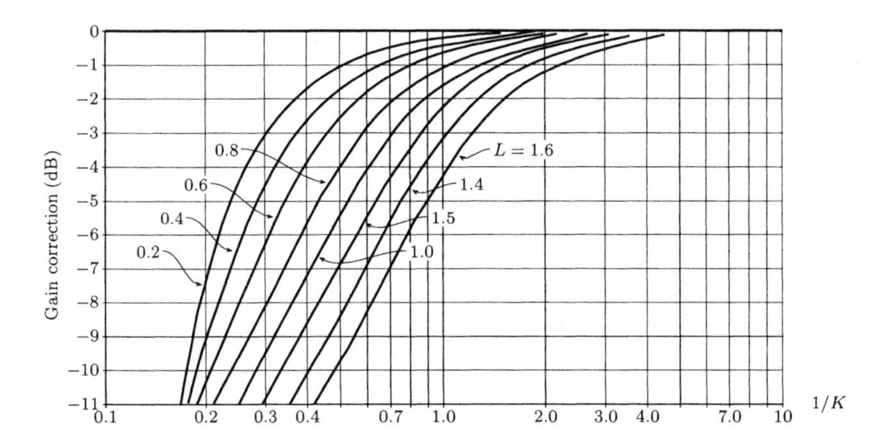

Fig. 2.62 Correction factor for the gain of a passive repeater in case the nearest terminal antenna is in the near field region (White [28]). Reproduced with permission from Valmont/Microflect.

The values given in Figure 2.62 are only for rectangular reflectors, by far the most common.

To avoid waveguide (or coaxial cable) attenuation between the transmitter (or the receiver) and the antenna, we may use a flat reflector together with a parabolic antenna in a set-up known as a periscope, represented schematically in Figure 2.63. With a correct choice of antenna and flat reflector dimensions a small gain (up to 6 dB) compared with the parabolic antenna on its own may be achieved.

The gain of the association of a parabolic antenna and a flat reflector, in a periscope set-up, may be calculated as for an antenna in the near field region of a flat reflector. Alternatively we may assume that the association of the antenna and the reflector is equivalent to the antenna by itself, with its gain corrected by a factor given in Figure 2.64, as a function of $1/K$ and L (previously defined).

Since large flat reflectors (with areas in excess of 10 m^2) have not only a higher aperture efficiency but also a lower cost than parabolic antennas with the same area, its use may be attractive even when the transfer angle is less than about 50 degrees. In these cases we may use two mirrors, in the near field of each other, as shown in Figure 2.65.

The gain of a set of two mirrors, in the near field region of each other is equal to the gain of the smaller mirror plus a corrective factor, given in Figure 2.65, as a function of the square root of the ratio between mirrors the effective

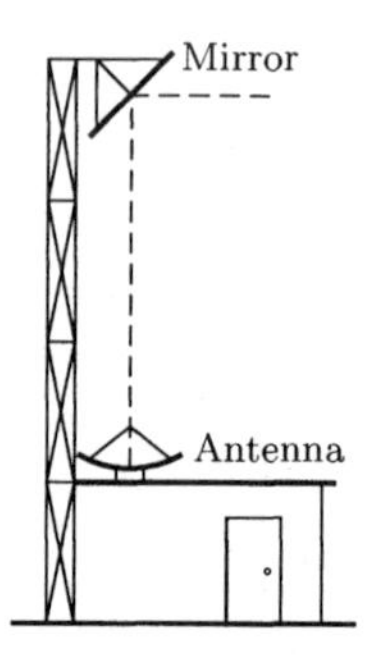

Fig. 2.63 The association of a parabolic antenna and a flat reflector, known as a periscope.

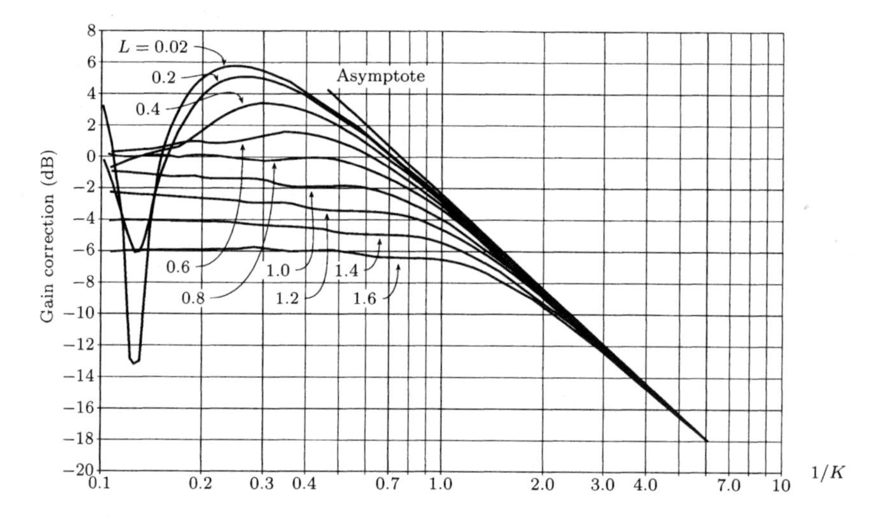

Fig. 2.64 Corrective factor for the association of a parabolic antenna and a flat reflector in a periscope set up (White [28]). Reproduced with permission from GTE Lenkurt.

areas $\sqrt{b_{ef}/a_{ef}}$ (where $b_{ef} > a_{ef}$) and the parameter $1/K$ defined as:

$$\frac{1}{K} = \frac{2\lambda d_1}{a_{ef}}, \tag{2.257}$$

where d_1 is the distance between the centers of the mirrors and a_{ef} is the smaller effective area of the two mirrors.

For $1/K < 0.1$ the corrective factor should be taken as -1 dB.

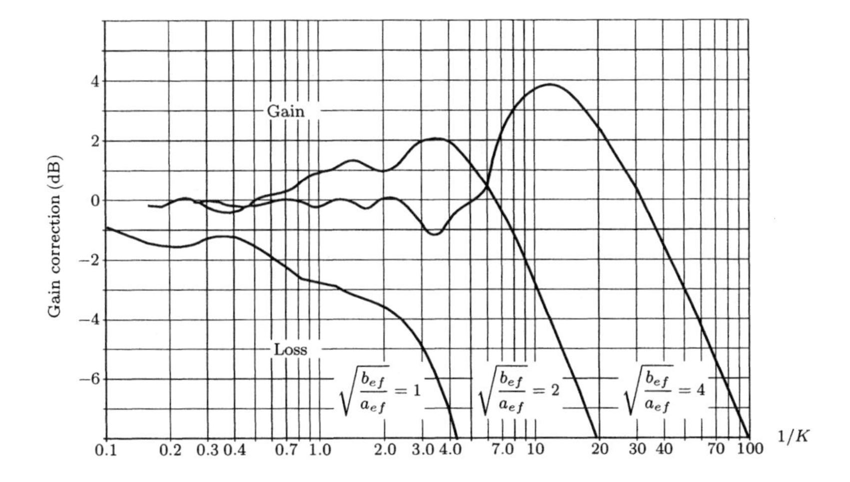

Fig. 2.65 Corrective factor for the gain of two mirrors in the near field region of each other (White [28]). Reproduced with permission from GTE Lenkurt.

When one of the link terminal antennas is in the near field region of the set of mirrors, this fact must be considered in the calculation of the path loss.

REFERENCES

1. Assis, M. S., A simplified solution to the problem of multiple diffraction over rounded obstacles. *IEEE Transactions on Antennas and Propagation*, Vol. AP-19, No. 3, pp. 292–295, 1971.

2. Barrick, D., Rough surface scattering based on the specular point theory. *IEEE Transactions on Antennas and Propagation*, Vol. AP-16, No. 4, pp. 449–454, 1968.

3. Blomquist, A. and Ladell, L., Prediction and calculation of transmission loss in different types of terrain. *AGARD Conference Proceedings*, No. 144, paper 32, 1975.

4. Boithias, L. and Battesti, J., Les faiceaux hertziens trans-horizon de haute qualité. *Annales des Tèlècommunications*, Vol. 20. No. 7, 8, 11, 12, 1965.

5. Boithias, L. and Battesti, J., Nouvelles expérimentations sur la baisse de gain d'antenne dans les liaisons transhorizon. *Annales des Tèlècommunications*, Vol. 22, Septembre–Octobre, 1967.

6. Bullington, K., Radio propagation at frequences above 30 megacycles. *Proceedings of the IRE*, Vol. 35, October, pp. 1122–1136, 1947.

7. Carlson, A. and Waterman, A., Microwave propagation over mountain-diffraction paths. *IEEE Transactions on Antennas and Propagation*, Vol. AP-14, No. 4, pp.489–496, 1966.

8. CCIR, *Reports of the CCIR, 1990*, Annex to Volume V, *Propagation in Non-Ionized Media.* ITU, Geneva, 1990

9. Chrysdale, J. H., Comparision of some experimental terrain diffraction losses with predictions based on Rice's theory for diffraction by a parabolic cylinder. *IRE Transactions on Antennas and Propagation*, Vol. AP-6, July, pp. 293–295, 1958.

10. Deygout, J., Multiple knife-edge diffraction of microwaves. *IEEE Transactions on Antennas and Propagation*, Vol. AP-14, pp. 480–489, 1966.

11. Dickson, F. H., Egli, J. J., Herbstreit, J. W. and Wickizer, G. S., Large reductions of VHF transmission loss and fading by the presence of a mountain obstacle in beyond-line-of-sight paths, *Proceedings of the IRE*, Vol. 41, pp. 967–969, 1953.

12. Epstein, J. and Pewterson, D. W., An experimental study of wave propagation at 850 Mc, *Proceedings of the IRE*, Vol. 41, May, pp. 595–611, 1953.

13. Figanier, J. *Aspectos de Propagação na Atmosfera*, editado por Fernandes, C., Associação dos Estudantes do Instituto Superior Técnico, Lisboa, 1993

14. Hartman, W. J. and Smith. D., Tilting antennas to reduce line-of-sight microwave link fading, *IEEE Transactions on Antennas and Propagation*, Vol. AP-25, No. 5, pp. 642–645, 1977.

15. ITT, *Reference data for radio-engineers.* 4th edition, ITT, New York, 1956.

16. ITU, *ITU-R Recommendations*, 1994 F Series Volume, Part 1, *Fixed service using radio-relay systems*, UIT, Geneva, 1995.

17. ITU, *ITU-R Recommendations*, 1994 PN Series Volume, *Propagation in Non-Ionized Media.* ITU, Geneva, 1995.

18. ITU, *ITU-R Recommendations on CD-ROM*, ITU, Geneva, 1997.

19. ITU, *ITU-R Recommendations on CD-ROM*, ITU, Geneva, 2000.

20. ITU, *ITU-R Recommendations on CD-ROM*, ITU, Geneva, 2001.

21. Kerr, D. E., *Propagation of short radio waves*, Dover Publications Inc., New York, 1965.

22. Laws, J. O. and Parsons, D. A., The relation of raindrop–size to intensity. *Transactions of the America Geophysics Union*, Vol. 24, pp. 452–460, 1943.

23. Millington, G., Hewitt, R. and Immirzé, F. S., Double knife-edge diffraction in field strength predictions. *IEE Monograph 507E*, March 1962, pp. 419–429.

24. Olsen, R. L., Rogers, D. V. and Hodge, D. B., The aR^b relation in the calculation of rain attenuation. *IEEE Transactions on Antennas and Propagation*, Vol. AP-26, Nr. 2, pp. 318–329, 1978.

25. Panter, P., *Communication Systems Design: Line-of-sight and Troposcatter Systems*, McGraw-Hill Inc., New York, 1972.

26. Rice, P. L., Longley, A. G., Norton, K. A. and Barsis, A. P., *Transmission Loss Predictions for Tropospheric Communication Circuits, I* e *II*, National Bureau of Standards, Boulder, Co., 1967.

27. Salema, C., Figanier, J. and Fernandes, C., Estimation of ground reflections in microwave links. *Proceedings of the IEE*, Vol. 128, Pt. H, No. 5, pp. 252–256, 1981.

28. White, R., *Engineering Considerations for Microwave Communications Systems*, 3rd edition, GTE Lenkurt Inc., San Carlos, CA.,1975.

29. Yeh, L. P., Simple method for designing troposcatter circuits. *IRE Transactions on Communication Systems*, September, 1960.

Problems

2:1 Calculate the free space attenuation for a link with 50 km at 6 GHz.

2:2 Calculate the gain of a 3 m diameter parabolic reflector antenna at 6 GHz, assuming an illumination efficiency of 50%.

2:3 Assume a 50-km-long microwave radio link in free space, operating at 6 GHz. This link uses 3 m diameter parabloic reflector antennas. Antennas are connected to the transmitter and receiver using 50 m EW52 elliptic waveguides (at each end) with an attenuation of 0.039 dB/m. Calculate the system attenuation and the global attenuation.

2:4 Calculate the receiver input power (in W and dBW) in the previous problem assuming that the transmitter output power is 2 W.

2:5 Calculate the maximum diameter of the first Fresnel ellipsoid for a 25 km link at 11 GHz.

2:6 Consider a 50-km microwave radio link over plane Earth ($\epsilon_s = 15\epsilon_0$, $\sigma_s = 0.005\Omega^{-1}m^{-1}$) at 4 GHz, with a transmitter antenna height equal to 80 m. Plot the receiver input power as a function of the receiver antenna height, varying between 80 and 140 m.

2:7 Plot the profile of 40-km-long microwave radio link over spherical Earth (radius equal to 6370 km), with antenna heights equal to 100 and 200 m, including the direct ray, the first Fresnel ellipsoid and the direction of the horizon at each terminal

2:8 For the link of the previous problem calculate the elevation angle at each terminal

2:9 Calculate the radio horizon for an antenna at 20 m height over spherical Earth (radius equal to 6370 km)

2:10 Assume a 45-km-long microwave radio link over spherical Earth (radius equal to 6370 km), with antenna heights equal to 100 and 200 m. Calculate the specular point position and the divergence factor.

2:11 Calculate the water vapor content at 30° C and 80% relative humidity.

2:12 Calculate the atmospheric attenuation, due to the presence of both oxygen and water vapor, at 1030 hPa, 25°C and 80% humidity, for the following path lengths (d) and frequencies (f):

(a) $d = 50$ km, $f = 4$ GHz;

(b) $d = 50$ km, $f = 8$ GHz;

(c) $d = 30$ km, $f = 12$ GHz;

(d) $d = 6$ km, $f = 20$ GHz.

2:13 Calculate the attenuation per unit length γ_r due to heavy rain (30 mm/h) at 4, 8, 12 and 16 GHz, both for vertical and horizontal polarization.

2:14 Calculate the excess attenuation due to rain intensity not exceed for more than 0.001%, 0.01% and 0.1% of the worst month in zone H, for the paths given in Problem 2:12.

2:15 Consider a path profile defined by the following pairs of of distances d (in km) to the terminal on the left and heights h (in m): Antenna heights

d (km)	0	8	15	22	25	31	40	45	48	50
h (m)	97	45	25	31	13	0	0	0	27	20

are 30 m (for $d = 0$) and 40 m (for $d = 50$ km), antenna gains are equal to 30 dBi, and the working frequency is 6 GHz. Estimate the ratio between the received ground scattered power and direct ray power for ground roughness equal to 0.0001, 0.01 and 0.1.

2:16 Consider a 200-km-long microwave radio link between two islands. Antenna height (above mean sea level) is 700 m and 300 m, respectively. For $k_e = 4/3$ calculate the radio horizons at each site and the basic transmission loss propagation attenuation for 200 MHz (assuming that the main propagation mechanism is diffraction by a spherical Earth).

2:17 Consider a 50-km-long microwave radio link at 4 GHz, with transmitter and receiver antenna heights equal to 50 and 70 m, respectively. The direct ray has two knife edge obstacles. One, located at 20 km from the transmitter, is 100 m high and the other, 40 km from the transmitter, is 80 m high. Calculate the excess path attenuation due to these obstacles.

2:18 Calculate the excess attenuation due to the canopy of a single tree (in leaf) located at 10 km from one terminal and 2 km from the other terminal, in good weather, for a 13 GHz link, using 2-m diameter parabolic antennas. Assume free space propagation and neglect excess losses due to atmospheric gases. The canopy may be taken as a 10-m high rectangular prism, with a $3 \times 3\ m^2$ base.

2:19 Calculate the median of the hourly average attenuation between isotropic antennas A_{50} for a 300-km troposcatter link, over the sea, in temperate maritime climate, assuming $k_e = 4/3$ and antenna heights of 300 m. The radio horizon is defined by mean sea level in both terminals.

2:20 Assume the use of a 7×5 m^2 mirror, with a transfer angle of 60° located at 3 km from one terminal and 27 km from the other terminal, of a 6-GHz microwave radio link. Check that the nearest terminal is inside the far distance of the mirror. Calculate the free space attenuation between terminal antennas and the minimum excess path attenuation that would make the use of this mirror worthwhile.

2:21 Compare the use of 50 m of elliptical waveguide (attenuation 0.056 dB/m) to feed a 3.0-m-diameter parabolic reflector, with a periscope using the same reflector in conjunction with a 3×4.5 m^2 mirror, at the distance of 50 m.

3

Fading

3.1 INTRODUCTION

In a point-to-point link, immersed in a time varying medium, we should expect the received signal power to vary in time even when the transmitter power remains constant. Microwave radio links in the troposphere experience (sometimes rather wild) received signal fluctuations above and below its long time median value. This phenomenon, usually known as fading, has a very strong influence in the quality of service. Thus it is essential to know as much as possible about fading and to try to predict or counteract its effects.

On a plot of the received power versus time it is possible to identify two types of fluctuations:

1. slow, with periods of a few hours;

2. fast, with periods ranging from a fraction of a second to a few minutes.

In line-of-sight links, slow changes in the received power are usually associated with changes in the refraction index of the atmosphere, that decrease the effective Earth radius and bring up obstacles (non-existent in normal propagation conditions). The probability of such fading, also known as sub-refractive fading, may be reduced by the adoption of prudent criteria for the clearance between the direct ray and the terrain. Criteria referred to in Section 2.5.5 lead to a probability of sub-refractive fading which may be neglected when compared to other types of fading.

For frequencies above about 8 – 10 GHz, particularly in longer paths, rainfall may cause an excess attenuation for periods, often longer than a few minutes, which reduce the link availability and may become the dominant factor in the design of such links.

Changes in the refracting index of the atmosphere, responsible for changes in the angle of arrival in the vertical plane (up to 0.7°), may cause a loss of a few decibels in antenna gains. Changes in the angle of arrival in the horizontal plane appear to be limited to about 0.1°. Even if changes in the angle of arrival seem to occur only a few places [4, 12], they raise questions about the use of antennas with very high directivities in the vertical plane.

In obstructed links, super standard propagation conditions[1] may significantly reduce path loss for up to a few hours. These conditions do not affect the performance of links where they occur, but they are a potential source of interference to other links operating in the same frequency.

In all cases mentioned so far, and except for the effect of rain, fading does not vary rapidly with frequency.

While normally associated with fast fading, there is another cause which is also capable of producing slow fluctuations in the level of the received signal – multipath. The main feature of multipath fading is its dependence, on the one hand, on the operating frequency and, on the other hand, on antenna location.

A link is subject to multipath when there is more than a single path between transmitter and receiver antennas. A typical case of multipath is a link where the specular point is in line-of-sight from both antennas and falls on smooth terrain or water. The existence of two paths with similar attenuations gives rise, in the receiver antenna, to interference of two fields with comparable amplitudes, whose relative phase depends on the difference in path length, and thus on propagation conditions, and on the wavelength (or frequency).

Experience [4] shows that reflections in atmospheric layers near the ground may occur, particularly when there is fog or morning mists in humid valleys, over swamps, or flood plains. These conditions are more frequent when there is no wind, at dawn or early morning.

Even without well-defined ground reflections, random changes in the atmospheric refraction index allow multiple paths between the transmitter and receiver antennas to exist. In such cases fading is often fast and dependent on frequency and antenna location.

Multipath fading in the atmosphere is not a permanent phenomenon. It occurs when there is no wind and the atmosphere is well stratified. It is more frequent at night and in the first morning hours and it is seldom felt at mid-day or during periods of intense rain. Its frequency is higher in tropical climates and in paths over extended water bodies.

Multipath fading has a pronounced effect in the performance of microwave radio links. Besides changes in the received power level, which are felt both in the signal to noise ratio and in the bit error ratio, multipath fading is frequency

[1] Corresponding to $k_e > 4/3$ or $k_e < 0$ or to the existence of ducts.

dependent and hence it introduces severe signal distortions in wideband links. In addition, in links where orthogonal polarizations are used, it may become the main cause of reduced isolation between polarizations.

In some cases we may get a superposition of the various fading mechanisms: for instance, a selective fast fading superimposed on non selective slow fading.

Line-of-sight paths with no deep multipath fading may experience fast changes in the amplitude of the received signal, known as scintillation. This phenomenon may occur in paths, irrespective of the length and frequency, but it becomes more obstructive for frequencies above 10 GHz. For frequencies below 40 GHz, the peak-to-peak amplitude of these fast fluctuations is below a few decibels and therefore they are usually not of concern in the design of microwave radio links.

After this, necessarily short, description of the main features of fading we continue by presenting theoretical fading models, due to Rayleigh, Nakagami and Rice that enable the prediction of fast fading amplitude distribution. Then we consider slow fading associated with paths with propagation by diffraction or tropospheric scattering. Finally we describe a method to forecast total fading (slow and fast). This chapter ends with a description and an analysis of diversity, a procedure used to counteract the effects of fading.

In the following we will assume that a path is not significantly affected by reflections, or has no reflections, when the power received via reflections is 10 dB, or more, below direct ray power.

3.2 FAST FADING: THEORETICAL MODELS

Consider an unobstructed line-of-sight path with no reflections. While usually there is only one path between the two antennas, during some periods of time we observe that a few distinct paths may coexist. If we assume that:

- the transmitted signal is sinusoidal,
- the number of paths is large,
- no path dominates the others,
- at the receiver antenna the signals phase distribution for each path is uniform,

then the received field $\vec{e}_R$ is the sum of the fields $\bar{e}_i$ due to each path

$$\vec{e}_R = \sum_{i=1}^{n} \vec{e}_i. \tag{3.1}$$

We will now consider separately the components of vector $\vec{e}_R$ along the X axis, x_R, and the Y axis, y_R. Let e_i and ϕ_i be the amplitude and the phase of vector $\vec{e}_i$

$$x_R = \sum_{i=1}^{n} e_i \cos(\phi_i), \tag{3.2}$$

$$y_R = \sum_{i=1}^{n} e_i \sin(\phi_i). \tag{3.3}$$

Variables x_R and y_R are random, independent, with zero average value. According to the central limit theorem, when the number of paths is large the distributions of x_R and y_R tend to be Gaussian, that is, their probability density functions may be written as

$$p(x_R) = \frac{1}{\sqrt{2\pi}\,\sigma} \exp\left(-\frac{x^2}{2\sigma^2}\right), \tag{3.4}$$

$$p(y_R) = \frac{1}{\sqrt{2\pi}\,\sigma} \exp\left(-\frac{y^2}{2\sigma^2}\right). \tag{3.5}$$

Since variables x_R e y_R are independent, the joint probability density $p(x_R, y_R)$ is the product of their probability density functions

$$p(x_R, y_R) = \frac{1}{2\pi\sigma^2} \exp\left(-\frac{x_R^2 + y_R^2}{2\sigma^2}\right). \tag{3.6}$$

Changing to polar coordinates r, θ expression (3.6) becomes

$$p(r, \theta) = \frac{1}{2\pi\sigma^2} \exp\left(-\frac{r^2}{2\sigma^2}\right). \tag{3.7}$$

The probability of receiving a field with amplitude below or equal to r_0, $\mathcal{P}(r \le r_0)$, may be obtained by integration of equation (3.7)

$$\begin{aligned} \mathcal{P}(r \le r_0) &= \int_0^{2\pi} d\theta \int_0^{r_0} p(r,\theta) r dr \\ &= \int_0^{r_0} \frac{1}{2\pi\sigma^2} \exp\left(-\frac{r^2}{2\sigma^2}\right) r dr \int_0^{2\pi} d\theta \\ &= 1 - \exp\left(-\frac{r_0^2}{2\sigma^2}\right). \end{aligned} \tag{3.8}$$

The probability distribution function defined in equation (3.8) is the negative exponential, also known as Rayleigh distribution.

The probability density function of the amplitude of the received field is obtained by differenting equation (3.8) with respect to the amplitude

$$p(r) = \frac{r}{\sigma^2} \exp\left(-\frac{r^2}{2\sigma^2}\right). \tag{3.9}$$

In telecommunication systems it is often handier to work with power densities, that is with the Poynting vector, than with field amplitudes. Taking

$$p_0 = \frac{r_0^2}{2Z_0}, \tag{3.10}$$

where Z_0 is the characteristic impedance of the transmission line, assumed to be real, (3.8) becomes

$$\mathcal{P}(p \leq p_0) = 1 - \exp\left(-\frac{p_0\, Z_0}{\sigma^2}\right). \tag{3.11}$$

From the definition of median of the received power p_m, we have

$$\mathcal{P}(p \leq p_m) = \frac{1}{2},$$

hence from (3.11)

$$p_m = \frac{\sigma^2 \log_e(2)}{Z_0}. \tag{3.12}$$

Substituting the value of σ^2 given in (3.12) in (3.11) we finally get:

$$\mathcal{P}(p \leq p_0) = 1 - \exp\left[-\frac{\log_e(2) p_0}{p_m}\right] \tag{3.13}$$

$$= 1 - 2^{-\frac{p_0}{p_m}}. \tag{3.14}$$

In Figure 3.1 we plot $\mathcal{P}(p \leq p_0)$ as a function of p_0/p_m as given in equation (3.14). The use of a logarithmic scale for $\mathcal{P}$ facilitates reading for very low values of probability, often used in the design of microwave radio links.

Expanding the exponential in (3.14) in series and keeping the first two terms, we get the following approximate expression for $\mathcal{P}(p \leq p_0)$, when $p_0/p_m \ll 1$:

$$\mathcal{P}(p \leq p_0) = \log_e(2)\, \frac{p_0}{p_m}, \tag{3.15}$$

$$= k\, \frac{p_0}{p_m}. \tag{3.16}$$

This result means that when $p_0/p_m \ll 1$ the plot of $\mathcal{P}(p_0/p_m)$ (using a logarithmic scale for the Y axis) is a straight line, with a slope of 10 dB per decade, that is $\mathcal{P}$ decreases to 1/10 when the received power decreases by 10 dB.

Before going any further we must define some parameters of the negative exponential (or Rayleigh) distribution, namely, the average $\bar{r}$ and the standard deviation $\bar{r^2}$. By definition we have:

$$\bar{r} = \int_0^\infty r\, p(r) dr. \tag{3.17}$$

$$\bar{r^2} = \int_0^\infty r^2 p(r) dr. \tag{3.18}$$

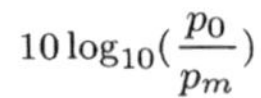

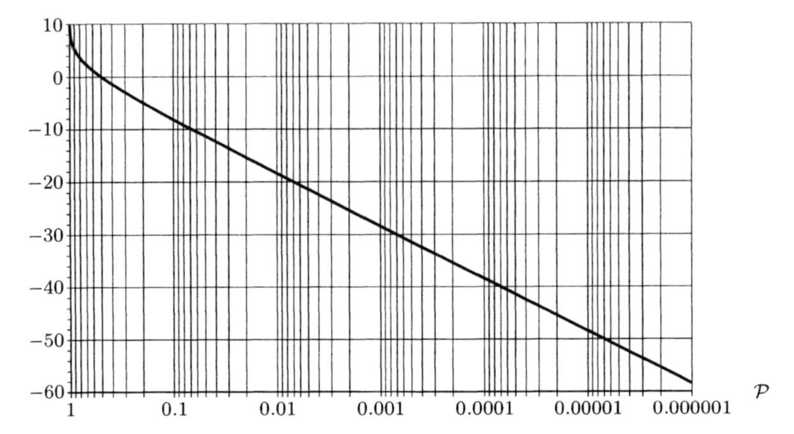

Fig. 3.1 Rayleigh fading. Probability $\mathcal{P}$ that the received power is less than or equal to p_0.

Substituting successively (3.9) in (3.17) and (3.18) and manipulating we have

$$\begin{aligned}\bar{r} &= \int_0^\infty \frac{r^2}{\sigma^2} \exp\left(-\frac{r^2}{2\sigma^2}\right) dr \\ &= \sigma\sqrt{\frac{\pi}{2}}, \qquad (3.19)\\ \bar{r^2} &= \int_0^\infty \frac{r^3}{\sigma^2} \exp\left(-\frac{r^2}{2\sigma^2}\right) dr \\ &= 2\sigma^2. \qquad (3.20)\end{aligned}$$

The assumptions used to derive the Rayleigh distribution are not always valid. Sometimes it is better to assume that between all the received signals there is a dominant one, with amplitude a whose phase is taken for the origin of phases. Using again the expression for the total received field (3.1) we get

$$\vec{e}_R = a + \sum_{i=1}^{n} \vec{e}_i. \qquad (3.21)$$

As we stated before, the components of $\vec{e}_R - a$ along the X axis x_R and the Y axis y_R are independent, random, variables with probability density functions:

$$p(x_R) = \frac{1}{\sqrt{2\pi}\,\sigma} \exp\left(-\frac{x_R^2}{2\sigma^2}\right),$$

$$p(y_R) = \frac{1}{\sqrt{2\pi}\,\sigma}\exp\left(-\frac{y_R^2}{2\sigma^2}\right).$$

With the following change of variables:

$$u = a + x_R, \quad (3.22)$$
$$v = y_R, \quad (3.23)$$

we get

$$p(u) = \frac{1}{\sqrt{2\pi}\,\sigma}\exp\left[-\frac{(u-a)^2}{2\sigma^2}\right], \quad (3.24)$$
$$p(v) = \frac{1}{\sqrt{2\pi}\,\sigma}\exp\left(-\frac{v^2}{2\sigma^2}\right). \quad (3.25)$$

Since variables x_R and y_R are independent, variables u and v are also independent, and the joint probability density function $p(u,v)$ is the product of the probability density functions of each variable, that is,

$$p(u,v) = \frac{1}{2\pi\sigma^2}\exp\left[-\frac{(u-a)^2+v^2}{2\sigma^2}\right]. \quad (3.26)$$

Changing (3.26) into polar coordinates r, θ we obtain

$$p(r,\theta) = \frac{1}{2\pi\sigma^2}\exp\left(-\frac{a^2+r^2}{2\sigma^2}\right)\exp\left[\frac{ar\cos(\theta)}{\sigma^2}\right]. \quad (3.27)$$

The probability of receiving a signal with an amplitude equal or less than r_0 will now be

$$\mathcal{P}(r \le r_0) = \frac{1}{2\pi}\int_0^{\pi}\exp\left[\frac{ar\cos(\theta)}{\sigma^2}\right]d\theta\int_0^{r_0}\frac{r}{\sigma^2}\exp\left(-\frac{a^2+r^2}{2\sigma^2}\right)dr. \quad (3.28)$$

Considering that

$$I_0(z) = \frac{1}{\pi}\int_0^{\pi}\exp[z\cos(\theta)]d\theta, \quad (3.29)$$

where $I_0(z)$ is the modified Bessel function of the first kind, order zero and argument z [1]. Substituting (3.29) in (3.28) and manipulating, we get

$$\mathcal{P}(r \le r_0) = \int_0^{r_0}\frac{r}{\sigma^2}\exp\left(-\frac{a^2+r^2}{2\sigma^2}\right)I_0\left(\frac{ar}{\sigma^2}\right)dr. \quad (3.30)$$

Equation (3.30) may be written in a more compact form, using function Q [10], described by Marcum and defined as:

$$Q(a,b) = \int_b^{\infty} x\exp\left(-\frac{a^2+x^2}{2}\right)I_0(ax)dx. \quad (3.31)$$

Function Q obeys the following:

$$Q(a,0) = 1, \tag{3.32}$$

$$Q(0,b) = \exp\left(-\frac{b^2}{2}\right). \tag{3.33}$$

Considering (3.31) and (3.32), (3.30) becomes

$$\mathcal{P}(r \leq r_0) = 1 - Q\left(\frac{a}{\sigma}, \frac{r_0}{\sigma}\right). \tag{3.34}$$

The distribution whose accumulated probability function is given in (3.30) and (3.34) is known as the Rice distribution, or the Nakagami-Rice distribution. Its probability density function may be obtained by differentiating (3.30) with respect to the amplitude

$$p(r) = \frac{r}{\sigma^2} \exp\left(-\frac{a^2 + r^2}{2\sigma^2}\right) I_0\left(\frac{ar}{\sigma^2}\right). \tag{3.35}$$

For the design of microwave radio links it is convenient to express the accumulated probability as a function of the median received power p_m and of the ratio between the power in the main component p_0 and in the other random components.

Since the average value of the received power density (the absolute value of the Poynting vector) p_a of an electromagnetic plane wave whose (sinusoidal) electric field has a peak amplitude a is

$$p_a = \frac{a^2}{2Z_0}, \tag{3.36}$$

where Z_0 is the characteristic impedance of medium where the wave is propagating, assumed to be real, and that the average value of the power density conveyed by the random components p_n is given, from (3.20), by

$$\begin{aligned} p_n &= \frac{\bar{r^2}}{2Z_0} \\ &= \frac{2\sigma^2}{Z_0}, \end{aligned} \tag{3.37}$$

we have:

$$\frac{a}{\sigma} = \sqrt{\frac{2p_a}{p_n}}. \tag{3.38}$$

Recalling that the received field is a sum of a dominant component, with constant peak amplitude a, and a large number of random components, the median received power density p_{mr} should be the sum of the power of the main component p_a plus the median power due to the random components p_m. From (3.36) and (3.12), we get

$$p_{mr} = \frac{a^2}{2Z_0} + \frac{\sigma^2 \log_e(2)}{Z_0}. \tag{3.39}$$

Denoting by p_0 the power density of an electromagnetic plane wave whose electric field has a peak amplitude r_0

$$p_0 = \frac{r_0^2}{2Z_0}, \tag{3.40}$$

from (3.40) and (3.37) the ratio r_0/σ becomes

$$\begin{aligned} \frac{r_0}{\sigma} &= \sqrt{\frac{2p_0}{p_n}}, \\ &= \sqrt{\frac{2p_0}{p_{mr}}\frac{p_{mr}}{p_n}}, \\ &= \sqrt{\frac{2p_0}{p_{mr}}\left(\frac{p_a}{p_n} + \log_e 2\right)}. \end{aligned} \tag{3.41}$$

Substituting the values of a/σ and r_0/σ, given in (3.38) and (3.41), in (3.34) we finally get

$$\mathcal{P}(p \leq p_0) = 1 - Q\left[\sqrt{\frac{2p_a}{p_n}}, \sqrt{\frac{2p_0}{p_{mr}}\left(\frac{p_a}{p_n} + \log_e 2\right)}\right]. \tag{3.42}$$

In Figure 3.2 we plot $\mathcal{P}(p \leq p_0)$ as a function of p_0/p_{mr}, expressed in decibels, taking as a parameter p_a/p_n, also expressed in decibels. The values of Q were obtained by numeric integration.

In the limit, when p_a/p_n tends to zero, the Nakagami-Rice distribution tends to the Rayleigh distribution. To ease the comparison, we also plot in Figure 3.2 the Rayleigh distribution.

Sometimes, as in ITU-R Recommendation P.1057 [6], the parameter used to plot the Nakagami-Rice fading is the fraction of the average received power conveyed by the random components p_n/p_r, rather than, p_a/p_n, as in figure 3.2. Since

$$p_r = p_a + p_n, \tag{3.43}$$

we get the relation between p_n/p_r and p_a/p_n

$$\frac{p_n}{p_r} = 1 - \frac{1}{1 + \frac{1}{\frac{p_a}{p_n}}}. \tag{3.44}$$

Although experience shows that in line-of-sight links fast fading follows the Nakagami-Rice model, its use is somewhat limited since, for a given path and operating frequency, it is not possible to predict the ratio p_a/p_n.

Assuming, with Boithias [2], that the contribution of the dominant component in the received power decreases with path length, we may obtain, at 4 GHz, a good fit between experimental fading data (in Western Europe, in

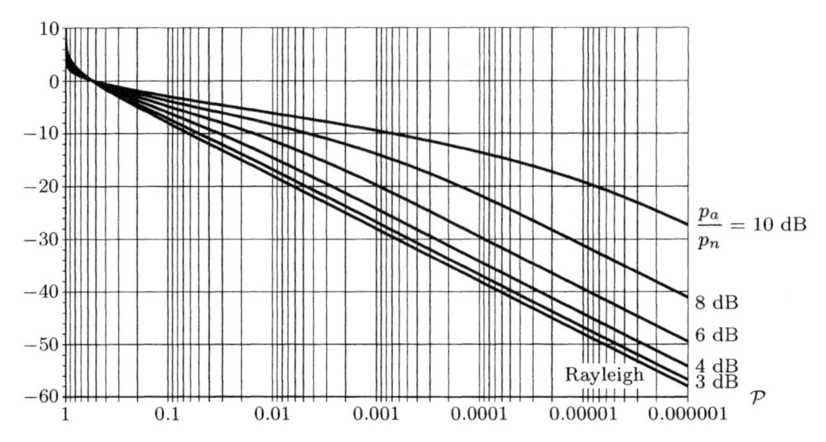

Fig. 3.2 Nakagami-Rice fading. Probability $\mathcal{P}$ that the received power p_0/p_{mr} (expressed in dB) is exceeded, taking p_a/p_n, in dB, as a parameter.

the worst month, for distances up to 100 kilometers) and the Nakagami-Rice model, by taking

$$\left(\frac{p_a}{p_n}\right)^2 = \frac{\exp(-d/265)}{1 - \exp(-d/265)}, \tag{3.45}$$

where d is the path length in kilometers.

For typical Western Europe climate conditions, the effect of using frequencies other than 4 GHz is equivalent to a change in path length from d to d_{eq}

$$d_{eq} = d\left(\frac{f}{4}\right)^{-0.25}. \tag{3.46}$$

3.3 FAST FADING: EMPIRICAL MODELS

For line-of-sight links, in temperate continental climate and undulating terrain, as often found in Western Europe, the probability that the received power p is less or equal to to p_0, in the worst month, may be estimated by the following expression also known as Morita's law

$$\mathcal{P}(p \leq p_0) = 1.4 \cdot 10^{-8}\ f\ d^{3.5}\ \frac{p_0}{p_n}, \tag{3.47}$$

where:

- f is the operating frequency in GHz;

- d is the path length in km;
- p_n is the received power with no fading.

Equation (3.47) is only valid for narrow band systems and when

$$\begin{aligned} P_0 - P_n &< -15 \text{ dB}, \\ \mathcal{P} &\leq 10^{-3}, \\ 15 \leq d &\leq 100, \\ 2 \leq f &\leq 37, \end{aligned}$$

where

$$P_0 - P_n = 10 \log_{10}\left(\frac{p_0}{p_n}\right).$$

The nomogram in figure 3.3 may be used to estimate $\mathcal{P}$ from $P_0 - P_n$ (in dB), f (in GHz) and d (in km).

ITU-R Recommendation P.530-8 [6] describes two methods to forecast fast fading, the first of which is only applicable to small percentages of time.

In the first method:

1. Calculate the geoclimatic factor K, for the average worst month, from the meteorologic data for the region. In the absence of such data, K may be estimated from the percentage of time P_L when the average refractivity gradient in the lower 100 m of the atmosphere is below -100 N/km, as given in Figures 3.4 to 3.7[2], and the type of path according to:

    - inland,
    - coastal, over or near large bodies of water,
    - coastal, over or near medium-sized bodies of water.

    Paths are considered inland:

    - when the entire path lies above the altitude of 100 m and further than 50 km from the nearest coastline;
    - when the entire path lies at less than 50 km from the nearest coast line and part of it is at an altitude of less than 100 m but there are obstacles higher than 100 m between the path and the coastline.

    Paths over small lakes or rivers are usually considered as inland paths.

    For inland paths the value of K is given by

$$K = 5.0 \cdot 10^{-7} \cdot 10^{-0.1(C_o - C_{Lat} - C_{Long})} P_L^{1.5}. \tag{3.48}$$

[2]For latitudes higher than 60°N or 60°S only Figures 3.5 and 3.6 should be used.

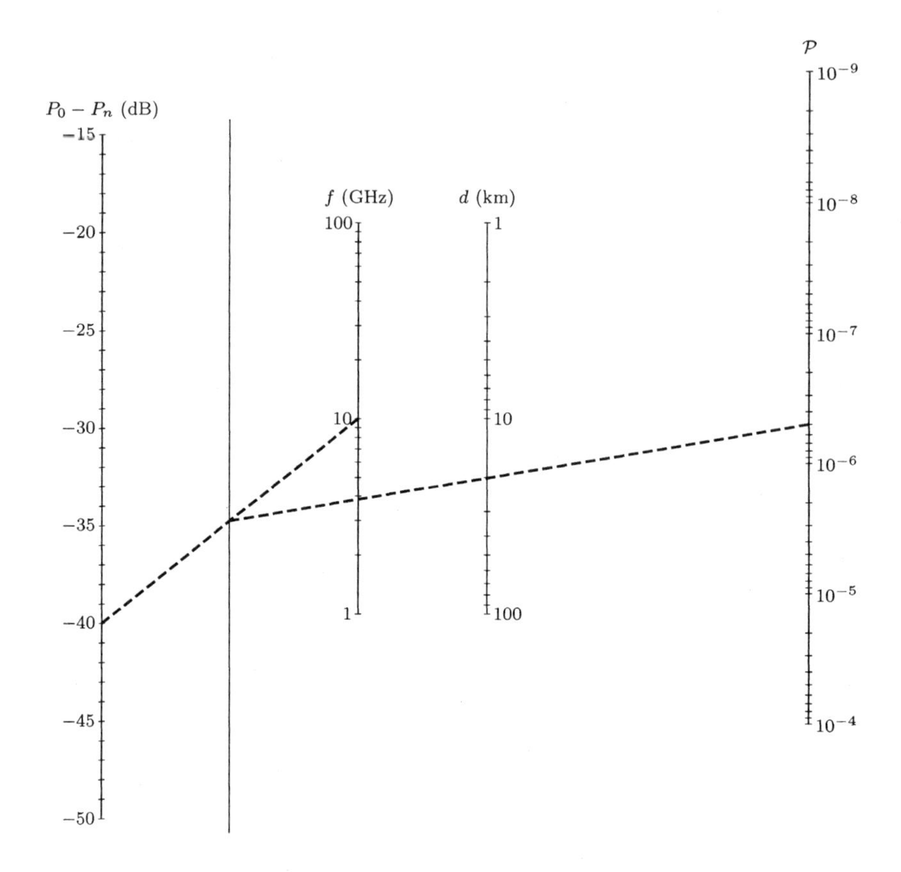

Fig. 3.3 Nomogram to estimate the fast fading given by (3.47), for the worst month. In the example we have $P_0 - P_n = -40$ dB, $f = 10$ GHz, $d = 20$ km and $\mathcal{P} \approx 5.10^{-7}$.

In (3.48) the value of C_o depends on the lower altitude antenna and the type of terrain, according to:

- lower altitude antenna between 0 and 400 m, plains

$$C_o = 0;$$

- lower altitude antenna between 0 and 400 m, hills

$$C_o = 3.5;$$

- lower altitude antenna between 400 and 700 m, plains

$$C_o = 2.5;$$

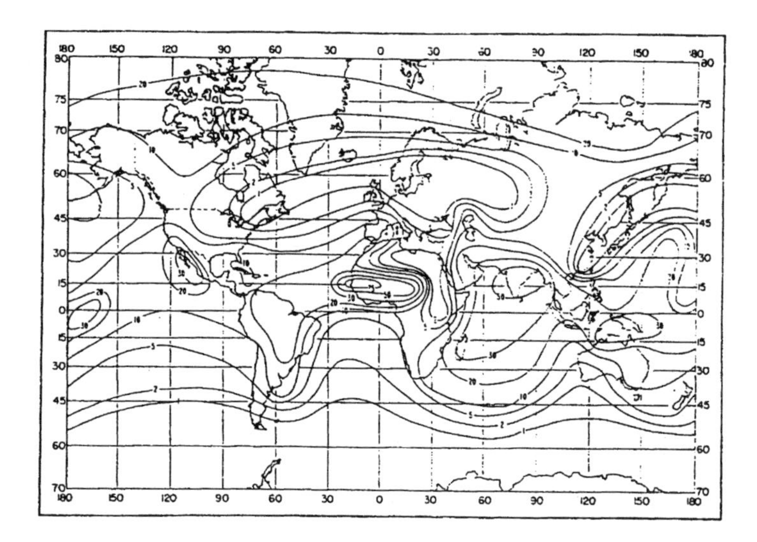

Fig. 3.4 World chart showing the percentage of time the average refractivity gradient is less than −100 N/km in the month of February (ITU-R Recommendation P.453-6 [6]). Reproduced with ITU permission.

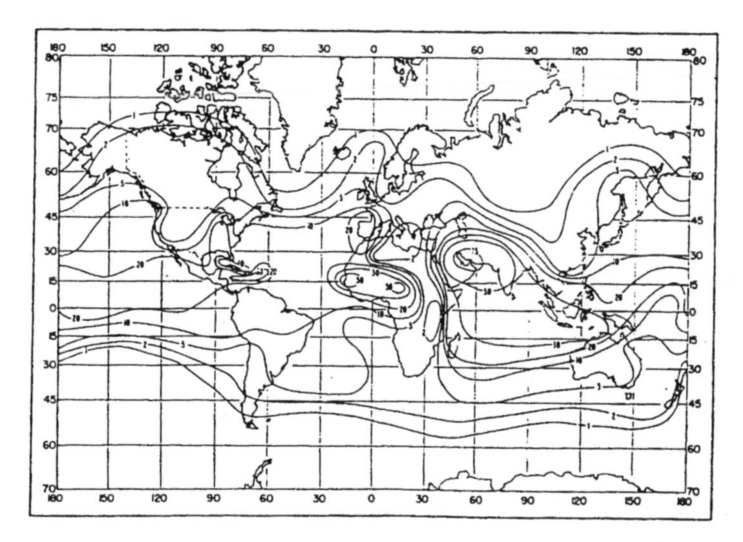

Fig. 3.5 World chart showing the percentage of time the average refractivity gradient is less than −100 N/km in the month of May (ITU-R Recommendation P.453-6 [6]). Reproduced with ITU permission.

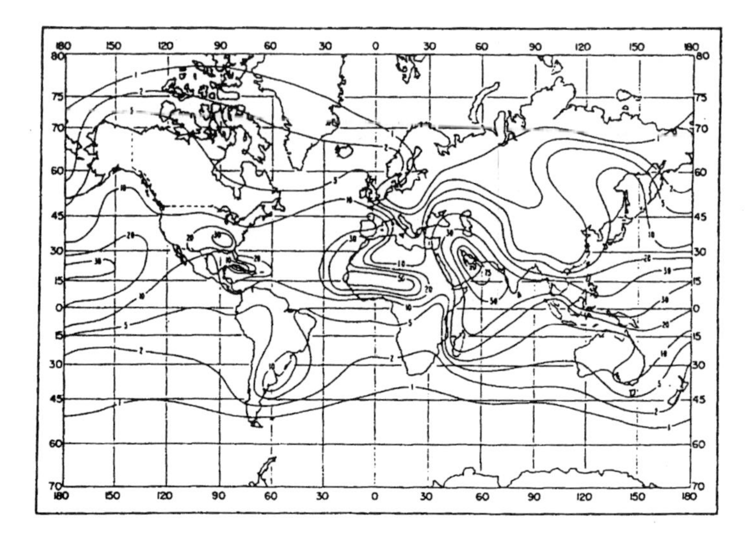

Fig. 3.6 World chart showing the percentage of time the average refractivity gradient is less than −100 N/km in the month of August (ITU-R Recommendation P.453-6 [6]). Reproduced with ITU permission.

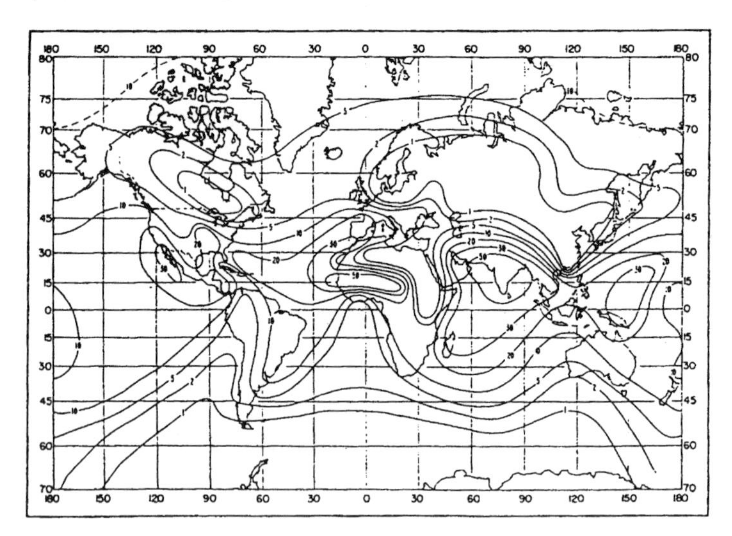

Fig. 3.7 World chart showing the percentage of time the average refractivity gradient is less than −100 N/km in the month of November (ITU-R Recommendation P.453-6 [6]). Reproduced with ITU permission.

- lower altitude antenna between 400 and 700 m, hills

$$C_o = 6;$$

- lower altitude antenna above 700 m, plains

$$C_o = 5.5;$$

- lower altitude antenna above 700 m, hills

$$C_o = 8;$$

- lower altitude antenna above 700 m, mountains

$$C_o = 10.5.$$

Constants C_{Lat} and C_{Long} are given as a function of the average path latitude η and longitude ξ by:

- for latitudes between 53° North and 53° South

$$C_{Lat} = 0;$$

- for latitudes between 53° and 60° (North or South)

$$C_{Lat} = -53 + \eta;$$

- for latitudes higher than 60° (North or South)

$$C_{Lat} = 7;$$

- for longitudes corresponding to Europe and Africa

$$C_{Long} = 3;$$

- for longitudes corresponding to America (North, Central and South)

$$C_{Long} = -3;$$

- elsewhere

$$C_{Long} = 0.$$

Paths are classified as coastal when part of the path has an altitude of less than 100 m and is at a distance up to 50 km from a average size or large body of water and if there is no obstacle with an altitude higher than 100 m between the path and the coastline.

For coastal paths over or near large bodies of water the value of K may be estimated as

$$K = K_l(r_c) = 10^{(1-r_c)\log_{10}(K_i)+r_c\log_{10}(K_{cl})}, \tag{3.49}$$

for $K_{cl} \geq K_i$, or as

$$K = K_i, \tag{3.50}$$

for $K_{cl} < K_i$, where r_c is the fraction of the path with an altitude less than 100 m above the median level of the water in the body of water in question and nearer than 50 km from the coastline. Factor K_i is given in (3.48) and K_{cl} is

$$K_{cl} = 2.3 \cdot 10^{-4} \cdot 10^{-0.1C_o-0.011|\eta|}, \tag{3.51}$$

where C_o is the same used for inland paths.

For coastal paths, over or near average sized bodies of water the value of K may be estimated as

$$K = K_m(r_c) = 10^{(1-r_c)\log_{10}(K_i)+r_c\log_{10}(K_{cm})}, \tag{3.52}$$

for $K_{cm} \geq K_i$, or as

$$K = K_i \tag{3.53}$$

for $K_{cm} < K_i$, where r_c is the fraction of the path with an altitude less than 100 m above the median level of the water in the body of water in question and nearer than 50 km from the coastline. The value of K_i is given in (3.48) and K_{cm} is:

$$K_{cm} = 10^{0.5[\log_{10}(K_i)+\log_{10}(K_{cl})]}. \tag{3.54}$$

The Gulf of Finland (about 300 km long and 30–40 km wide) is considered as an average sized body of water while the English Channel, the North Sea and the Strait of Hudson are considered as large bodies of water. When in doubt the value of K should be estimated as

$$K = 10^{(1-r_c)\log_{10}(K_i)+0.5[\log_{10}(K_{cm})+\log_{10}(K_{cl})]}, \tag{3.55}$$

where r_c is defined in the following, K_i is the value of K in (3.48), K_{cl} in (3.51) and K_{cm} em (3.54).

Inland paths, in regions with many lakes, seem to behave as coastal path and we should take

$$K = 10^{0.5[(2-r_c)\log_{10}(K_i)+r_c\log_{10}(K_{cm})]}. \tag{3.56}$$

2. Compute the absolute value of the path slope ε_p, in milliradians, from the altitudes of the transmitter and the receiver antennas, h_T and h_R, in meters, and the path length d in kilometers

$$|\varepsilon_p| = \frac{|h_R - h_T|}{d}. \tag{3.57}$$

3. The probability that the received power p is less or equal to p_0, at the frequency f (in GHz), in the worst month, is given by

$$\mathcal{P}(p \leq p_0) = K\, d^{3.6}\, f^{0.89}\, (1 + |\varepsilon_p|)^{-1.4}\, \frac{p_0}{p_n} \times 10^{-2}. \tag{3.58}$$

Equation (3.58) was obtained from data for paths with lengths between 7 and 95 km, frequencies between 2 and 37 GHz, path slopes between 0 and 24 milliradians and angles between the direct and the reflected ray in range of 1 to 12 milliradians. Some data suggest that its validity may be extended with respect to the path length and the frequency. The lower limit for the frequency range appears to decrease with the path length as follows:

$$f_{[GHz]} = \frac{1.5}{d_{[km]}}.$$

The second method described in ITU-R Recommendation P.530-8 [6] enables us to calculate the fraction of time when a given fading depth A (higher than 0 dB) is exceeded. As before, this method will be presented in a step-by-step mode:

1. Compute the multipath occurrence factor $\mathcal{F}$

$$\mathcal{F} = K\, d^{3.6}\, f^{0.89}\, (1 + |\varepsilon_p|)^{-1.4} \times 10^{-2}, \tag{3.59}$$

   where K is the geoclimatic factor previously defined. The method is valid for $\mathcal{F} \leq 20$.

2. Calculate the fading depth A_t at which occurs the transition between shallow and deep fading

$$A_t = 25 + 1.2 \log_{10}(100 p_0). \tag{3.60}$$

   The procedure now depends on whether the required fading depth A is higher or lower than A_t.

3. If $A \geq A_t$ then use equation (3.58) to compute $\mathcal{P}_A$.

4. If $A < A_t$ then:

   (a) Compute $\mathcal{P}_t$ as

$$\mathcal{P}_t = \mathcal{F} \times 10^{-A_t/10}; \tag{3.61}$$

   (b) Compute q'_a given by:

$$q'_a = \frac{-20 \log_{10}[-\log_e(1 - \mathcal{P}_t)]}{A_t}; \tag{3.62}$$

(c) Compute q_t given by:

$$q_t = \frac{q'_a - 2}{\left[\left(1 + 0.3 \times 10^{-A_t/20}\right) 10^{-0.016A_t}\right]} - 4.3\left(10^{-A_t/20} + \frac{A_t}{800}\right); \tag{3.63}$$

(d) Compute q_a as

$$q_a = 2 + \left(1 + 0.3 \times 10^{-A/20}\right) 10^{-0.016A} \times \left[q_t + 4.3\left(10^{-A/20} + \frac{A}{800}\right)\right]; \tag{3.64}$$

(e) The probability $\mathcal{P}_A$ that the fading depth A is exceeded is now given by

$$\mathcal{P}_A = 1 - \exp\left(-10^{-q_a A/20}\right). \tag{3.65}$$

In Figure 3.8 we plot $\mathcal{P}_A$, the probability that the fading depth exceeds A (in dB), taking $\mathcal{F}$ as a parameter.

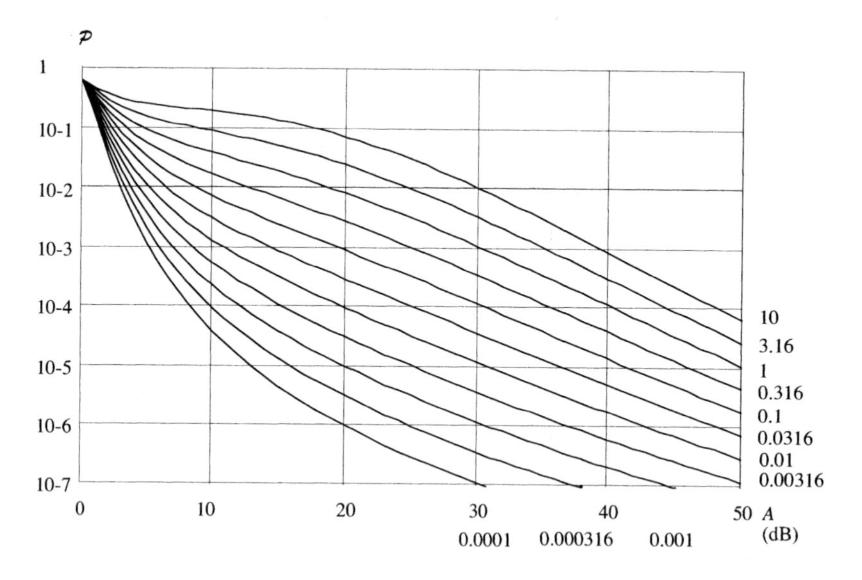

Fig. 3.8 Probability $\mathcal{P}_A$ that the fading depth exceeds A (in dB), taking $\mathcal{F}$ as a parameter.

For paths with large reflections, fading is more dependent on reflections than on tropospheric multipath and its characteristics are quite different from those described before. There are little data for these paths so forecasts are rather unreliable.

It is possible to summarize the different methods, both theoretical and empirical, used to estimate fast fading probability in a simple expression, valid for deep fades (over about 15 dB):

$$\mathcal{P} = \mathcal{F} \cdot \frac{p_0}{p_n}, \tag{3.66}$$

where $\mathcal{F}$ is a constant, dependent on the method chosen, often known as the deep fade occurrence factor (ITU-R [6] Recommendation P.530-8).

As an alternative to the second method presented in ITU-R Recommendation P.530-8 [6], we may get a rather crude estimation of the fading depth not exceeded for more than 20 percent of the time, in the worst month F_{20} (in dB), in average Western Europe conditions, from:

$$F_{20} = 0.03d\sqrt{f}, \tag{3.67}$$

where d is the path length and f the operating frequency in GHz.

In Figure 3.9 we plot the fading depth F_{20} (in dB) given by (3.67), as a function of the path length, taking the operating frequency, in the range 0.5 to 20 GHz, as a parameter.

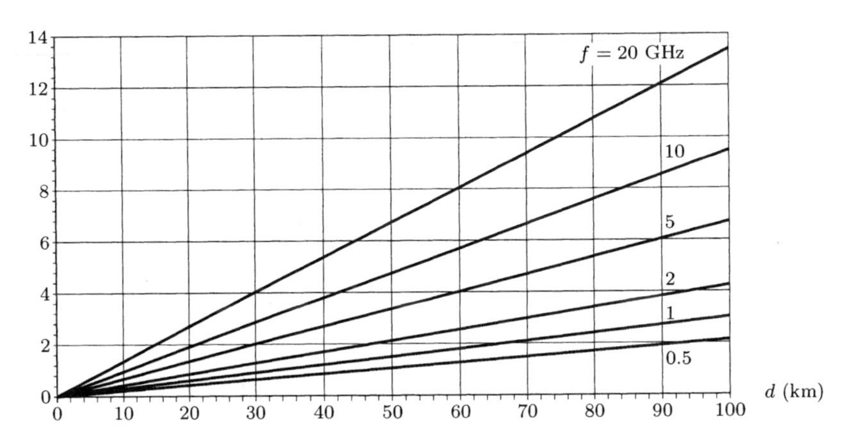

Fig. 3.9 Fading depth F_{20}, in dB, as a function of the path length, taking the operating frequency, in the range 0.5 to 20 GHz, as a parameter.

Stating that the fading depth F_p is not exceeded for more than p percent of the time is equivalent to stating that the probability $\mathcal{P}$ that the received power is F_p dB below the received power without fading is $100 - p$ percent, that is:

$$\mathcal{P}_{\%}(P < P_n - F_p) = 100 - p.$$

It is instructive to compare the values of F_{80} given by the empirical expression (3.67) with those obtained from the Rayleigh distribution. Taking $\mathcal{P} = 0.2$ in (3.14) we get $p_0/p_n \approx 0.32$ which corresponds to $F_{80} \approx 4.9$ dB. From Figure 3.9 we find that this valor is higher than the one obtained from (3.67) for paths with $d < 50$ km and $f < 10$ GHz, which confirms the pessimistic character of the Rayleigh distribution.

Experience shows that on adjacent paths fading depths higher than about 20 dB are almost completely uncorrelated and thus we may state that the probability of such events in a multihop link is the sum of the probabilities for each hop. According with ITU-R Recommendation P.530-8 [6] for fading depths not exceeding 10 dB the probability $\mathcal{P}_{12}$ that a given fading depth is exceeded simultaneously in two adjacent hops is:

$$\mathcal{P}_{12} = (\mathcal{P}_1 \mathcal{P}_2)^{0.8}, \tag{3.68}$$

where $\mathcal{P}_1$ and $\mathcal{P}_2$ are the probabilities that the given fading depth is exceeded in each of the adjacent hops.

Besides fading depth, for link design, particularly for digital links where performance indicators includes, besides other parameters, the number of seconds without errors, we require data on fade duration.

Let n be the number of times, per unit time, that fade depths F higher than 15 dB, are exceeded, in a link at the frequency f, in GHz. The fade duration follows a log-normal distribution with average value t_m. Empirically it is found that

$$n = C_1 \, (10^{-F/10})^{\alpha_1} \, f^{\beta_1}, \tag{3.69}$$

$$t_m = C_2 \, (10^{-F/10})^{\alpha_1} \, f^{\beta_1}, \tag{3.70}$$

where, according to measurements referred to in CCIR Report 338-6 [4], we have:

$$\begin{aligned} \alpha_1 &= 0.5, \\ \alpha_2 &= 0.5, \\ \beta_1 &= 1.32, \\ \beta_2 &= -0.50, \\ C_2 &= 56.6\sqrt{d}. \end{aligned}$$

In Figure 3.10 we plot t_m, as a function of d/f, taking F as a parameter.

Some other data appear to indicate that t_m decreases, rather than increases, as in (3.70), with path length.

Some measurements, relative to narrow band systems and referred to in CCIR Report 338-6 show that in fast fading the rate of change of attenuation

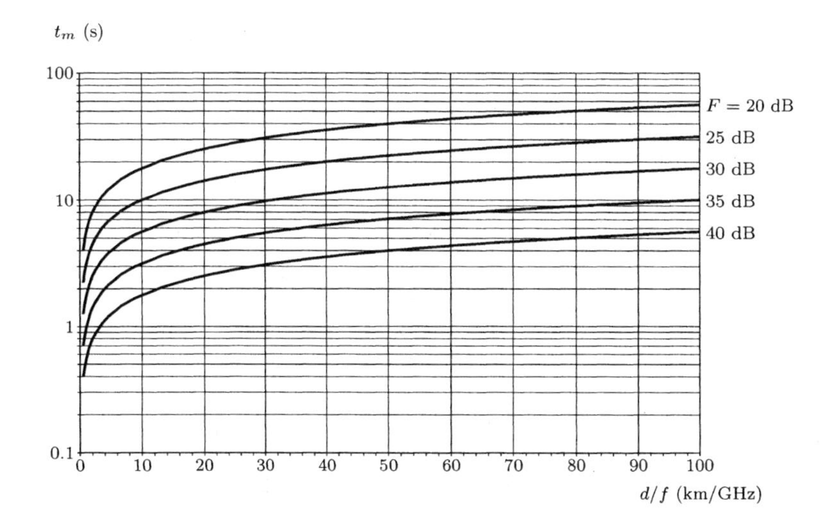

Fig. 3.10 Median fade duration t_m, as a function of d/f, taking the fade depth F (in dB) as a parameter.

has a log-normal distribution with median of the order of 5 dB/s for fade depths of 20 dB and 20 dB/s for fade depths of 30 dB.

In line-of-sight microwave radio links we usually consider only fast fading. Since the Rayleigh model leads to pessimistic results we should rather use the Nakagami-Rice model or the empirical models applicable to the climatic region, the dominant type of terrain and the slope of the direct ray.

Paths employing passive repeaters are taken as a series of hops provided the passive repeaters are at the far distance from each other or of the terminal antennas. Passive repeaters close to terminal antennas, as in periscopes, are taken as part of the terminal antenna. Two close-by flat reflectors are considered as a single passive repeater.

3.4 SLOW FADING: LOG-NORMAL MODEL

Experience shows that in a troposcatter link, the distribution of the hourly medians of the attenuation is log-normal[3], with a standard deviation between 4 and 8 dB, depending on the path length and the climate.

[3]In a log-normal distribution the logarithms of the observed values (in this case the attenuation values expressed in dB) follow a normal distribution.

In order to write down the probability density function of a log-normal distribution, we note that if $f(x)$ is the probability density function of the received signal, the probability that this signal is lower than x_0 is

$$\mathcal{P}(x \leq x_0) = \int_0^{x_0} f(x)dx. \tag{3.71}$$

Taking $x = \log_e(y)$, (3.71) becomes:

$$\mathcal{P}(y \leq y_0) = \int_0^{y_0} \frac{f[\log_e(y)]}{y} dy \tag{3.72}$$

$$= \int_0^{y_0} g(y)dy. \tag{3.73}$$

If $f(x)$ is a normal distribution than $g(y)$ is a log-normal distribution:

$$g(y) = \frac{1}{\sqrt{2\pi}\sigma} \exp\left\{-\frac{[\log_e(y) - m]^2}{2\sigma^2}\right\}. \tag{3.74}$$

Let c be the hourly median of the received signal. According to the log-normal distribution, the probability density of this value is:

$$f(c) = \frac{1}{\sqrt{2\pi}\sigma} \frac{1}{c} \exp\left\{-\frac{[\log_e(c/c_0) - m]^2}{2\,\sigma^2}\right\}, \tag{3.75}$$

where c_0 is an arbitrary signal, m and σ are the average and standard deviation of $\log_e(c/c_0)$ expressed in Neper.

The probability that the received signal is lower than c_0 is:

$$\mathcal{P}(c \leq c_0) = \int_0^{c_0} f(c)\, dc \tag{3.76}$$

$$= \frac{1}{2}\left\{1 + \text{erf}\left[\frac{\log_e(c_0) - m}{\sqrt{2}\sigma}\right]\right\}, \tag{3.77}$$

where erf is the error function, defined as:

$$\text{erf}(x) = \frac{2}{\sqrt{\pi}} \int_0^x \exp(-t^2)dt. \tag{3.78}$$

In Figures 3.11 and 3.12 we plot the probability that the received power p is above a given threshold p_0. In these figures the reference power p_0 is the average of the hourly median received power. For ease of use both p/p_0 and σ are expressed in dB.

In Figure 3.11 we use a scale for the X axis such that the normal (or the log-normal) distribution is plotted as a straight line. This type of plot is often known as a Gaussian plot.

In figure 3.12 we apply in the X axis the same logarithmic scale used for the other distribution to simplify comparisons.

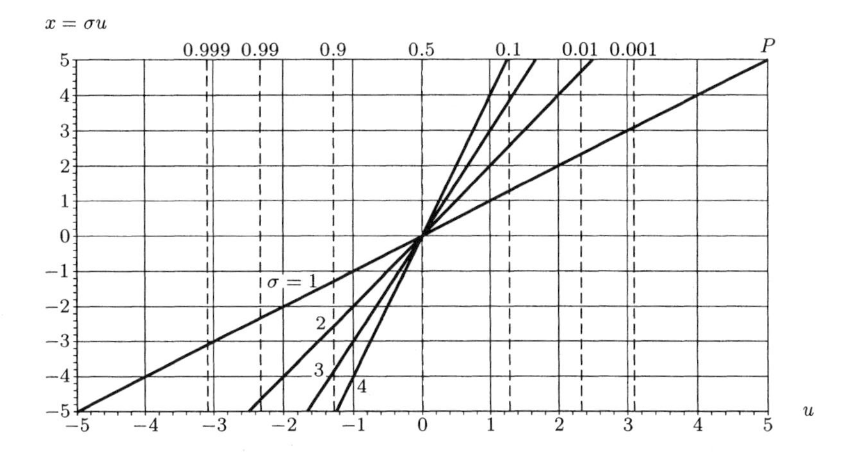

Fig. 3.11 Gaussian plot of the log-normal distribution, taking the standard deviation as a parameter.

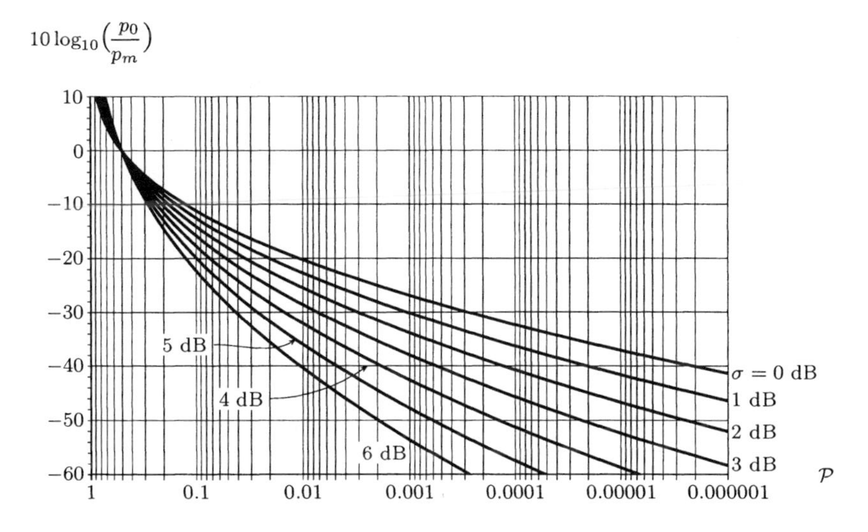

Fig. 3.12 Log-normal distribution, taking the standard deviation as a parameter.

3.5 SLOW FADING: EMPIRICAL MODELS

In this section we describe the method for predicting the variation of the attenuation (slow fading) in troposcatter links referred to in ITU-R Recommendation P.617-1 [6].

q	50	90	99	99.9	99.99
c(q)	0	1	1.82	2.41	2.90

Table 3.1 Values of $c(q)$ for the most frequently used values of probability q, in percent, (ITU-R Recommendation P.617-1 [6]).

The hourly median attenuation exceeded during $100 - p$ percent, of the hour in a year, $A(q)$ is given by

$$A(q) = A(50) - c(q)\,Y(90), \tag{3.79}$$

where $A(50)$ is the hourly median attenuation exceeded 50 percent of the hours on a year, calculated according to Section 2.10.

The value $Y(90)$ is a function of the distance between radio horizons d_{de} (defined in Section 2.9) and of the climate, plotted in Figure 3.13 for climates 1 (equatorial), 3 (maritime subtropical) and 4 (desert).

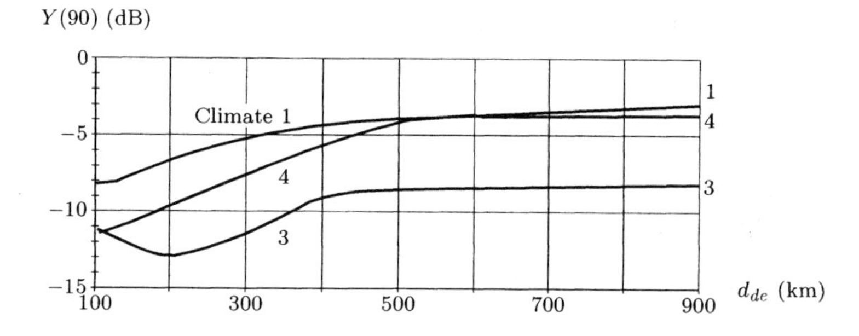

Fig. 3.13 Function $Y(90)$ for climates 1, 3 and 4 (ITU-R Recommendation P.617-1 [6]). Reproduced with ITU permission.

For climates 2 (continental subtropical), 6 (continental temperate) and 7a (maritime temperate, over land) we have

$$Y(90) = -2.2 - (8.1 - 2.3 \cdot 10^{-4} f)\ \exp(-0.137h), \tag{3.80}$$

where f is the operating frequency in GHz and h is given in (2.240).

Finally, for climate 7b (maritime temperate, over the sea) we have

$$Y(90) = -9.5 - 3.0\ \exp(-0.137h), \tag{3.81}$$

where h is also calculated from (2.240).

The values of $c(q)$ are given in Table 3.1 for the usual values of q.

The attenuation for the worst month may be estimated by adding to the annual value a corrective term, given in Figure 3.14 for equatorial, subtropical humid, desert and temperate climates.

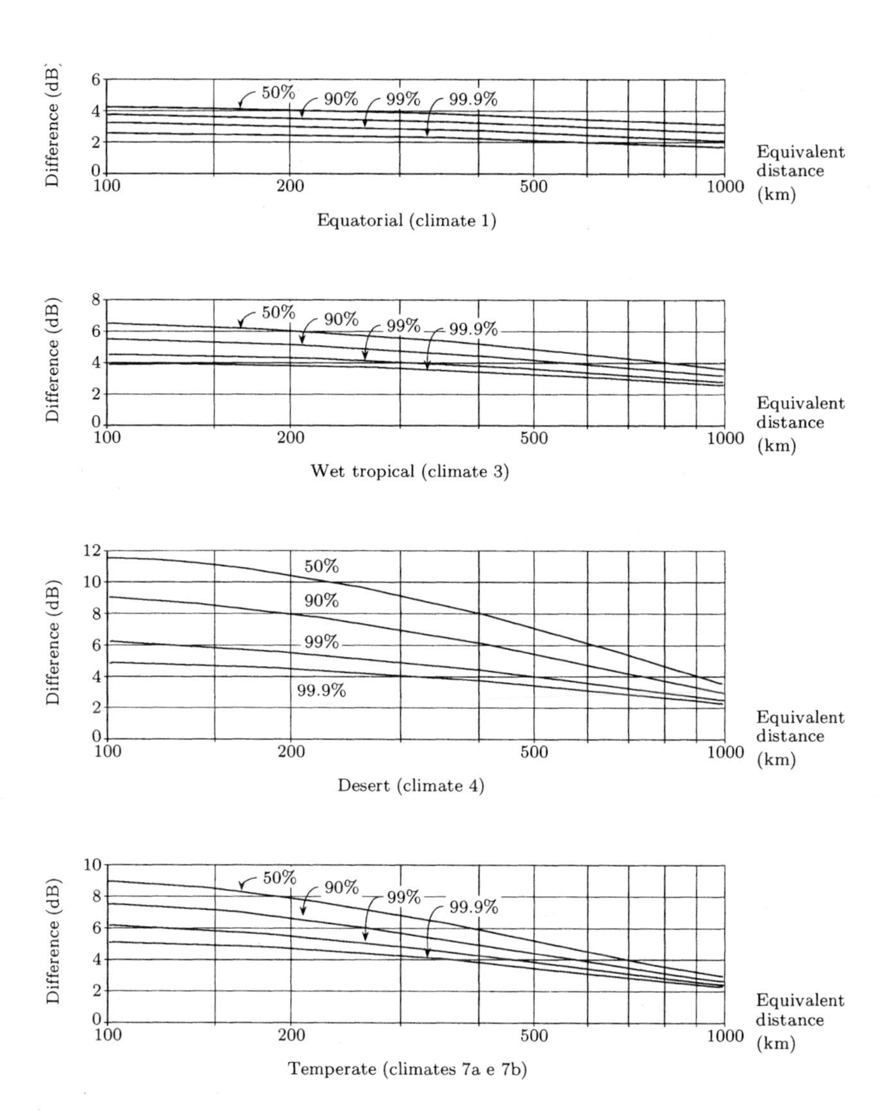

Fig. 3.14 Difference between the attenuation on the worst month and the annual attenuation (ITU-R Recommendation P.617-1 [6]). Reproduced by permission of ITU.

3.6 SIMULTANEOUS SLOW AND FAST FADING

In links with propagation by diffraction or by tropospheric scattering slow and fast fading exist simultaneously.

The joint distribution of slow and fast fading may be obtained [3, 8] if we consider that the received signal power P_R, in dB, is the result of adding two, supposedly uncorrelated, terms. Let the first one, P_{sf}, be due to slow fading and the second one, P_{ff} be due to fast fading

$$P_R = Psf + P_{ff}. \tag{3.82}$$

Let f_{sf} and f_{ff} be the probability density functions corresponding to the slow and the fast varying terms of the received power. The probability density function f_R is the convolution of f_{sf} and f_{ff}

$$f_R(P_R) = \int_{-\infty}^{\infty} f_{sf}(P_{sf})\, f_{ff}(P_R - P_{sf})\, dP_{sf}, \tag{3.83}$$

or

$$f_R(P_R) = \int_{-\infty}^{\infty} f_{sf}(P_R - P_{ff})\, f_{ff}(P_{ff})\, dP_{ff}. \tag{3.84}$$

The probability $\mathcal{P}_r$ that the received power is less or equal to P_0 is

$$\mathcal{P}_r(P_R \le P_0) = \int_{-\infty}^{P_0} dP_R \int_{-\infty}^{\infty} f_{sf}(P_{sf})\, f_{ff}(P_R - P_{sf})\, dP_{sf}. \tag{3.85}$$

Let $\mathcal{P}_{ff}(P \le P_0)$ be the cumulative distribution function of fast fading. From (3.85) we get

$$\mathcal{P}_r(P_r \le P_0) = \int_{-\infty}^{\infty} f_{sf}(P_{sf})\, \mathcal{P}_{ff}(P \le P_0 - P_{sf})\, dP_{sf}. \tag{3.86}$$

Assume now that slow fading follows a log-normal distribution and fast fading a Rayleigh distribution. Let c_{sf} be the received field intensity. From (3.75) we get

$$f_{sf}(c_{sf}) = \frac{1}{\sqrt{2\pi}\sigma c_{sf}} \exp\left[-\frac{(c_{sf} - m)^2}{2\sigma^2}\right], \tag{3.87}$$

and, from (3.14)

$$\mathcal{P}_{ff}(P \le P_0 - P_{sf}) = 1 - 2^{-(c_0/c_{sf})^2}, \tag{3.88}$$

where m and σ are the average and the standard deviation of the logarithm of the received field intensity, respectively.

Taking for the reference power the average received power assuming that only slow fading exists, we have $m = 0$. Considering (3.87) and (3.88) and noting that the fields are expressed in linear units, (3.86) may be written as

$$\mathcal{P}_R(P_R \leq P_0) = \int_0^\infty \frac{1}{\sqrt{2\pi}\sigma P_{sf}} \exp\left[-\frac{\log_e^2(c_{sf})}{2\sigma^2}\right] \left[1 - 2^{-(c_0/c_{sf})^2}\right] dc_{sf}. \tag{3.89}$$

In Figure 3.15 we plot $\mathcal{P}_R(P_R \leq P_0)$, as obtained by numeric integration of (3.89), for σ between 0 and 16 dB. As shown when σ decreases the resulting distribution tends to the Rayleigh distribution.

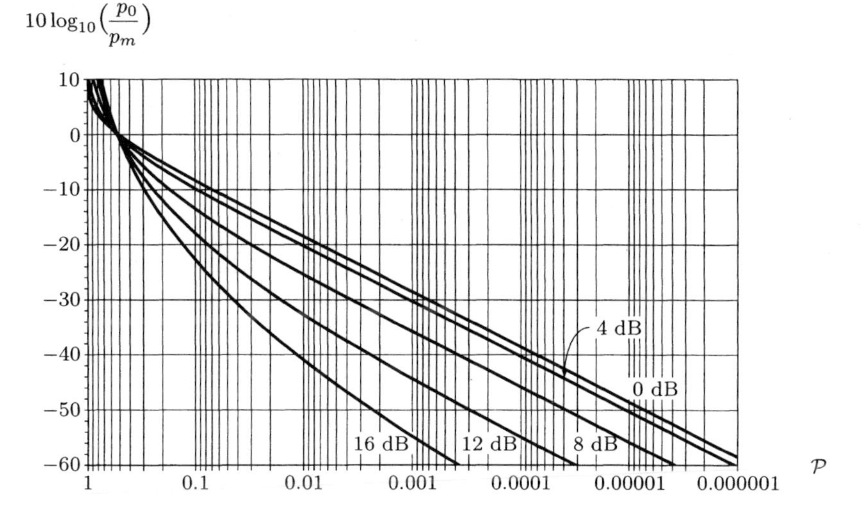

Fig. 3.15 Simultaneous slow (log-normal) and fast (Rayleigh) fading.

3.7 DIVERSITY

Increasing the transmitter power and (or) the antenna gains to obviate the effects of fading and to ensure a given power at the receiver may not always be economical or even possible. There is, however, an alternative solution, both elegant and efficient: diversity.

Experience shows that multipath fading is imperfectly correlated in receivers whose antennas are sufficiently apart (a few tens of meters), or that manage to discriminate between very close arrival angles (some tenths of degrees), or that use different frequencies (separated by some tens of MHz) or that employ orthogonal polarizations. By proper processing of the received signals, in an operation which is known as combination, it is possible to obtain a signal in which fading is far less intense than in any of the signals in isolation.

When the desired signal is obtained through n different signals there is nth order diversity. If these signals come from separate antennas we have space diversity. A variant of space diversity, known as angle diversity, is obtained with highly directive (in the vertical plane) receiver antennas pointed to different arrival angles. With the same antenna and different frequencies we get frequency diversity. The two basic diversity techniques of order 2, or double diversity, are illustrated schematically in Figure 3.16.

In general, space diversity is to be preferred, for economy in the use of the radio spectrum. This type of diversity also has the lowest cost, except when very large antennas are employed, since we only need to duplicate the receivers at each terminal.

For line-of-sight or for links with propagation by diffraction double diversity is usually sufficient.

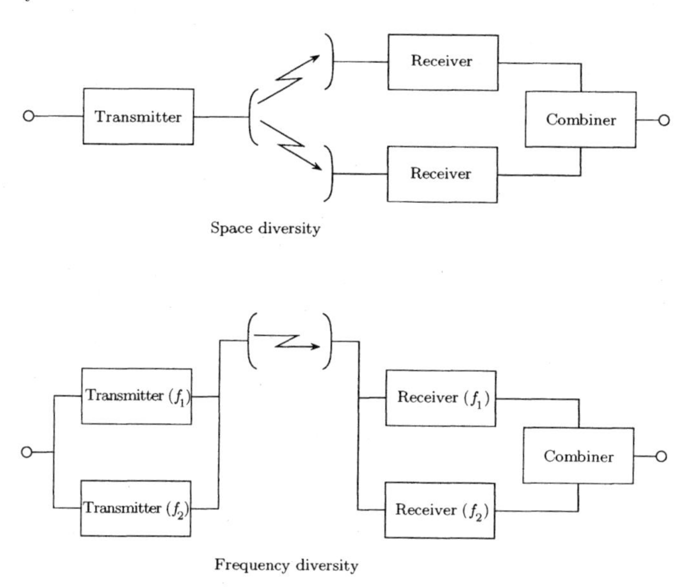

Fig. 3.16 Schematic diagram of double diversity.

Troposcatter links, on the other hand, often require quadruple diversity, both space and frequency (Figure 3.17).

The basic signals may be combined to obtain the final signal both before demodulation, that is, at radio frequency, or at intermediate frequency, or after demodulation. However, to enable combination before demodulation, we must synchronize the basic signals, because propagation delays, although small compared to the modulated signals, may reach quite few periods at radio frequency.

Combiners may use selection, simple addition or optimal addition.

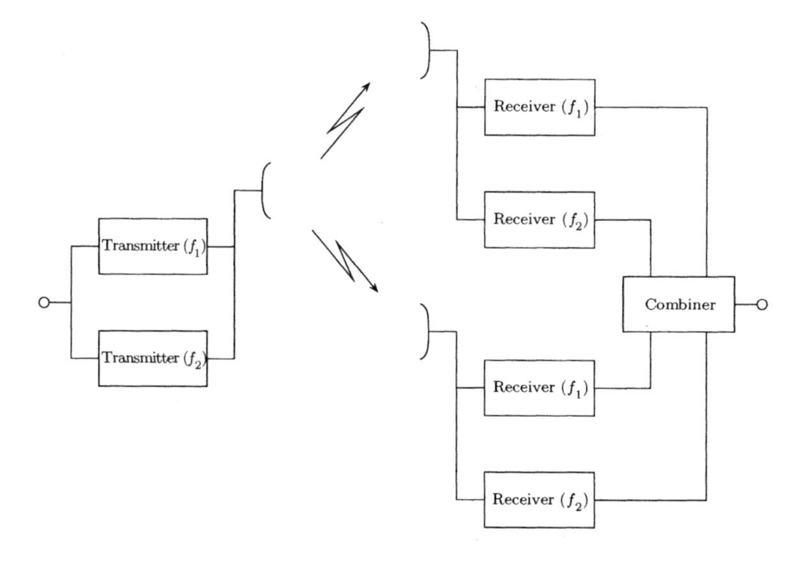

Fig. 3.17 Schematic diagram of quadruple diversity, both frequency and space.

In selection combiners the output signal is, at each moment, the strongest of the basic signals, or the one with the signal-to-noise ratio. This process may be performed before or after detection.

In equal gain combiners or addition combiners, the n basic signals are synchronized before demodulation and then added. If these signals have identical amplitudes the power of the output signal increases with n^2. On the other hand, assuming that noise associated with each of the basic signals is uncorrelated, the output noise power increases with n and thus the output signal to noise ratio increases with n.

In maximal ratio or optimizing combiners the output signal is a linear combination of the (previously synchronized) basic signals, such that it maximizes the output signal to noise ratio.

3.7.1 Selection combiner

We will start with the double diversity selection combiner, where the output signal is the strongest of two basic signals. Let $f_1(p)$ and $f_2(p)$ be the probability density functions of the basic signals which are to be combined. The output probability density function will be

$$f_s(p) = f_1(p) \int_0^p f_2(x)\, dx + f_2(p) \int_0^p f_1(x)\, dx. \tag{3.90}$$

The output cumulative distribution function $\mathcal{P}_s$ is obtained by integrating $f_s(p)$:

$$\begin{aligned} \mathcal{P}_s(p \leq p_0) &= \int_0^{p_0} f_s(p)\, dp \\ &= \int_0^{p_0} f_1(p) \int_0^{p} f_2(x)\, dx\, dp \\ &+ \int_0^{p_0} f_2(p) \int_0^{p} f_1(x)\, dx\, dp. \end{aligned} \tag{3.91}$$

If $f_1(p)$ and $f_2(p)$ are Rayleigh distributions with the same median power p_m, will be from (3.13) and (3.91)

$$f_s(p) = \frac{2\,k}{p_m} \exp\left(-\frac{k\,p}{p_m}\right)\left[1 - \exp\left(-\frac{k\,p}{p_m}\right)\right], \tag{3.92}$$

$$\mathcal{P}_s(p \leq p_0) = \left[1 - \exp\left(-\frac{k\,p_0}{p_m}\right)\right]^2, \tag{3.93}$$

where $k = \log_e(2)$.

For the more general case of an nth diversity selection combiner we may write:

$$\mathcal{P}_s(p \leq p_0) = \left[1 - \exp\left(-\frac{k\,p_0}{p_m}\right)\right]^n. \tag{3.94}$$

In Figure 3.18 we plot the function $\mathcal{P}_s(p \leq p_0)$ for $n = 1$, 2, 3 and 4. As shown in this figure, diversity, even only double diversity, strongly reduces the deepest fades.

For $p_0/p_m \ll 1$, we may obtain from (3.94), a simplified expression, by expanding the exponential in series and taking only the first two terms

$$\mathcal{P}_s(p \leq p_0) = \left(k\frac{p_0}{p_m}\right)^n. \tag{3.95}$$

In these conditions the plot of $\mathcal{P}_s(p \leq p_0)$ in Figure 3.18, is linear with p_0/p_m (expressed in logarithmic units) with a slope of $10/n$ dB per decade.

The average output power in selection combiners may be derived from the density probability function $f_s(p)$:

$$\overline{p}_{s_n} = \int_0^{\infty} p\, f_s(p)\, dp. \tag{3.96}$$

Taking $n = 2$ and assuming that the basic signals follow Rayleigh distributions we have, from (3.92) and (3.96) and manipulating

$$\begin{aligned} \overline{p}_{s_2} &= \frac{2\,p_m}{k}\left(\int_0^{\infty} x\,\exp(-x)\,dx - \int_0^{\infty} x\,\exp(-2x)\,dx\right) \\ &= \frac{3}{2}\,\frac{p_m}{k}. \end{aligned} \tag{3.97}$$

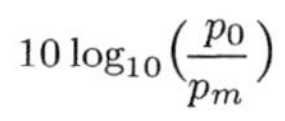

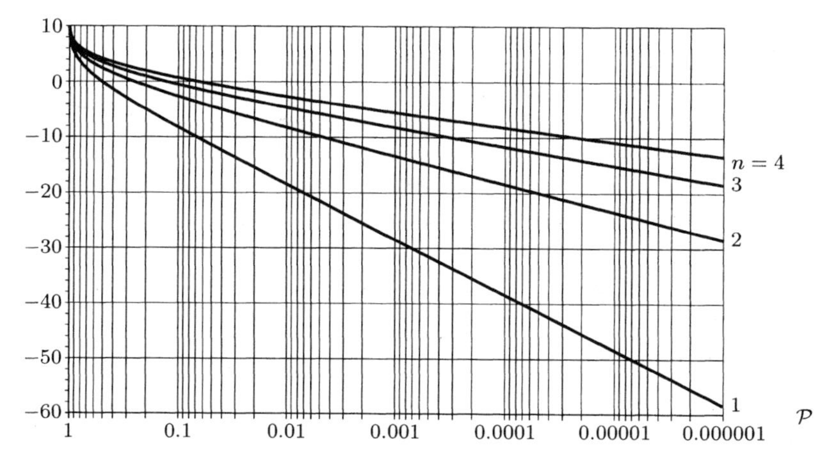

Fig. 3.18 Rayleigh fading without diversity and with selection diversity (stronger signal selection) of order $n = 1$, 2, 3 and 4.

Similarly for the case of no diversity (n=1),

$$\overline{p}_{s_1} = \frac{p_m}{k}. \tag{3.98}$$

Double diversity with selection combiners increases the average output power 1.5 times (1.76 dB). This is less than what is achieved either with equal gain combiners or with maximal ratio combiners, as will be shown later.

For nth order diversity selection combiner we get

$$\overline{p}_{s_n} = \overline{p}_{s_1} \left(\sum_{i=1}^{n} \frac{1}{i} \right). \tag{3.99}$$

If we define diversity gain g_{av} for the average power as the ratio between the average power with and without diversity

$$g_{av}(n) = \frac{\overline{p}_{s_n}}{\overline{p}_{s_1}}, \tag{3.100}$$

we get, in logarithmic units

$$G_{av}(n) = 10\log_{10} \left(\sum_{i=1}^{n} \frac{1}{i} \right). \tag{3.101}$$

In Figure 3.19 we plot the diversity gain G_{av} as a function of the order of diversity n. Even if only integer values of n make sense G_{av} is plotted as a continuous curve, joining with straight lines the points corresponding to integer values of n.

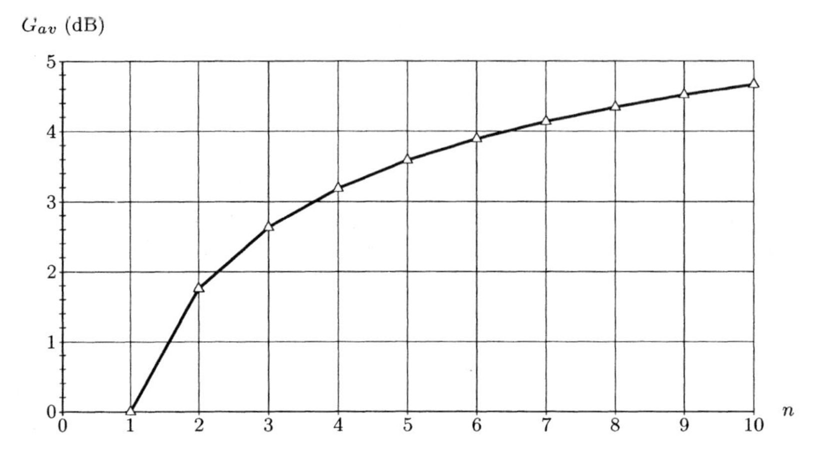

Fig. 3.19 Diversity gain for the average power in a selection combiner G_{av} as a function of the order of diversity n.

3.7.2 Equal gain combiner

Take, now, a double diversity equal gain or addition combiner, that is, a combiner where the output signal is the sum of the two input signals. Let e_1 and e_2 be the input signals and e_r the output signal,

$$e_r = e_1 + e_2. \tag{3.102}$$

Assume now that the input signals are incorrelated. Denoting the probability density functions of the input signals by $f_1(e_1)$ and $f_2(e_2)$ and the density probability function of the output signal by $f_{eq}(e_r)$ we have

$$f_{eq}(e_r) = \int_0^{e_r} f_1(e_1)\, f_2(e_r - e_1)\, de_1. \tag{3.103}$$

The probability that the amplitude of the output signal is less or equal to e_0 is

$$\mathcal{P}_{eq}(e_r \leq e_0) = \int_0^{e_0} de_r \int_0^{e_r} f_1(e_1)\, f_2(e_r - e_1)\, de_1. \tag{3.104}$$

Assuming that the input signals follow a Rayleigh distribution, whose cumulative distribution function, expressed in terms of power was given in equation (3.14) and is repeated here for easier reference

$$\mathcal{P}(p \leq p_0) = 1 - \exp\left(-\frac{k\, p_0}{p_m}\right).$$

In terms of field intensities we have

$$\mathcal{P}(e \leq e_0) = 1 - \exp\left(-\frac{k\, e_0^2}{e_m^2}\right). \tag{3.105}$$

When we differentiate the previous expression with respect to e_0 we obtain the probability density function of the Rayleigh distribution in terms of field intensities

$$f(e_0) = 2\,k\,\frac{e_0}{e_m^2}\,\exp\left(-\frac{k\, e_0^2}{e_m^2}\right). \tag{3.106}$$

Substituting (3.106) in (3.104) and manipulating

$$\mathcal{P}_{eq}(e_r \leq e_0) = 4\,k^2 \int_0^{e_0/e_m} dx_2$$

$$\times \int_0^{x_2} x_1(x_2 - x_1) \exp\left\{-k[x_1^2 + (x_2 - x_1)^2]\right\} dx_1, \tag{3.107}$$

or, in terms of power

$$\mathcal{P}_{eq}(p_r \leq p_0) = 4\,k^2 \int_0^{\sqrt{p_0/p_m}} dx_2$$

$$\times \int_0^{x_2} x_1(x_2 - x_1) \exp\left\{-k[x_1^2 + (x_2 - x_1)^2]\right\} dx_1. \tag{3.108}$$

To compare the performance of equal gain and selection combiners we must recall that in the selection combiner the average output noise power is equal to the average input noise power, while in the equal gain combiner it is the double of the later value. Thus for the signal to noise ratio we must substitute p_0/p_m by $2p_0/p_m$ in equation (3.108)

$$\mathcal{P}_{eq}(p_r \leq p_0) = 4\,k^2 \int_0^{\sqrt{2p_0/p_m}} dx_2$$

$$\times \int_0^{x_2} x_1(x_2 - x_1) \exp\left\{-k[x_1^2 + (x_2 - x_1)^2]\right\} dx_1. \tag{3.109}$$

In Figure 3.20 we plot $\mathcal{P}_{eq}(p \leq p_0)$, obtained by numeric integration of (3.109). In the same figure we also plot the cumulative distribution function for the selection combiner as given in equation (3.93). As shown the two curves are almost parallel at least for the usual probability values (less than 0.1).

If $p_0/p_m \ll 1$ we may get a simplified expression for $\mathcal{P}_r(p \leq p_0)$. From (3.109) neglecting the exponential term we get

$$\begin{aligned}
\mathcal{P}_{eq}(p \leq p_0) &= 4\,k^2 \int_0^{\sqrt{2\,p_0/p_m}} dx_2 \int_0^{x_2} x_1(x_2 - x_1)\, dx_1 \\
&= \frac{2}{3}\left(k\frac{p_0}{p_m}\right)^2.
\end{aligned} \tag{3.110}$$

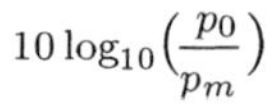

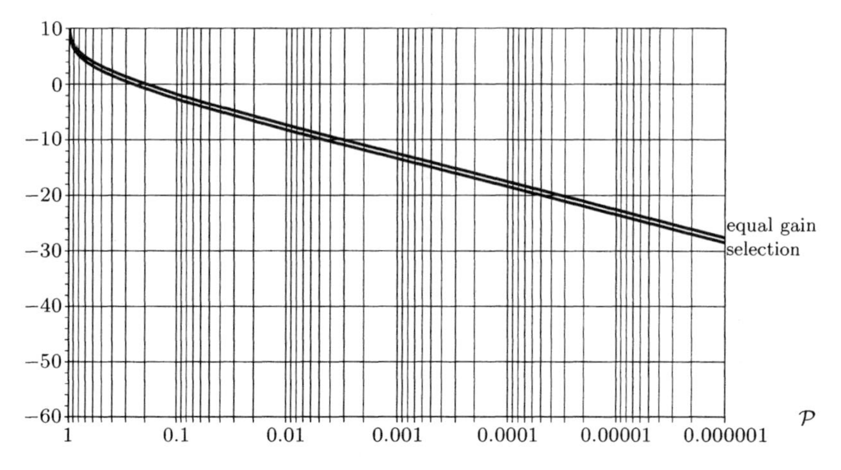

Fig. 3.20 Double diversity equal gain combiner versus selection combiner with Rayleigh fading.

Comparing (3.110) with (3.95), we note that for small signals, that is, when $p_0/p_m \ll 1$, the double diversity equal gain combiner shows a relative gain of $gr_s = \sqrt{1.5}$ in relation with the selection combiner.

The results obtained for double diversity may be generalized for nth diversity. Let $f_1(e_1)$, $f_2(e_2)$, $\cdots$, $f_n(e_n)$ be the probability density functions of the input signals (voltages). The probability density functions of the output signal $f_a(e_r)$ will be

$$f_{eq}(e_r) = \int_0^{e_r} de_1 \int_0^{e_r - e_1} de_2 \cdots$$

$$\cdots \int_0^{e_r - \sum_{i=1}^{n-2} e_i} \left[\prod_{j=1}^{n-1} f_j(e_j)\right] f_n\left(e_r - \sum_{k=1}^{n-1} e_k\right) de_{n-1} \qquad (3.111)$$

and the cumulative distribution function

$$\mathcal{P}_{eq}(e_r \leq e_0) = \int_0^{e_0} de_r \int_0^{e_r} de_1 \int_0^{e_r - e_1} de_2 \cdots$$

$$\cdots \int_0^{e_r - \sum_{i=1}^{n-2} e_i} \left[\prod_{j=1}^{n-1} f_j(e_j)\right] f_n\left(e_r - \sum_{k=1}^{n-1} ek\right) de_{n-1}. \qquad (3.112)$$

If the input signals follow a Rayleigh distribution, from (3.106) and (3.112), manipulating we have

$$\mathcal{P}_{eq}(e_r \leq e_0) = (2k)^n \int_0^{e_0/e_m} dx_n \int_0^{x_n} dx_1 \int_0^{x_n - x_1} dx_2 \cdots$$

$$\cdots \int_0^{x_n - \sum_{i=1}^{n-2} x_i} \left(\prod_{j=1}^{n-1} x_j \right) \left(x_n - \sum_{j=1}^{n-1} x_j \right)$$

$$\times \exp \left\{ -k \left[\sum_{j=1}^{n-1} x_j^2 + \left(x_n - \sum_{j=1}^{n-1} x_j \right)^2 \right] \right\} dx_{n-1}. \tag{3.113}$$

Taking again (3.113) in terms of power and recalling that in order to compare performance of equal gain and selection combiners we must substitute p_0/p_m for np_0/p_m

$$\mathcal{P}_{eq}(p_r \leq p_0) = (2k)^n \int_0^{\sqrt{\sqrt{n p_0/p_m}}} dx_n \int_0^{x_n} dx_1 \int_0^{x_n - x_1} dx_2 \cdots$$

$$\cdots \int_0^{x_n - \sum_{i=1}^{n-1} x_i} \left(\prod_{j=1}^{n-1} x_j \right) \left(x_n - \sum_{j=1}^{n-1} x_j \right)$$

$$\times \exp \left\{ -k \left[\sum_{j=1}^{n-1} x_j^2 + \left(x_n - \sum_{j=1}^{n-1} x_j \right)^2 \right] \right\} dx_{n-1}. \tag{3.114}$$

Approximating the exponential by unity, in the range $\mathcal{P}_a(p \leq p_0) \ll 1$ and $p_0/p_m \ll 1$, we get the following simplified expression:

$$\mathcal{P}_{eq}(p_r \leq p_0) = \frac{(2n)^n}{(2n)!} \left(k \, \frac{p_0}{p_m} \right)^n. \tag{3.115}$$

For $p_0/p_m \ll 1$, comparing (3.115) and (3.95), we may define the small signal gain gr_s of the equal gain combiner in relation to the selection combiner as the inverse of the power ratio that leads to the same value of $\mathcal{P}$

$$gr_s = \frac{[(2n)!]^{1/n}}{2n}. \tag{3.116}$$

In Figure 3.21 we plot $Gr_s = 10 \log_{10}(gr_s)$ as a function of the order of diversity n. Please note that Gr_s increases with n, from 0.9 dB for $n = 2$ up to 5.4 dB for $n = 10$. This is the reason for using equal gain combiners only when $n > 2$.

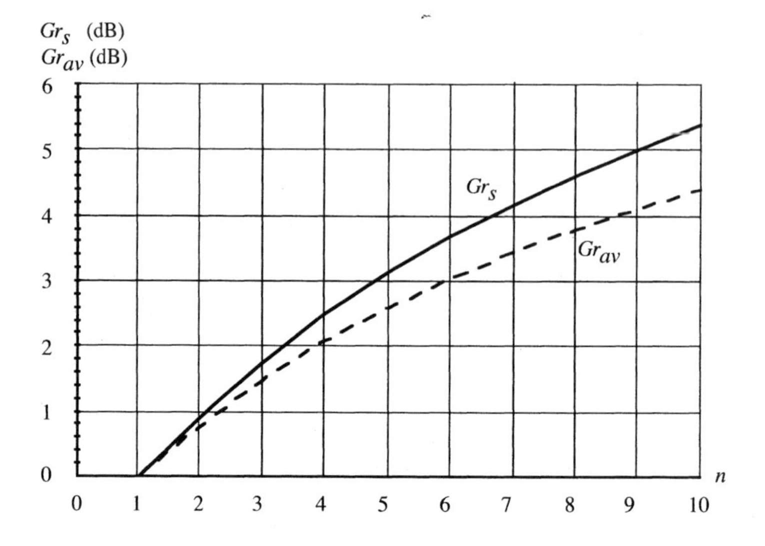

Fig. 3.21 Gains Gr_s and Gr_{av} of the equal gain combiner in relation to the selection combiner.

Assume again that e_i, with $i = 1, \ldots, n$ are the n the input signals to an equal gain combiner. The output signal will be

$$e_r = \sum_{i=1}^{n} e_i.$$

Assume also that all the input signals are uncorrelated, that is,

$$\int_0^{\infty} e_i e_j de = 0,$$

with $i \neq j$.

The average power output of the equal gain combiner will proportional to $\overline{e_d^2}$. If the input signals are uncorrelated, the average of the squared output signal may be calculated as

$$\begin{aligned} \overline{e_d^2} &= \overline{\left(\sum_{i=1}^{n} e_i\right)^2} \\ &= \sum_{i,j=1}^{n} \overline{(e_i e_j)} \\ &= n\overline{e_i^2} + n(n-1)\overline{e_i e_j}. \end{aligned}$$

If the input signals follow Rayleigh distributions we have, from (3.17) and (3.18)

$$\begin{aligned}\overline{e_d^2} &= 2n\sigma^2 + n(n-1)\sigma^2\frac{\pi}{2} \\ &= n\overline{e_1^2}\left[1 + (n-1)\frac{\pi}{4}\right].\end{aligned} \tag{3.117}$$

For n equivalent input signals the average noise power at the combiner output is equal to n times the average noise power at any of the combiner inputs. Hence the average signal-to-noise ratio at the combiner output $\frac{s_0}{n_0}$ as a function of the average signal-to-noise ratio at any of the combiner inputs $\frac{s_i}{/}n_i$ is

$$\frac{s_0}{n_0} = \left[1 + (n-1)\frac{\pi}{4}\right]\frac{s_i}{n_i}. \tag{3.118}$$

Comparing the performance of the equal gain and the selection combiner we may define a relative gain gr_{av} for the average output power, for the same average output noise power

$$gr_{av} = \frac{\left[1 + (n-1)\frac{\pi}{4}\right]}{\sum_{i=1}^{n}\frac{1}{i}}. \tag{3.119}$$

We also plot in Figure 3.21, $Gr_{av} = 10\log_{10}(gr_{av})$ Gr_{av} as a function of the order n of diversity.

For the usual values of n ($n \leq 10$) the difference between the values of Gr_s and Gr_{av} is always less than 1 dB. Thus, it is possible to obtain the plots of $\mathcal{P}_r(p \leq p_0)$, represented in Figure 3.20, by shifting the plot corresponding to the selection combiner (figure 3.18), parallel to the Y axis.

3.7.3 Maximal ratio combiner

In the maximal ratio (or optimizing) combiner the output signal is obtained by adding the input signals which are previously synchronized and multiplied by a constant, proportional to the signal power and inversely proportional to its noise power. It is possible to demonstrate [3] that the output signal-to-noise power s_0/n_0 may be obtained from the input signal-to-noise powers by

$$\frac{s_0}{n_0} = \sum_{i=1}^{n}\frac{s_i}{n_i}. \tag{3.120}$$

It may also be demonstrated that for an average output noise power equal to one of the input signals noise power, the output signal power p_{mr} is equal to the sum of the powers of the input signals p_i

$$p_{mr} = \sum_{i=1}^{n} p_i. \tag{3.121}$$

Let $f_1(p_1)$, $f_2(p_2)$, $\cdots$, $f_n(p_n)$ be the probability density functions of the input signals. The output signal probability density function $f_0(p_n)$ will be, from equation (3.111)

$$f_{mr}(p_r \leq p_0) = \int_0^{p_1} dp_1 \int_0^{p_r - p_1} dp_2 \cdots$$

$$\cdots \int_0^{p_r - \sum_{i=1}^{n-2} p_i} \left[\prod_{j=1}^{n-1} f_j(p_j)\right] f_n\left(p_r - \sum_{j=1}^{n-1} p_j\right) dp_{n-1}. \tag{3.122}$$

The output cumulative distribution density function is

$$\mathcal{P}_{mr}(p_r \leq p_0) = \int_0^{p_0} dp_r \int_0^{p_1} dp_1 \int_0^{p_r - p_1} dp_2 \cdots$$

$$\cdots \int_0^{p_r - \sum_{i=1}^{n-2} p_i} \left[\prod_{j=1}^{n-1} f_j(p_j)\right] f_n\left(p_r - \sum_{j=1}^{n-1} p_j\right) dp_{n-1}. \tag{3.123}$$

If input signals follow a Rayleigh distribution

$$\mathcal{P}_{mr}(p_r \leq p_0) = \int_0^{k\, p_0/p_m} dx_n \int_0^{x_n} dx_1 \int_0^{x_n - x_1} dx_2 \cdots$$

$$\cdots \int_0^{x_n - \sum_{i=1}^{n-2} x_i} \exp(-x_n)\, dx_{n-1}, \tag{3.124}$$

thus:

$$\mathcal{P}_{mr}(p_r \leq p_0) = 1 - \exp\left(-k\frac{p_0}{p_m}\right) \sum_{i=0}^{n-1} \left(k\, \frac{p_0}{p_m}\right)^i \left(\frac{1}{i!}\right). \tag{3.125}$$

In Figure 3.22 we plot $\mathcal{P}_{mr}(p_r \leq p_0)$ as a function of p_0 taking as a parameter the order n of diversity.

For $p_0/p_m \ll 1$ we may derive from (3.125) the following approximate expression:

$$\mathcal{P}_{mr}(p_r \leq p_0) = \frac{1}{n!}\left(k\, \frac{p_0}{p_m}\right)^n. \tag{3.126}$$

Comparing (3.126) and (3.95) we get the gain gr_s of the maximal ratio combiner in relation to the selection combiner for small signals ($p_0/p_m \ll 1$)

$$gr_s = (n!)^{1/n}. \tag{3.127}$$

We already stated that, in an maximal ratio combiner, the average output power is equal to the sum of the average powers of the input signals. If we compare this result (3.121) with the selection combiner (3.99), we may define

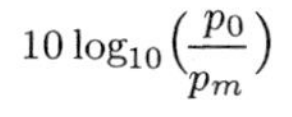

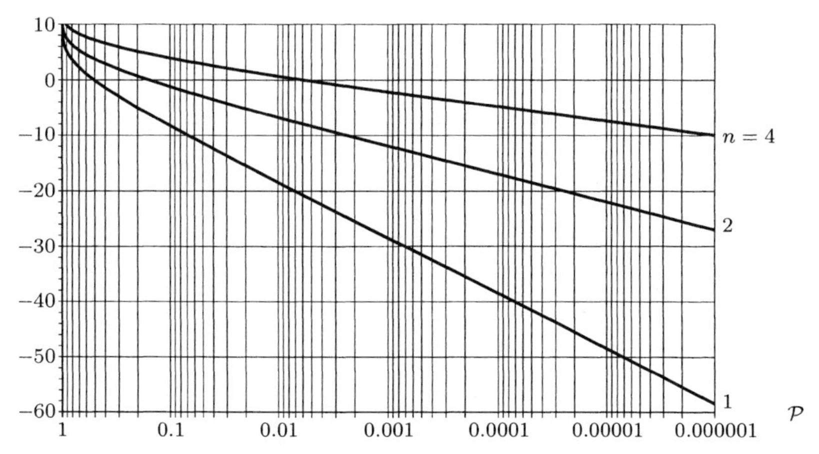

Fig. 3.22 Performance of maximal ratio combiner under Rayleigh fading.

a gain gr_{av} for the average power output of the maximal ratio combiner in relation to the selection combiner for the same output noise power. For the same input signals in both cases we have

$$gr_{av} = \frac{n}{\sum_{i=1}^{n} \frac{1}{i}}. \tag{3.128}$$

In Figure 3.23 we plot $Gr_s = 10\log_{10}(gr_s)$ and $Gr_{av} = 10\log_{10}(gr_{av})$ for the maximal ratio combiner taking as a parameter the order of diversity ($n \leq 10$).

Comparing the performance of the maximal ratio and the equal gain combiner, using Figures 3.21 and 3.23 or the relevant expressions, we may conclude that the maximal ratio combiner performs slightly better that the equal gain combiner both for the average output power and for for small signals. However this difference is small enough (less than 1 dB for up to $n = 10$) to justify, in most cases, the added complexity.

3.8 DIVERSITY AND FADING

In our analysis of diversity we have assumed that input signals were, at least, partly uncorrelated. This hypothesis is correct for line of sight, or slightly obstructed paths, when fading is mainly due to multipath. For slow fading,

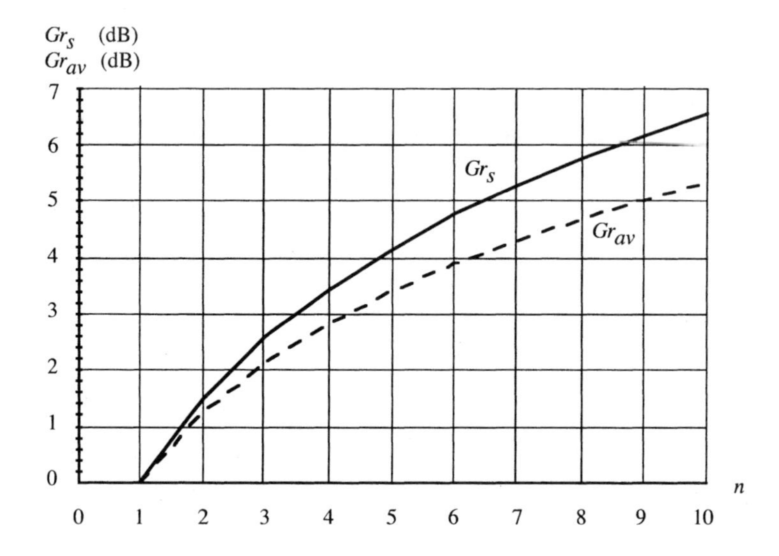

Fig. 3.23 Gains Gr_s and Gr_{av} of the maximal ratio combiner compared to the selection combiner as a function of the order of diversity n.

even when associated with fast fading, observed in heavily obstructed paths, this is not true. However, diversity, by modifying the average output power has an indirect influence on the distribution of slow fading.

Assuming that slow fading follows a log-normal distribution, the probability $\mathcal{P}_r$ that the output power in a selection combiner is less or equal to P_0 (expressed in dB) is, from (3.86) and (3.87)

$$\mathcal{P}_r(P_r \leq P_0) = \int_{-\infty}^{\infty} \frac{1}{\sqrt{2\pi}\sigma c_{sf}} \exp\left[-\frac{\log_e(c_{sf})^2}{2\sigma^2}\right] \mathcal{P}_{ff}(c \leq c_0 - c_{sf})\, dc_{sf}, \tag{3.129}$$

where σ is the standard deviation of the received power (expressed in Neper) and $\mathcal{P}_{ff}(c \leq c_0)$ is the probability that the received signal (voltage) is less or equal to c_0 due to fast fading.

For the order n selection combiner, from (3.129) and (3.94), we get

$$\mathcal{P}_r(P_r \leq P_0) = \int_{0}^{\infty} \frac{1}{\sqrt{2\pi}\sigma c_{sf}} \exp\left[-\frac{\log_e^2(c_{sf})}{2\sigma^2}\right] \left[1 - 2^{(c_0/c_{sf})^2}\right]^n dc_{sf}. \tag{3.130}$$

In Figure 3.24 we plot $\mathcal{P}_r$ for a double diversity (n) selection combiner taking σ as a parameter.

Had we opted for an equal gain or an maximal ratio combiner instead of a selection combiner, we would get sets of curves similar to those in Figure

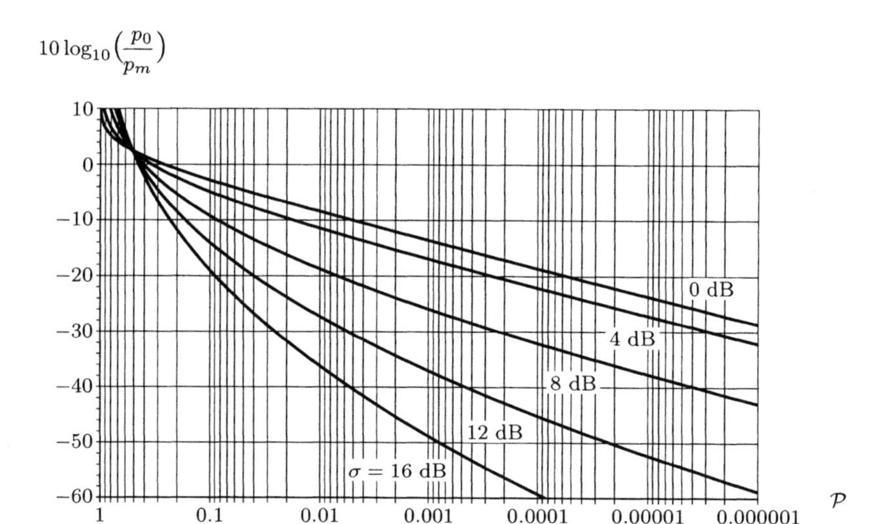

Fig. 3.24 Performance of a double diversity selection combiner under simultaneous slow and fast fading, taking as a parameter the value of σ.

3.24, shifted vertically by the difference of gains between the chosen combiner and the selection combiner.

3.9 DIVERSITY IN LINK DESIGN

We have already stated that in line-of-sight paths fading tends to follow a Nakagami-Rice distribution, which is less severe than the Rayleigh distribution.

When applying diversity to the design of real links we not only need to make use of the most realistic fading distributions but we also require to know how far apart should receiver antennas (or operating frequencies) be to achieve effective diversity. The method commonly used to design analog or narrow band digital links and which will be described in the following is due to Vigants [11].

As we have shown in Section 3.3, in analog or narrow band digital links, in the worst month, the probability that the received power, without diversity, is equal or less than p_0, where $p_0 \ll p_n$ is given by (3.66), which for convenience is repeated here:

$$\mathcal{P}(p \leq p_0) = \mathcal{F}\left(\frac{p_0}{p_n}\right).$$

Using double space diversity, with a selection combiner and receiver antennas vertically spaced, the probability $\mathcal{P}_r$ that the output signal is equal or less than p_0 is

$$\mathcal{P}_r(p_r \leq p_0) = \frac{\mathcal{P}(p \leq p_0)}{i_0}, \tag{3.131}$$

where i_0 is the space diversity improvement factor:

$$i_0 = 1.21 \cdot 10^{-3} \frac{d_c^2 f}{d} \frac{g_s}{g_p} \left(\frac{p_n}{p_0}\right) \tag{3.132}$$

and:

- g_s and g_p are the gains of the main and the secondary antennas;
- f is the operating frequency in GHz;
- d is the path length in kilometers;
- d_c is the distance between antenna centers, in meters.

Equation (3.132) is valid for the following range of parameters:

- $1 \geq g_s/g_p \geq 0.25$;
- $11 \geq f \geq 2$;
- $65 \geq d \geq 22.5$;
- $25 \geq d_c \geq 5$;
- $10^{-3} \geq (p_0/p_n) \geq 10^{-5}$;
- $200 \geq i_0 \geq 10$.

The reference to a main antenna and a secondary antenna stresses the possibility that both antennas may be different. This is a rather usual situation, when diversity is added to a pre-existing link and the mast does not allow for the use of another identical antenna.

As it was stated in Section 2.5 in paths with space diversity it is not necessary to provide the same clearance for both antennas. While for the main antenna we take the usual clearance, for the secondary one the clearance may be smaller. Thus, if the main antenna is the higher one, additional height is not required to achieve an adequate separation between antennas.

The main limiting factor in (3.132) stems from the small number of climatic regions and types of terrain used to derive it. ITU-R Recommendation P.530-8 [6] presents an alternative expression for the improvement factor i_0

$$i_0 = \left\{1 - \exp\left\{-3.34 \cdot 10^{-4}\, d_c^{0.87}\, f^{-0.12}\, d^{0.48} \left[\mathcal{P}(p \leq p_0)\frac{p_n}{p_0}\right]^{-1.04}\right\}\right\} \frac{g_s}{g_p}\frac{p_n}{p_0}. \tag{3.133}$$

The domain of validity for (3.133) is the following:

- $P_0 - P_n < -15$ dB;
- $\mathcal{P}(p \leq p_0) < 10^{-3}$;
- $11 \geq f \geq 2$;
- $243 \geq d \geq 43$ and, with lesser confidence, $43 > d > 25$;
- $23 \geq d_c \geq 3$.

If, in analog or narrow band digital systems, instead of double space diversity we use double frequency diversity (also with a selection combiner) the improvement factor becomes

$$i_0 = \frac{80}{fd} \frac{\Delta f}{f} \frac{g_s}{g_p} \frac{p_n}{p_0}, \tag{3.134}$$

where Δf is the frequency difference between carriers, in GHz. When $\Delta f >$ 0.5 GHz then one should use $\Delta f = 0.5$ GHz.

Expression (3.134) is valid for:

- $1 \geq g_s/g_p \geq 0.25$;
- $11 \geq f \geq 2$;
- $70 \geq d \geq 30$;
- $\Delta f/f \leq 0.05$;
- $i_0 \geq 5$.

The problems involved in the design of digital links with space, frequency and simultaneous space and frequency diversity are discussed in Section 5.9.

If fading followed a Rayleigh distribution, if input signals to the combiner were equivalent ($g_s/g_p = 1$) and uncorrelated, if $p_0/p_n \ll 1$, and if the median received power p_m coincided with the received power without fading p_n, then from (3.131) and (3.95) we would get:

$$i_0 = \frac{1}{k} \frac{p_n}{p_0}, \tag{3.135}$$

where $k = \log_e(2)$.

Comparing the theoretical value of i_0, given in (3.135), with its empirical value given in (3.132), (3.133) or (3.134) shows that in practice diversity improvement factors are much lower than theory would lead us to expect. This fact is mainly due to the existence of a non-negligible correlation between input signals to the combiner.

Empirical values given for the improvement factors assume ideal selection combiners, that is, combiners where the output signal is always the strongest of the input signals. Some combiners, in order to reduce selection frequency, switch only when one of the input signals exceeds the other by B dB. Under these conditions the value of i_0 is reduced to i_e

$$i_e = i_0 \frac{2}{b + \frac{1}{b}}, \tag{3.136}$$

where:

$$b = 10^{-B/10}. \tag{3.137}$$

In another type of selection combiner, known as a blind combiner, threshold combiner, or switching combiner selection only takes place when the input signal decreases below a given threshold. Thus, for an input power above the threshold, the system behaves as if there was no diversity and, for an input power below the threshold, as if there was double diversity.

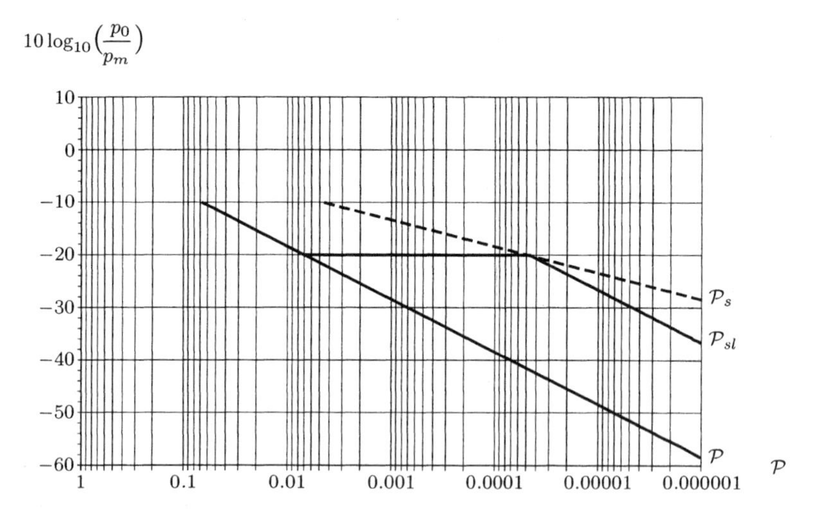

Fig. 3.25 Performance of a double diversity blind combiner.

Let $\mathcal{P}(p \leq p_n)$ and $\mathcal{P}_s(p \leq p_0)$ be, respectively, the probabilities that signal power be less or equal to p_0 without and with selection diversity (see Figure 3.25) and let p_l be the power corresponding to the threshold. The probability $\mathcal{P}_{bc}(p \leq p_0)$ that the output power is equal or less than p_0 for a blind combiner may be derived approximately from $\mathcal{P}$ and $\mathcal{P}_s$ as follows:

- If the power of any of the signals is higher than the threshold, $\mathcal{P}_{bc}$ coincides with $\mathcal{P}$, since there is no selection;

- If the power of the selected input signal falls below the threshold, $\mathcal{P}_{bc}$ changes, more or less abruptly, from $\mathcal{P}$ to $\mathcal{P}_s$;

- If both input signal powers fall below the threshold the combiner switches, successively from one to the other.

In the last case, let p_1 and p_2 be the input signal power, that we assume to have the same characteristics and to be uncorrelated. Then $\mathcal{P}_{bc}$ will be

$$\mathcal{P}_{bc}(p \leq p_O) = \frac{1}{2}\,\mathcal{P}(p_1 \leq p_O)\mathcal{P}(p_2 \leq p_l) + \frac{1}{2}\,\mathcal{P}(p_1 \leq p_l)\mathcal{P}(p_2 \leq p_O). \quad (3.138)$$

For $p_O/p_n \ll 1$ and $p_l/p_n \ll 1$, taking for $\mathcal{P}$ the approximated expression of the Rayleigh distribution given in (3.15), we get

$$\mathcal{P}_{bc}(p \leq p_O) = \left(k\,\frac{p_O}{p_n}\right)\left(k\,\frac{p_l}{p_n}\right). \quad (3.139)$$

Similarly, taking for $\mathcal{P}_s(p \leq p_O)$ the approximated expression (3.95), we get

$$\mathcal{P}_s(p \leq p_O) = \left(k\,\frac{p_O}{p_n}\right)^2. \quad (3.140)$$

Thus:

- for $p_O = p_l$, $\mathcal{P}_{bc}$ coincides with $\mathcal{P}_s$;

- for $p_O < p_l$, $\mathcal{P}_{bc}$ behaves as $\mathcal{P}$, and passes by the point corresponding to $\mathcal{P}_s(p \leq p_l)$.

Since empirical distributions used to compute $\mathcal{P}$ and $\mathcal{P}_s$ have the same dependence on p_O/p_n as the approximated theoretical expression, the process may also be applied to them.

Assuming that the input signals follow a Rayleigh distribution, with a correlation coefficient compatible with the empirical value of i_O, it is possible to compute $\mathcal{P}_{bc}$. We then find out that for $p_O > p_l$, $\mathcal{P}_{bc}$ does not change abruptly from $\mathcal{P}_s(p \leq p_l)$ to $\mathcal{P}(p \leq p_l)$, but instead follows a transition curve such that $\mathcal{P}_{bc}$ practically coincides with $\mathcal{P}_s$ for $p_O > 10\,p_l$.

For the design of microwave radio links it is sufficient to use the approximation to $\mathcal{P}_{bc}$ as previously done.

Since the improvement factor for a blind combiner is constant for input powers below the threshold and, under the same circumstances, the higher the lower the threshold, one should choose as low a value for the threshold as possible. In practice, the threshold is usually set about 2 dB above the receiver threshold (which will be referred to in Chapter 4).

3.10 DIVERSITY AND PASSIVE REPEATERS

Space diversity assumes the existence of n independent paths between terminal antennas. In paths with a passive repeater this fact implies the use of n passive repeaters. Moreover terminal antennas must have high enough directivity to discriminate between passive repeaters, which may be quite difficult in many cases. These two factors strongly reduce the appeal of space diversity in paths with passive repeaters.

3.11 DIVERSITY IN PATHS WITH REFLECTIONS

In some paths, particularly those that cross extended bodies of water, it is impossible to apply the procedures described in Section 2.7.1 to avoid reflections and the associated fading. We must thus make use of space diversity.

For a plane Earth ($k_e = \infty$) and a path length d, the vertical distance d_v between the a maximum and a minimum of the received signal may be derived from (2.39) as

$$d_v = \frac{\lambda\, d}{4\, h_E}, \tag{3.141}$$

where h_E is the height of the transmitter antenna above the ground and λ is the wavelength. The use of antennas with a vertical distance d_v enables us to guarantee that the two signals are never simultaneously zero. For other values of k_e we must verify that the distance d_v given in (3.141) ensures that the signals received by both antennas have a phase difference, given by (2.65) or (2.70), sufficiently close to π radians to get the same result.

A more economic alternative to space diversity is described by Fabri [5]. This alternative uses two antennas with a vertical distance d_v such that the path difference for the reflected ray is a odd multiple of a half wavelength.

Consider Figure 3.26. Let φ be the angle between the reflected ray (calculated as per Section 2.4), α be the angle between the direct ray and the horizontal, and λ the wavelength. If

$$d_v[\sin(\varphi + \alpha) - \sin(\alpha)] = (2n + 1)\frac{\lambda}{2},$$

adding (for instance using a hybrid) the two antenna outputs we double the power received via the direct ray (already accounting for the 3 dB hybrid loss) and we cancel the effect of the reflected ray.

The exact cancellation of the effects of the reflected ray occurs only for the value of k_e that corresponds to φ and α. In many cases, however it is possible to achieve a high enough attenuation of the reflected ray, for a wide range of k_e.

If, in addition to two receiver antennas, we use two transmitter antennas, the received power becomes double (relative to the case of one transmitter

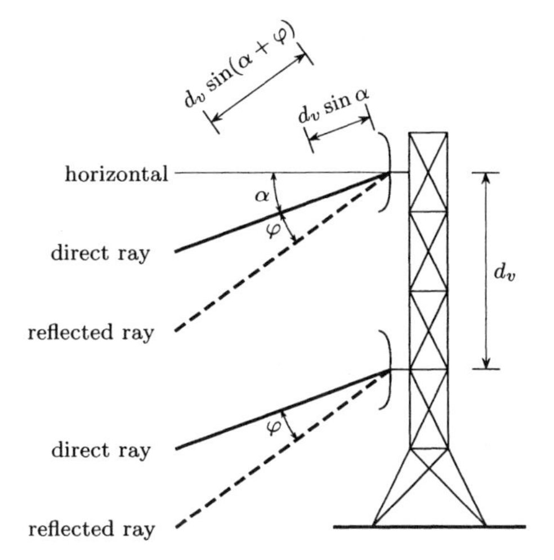

Fig. 3.26 Vertical distance between antennas to avoid fading due to specular reflection.

antenna and two receiver antennas) or quadruple (relative to the case of one transmitter and one receiver antenna).

This system, with a single receiver and two receiver antennas, is more economical than double space diversity, but it has an important setback: it introduces an additional loss, which may be rather high, when the differences between the propagation delays for the direct ray for the two antennas reaches or exceeds half a cycle at the operating frequency (0.125 ns at 4 GHz). Such a scheme must adopted rather carefully and only for frequencies above 2 GHz (for which there is experience).

REFERENCES

1. Abramowitz, M. and Stegun, I., *Handbook of Mathematical Functions*, Dover, New York, 1965.

2. Boithias, L., Multipath propagation in line-of-sight links. *Electronics Letters*, Vol. 15, No. 7, 29th March, pp. 209–210, 1979.

3. Boithias, L. and Battesti, J., Les faisceaux hertziens trans-horizon haute qualité. *Annales des Télécommunications*, Vol. 20, No. 7, 8, 11 and 12, July, August, November and December, 1965.

4. CCIR, *Reports of the CCIR, 1990*, Annex to Volume V, *Propagation in Non-Ionized Media*, UIT, Geneva, 1990.

5. Fabri, F. , Anti-reflecting system for 2 GHz over-sea radio links. *Bolletino Tecnico d'informazione Telettra (English Edition)*, Vol. 25, March, pp. 3–8, 1974.

6. ITU, *ITU-R Recommendations on CD-ROM*, UIT, Geneva, 2000.

7. Jakes, W. (Editor), *Microwave Mobile Communications*, IEEE Press, Piscataway, NJ, 1993

8. Rice, P. L., Longley, A. G., Norton, K. A. and Barris, A. P., *Transmission Loss Predictions for Tropospheric Communication Circuits, I* and *II*, National Bureau of Standards, 1967.

9. Schwartz, M., *Information Transmission, Modulation and Noise*, McGraw-Hill Kogakusha Ltd, Tokyo, 1970.

10. Schwartz, M., Bennett, W. and Stein S., *Communication Systems and Techniques*, McGraw-Hill Book Company, New York, 1966.

11. Vigants, A., Space diversity engineering. *Bell System Technical Journal*, Vol. 54, No. 1, pp. 103-142, 1975.

12. Webster, A. R. and Neno, T., Tropospheric Microwave Propagation – An X Band Diagnostic System. *IEEE Transactions on Antennas and Propagation*, Vol. AP - 28, No. 5, September, pp. 693–699, 1980.

Problems

3:1 Calculate the fade depth not exceeded for more than 0.0001 of the time for Raleigh fading and for Ricean fading ($10 \log_{10}(p_a/p_m) = 6$ dB.

3:2 Assuming a 50-km-long, microwave radio link, operating at 4 GHz, with a fading margin of 30 dB for a given quality of service (QoS). Assuming fading depth obeys Morita's law, for what fraction of time would the required quality of service be met?

3:3 What is the maximum link length that can achieved at 10 GHz with equipment that copes with up 170 dB system loss for a given quality of service?

3:4 Calculate fading depth not exceeded for more than 0.001% of the time for a 40 km-long-link, with a path slope of 10 milliradians, at 6 GHz, over hills, below the altitude of 400 m, in Wyoming, in the worst month. Compare this value with the one given by Morita's law.

3:5 Calculate the fade depth not exceeded for more than 0.001 of the time for a simultaneous uncorrelated log-normal slow fading with $\sigma = 8$ dB and Rayleigh fast fading.

3:6 Calculate the Rayleigh fade depth not exceeded for more than 0.001 of the time, and the diversity gain for the average power for double and quadruple diversity selection combiners.

3:7 In the previous problem, what would be the difference if we had chosen equal gain combiners or maximal ratio combiners, instead of selection combiners?

3:8 Compare the diversity gain for the power exceeded more than 0.9999 of the time of a double diversity selection combiner under Rayleigh fading and under simultaneous slow and fast fading with $\sigma = 8$ dB.

3:9 Calculate the (double) space diversity improvement factor for a 30-dB fade, in a 50-km-long line-of-sight microwave radio link, at 6 GHz, with two equal sized antennas 10 m apart, using both the Vigants and the ITU-R expressions.

3:10 Compare the previous (double) diversity improvement factor with expected value for uncorrelated Rayleigh fading.

3:11 Calculate the (double) frequency improvement factor for a 30-dB fade, in a 50-km-long line-of-sight microwave radio link, at 6 GHz, with a 4×29.65 MHz difference between carrier frequencies.

3:12 Consider a path where the reflected ray makes a 15 milliradian angle with the direct ray, assumed to be horizontal. Calculate the minimum distance between terminal antenna centers required to cancel reflections at 4 GHz when adding antenna outputs.

4 Analog Links

4.1 INTRODUCTION

Microwave radio links may be used to transmit either analog or digital signals, Obviously, modulators and demodulators, as well as low-frequency stages, are different in both cases. However, apart from bandwidth and distortion requirements, high (or radio) frequency, frequency converter and intermediate frequency stages are quite similar.

In the design of analog radio links, the most important variable to watch is the signal-to-noise ratio (or the noise power) measured at the modulating signal (base band) level, whereas for digital radio links that variable is basically the bit error ratio, sometimes defined through block or frame error ratios.

This chapter starts with a description of the most frequently used analog signals:

- single-channel telephony;
- frequency division multiplex telephony,;
- high-quality sound broadcast, both monophonic and stereophonic;
- black-and-white and color television.

The calculation of the base band signal to thermal noise ratio and thermal noise power level, for the usual modulations, taking into account noise weighting factors follows. Then intermodulation noise, due to mismatches in antenna feeders, to the bandwidth limitations, and to imperfections in oscillators, filters and other components is accounted for. After looking into the chain connection of radio links the chapter concludes with an analysis of the applicable ITU-R quality recommendations.

We assume that the reader is familiar with the theory of modulation and demodulation in the presence of thermal noise and thus, in most cases, results are presented with no demonstrations. In Annex B we derive expressions for the output signal to noise ratio as a function of the input signal-to-noise for phase and for frequency demodulators, using sinewave modulating signals and white, Gaussian, band limited noise.

4.2 ANALOG SIGNALS

4.2.1 Telephony

Even if single channel telephony radio links are nowadays seldom used, it is appropriate to start the description of analog signals by the telephone signal (also known as the voice signal) given its importance in the making up the multiplex telephone signal.

ITU-T [13] Recommendation G.132 standardized the bandwidth of the voice channel to 3100 Hz, between 300 and 3400 Hz, even if the human voice spectrum extends well over these limits. The center frequency at which all measurements are performed is 800 Hz. Frequency multiplexing of voice channels uses 4 kHz channel spacing.

ITU-T sets quality criteria for voice signals. The most important are amplitude and group delay distortion, propagation delay and difference between transmitted and received signal frequencies.

Limits for the variation of the channel attenuation with frequency for a 12 channel multiplex, specified in ITU-T[13] Recommendation G.132 are shown in Figure 4.1.

To achieve these objectives amplitude response of transmitting plus receiving equipment for 12 voice channels should be within the limits shown in Figure 4.2 (Recommendation G.232 [13]).

Oscillator stability should be such that the difference between transmitted and received frequencies in any channel does not exceed 2 Hz (Recommendation G.225 [13]).

Transmission time, including propagation and equipment delay should never exceed 400 ms (Recommendation G.114 [13]). Values between 150 and 400 ms are acceptable with special precautions, such as echo suppressors, appropriate for large delay circuits.

The differences between group delay at the lowest and the highest frequency of the transmitted bandwidth should not exceed 60 and 30 ms, for the whole 4 wire circuit of an international link with 12 circuits each with 12 voice channels ITU-T [13]) (Recommendation G.133).

ITU-T [13] Recommendation G.223 defines the average power of a voice channel, during the busy hour, at the zero level point of a unidirectional

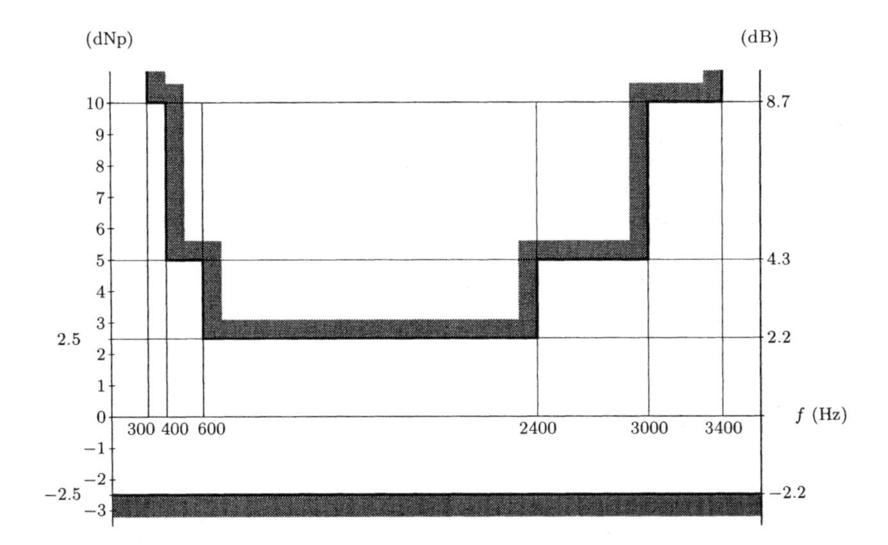

Fig. 4.1 Limits for channel attenuation with frequency (Recommendation G.132 [13]).

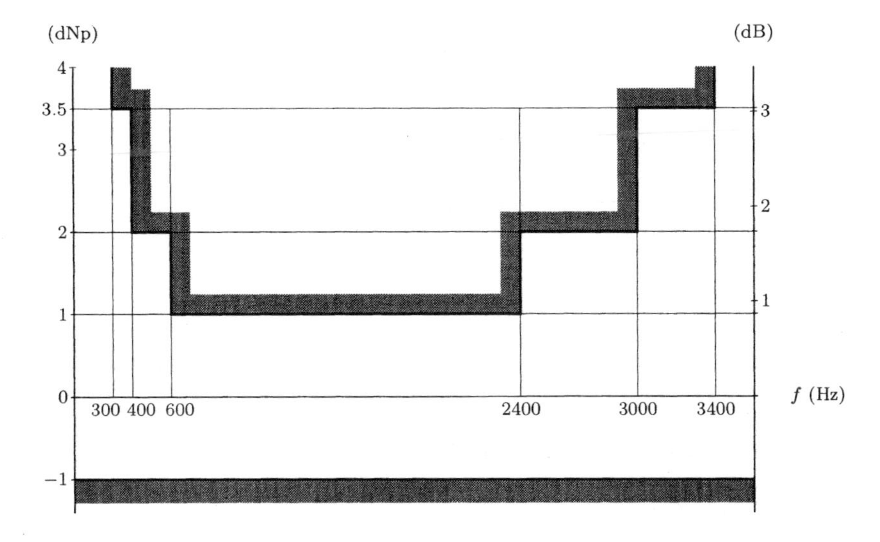

Fig. 4.2 Amplitude response limits for transmitting plus receiving equipment (Recommendation G.232 [13]).

circuit, as −15 dBm0 (32 μW). This value includes signaling (10 μW) and voice, including echoes (22 μW).

Since the test signal for voice circuits is usually a 800 Hz sine wave, with a power of 0 dBm0 (1 mW), at the zero point level, the ITU-T recommended value may appear to be rather low. A simple reasoning will show that this is not the case.

Assume that the power of the voice signal of an active user in a unidirectional circuit is 1 mW. If a set of n circuits ($n > 12$ is to provide a reasonably good level of service, during the busy hour, that is, for a low probability that all circuits are simultaneously occupied, Erlang's formula indicates that the number of occupied circuits, on average, should not exceed about 70 percent of the total number of circuits. On the other hand, in a telephone call, on average, each partner speaks for about half of the time (and listens to, during the other half). Thus we have an activity index of 0.7 times 0.5, that is, 0.35 or 35 percent. If now we consider the inter-syllabic pauses we arrive at an activity index of 0.25, as used by the ITU-T. To complete our reasoning we must now consider the attenuation between the user equipment and the network zero level point. Taking a 10 dB attenuation, which corresponds to about 9 km, of 4-wire 0.6 mm diameter, paper isolated, telephone cable, with an attenuation of about 1.1 dB/km, the average power level Pm of the voice signal at the zero point becomes

$$\begin{aligned} Pm &= 10\log_{10}(1) + 10\log_{10}(0.25) - 10 \\ &= -16 \text{ (dBm0)}, \end{aligned}$$

or:

$$pm = 25 \ (\mu\text{W}).$$

Experimental results [10] show that the maximum power level of the unidirectional voice signal is, on average, 18.6 dB above the average power level.

4.2.2 Frequency division multiplexing

Frequency division multiplex, or FDM, is obtained by grouping unidirectional voice channels after a frequency shift such that they occupy successive 4 kHz slots. This signal, as well as any other signal used to modulate a radio link, is also known as the base band signal.

According to ITU-R Recommendation F.380-4 [12] the main features of this base band signal are:

- number of telephone channels;
- frequency limits (lower and upper) of the band occupied by the telephone channels;
- frequency limits (lower and upper) of the base band, including pilots or frequencies that might be transmitted;

- relative input and output power levels and nominal impedance at the interconnection points.

The preferred values for all these variables are shown in Table 4.1. The international interconnection points of the base band circuits, known as $R\prime$ and R, represent the input and the output of a radio link. At these points:

- all telephone channels and pilots should be at the spectral position at which they are to be transmitted, according to the relevant ITU-T and ITU-R Recommendations;

- continuity pilots and any other signals that may be transmitted in a radio relay link outside the telephony band have been suppressed;

- any radio relay protection switching is considered as part of the radio relay;

- in case of diversity, the output of the combiner output (after suppressing the continuity or any other pilots) corresponds to point R;

- pre-emphasis and de-emphasis circuits are taken as integral part of receivers and/or transmitters.

For a large number ($n \geq 240$) of telephone channels the average unidirectional multiplex signal power pm, during the busy hour, at the zero level point, is quite simply the product of the number of channels by the average unidirectional signal power. According to the ITU-T [13] Recommendation G.223 for $n \geq 240$ we have

$$pm_{\ [mW]} = n \cdot 10^{-15/10}.$$

Expressing pm in logarithmic units

$$Pm_{\ [dBm0]} = 10 \log_{10}(pm),$$

or:

$$Pm_{[dBm0]} = -15 + 10 \log_{10}(n). \tag{4.1}$$

The average noise power used to simulate the multiplex signal is known as the conventional load, or simply, the system load.

Introduction of new telephone sets, changes in the activity factor brought about by new operating methods, differences in the average loss between the user set and the system zero level point and the increasing use of voice channels to transmit data may affect the power level of the multiplex signal. As an example we may refer that for systems used in the United States of America [17]

$$Pm_{\ [dBm0]} = -16 + 10 \log_{10}(n), \tag{4.2}$$

Number of telephone channels	Frequency limits of the band occupied by telephone channels (kHz)	Frequency limits of the base band (kHz)	Nominal impedance at interconnection points (Ohm)
12	12 – 60	12 – 60	—
	60 – 108	60 – 108	—
24	12 – 108	12 – 108	150 sym.
60	12 – 252	12 – 252	150 sym.
	60 – 300	60 – 300	75 assym.
120	12 – 552	12 – 552	150 sym.
	60 – 552	60 – 552	75 assym.
300	60 – 1300	60 – 1364	75 assym.
	60 – 1296	60 – 1364	75 assym.
600	60 – 2540	60 – 2792	75 assym.
	64 – 2660	60 – 2792	75 assym.
960	60 – 4028	60 – 4287	75 assym.
	60 – 5636		
1260	60 – 5564	60 – 5680	75 assym.
	316 – 5564		
	312 – 8204		
1800	326 – 8204	300 – 8248	75 assym.
	312 – 8120		
	312 – 12388		
2700	316 – 12388	300 – 12435	75 assym.
	312 – 12336		

Table 4.1 Preferred characteristics for frequency division multiplex signals according to ITU-R Recommendations F.380-4, F.398-3 and F.401-2 [12]).

and, for some military systems, where a high proportion of telephone channels are occupied with telegraphic or data signals

$$Pm_{\ [dBm0]} = -10 + 10\log_{10}(n). \tag{4.3}$$

Theoretically, the peak power *pmx* of the multiplex signal occurs when, simultaneously, all channels reach peak power, 18.6 dB above average power. Assuming an average power per channel of −15 dBm0 and a voltage addition, we would have, for n channels

$$Pmx_{\ [dBm0]} = -15 + 20\log_{10}(n \cdot 10^{18.6/20}).$$

This peak power level is statistically less probable as the number of channels increases. Tests, carried out in 1938 by Holbrook and Dixon [11] showed that multiplex signal peaks could be allowed to exceed the overload point of the (vacuum tube) amplifiers for about 0.1 percent of the time without serious degradation of system performance. To enable the use of different types of amplifiers is has become usual to define the multiplex signal peak power at the zero level, as the peak power of a sine wave signal such that its peak amplitude is not exceeded by the amplitude of the multiplex signal during more than 0.001 percent of the time, during the busy hour.

By definition (ITU-T [13]Recommendation G.223) the overload point of an amplifier is the output power level for which the absolute power of the third harmonic increases by 20 dB when the input signal, assumed to be sinusoidal, increases 1 dB. This definition is not applicable when the input frequency is such that the third harmonics falls outside the useful amplifier bandwidth. In this case the overload point is defined as being 6 dB above the output power level, expressed in dBm, of each of two signals with frequencies f_A and f_B such that when its input level increases simultaneously 1 dB, the intermodulation product at the frequency $2f_A - f_B$ increases 20 dB.

The peak power level Pmx of an n channel multiplex signal is defined as the (r.m.s) power of a sinewave signal with a peak amplitude equal to the peak amplitude of the multiplex signal. Table 4.2 and Figure 4.3 display the value of Pmx calculated from Holbrook [11], for an average channel power of −15 dBm0 (instead of −16 dBm0) as given in ITU-T [13] Recommendation G.223. For $n > 1000$ the same recommendation states that the peak power of the multiplex signal may be calculated from

$$Pmx_{\ [dBm0]} = -5 + 10\log_{10}(n) + 10\log_{10}\left(1 + \frac{15}{\sqrt{n}}\right). \tag{4.4}$$

Alternatively, we may calculate the multiplex peak power from the average power, given in equation (4.1), by adding the multiplex peak factor Δc

$$Pmx = Pm + \Delta c. \tag{4.5}$$

Number of telephone channels	Multiplex peak power (dBm0)
12	19
24	19.5
36	20
48	20.5
60	20.8
120	21.2
300	23
600	25
960	27

Table 4.2 Multiplex peak power as a function of the number of telephone channels, according to ITU-T [13] Recommendation G.223.

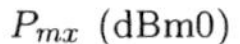

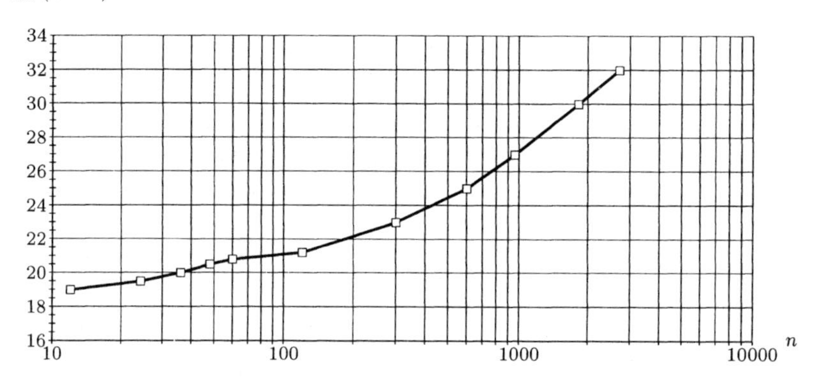

Fig. 4.3 Peak power of the multiplex signal as a function of the number of telephone channels, according to ITU-T [13] Recommendation G.223.

If we assume that the power of the n individual channels, that make up the multiplex signal, have a normal distribution, the multiplex peak factor Δc is approximately [1] given by:

$$\Delta c(n)_{[dB]} = 10.5 + \frac{40\sigma}{n\tau_L + 5\sqrt{2\sigma}}, \tag{4.6}$$

where τ_L is the activity index. Taking, as usual, $\tau_L = 0.25$ and $\sigma = 5.8$ dB we get $\Delta c(n)$ as plotted in Figure 4.4, with the label Bell. For easy referencing we plot, in the same Figure 4.4, the multiplex peak factor derived from ITU-T Recommendation G.223, the one given in equation (4.6) and the test signal referred to later. Please note the similarity between the two values of the multiplex peak factor, particularly for a number of channels between 24 and 960. For less than 12 or more than 960 channels the peak multiplex factor derived from the ITU-T recommendation is always about 1 to 2 dB higher than the one given by (4.6).

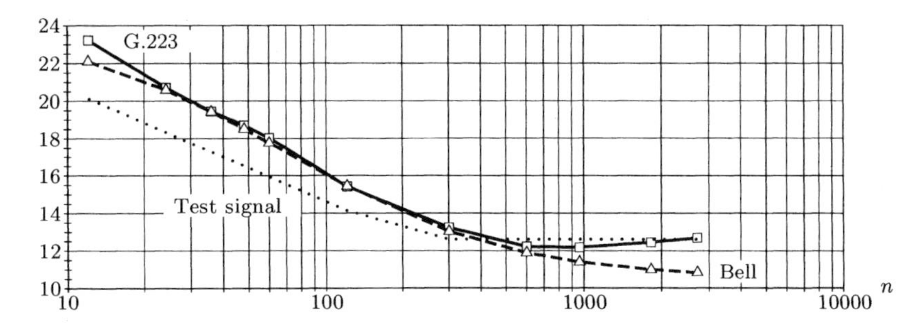

Fig. 4.4 Peak multiplex factor as a function of the number of channels.

Tests of radio links require the simulation of multiplex signals. Fortunately, for a large number of channels (in practice more than 60), the multiplex signal may be substituted by white Gaussian noise with the same bandwidth and average power.

For a Gaussian noise with unit average power ($\sigma = 1$) the probability to exceed the amplitude of 4.265 (i.e., 12.6 dB above average power) is 0.00001 (or 0.001 percent). This is the reasoning behind the ITU-R[12] Recommendation F.399-3 which asks for a peak Gaussian noise power 12 dB above the average power and for the peak multiplex factor calculated from Table 4.2.

ITU-R[3] Report 836-1 uses a peak multiplex factor of 11.5 for 12 or more telephone channels and 13 dB between 3 and 12 telephone channels.

The assumption that the multiplex amplitude distribution is Gaussian is not appropriate for a small number of telephone channels ($n < 60$). For $240 > n \geq 12$, ITU-T suggests a compromise solution where the simulated multiplex signal continues to have a Gaussian amplitude distribution (i.e., has the same 12.6 dB ratio between peak and average power) but its average power is increased to

$$Pm_{\ [dBm0]} = -1 + 4\log_{10}(n). \tag{4.7}$$

As a partial alternative to the previous equation, ITU-R [3] Report 836-1 suggests that the multiplex average power for $60 > n \geq 12$ should be given by

$$Pm_{\ [dBm0]} = 2.6 + 2\log_{10}(n). \tag{4.8}$$

The difference between the average power values given by equations (4.7) and (4.1) is quite small (−0.3 dB) for 240 telephone channels and increases as the number of channels decreases, reaching 7.5 dB for 12 channels. The difference between the average power values given by (4.7) and (4.8) is always less than 1.5 dB.

Even after increasing the average power value, the Gaussian signal has a lower peak power than the multiplex signal, particularly for a small number of telephone channels. Thus for $n = 12$ the peak multiplex power, in Table 4.2, is 19 dBm0, while for the Gaussian signal we have, from (4.7) an average power of 3.3 dBm0, which after adding 12.6 dB, adds up to a peak power of 15.9 dBm0. Making use of (4.8) we would have $Pm = 4.8$ dBm0 which for the same 12.6 dB peak multiplex factor, leads to a peak power of 17.4 dBm.

4.2.3 Sound broadcasting

According to ITU-T [13] Recommendation J.12 radio circuits used to transmit sound broadcast signals may be classified as:

- average quality, 5 or 7 kHz monophonic circuits, used both for temporary and permanent voice links, or to interconnect broadcast studios to long-medium- and short-wave transmitter centers;
- quality, 10 kHz circuits, often used in the past, but no longer recommended;
- high-quality, 15 kHz circuits, used both for monophonic and stereophonic transmissions.

The hypothetical reference circuit for ground links, as defined in ITU-T [13] Recommendation J.11 is 2500 km long between audio frequency terminals. Between these terminals there are two intermediate audio frequency points that divide the circuit in three equal length parts.

For the hypothetical reference circuit, the maximumfrequency response deviation for the 7 kHz and the 15 kHz circuits, according to ITU-T [13] Recommendations J.21 and J.23, is given in Table 4.3.

In the studio, signals used for sound broadcasting may be subject to one or more of the following procedures:

- emphasis[1];
- selective and adjustable emphasis, for artistic reasons;
- compression to decrease signal dynamics and to improve signal to noise ratio for low level signals;
- clamping to decrease the peak power level.

[1]Emphasis is a procedure that modifies the characteristic amplitude as a function of frequency of a signal. The words pre-emphasis and de-emphasis indicate that emphasis is applied before and after transmission, respectively.

Frequency (kHz)	Frequency response (dB)	
	7 kHz	15 kHz
0.05 – 0.1	+1 to −3	
0.1 – 6.4	+1 to −1	
6.4 – 7.0	+1 to −3	
0.04 – 0.125		+0.5 to −2.0
0.125 – 10		+0.5 to −0.5
10 – 14		+0.5 to −2.0
14 – 15		+0.5 to −3.0

Table 4.3 Frequency response, given by the maximum frequency response deviation, in dB, for the transmission of broadcast signals in different types of circuits. The reference frequency is 1000 Hz, except for the 15 kHz circuits, where it may also be 800 Hz.

At the zero level, the quasi peak power of the sound broadcast signal, measured with an integration time of 20 ms, must be 9 dBm0 (ITU-T [13] Recommendation J.14).

Since clampers operate as linear amplifiers, whose gain decreases when the power of the sinewave, equivalent to the broadcast signal, is above a peak power of 9 dBm0, and since the time constant for gain decrease is 0.5 ms, it is possible, although rather improbable, to have peak power levels in excess of 9 dBm0, as shown in Table 4.4.

Probability that the signal power is exceeded	Equivalent sine wave power (dBm0)
0.001	9.5
0.0001	11.2
0.00001	12.0

Table 4.4 Probability that the peak power level of a broadcast signal is exceeded (ITU-R [8] Report 491-3).

Unlike the multiplex signal, the power level distribution of the sound broadcast signal is not Gaussian. According to CCIR [8] Report 491-3 the average power level is:

- −8.9 dBm0, for a non-processed signal;
- −6.4 dBm0, for a amplitude limited, partly compressed signal;
- −2.4 dBm0, for a fully amplitude limited and compressed signal.

A typical value of −4 dBm0 has been suggested, in CCIR [8] Report 491-3, for all broadcasting signals.

In microwave radio links, broadcasting signals are transmitted in primary groups (12 telephone channels), at the rate of six telephone channels for each 15 kHz channel (ITU-T [13] Recommendation J.31). Since, as we have shown in Section 4.2.2, the power levels for a primary group are:

- average rated power in the busy hour

$$\begin{aligned} Pm &= -15 + 10\log_{10}(12) \\ &= -4.2 \text{ dBm0}, \end{aligned}$$

- maximum power

$$Pmx = 19 \text{ (dBm0)},$$

it is clear that these levels are compatible with the requirements of a 15 kHz circuit.

4.2.4 Video

In international circuits, the same microwave radio link may transmit, in different instances, video signals according to various standards (both 525 and 625 lines). Thus, the link characteristics should be, as far as possible, the same (ITU-T [13] Recommendation J.61). For new systems, ITU-R [7] recommends 525 or 625 lines, according to the NTSC, PAL or SECAM[2] (Recommendation BT.470-6)

At video interconnection points, where impedance should be 75 Ohm asymmetric or 124 Ohm symmetric, the nominal signal level, here defined as the peak-to-peak amplitude of the monochromatic signal (the sum of the luminance signal plus the synchronism pulse) is 1 Volt (ITU-R [12] Recommendation F.270-2). Maximum white level corresponds to the maximum signal level (1 V), synchronism corresponds to the minimum signal level (0 V) and suppression (black) to 0.3 V.

Composite video signal, resulting from the sum of the monochromatic signal with a subcarrier modulated by the chromatic signals, may exceed the nominal amplitude for high-luminance saturated colors, such as yellow.

The nominal bandwidth of the video signal varies between 5 and 6 MHz for the usual standards.

Besides video signal, television requires one or more audio channels with about 10 kHz bandwidth. In television broadcast the sound signal usually modulates in frequency a subcarrier, 5.5 to 6.5 MHz above the video carrier, with a maximum frequency deviation of 50 kHz. In television links sound signals may be transmitted in one of the following ways:

[2]The acronyms NTSC (National Television Systems Committee), PAL (Phase Alternating Line) and SECAM (Séquentiel à Mémoire) identify the color television systems originated in the United States of America, Germany and France, respectively.

- Frequency modulation, with a 140 kHz root-mean-square frequency (r.m.s) deviation, of a 7.5 MHz subcarrier (ITU-R [12] Recommendation F.402-2). The amplitude of this subcarrier is such that, when unmodulated, it leads to a r.m.s. deviation of 300 kHz (or to a peak deviation of 420 kHz).
- Sampling the sound signal at twice the line frequency (i.e. $2 * 15625$ Hz for the 625 line systems), coding the samples, after adequate compression, with 10 bits per sample, grouping the 20 bits corresponding to 2 samples, adding 1 reference bit and superimposing the resulting 21 bits with the synchronizing pulse in such a way that the reference bit and the least significant bits are placed on the forward pulse front (ITU-T [13] Recommendation J.66). The amplitude of these pulses extends from the levels of synchronism to maximum white. Since the sampling rate is quite close to the Nyquist rate it may be difficult to guarantee a frequency response up to 15 kHz. In this case the maximum frequency for which the frequency response of the hypothetical reference circuit is specified may be reduced from 15 to 14 kHz.

When sound signals are transmitted using a different channel from the video signal, the difference between propagation delays must be kept small. A sound delay of 140 ms, or an advance of 70 ms, produces a "just perceptible" anomaly for 50 percent of the listeners (ITU-R [8] Report 412-3).

4.3 RADIO FREQUENCY BANDWIDTH

In analog microwave radio links frequency modulation is used almost exclusively, with the exception of single channel systems, where phase modulation is also employed,

According to ITU-R [3] Report 836-1 radio frequency bandwidth b_{rf} for frequency modulated signals must, at least, be equal to the value provided by Carson's (approximate) formula b_{cr}

$$\begin{aligned} b_{cr} &= 2(\Delta f_{max} + f_{max}) \\ &= 2(k_f + 1) f_{max}, \end{aligned} \tag{4.9}$$

where Δf_{max} is the maximum carrier deviation frequency, f_{max} is the highest frequency of the base band signal and k_f is the ratio $\Delta f_{max}/f_{max}$. Please note that, in some cases, this formula may underestimate the required bandwidth [16].

In phase modulation, with a maximum phase deviation k_p (in radians) Carson's bandwidth is given by

$$b_{cr} = 2(k_p + 1) f_{max}. \tag{4.10}$$

Besides the highest frequency of the modulating signal, the use of the two formulas above requires the maximum frequency, or phase, deviation.

For single channel radio links there are no general standards but Δf_{max} is often between 10 and 30 kHz, that is, k_f between 3 and 9 and k_p between 2 and 5.

In radio links used for television signals, a 1 V signal at the video interconnection points, in the absence of pre-emphasis (later referred to), should produce a 4 MHz maximum frequency deviation (ITU-R [12] Recommendation F.276-2).

In frequency division multichannel links, the r.m.s. deviation per channel Δf_{rms} for a 1 mW, 800 Hz, signal at the zero level point, is given in Table 4.5.

Number of channels	**R.m.s. frequency deviation per channel (kHz)**
12	35
24	35
60	50, 100, 200
120	50, 100, 200
300	200
600	200
960	200
1260	140, 200
1800	140
2700	140

Table 4.5 R.m.s. frequency deviation per channel as a function of the number of channels, in frequency division multichannel radio links (ITU-R [12] Recommendation F.404-2.)

Since frequency modulators produces a carrier frequency deviation proportional to the input voltage, if all input signals but one in a frequency division multiplex signal are zeroed and the remaining one has a 800 Hz sinewave with 1 mW (0 dBm), at the zero level point, then the output signal frequency deviation is also a sinewave with a r.m.s. value given in Table 4.5 and a peak value $\sqrt{2}$ times higher. If pm is the average power of the multiplex signal, in mW, the r.m.s. frequency deviation Δf_{rms}^n will be

$$\begin{aligned}\Delta f_{rms}^n &= \Delta f_{rms} \frac{u_{rms}^n}{u_{rms}^1} \\ &= \Delta f_{rms}^1 \sqrt{pm},\end{aligned}$$

where:

- Δf_{rms}^1 is the r.m.s. frequency deviation per channel;
- u_{rms}^n is the r.m.s. voltage corresponding to the multiplex signal;

- u_{rms}^1 is the r.m.s. voltage of the test signal for which the r.m.s. frequency deviation is defined.

Using logarithmic units

$$\Delta f_{rms}^n = \Delta f_{rms}^1 \, 10^{Pm/20},$$

and the maximum frequency deviation becomes

$$\Delta f_{max} = \Delta f_{rms}^1 \, 10^{Pmx/20}. \tag{4.11}$$

Pmx may be obtained from Table 4.2 or from equations (4.1) and (4.5) or adding to the average power given in (4.1), (4.7) or(4.8) a constant peak factor of about 12.6 dB.

4.4 THERMAL NOISE

4.4.1 Introduction

The quality of an analog link may be defined in terms of:

- uniformity of the modulus of the transfer function,
- linearity of the argument of the transfer function,
- signal-to-noise ratio,

throughout the bandwidth.

Except for the existence of multipath with differing propagation delays, link transfer function is mainly the responsibility of equipment designers and manufacturers. On the other hand, signal-to-noise ratio depends mainly on link design, thus justifying a more detailed analysis.

The base band thermal signal-to-noise ratio is dependent on the signal-to -noise ratio at the demodulator input (i.e., at the intermediate frequency level) and on the modulation parameters. For angle modulations the expression *radio frequency signal-to-thermal noise ratio* is often substituted by the equivalent expression *carrier-to-thermal noise ratio* or simply *carrier-to-noise ratio.*

4.4.2 Carrier-to-thermal noise ratio

At the usual microwave radio frequencies (well above 100 MHz) noise at the antenna terminals is mostly thermal noise with a power n_0 given by

$$n_0 = K_B T \, b_{rf}, \tag{4.12}$$

where:

- K_B is the Boltzman constant ($K_B = 1.38 \times 10^{-23} J/K$);
- T is temperature seen by the antenna, in K;
- b_{rf} is the effective noise bandwidth, in Hz.

In the large majority of terrestrial microwave radio links, antennas "see" the Earth at a temperature of about 293 K (or 20°C) and thus, the thermal noise power at the antenna terminals, in logarithmic units, is given by

$$N_{0\ [dBW]} = -204 + 10 \log_{10}(b_{rf}), \tag{4.13}$$

or

$$N_{0\ [dBm]} = -174 + 10 \log_{10}(b_{rf}). \tag{4.14}$$

A more rigorous analysis requires the integration of the temperature seen by the antenna for each direction. This leads to a dependence of the result not only on the antenna but also on its location. In general, we would obtain a temperature somewhat lower than 293 K, since only part of the antenna radiation pattern "sees" the Earth. The approximation used above is acceptable, because equipment noise is usually much larger than thermal noise at the antenna terminals.

Thermal noise power N at the demodulator input, referred to the antenna terminals, may be calculated by adding the receiver noise factor[3] Nf (in dB) to N_0

$$N = N_0 + Nf. \tag{4.15}$$

In general, the receiver noise factor is mostly due to the input stages, that is, radio frequency amplifier and mixer.

When, instead of the receiver noise factor, we would rather use the receiver equivalent noise temperature Tr (in K), for an ambient temperature of 293 K

$$Nf = 10 \log_{10} \left(1 + \frac{Tr}{293}\right). \tag{4.16}$$

After calculating the receiver input power P_R (in logarithmic units), using the procedures described in Chapter 2, we may proceed to calculate the radio frequency signal-to-noise ratio S/N_{rf}, better referred to as the carrier-to-noise ratio C/N

$$\begin{aligned} S/N_{rf} &= C/N \\ &= P_R - N. \end{aligned} \tag{4.17}$$

Substituting in (4.17) the value of P_R from the Friis formula in logarithmic form (2.8) we get

$$C/N = P_T - A_T + G_T + G_R - A_R - A - N,$$

[3] For a more detailed analysis of the noise factor see Section 4.8.

where A, is the transmission loss. Introducing the concept of equivalent isotropic radiated power $EIRP$ we get

$$C/N = EIRP + G_R - A_R - A - N. \tag{4.18}$$

Noting that the thermal noise power N may be written as

$$N = 10\, log_{10}(KTr\, b_{rf}),$$

we may write from (4.18)

$$C/N = EIRP + G_R - A_R - 10\ \log_{10}(Tr) - 10\ \log_{10}(Kb_{rf}) - A. \tag{4.19}$$

In satellite communications $G_R - A_R - 10\ \log_{10}(Tr)$ is known as G/T, or figure of merit, of the receiver station and is expressed in dB/K. Introducing G/T in (4.19) we finally get

$$C/N = EIRP + G/T - A - 10\ \log_{10}(Kb_{rf}), \tag{4.20}$$

which clearly shows that the carrier-to-termal noise at the demodulator input depends on the transmitter, the receiver, the transmission attenuation and the occupied bandwidth.

4.4.3 Demodulation

The base band signal-to-noise ratio depends on:

- the carrier-to-noise ratio, at the demodulator input;
- the modulating signal;
- the type of modulation.

As shown in Appendix B, for a frequency modulated sinewave carrier with a maximum frequency deviation Δf_{max} the base band signal-to noise ratio s/n, measured between a minimum frequency f_m and a maximum frequency f_M, is given as a function of the carrier signal-to-noise ratio c/n as

$$\frac{s}{n} = \frac{3}{2} \frac{\Delta f_{max}^2}{f_M^2 + f_M f_m + f_m^2} \frac{b_{rf}}{b_s} \frac{c}{n}, \tag{4.21}$$

where $b_s = f_M - f_m$.

If $f_M \gg f_m$ and thus $b_s \approx f_M$, which is true both for sound and video broadcast signals and, to a lesser extent, also for single channel telephony, (4.21) simplifies to

$$\frac{s}{n} = \frac{3}{2} \frac{\Delta f_{max}^2}{f_M^2} \frac{b_{rf}}{b_s} \frac{c}{n}. \tag{4.22}$$

For a single channel of a frequency division multiplex signal

$$f_M = f_c + \frac{b_s}{2}, \quad (4.23)$$

$$f_m = f_c - \frac{b_s}{2}, \quad (4.24)$$

where the channel center frequency f_c is given by:

$$f_c = \frac{f_M + f_m}{2}. \quad (4.25)$$

Substituting (4.23) and (4.24) into (4.21), we get

$$\frac{s}{n} = \frac{3}{2} \frac{\Delta f_{max}^2}{3\left(f_c^2 + \frac{b_s^2}{12}\right)} \frac{b_{rf}}{b_s} \frac{c}{n}, \quad (4.26)$$

where Δf_{max} is the maximum deviation due to the test signal, assumed to be at a frequency such that, for the base band signal, we get a channel center frequency f_c.

Assuming $f_c \gg b_s$, with is true for most channels and, in particular for the highest frequency channel

$$\frac{s}{n} = \frac{1}{2} \frac{\Delta f_{max}^2}{f_c^2} \frac{b_{rf}}{b_s} \frac{c}{n}. \quad (4.27)$$

As referred to before, for multiplex signals, the r.m.s. rather than the peak frequency deviation is specified. Taking into account that for a sine wave test signal we have

$$\Delta f_{rms} = \frac{\Delta f_{max}}{\sqrt{2}},$$

equation (4.27) may be re-written in terms of Δf_{rms} as

$$\frac{s}{n} = \frac{\Delta f_{rms}^2}{f_c^2} \frac{b_{rf}}{b_s} \frac{c}{n}. \quad (4.28)$$

4.4.4 Sensitivity threshold

The expressions for the signal-to-noise ratio in the base band as a function of the carrier signal-to-noise assume that

$$c \geq st, \quad (4.29)$$

where st, the sensitivity threshold is given, in logarithmic units, for the usual demodulators found in microwave radio links as

$$St = N + 11, \quad (4.30)$$

where N is given in (4.15).

4.5 EMPHASIS

4.5.1 Introduction

Expressions for the base band signal-to-noise ratio show that, for the same frequency deviation, the signal-to-noise ratio decreases with the increase of the modulating signal frequency. This fact is particularly serious for frequency division multiplex signals because the quality of a given telephony channels (as far as its signal-to-noise ratio) depends on the channel position in the base band. To minimize this effect, the amplitude of the higher frequency signals in the base band are often increased before modulation and decreased after demodulation. These operations are known as pre-emphasis and de-emphasis, respectively.

Although later on we may simply refer to emphasis, it must be understood that we mean pre-emphasis on the transmitter side and de-emphasis on the receiver side, so that the overall transfer function is linear.

4.5.2 Single channel telephone signal

Emphasis for the single channel telephone signal is not standardized internationally because single channel microwave radio links are not usually part of international circuits.

4.5.3 Multiple channel telephone signal

To allow for interconnection of microwave radio links from different makers, emphasis should conform to ITU-R [12] Recommendation F.275-3. In this recommendation emphasis corresponds to a variable frequency emphasis gain $G(f)$ given, in dB, by

$$G(f) = 5 - 10\log_{10}\left[1 + \frac{6.90}{1 + \dfrac{5.25}{(\frac{fr}{f} - \frac{f}{fr})^2}}\right], \tag{4.31}$$

where the resonant frequency of the emphasis circuit, fr, is equal to $1.25 f_{max}$ and f_{max} is the highest base band frequency in Table 4.1. Function $G(f)$ is plotted in Figure 4.5.

Emphasis gain is such that the r.m.s. frequency deviation due to the multiplex signal is the same with or without emphasis.

The use of emphasis approximates (but does not exactly equals) the signal-to-noise ratios of telephone channels in a multiplex signal. As shown in detail in Appendix B, equal signal-to-noise ratios require a 6 dB/octave emphasis (which correspond to a change from frequency to phase modulation). Such an emphasis is inconvenient because is makes the lower telephone channels more

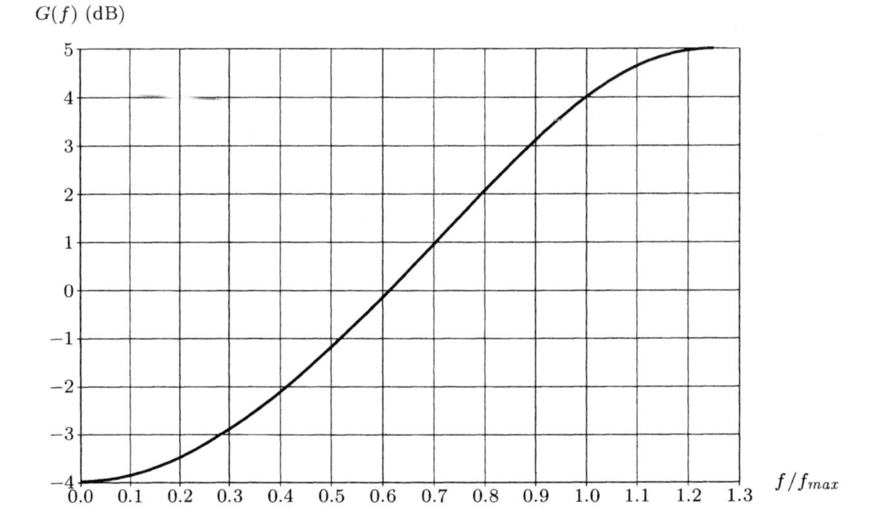

Fig. 4.5 Emphasis gain for multiplex signals (ITU-R [12]) Recommendation F. 275-3).

susceptible to intermodulation noise. The recommended emphasis curve aims at equalizing all channels in terms of the total (thermal plus intermodulation) noise power .

If we substitute f for f_{max} (highest base band frequency used by telephone channels) in (4.31) we get a 4 dB emphasis for the least favored channel. To be precise, the emphasis should have been calculated for the center frequency of the highest channel (f_{max} − 2 kHz), making the emphasis dependent on f_{max}, that is, on the number of channels of the multiplex signal. The error, which in the worst case is less than a 0.30 dB, may be safely neglected.

With the recommended emphasis, the signal-to-noise ratio in any voice channel with 3100 Hz bandwidth (from 300 to 3400 Hz) becomes, from (4.28) and (4.31)

$$
\begin{aligned}
S/N &= C/N + 20\log_{10}\left(\frac{\Delta f_{rms}}{f_c}\right) + 10\log_{10}\left(\frac{b_{rf}}{3100}\right) \\
&\quad +5 - 10\log_{10}\left[1 + \frac{6.90}{1 + \frac{5.25}{\left(\frac{fr}{fc} - \frac{fc}{fr}\right)^2}}\right],
\end{aligned}
\tag{4.32}
$$

where:

- S/N is the signal-to-thermal noise in the required expressed in logarithmic units

$$S/N = 10\log_{10}\left(\frac{s}{n}\right);$$

- C/N is the ratio between the carrier power and the thermal noise power, at the demodulator input, expressed in logarithmic units

$$C/N = 10\log_{10}\left(\frac{c}{n}\right);$$

- b_{rf} is the modulated signal bandwidth;
- Δf_{rms} is the r.m.s. frequency deviation given in Table 4.5;
- f_c is the channel center frequency;
- f_r is the resonant frequency of the emphasis circuit;
- f_{max} is the maximum frequency used by telephone channels in the base band;
- $f_r = 1.25 f_{max}$.

For the highest frequency of the multiplex signal , where $f_r = f_{max}$, equation (4.32) simplifies to

$$S/N = C/N + 20\log_{10}\left(\frac{\Delta f_{rms}}{f_c}\right) + 10\log_{10}\left(\frac{b_{rf}}{3100}\right) + 4. \qquad (4.33)$$

4.5.4 Sound broadcast signal

The recommended emphasis for sound broadcast signals is defined in ITU-T [13] Recommendation J.17. At the angular frequency $\omega = 2\pi f$, it corresponds to a gain $G(\omega)$, in dB, given by

$$G(\omega) = 10\log_{10}\left[\frac{75 + (\frac{\omega}{3000})^2}{1 + (\frac{\omega}{3000})^2}\right]. \qquad (4.34)$$

With the recommended pre-emphasis and de-emphasis, the noise power reduction calculated from CCIR [8] Report 496-4 (with the 1.5 dB correction referred to in item b of the notes to Table II in this report) is given in Table 4.6. The values related to white noise power reduction correspond to situations such as base band transmission or amplitude modulation, where noise power density is constant. In the case of frequency modulation where noise power density at the demodulator output increases with the square of the frequency, the values for triangular noise apply.

Broadcast signal bandwidth (Hz)	White noise power reduction (dB)	Triangular noise power reduction (dB)
70 – 5000	–	–
50 – 7000	1.3	7.2
50 – 10000	3.8	9.9
40 – 15000	5.2	10.8

Table 4.6 Unweighted noise power reduction using the recommended pre-emphasis and de-emphasis for sound broadcast (CCIR [8] Report 496-4).

4.5.5 Video broadcast signal

Unlike multiplex telephony signals, video broadcast signals do not have a flat power spectrum; most of the power is concentrated in the lower frequencies. To allow for the same equipment to be used in both cases the emphasis circuit must:

- attenuate the lower and increase the higher frequencies in the video spectrum;
- equalize, as much as possible, the r.m.s. frequency deviations for the middle frequencies both for multiplex and video signals.

The recommended emphasis circuit for video signal, as defined in ITU-R [12] Recommendation $G(f)$ F.405-1, corresponds to a gain $G(f)$, in dB, as a function of the signal frequency f (in MHz), given by

$$G(f) = 10\log_{10}\left(\frac{1+Cf^2}{1+Bf^2}\right) - A_{f=0}, \tag{4.35}$$

where:

- $A_{f=0}$ is the attenuation for a low signal frequency (less than 0.01 MHz);
- B and C are constants.

The values of $A_{f=0}$, B e C depend on television standards (525 or 625 lines per image), according to Table 4.7. In this table we also indicate the transition frequency f_e, defined as the signal frequency at which the frequency deviation is the same with or without emphasis. Emphasis curves defined in (4.35) together with the constants in Table 4.7 are plotted in Figure 4.6.

With the recommended pre-emphasis and de-emphasis the signal-to noise ratio is slightly improved as shown in Table 4.8.

In television it is usual to define the signal-to-noise ratio at a given point, as the ratio between the peak-to-peak image voltage at this point and the

Number of lines per image	525	625
$A_{f=0}$	10.0	11.0
B	1.306	0.4038
C	28.58	10.21
f_e (MHz)	0.7616	1.512
Peak frequency deviation at low frequencies (MHz)	2.530	2.255

Table 4.7 Recommended constant values for video emphasis (ITU-R [12] Recommendation F.405-1).

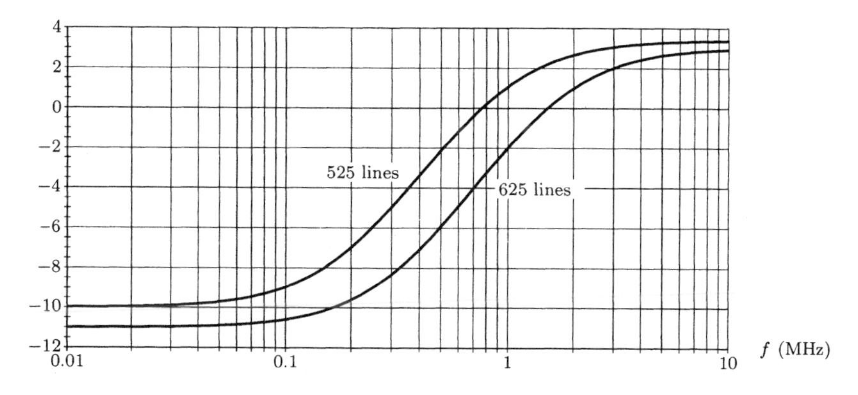

Fig. 4.6 Emphasis gain for video signals (ITU-R [12] Recommendation F.405-1).

r.m.s. noise voltage at the same point. Since the image signal, at the video interconnection points, has a 0.7 V peak-to-peak, the signal-to-noise ratio at that point will be

$$SN_{TV} = 20 \log_{10} \left(\frac{0.7}{u_n} \right), \tag{4.36}$$

where u_n represents the r.m.s. noise voltage.

We should now recall that the expressions for the signal-to-noise ratio assume that the modulating signal is a sinewave with an amplitude such that it causes carrier peak deviation. For the television signal, at the video interconnection points, a 1 V peak-to-peak signals causes maximum frequency deviation and thus

$$S/N = 10 \log_{10} \left(\frac{0.5^2}{2u_n^2} \right)$$

Number of lines per image	525	625		
Maximum frequency of the video signal (in MHz)	4.2	5.0	5.5	6.0
Signal-to noise ratio improvement (in dB)	2.9	2.0	2.1	2.3

Table 4.8 Signal-to-noise ratio improvement using the ITU-R recommended pre-emphasis and de-emphasis.

$$= 20\log_{10}\left(\frac{0.5}{\sqrt{2}u_n}\right), \tag{4.37}$$

hence:

$$S/N_{TV} = S/N + 6. \tag{4.38}$$

4.6 NOISE WEIGHTING

4.6.1 Introduction

In most cases microwave radio link specifications make use of weighted signal-to-noise ratios (or weighted noise powers at the zero level point) since these relate better to circuit quality than the corresponding unweighted values. In the following paragraphs we describe the effects of noise weighting in telephone, sound and video broadcast signals.

4.6.2 Telephone signal

Noise measurements in telephone circuits should be made with a special measuring apparatus – psophometer – which weighs noise contributions at different frequencies according to ITU-T [13] Recommendation G.223, reproduced in Table 4.9 and Figure 4.7.

When a 0 dBm white noise power in the frequency band 300 to 3400 Hz is applied at the weighting circuit input, the output (weighted) noise power is −2.5 dBm. When the white noise frequency band is increased to 0 to 4000 Hz the (weighted) noise power output changes to −3.6 dBm.

Frequency (Hz)	Weighting coefficient (dB)
50	−63.0
100	−41.0
150	−29.0
200	−21.0
300	−10.6
400	−6.3
500	−3.6
600	−2.0
800	0.0
1000	+1.0
1200	0.0
1500	−1.3
2000	−3.0
2500	−4.2
3000	−5.6
3500	−8.5
4000	−15.0
5000	−36.0

Table 4.9 Noise weighting coefficients defined in ITU-T [13] Recommendation G.223.

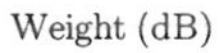

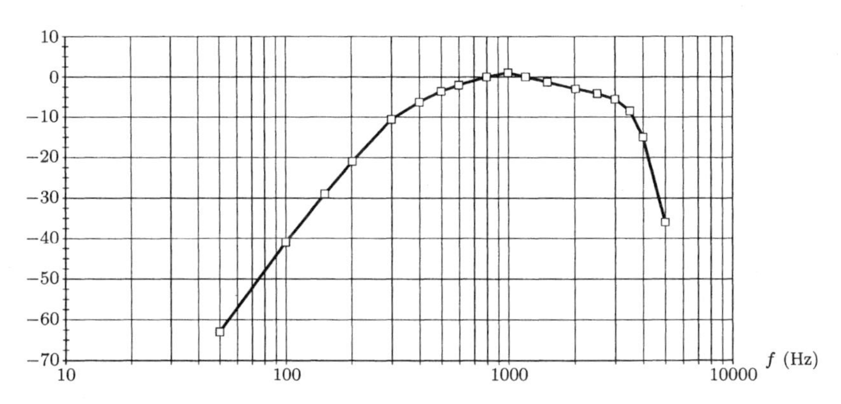

Fig. 4.7 Noise weight coefficients defined in ITU-T [13] Recommendation G.223.

4.6.3 Sound broadcasting

According to ITU-R [12] Recommendation BS.468-4 noise power in circuits used to transmit sound broadcast signals should be measured with a quasi-peak device, preceded by a weighting circuit. Frequency response of the weighting circuit is given in Table 4.10 and plotted in Figure 4.8.

Frequency (Hz)	Response (dB)
31.5	−29.9
63	−23.9
100	−19.8
200	−13.8
400	−7.8
800	−1.9
1000	0.0
2000	+5.6
3150	+9.0
4000	+10.5
5000	+11.7
6300	+12.2
7100	+12.0
8000	+11.4
9000	+10.1
10000	+8.1
12500	0.0
14000	−5.3
16000	−11.7
20000	−22.2
31500	−42.7

Table 4.10 Frequency response of the weighting circuit used for sound broadcasting signals according to ITU-R [12] Recommendation BS.468-4.

For calibration, a 0.775 V sinewave at 1000 Hz applied to the input (600 Ω impedance) of the measuring device should produce a reading of 0 dB. The noise power measured in this way is expressed in dBqps

The quasi peak noise power, irrespective of the weighting circuit, expressed in logarithmic units, is equal to the average noise power (expressed in the same units) plus 5 dB.

A signal level of 9 dBm (which corresponds to a 2.18 V sinewave voltage on 600 Ω), at the zero level point, and a noise power N, expressed in dBqps, correspond to a weighted signal-to-noise ratio S/N, expressed in dB, given by

$$S/N_{[\mathrm{dB}]} = 9 - N_{[\mathrm{dBqps}]}. \tag{4.39}$$

The use of the recommended weighting circuit increases the noise power as shown in Table 4.11, obtained from CCIR [8] Report 496-4 for the 5, 10 and 15 kHz circuits and from CCIR [4] Report 375-3 for the 7 kHz circuit. In

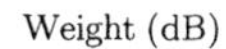

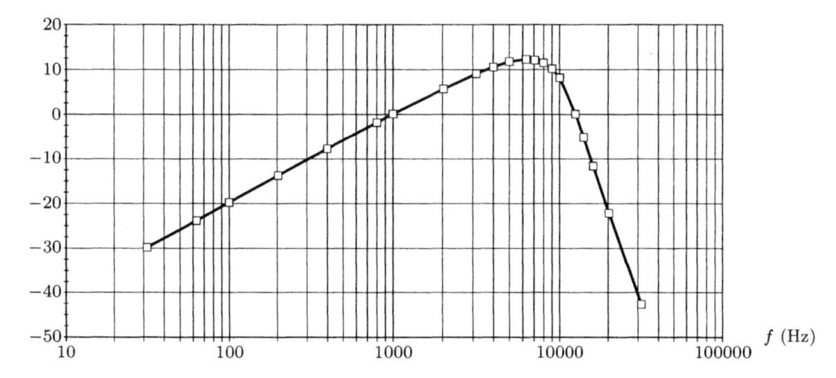

Fig. 4.8 Frequency response of the weighting circuit used for sound broadcasting signals according to ITU-R [12] Recommendation BS.468-4.

Table 4.11 the triangular noise corresponds to the noise output of a frequency demodulator.

Signal bandwidth (Hz)	White noise (dBm)	Triangular noise (dBm)
70 – 5000	2.3	9.1
50 – 7000	3.9	11.5
50 – 10000	4.8	10.6
40 – 15000	4.7	7.1

Table 4.11 Weighting circuit output noise power assuming an input noise power 0 dBm (ITU-R [8] Report 496-4 for the 5, 10 and 15 kHz circuits and ITU-R [4] Report 375-3 for the 7 kHz circuit).

4.6.4 Video signal

The weighting curve for the video signal, defined in CCIR [8] Recommendation 567-2, corresponds to an attenuation which, at the angular frequency ω, is given by

$$a(\omega) = \frac{1 + \left[(1 + \frac{1}{a})\omega\right]^2}{1 + \left(\frac{\omega\tau}{a}\right)^2}, \tag{4.40}$$

where:

- $a = 4.5$;

- $\tau = 245$ ns.

The use of the recommended weighting circuit decreases power noise as shown in Table 4.12 (for 625 line systems).

Video signal Frequency limits	White noise (dBm)	Triangular noise (dBm)
50 Hz and 5.0 MHz	-7.4	-12.2
50 Hz and 5.5 MHz	-7.7	-12.5
50 Hz and 6.0 MHz	-8.0	-12.8

Table 4.12 Video weighting circuit output noise for an input noise power of 0 dBm.

4.7 INTERMODULATION NOISE

In real systems, besides thermal noise we must consider all circuit imperfections which give rise to signal distortion. When the base band is occupied by a multiplex signal, distortion manifests itself in telephone channels as intermodulation noise. The main responsibles for intermodulation noise are:

- bandwidth limitation;
- modulator and demodulator non-linearity;
- changes in attenuation or group delay with frequency, within the signal bandwidth.

Most of these factors are outside the control of link designers and will not be further referred to here. The exception are changes in attenuation and group delay, caused by impedance mismatches between transmitter, receiver, antennas and waveguides which, given their importance in link design and installation, deserve a more detailed approach.

The analysis of intermodulation noise radio links loaded with multiplex signals is complex and we refer the interested reader to Medhurst[14] and Müller [15], cited by Carl [2]. Here we limit ourselves to the description of a procedure to calculate intermodulation noise due to linear and sinewave changes in the group delay given in Carl [2].

For a linear variation of the group delay $\Delta\tau$ (in ns) with a total delay $\Delta\tau$ (in ns) within the bandpass, the intermodulation noise n_i (in pW) is

$$n_i = 26\,\Delta\tau^2\,\Delta f^2\,y_{kl}, \tag{4.41}$$

where:

- Δf is the equivalent frequency deviation in MHz;
- y_{kl} is a coefficient between 0 and 1, which is a function of the modulating signal frequency.

The value of y_{kl} is dependent not only on the modulating signal frequency but also on the applied emphasis. For the ITU-R recommended emphasis we have

$$y_{kl} \approx 0.3 \frac{f}{f_{max}}, \tag{4.42}$$

where:

- f is the modulating frequency, that is, the center frequency of the telephone channel under consideration;
- f_{max} is the base band highest frequency.

For a sinewave variation of the group delay, with n oscillations with a peak-to-peak amplitude $\Delta\tau$, intermodulation noise is given by

$$n_i = \frac{41.5}{n^2} \Delta\tau^2 \, \Delta f^2 \, y_{kls}, \tag{4.43}$$

where:

- Δf is the equivalent frequency deviation in MHz;
- y_{kls} is a coefficient, function of n and of the modulating signal frequency.

In many cases y_{kls} obeys the following:

$$y_{kls} \leq 1.5 - 0.75 \frac{f}{f_{max}}. \tag{4.44}$$

Besides ground reflections and multipath fading, the main causes for changes in attenuation and group delay with frequency are impedance mismatches between transmitter, receiver, antenna and waveguides (or cables). This type of intermodulation noise is a function of:

- the number of telephone channels of the multiplex signal;
- the r.m.s. frequency deviation per channel;
- the waveguide length between the transmitter (or the receiver) and the antenna.

This type intermodulation noise may be estimated by:

$$n_{i_{[mW]}} = 10^{\frac{N_{it}}{10}} + 10^{\frac{N_{ir}}{10}}, \tag{4.45}$$

$$N_{it_{[dBm0p]}} = N_{io_T} - 2\, Ag_T - Ar_T, \tag{4.46}$$

$$N_{ir_{[dBm0p]}} = N_{io_R} - 2\, Ag_R - Ar_R, \tag{4.47}$$

where:

- N_{io_T} e N_{io_R} (in dBm0p) are given in Figure 4.9;
- Ag_T and Ag_R are the waveguide attenuation, in dB, between the transmitter and the antenna and the receiver and the antenna, respectively;
- Ar_T is the sum of the return attenuations, in dB, in the transmitter-waveguide and waveguide-antenna interfaces;
- Ar_R is the sum of the return attenuations, in dB, in the receiver-waveguide and waveguide-antenna interfaces.

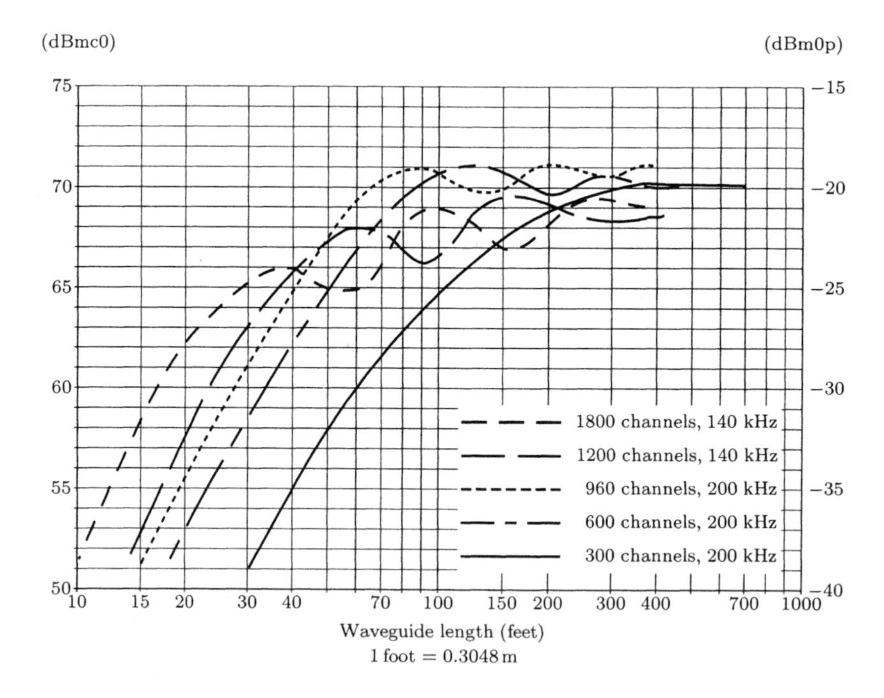

Fig. 4.9 Intermodulation noise due to mismatch between transmitter, receiver or antenna and waveguide, to be used in (4.41) (White [17]). Reproduced with permission from Valmont/Microflet.

The return loss A_r at an interface, due to a single impedance mismatch is given by

$$A_{r_{[dB]}} = -20 \log_{10} \left(\frac{vswr - 1}{vswr + 1} \right), \tag{4.48}$$

where

$$vswr = \frac{1+\mid\Gamma\mid}{1-\mid\Gamma\mid}, \tag{4.49}$$

$$\Gamma = \frac{z_1 - z_2}{z_1 + z_2}, \tag{4.50}$$

z_1 and z_2 being the impedances seen from the interface and $vswr$ the voltage standing wave ratio. If we substitute (4.49) and (4.50) in (4.48) we obtain

$$A_r = -20\log_{10}\left(\frac{z_1 - z_2}{z_1 + z_2}\right). \tag{4.51}$$

In large capacity (equal to or more than 300 voice channels) systems with long (more than about 10 to 20 meters) waveguides , intermodulation noise requirements lead to very strict mismatch specifications for the transmitter/receiver connection to the waveguide and from the latter to the antenna. As an example take the case of a microwave radio link for 960 telephone channels and a r.m.s. frequency deviation per channel $\Delta^1_{rms} = 200$ kHz, using a 60-m- long waveguide, where $Ag_T = Ag_R = 1.3$ dB and $vswr$=1.1. From Figure 4.9 we get $N_{io_T} = N_{io_R} = -19$ dBm0p. From (4.48) we have $A_r = 26.4$ dB which, from (4.46) and (4.47), lead to $N_{it} = -74.5$ dBm0p and $N_{ir} = -74.5$ dBm0p, respectively, and finally from (4.45) $n_i = 71.2$ pW0p. As will be shown later, this value is unacceptable for links with less than about 95 km, that is, for most links. Hence the importance of ensuring very low $vswr$ values.

4.8 INTERCONNECTION OF MICROWAVE RADIO LINKS

Interconnection of microwave radio links may be performed at three levels:

1. telephone channel;
2. base band;
3. intermediate frequency.

To facilitate interconnection at intermediate frequency level the center frequency of the intermediate frequency should follow the ITU-R [12] Recommendation F.403-3, reproduced in Table 4.13.

4.9 NOISE IN INTERCONNECTED LINKS

Most real-life microwave radio links have more than one hop, sometimes up to 50 hops or even more. Thus, in practice, we must consider interconnection of microwave radio links.

Link capacity expressed in telephone channels	Intermediate frequency center value (MHz)
12, 24, 60, 120	35 (for carrier frequencies below about 1.7 GHz) 70 (for carrier frequencies above about 1 GHz
300, 600, 960, 1260, 1800	70
2700	140

Table 4.13 Intermediate frequency center value as a function of link capacity (ITU-R [12] Recommendation F.403-3).

The quality of an analog link, taken as a quadrupole, within a specified bandwidth, derives from:

- the uniformity of the modulus of the transfer function;
- the linearity of the argument of the transfer function;
- the added noise.

It is well known that the transfer function of a quadrupole chain is the product of the transfer functions of the individual quadrupoles. If, as usual, the modulus of the individual transfer functions are expressed in logarithmic units, then the modulus of the overall transfer function is equal to the sum of the modulus of the individual transfer functions. Thus high quality in a long chain of quadrupoles implies very high quality in individual quadrupoles.

Assume that each of the m hops of a microwave radio link may be modelled by a quadrupole with a power gain g_i and a noise factor nf_i. Assume also that the quadrupole input and output impedances are matched. The overall power gain g is

$$g = \prod_{i=1}^{m} g_i. \tag{4.52}$$

When the input and the output are loaded with matched impedances, output power noise n_o g is

$$n_o = g\,(n_i + n_{int}), \tag{4.53}$$

where:

- n_i is the input noise power;
- n_{int} is the internal noise power, assumed at the input.

Manipulating (4.53) we get

$$n_o = g\, n_i \left(1 + \frac{n_{int}}{n_i}\right), \tag{4.54}$$

or

$$n_o = g\, n_i\, nf, \tag{4.55}$$

with

$$nf = 1 + \frac{n_{int}}{n_i}, \tag{4.56}$$

where nf is the quadrupole noise factor.

Equation (4.56) shows that the noise factor is not a constant and depends on the input noise power. For meaningful comparisons, and when nothing else is specified, input noise power n_i is taken as white thermal noise, at ambient temperature ($T_i = 293K$), given in (4.12) for a quadrupole bandwidth b.

An equivalent expression for the noise factor may be obtained if, in (4.56), we express the internal noise in terms of the equivalent noise temperature T_{int}. From (4.12) we get

$$n_{int} = K_B\, T_{int}\, b, \tag{4.57}$$

and

$$n_i = K_B\, T_i\, b, \tag{4.58}$$

and, substituting (4.57) and (4.58) in (4.56),

$$nf = 1 + \frac{T_{int}}{T_i}. \tag{4.59}$$

Noise factor may be used to interrelate input and output signal-to-noise ratios. Let s_i and s_o be the input and the output signal power, respectively. By definition of power gain g

$$s_o = g\, s_i. \tag{4.60}$$

From (4.55) and (4.60) output signal-to-noise ratio $(s/n)_o$ is

$$(s/n)_o = \frac{g\, s_i}{g\, nf\, n_i},$$

or, if we represent input signal-to-noise ratio as $(s/n)_i$,

$$(s/n)_o = \frac{(s/n)_i}{nf}. \tag{4.61}$$

As it will be shown later on, we often have to calculate the noise power from the signal-to-noise ratio. Let s be the signal power at a given point. The noise power n, at the same point, may be written in term of the signal to noise ratio s/n as

$$n = \frac{s}{s/n}, \tag{4.62}$$

or, if the signal-to-noise ratio is expressed in logarithmic units,

$$n = s \times 10^{-\frac{S/N}{10}}. \tag{4.63}$$

When, as is the case for telephone channels, the reference power is equal to 1 mW, equation (4.63) may be rewritten as

$$n_{pW} = 10^{\frac{90-S/N}{10}}. \tag{4.64}$$

If we chain-link two quadrupoles with power gains g_1 and g_2 and noise factors nf_1 and nf_2 the noise power at the output of the first quadrupole will be

$$n_{o_1} = n_i\, g_1\, nf_1, \tag{4.65}$$

where the reference input noise power for the first quadrupole n_i is given in (4.12).

Assuming that the internal noise of the second quadrupole is uncorrelated with the output noise of the first quadrupole, at the input of the second quadrupole, including its own internal noise, we will have

$$n_{i_2} = n_{o_1} + n_{int_2}. \tag{4.66}$$

If the quadrupoles have the same bandwidth, the internal noise power of the second quadrupole may be written as a function of its noise factor. Thus from (4.57) and(4.59) we get

$$n_{int_2} = n_i(nf_2 - 1). \tag{4.67}$$

Substituting (4.65) and (4.67) in (4.66) we get

$$n_{i_2} = n_i\, g_1\, nf_1 + n_i(nf_2 - 1),$$

or

$$n_{i_2} = n_i\, g_1 \left(nf_1 + \frac{nf_2 - 1}{g_1} \right). \tag{4.68}$$

From (4.68) the output noise power of the second quadrupole becomes

$$\begin{aligned} n_{o_2} &= g_2\, n_{i_2} \\ &= n_i\, g_1\, g_2 \left(nf_1 + \frac{nf_2 - 1}{g_1} \right). \end{aligned} \tag{4.69}$$

Since the gain of a chain with two quadrupole is the product of the gains of the individual quadrupoles, from (4.65) and (4.69) the noise factor nf of the chain becomes

$$nf = nf_1 + \frac{nf_2 - 1}{g_1}. \tag{4.70}$$

The previous expression is easily extended to a chain with m quadrupoles

$$nf = nf_1 + \frac{nf_2 - 1}{g_1} + \frac{nf_3 - 1}{g_1 g_2} + ... + \frac{nf_m - 1}{g_1 g_2 ... g_{m-1}}. \tag{4.71}$$

Since for all quadrupoles that represent hops in a microwave radio link we have $g_i = 1$ for all $i = 1, \ldots, m$,

$$\begin{aligned} g &= 1, \\ nf &= 1 + \sum_{i=1}^{m} (nf_i - 1). \end{aligned} \tag{4.72}$$

When we chain connect microwave radio links we usually do not work with individual output signal-to-noise ratios $(s/n)_o$, nor with noise factors, but rather with the output noise powers n_o of each link, assumed to be on its own.

For a unity gain microwave radio link output noise power is equal to the reference input noise power plus the internal noise power. In most cases the internal noise power is many orders of magnitude higher than the reference input noise power.

The previous statement may be easily demonstrated with a simple example. Assume a 3.1 kHz bandwidth telephone channel, with a very high output signal-to-noise ratio, say 100 dB. Since the test signal power is 1 mW, a 100 dB signal-to-noise ratio leads to a noise power of 0.1 pW. The reference (thermal) white noise power, at ambient temperature (293 K) in a 3.1 kHz bandwidth is 0.0000123 pW, four orders of magnitude below the internal noise. For more realistic values of the signal-to-noise ratio, that is lower than the assumed value of 100 dB, the internal noise power would be even higher than 0.1 pW.

Stating that the internal noise power is much higher than the reference noise power is equivalent to stating that the noise factor is much larger than unity and thus (4.72) may be simplified to

$$nf = \sum_{i=1}^{m} nf_i. \tag{4.73}$$

Taking into account the relation between noise factor and input and output signal-to-noise ratios (4.61) we have

$$(s/n)_o = \frac{(s/n)_i}{\sum_{i=1}^{m} nf_i}, \tag{4.74}$$

or, recalling that the input and output signal powers are the same and that the reference power n_i is the same for all elements of the chain (which have the same signal bandwidth),

$$n_{o_n} = n_i \left(\sum_{i=1}^{m} nf_i \right). \tag{4.75}$$

From (4.55) the output noise power for the ith link is

$$n_{o_i} = n_i \, nf_i, \tag{4.76}$$

and thus (4.75) may be written as

$$n_{o_m} = \sum_{i=1}^{m} n_{o_i}. \tag{4.77}$$

Expression (4.77) may be expressed as follows: *In a microwave radio link chain the output noise power is equal to the sum of the output noise power of each element in the chain, assumed to be alone.*

Since signal power obeys

$$s_{o_m} = s_{o_{m-1}} = \dots = s_{o_2} = s_{o_1}, \tag{4.78}$$

from (4.77) we get

$$\left(\frac{n}{s}\right)_{o_m} = \sum_{i=1}^{m} \left(\frac{n}{s}\right)_{o_i},$$

hence

$$\frac{1}{(s/n)_{o_m}} = \sum_{i=1}^{m} \frac{1}{(s/n)_{o_i}}, \tag{4.79}$$

which may be stated as: *In a microwave radio link chain, the inverse of the output signal-to-noise ratio is equal to the sum of the inverses of the signal to noise ratios of each elements, assumed to be alone.*

It should now be clear from (4.77) and (4.79) that, in a microwave radio link chain, it is much simpler to work with noise powers than with signal-to-noise ratios.

4.10 THE HYPOTHETICAL REFERENCE CIRCUIT

4.10.1 Introduction

As shown in the previous section, the output signal to noise ratio, or the output noise power of a microwave radio link is dependent on its nature. Thus ITU-R, before prescribing the quality criteria for international links[4],

[4]Even if the ITU-R recommendations are only strictly applicable to international links many administrations, and telecom operators, particularly in Europe, also enforce them in the national links.

defined a hypothetical reference circuit, whose layout depends on the nature of the transmitted signal.

4.10.2 Telephone signal

For line-of-sight, or near line-of-sight, links using frequency division multiplex, the hypothetical reference circuit is 2500 km long and its layout varies, according to link capacity (expressed by the number of telephone channels).

For capacities between 12 and 60 telephone channels, in each direction, ITU-R Recommendation F.391 defines a hypothetical reference circuit composed of:

- 3 telephone channel modulator pairs;
- 6 primary group modulator pairs;
- 6 secondary group modulator pairs;
- 6 base band signal modulator pairs, that divide the circuit in 6 equal length homogenous sections;

which is schematically represented in Figure 4.10. A modulator pair is defined as the set consisting of a modulator and a demodulator.

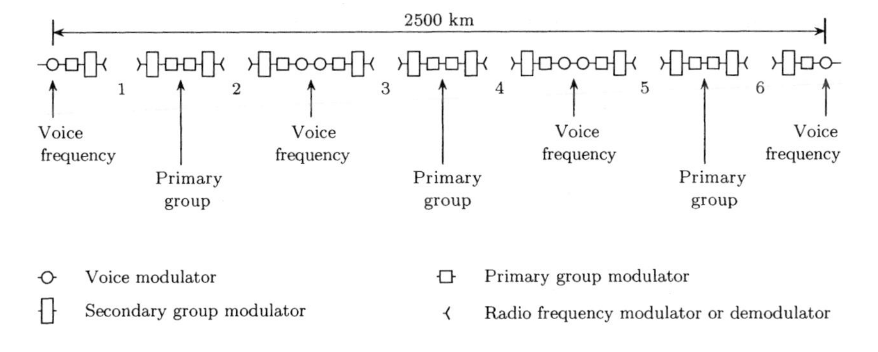

Fig. 4.10 Hypothetical reference circuit for microwave links with capacity from 12 to 60 telephone channels (ITU-R [12] Recommendation F.391).

For capacities in excess of 60 telephone channels in each direction, ITU-R[12] Recommendation F.392 defines a hypothetical reference circuit (see Figure 4.11) composed of:

- 3 telephone channel modulator pairs;
- 6 primary group modulator pairs;
- 9 secondary group modulator pairs;

- 9 base band signal modulator pairs, that divide the circuit in 9 equal length homogenous sections.

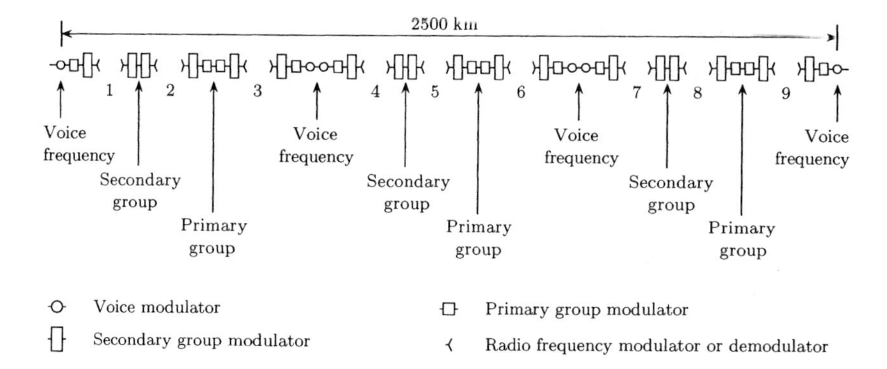

Fig. 4.11 Hypothetical reference circuit for microwave links with a capacity higher than 60 telephone channels (ITU-R [12] Recommendation F.392).

For transhorizon radio links (with propagation by diffraction or troposcatter) the hypothetical reference circuit, defined in ITU-R [12] Recommendation F.396-1 is 2500 km long, undivided, because hops may have very different lengths. If d is a hop length, the hypothetical reference circuit should be made up of the nearest integer of $\frac{2500}{d}$ identical sections. According to this Recommendation in each direction the hypothetical reference circuit should be composed of:

- 3 telephone channel modulator pairs;
- 6 primary group modulator pairs;
- 6 secondary group modulator pairs.

4.10.3 Sound broadcasting

The hypothetical reference circuit for ground links, as defined in ITU-T [13] Recommendation J.11 is 2500 km long between audio frequency terminals. Between these terminals there are two intermediate audio frequency points that divide the circuit in three equal length parts.

4.10.4 Television broadcasting

The hypothetical reference circuit for television for terrestrial links is 2500-km-long between video terminals (Recommendation J.61). Between these terminal points there are two intermediate points, at video frequency, that divide the reference circuit in there equal length paths. The reference circuit does not include inter-standard converters, nor regeneration of synchro pulses, nor devices to insert signals in line or in frame suppression periods.

4.11 NOISE IN THE HYPOTHETICAL REFERENCE CIRCUIT

4.11.1 Telephone signal

For frequency division multiplex microwave radio links the allowable noise on the hypothetical reference circuit is defined in ITU-R [12] Recommendation F.393-4 as follows:

- Noise power, at the zero level point, in the worst channel, should not exceed:

    1. 7500 pW0p, one minute averaged weighted noise power, for more than 20 percent of the worst month;
    2. 47 500 pW0p, one minute averaged weighted noise power, for more than 0.1 percent of the worst month;
    3. 1 000 000 pW0, unweighted noise power, with an integration time of 5 ms, for more than 0.01 percent of the worst month;

- For one or more homogeneous sections of the hypothetical reference circuit the one minute averaged weighted noise power, not exceeded for more than 20 percent of the worst month, should be proportional to the number of homogeneous sections;

- For one or more homogeneous sections of the hypothetical reference circuit, the percentage of time during which the one minute averaged weighted noise power should not exceeded 47 500 pW0p and the unweighted noise power, with an integration time of 5 ms, should not exceed 1 000 000 pW0, should be proportional to the number of homogeneous sections.

ITU-R [12] Recommendation F.393-4 , a ITU-R draws the attention to the following points:

- Noise power referred to above does not include the multiplex noise, which for the hypothetical reference circuit should not exceed 2500 pW0p;

- Noise from different sections of the hypothetical reference circuit add up in power;

- Recommendation F.393-4 only applies when the system is available;

- A system is unavailable when, in any channel, noise power exceeds 1 000 000 pW0, for more than 10 consecutive seconds (ITU-R [12] Recommendation F.557-4). According to the same Recommendation the hypothetical reference circuit should be available for more than 99.7 percent of the time.

Before proceeding to the noise objectives for sound and video broadcasting links, it is appropriate to explain the reasons behind ITU-R Recommendation F.393-4 [12]:

- The one minute averaged weighted noise power was chosen by the CCITT Study Group XII (in charge of telephone quality), because it is well correlated with the degree of inconvenience felt by users as regards to noise;

- To divide the noise power, not exceed for more than 20 percent of the time, proportionally to the number of homogenous sections, is equivalent to assuming that the 20 percent of the time in different sections may occur simultaneously, which is a reasonable assumption for links up to a few hundred kilometers, but it is, probably, pessimistic for a 2500 km link;

- For the noise power, not exceeded for more than 0.1 percent of the time, taking into account that this noise power is due to fast fading, it is unreasonable to assume simultaneous fading in consecutive adjacent noise sections;

- Given that 47 500 pW0p is a much higher value than 7500 pW0p, it is acceptable to approximate the total noise, under these conditions, to the noise in the section under deep fading, and to neglect the contribution of the other sections and thus, for the hypothetical reference circuit, the total noise is the same (47 500 pW0p) as the section under deep fading, and the periods of occurrence of deep fading for each section should be add up.

For transhorizon links, the maximum allowable noise in the hypothetical reference circuit (ITU-R [12] Recommendation F.397-3) depends on the feasibility of alternate solutions. If these are deemed possible, then Recommendation F.393-4 applies. Otherwise:

1. The one minute averaged weighted noise power should not exceed 25 000 pW0p, during more than 20 percent of the time, during the worst month;

2. The one minute averaged weighted noise power should not exceed 63 000 pW0p, for more than 0.5 percent of the time, during the worst month;

3. The unweighted noise power, with an integration time of 5 ms, should not exceed 1 000 000 pW0, during more than 0.05 percent of the time, during the worst month.

4.11.2 Sound broadcasting

Noise objectives for microwave radio links used to transmit sound broadcasting signals, measured according to ITU-R [12] Recommendation BS.468-4, that shall me met during at least 80 percent of the time during the worst month, are given in Table 4.14. These noise values may be exceeded by +4 dB and +12 dB during no more than 1 and 0.1 percent of the worst month, respectively.

Type of circuit	ITU-T Recommendation	Noise objective (dBq0ps)
7 kHz	J.23	−44
15 kHz	J.21	−42

Table 4.14 Noise objectives for international circuits used to transmit sound broadcasting signals.

For the one-minute weighted noise levels (including multiplex noise) in a telephone channel, which should not be exceeded for more than 20 percent and 0.1 percent of the worst month, that is, 10 000 and 50 000 pW0p, respectively, the corresponding noise levels in a sound broadcasting circuit are given in Table 4.15.

Analysis of Tables 4.14 and 4.15 shows that it is not possible to meet the quality specifications for sound broadcasting signals without the use of a compressor-expander

For circuits using a subcarrier of a microwave television link, according to ITU-R [12] Recommendation F.402-2, CCIR [9] Report 375-3 suggests the noise objectives given in Table 4.16

Type of circuit (kHz)	One-minute averaged weighted noise, 20 % of worst month		One-minute averaged weighted noise, 0.1 % of worst month	
	without ce (dBq0ps)	with ce (dBq0ps)	without ce (dBq0ps)	with ce (dBq0ps)
7	−34.6	−46.6	−27.6	−39.6
15	−36.2	−48.2	−29.2	−41.2

Note: ce - compressor expander

Table 4.15 One-minute mean value of weighted noise levels in a sound broadcasting circuit established on a analog microwave radio link, for which the one-minute weighted noise levels (including multiplex noise) in a telephone channel, correspond to those which should not be exceeded for more than 20 percent and 0.1 percent of the worst month, that is, 10 000 and 50 000 pW0p, according to CCIR [9] Report 375-3.

According to ITU-R [12] Recommendation F.402-2 the main features of television microwave radio links using a subcarrier for the transmission of sound broadcasting signals are as follows:

- sound subcarrier frequency: 7.5 MHz;
- maximum level of the sound signal at the zero level point (relative to 0.775 V on 600 Ohm): 9 dB;
- sound bandwidth: 30 to 10000 Hz (or more, by agreement between the administrations);
- subcarrier r.m.s. frequency deviation: 140 kHz (for a sinewave signal with a level corresponding to the maximum sound signal level);
- pre-emphasis, to be agreed between administrations;
- unmodulated subcarrier amplitude should be such that it leads to a carrier r.m.s. frequency deviation (at intermediate or radio frequency) of 300 kHz.

4.11.3 Video signal

For the hypothetical reference circuit ITU-R [12] Recommendation F.555-1 specifies that the ratio, in dB, between the nominal amplitude of the luminance signal and the weighted (according to ITU-T [13] Recommendation J.61) r.m.s. noise should not be less than:

1. 57 dB, during more than 20 percent of the time, during the worst month
2. 45 dB, during more than 0.1 percent of the time, during the worst month.

The division of noise among the sections of the hypothetical reference circuit is performed as per ITU-R [12] Recommendation F.393-4.

Type of circuit (kHz)	**One-second averaged weighted noise level, in dBq0ps, exceed during the fraction of time, during the worst month**		
	20 %	**1 %**	**0.1 %**
10	−39	−35	−27
15	−42	−38	−30

Table 4.16 Hypothetical reference circuit noise objectives for sound broadcasting signals transmitted in a subcarrier of a television microwave radio link (CCIR [9] Report 375-3.

In a note annex to Recommendation F.555-1 it is mentioned that the value of 57 dB, for 20 percent of the time, should be the objective in the absence of fading, a different position from the one adopted in Recommendation F.393-4, which goes some way to justify that 20 percent simultaneous fading in all sections is indeed pessimistic.

4.12 NOISE IN REAL CIRCUITS

One minute averaged, weighted, noise power, at the zero level point, in the worst channel of a frequency division multiplex line-of-sight microwave radio link, should not exceed the following limits (ITU-R [12] Recommendation F.395-2):

- In links with length d (in km), between 280 and 2500 km, that are not very different from the hypothetical reference circuit:
  1. $3d$ pW0p, for more than 20 percent of the time, during the worst month,
  2. 47 500 pW0p, for more than $(d/2500) \times 0.1$ percent of the time, during the worst month;
- In links that are considerably different from the hypothetical reference circuit:
  - For $50 \leq d \leq 840$:
    1. $3d + 200$ pW0p, for more than 20 percent of the time, during the worst month,
    2. 47 500 pW0p, for more than $(280/2500) \times 0.1$ percent of the time, during the worst month, when $d < 280$ or $(d/2500) \times 0.1$ percent during the worst month when $d > 280$;
  - For $840 \leq d \leq 1670$:
    1. $3d + 400$ pW0p, for more than 20 percent of the time, during the worst month,
    2. 47 500 pW0p, for more than $(d/2500) \times 0.1$ percent of the time, during the worst month;
  - For $1670 \leq d \leq 2500$:
    1. $3d + 600$ pW0p, for more than 20 percent of the time, during the worst month,
    2. 47 500 pW0p, for more than $(d/2500) \times 0.1$ percent of the time, during the worst month.

Noise powers referred to above include all types of noise, except multiplex noise which, for the hypothetical reference circuit, should not exceed 2500 pW0p.

For transhorizon microwave radio links, ITU-R [12] Recommendation F.593 specifies that, for a d km long link, one minute averaged, weighted, noise power, at the zero level, on the worst channel, should not exceed:

1. $10d$ pW0p for more than 20 percent of the time during the worst month;

2. 63 000 pW0p for more than $(d/2500) \times 0.5$ percent of the time during the worst month.

As in line-of-sight links, noise powers in troposcatter links include all types of noise except multiplex noise which should not exceed 2500 km, for the hypothetical reference circuit.

4.13 NOISE DISTRIBUTION

Distribution of noise powers among the different origins, also known as noise budget, is a link designer responsibility. For multiplex telephony the 7500 pW0p pertaining to the 20 percent clause may be splitted as follows:

- 50 percent, for thermal noise;

- 50 percent, for intermodulation noise.

In very long hops, or when particularly difficult propagation conditions are met, thermal noise contribution may be increased at the expense of intermodulation noise.

At the predesign stage the following intermodulation noise budget may be used:

- 25 percent, for modulators and demodulators;

- 25 percent, for varying group delay;

- 50 percent, for impedance mismatches.

At the design or at the implementation stages intermodulation noise budget may change according to the equipment chosen.

We should note that since the intermodulation noise in independent of fading[5] in the 0.1 percent clause, with the noise budget above, we have 43750 pW0p available for thermal noise.

[5]Neglecting the influence of changes in the group delay, which may be unacceptable in some cases.

4.14 MEASURING NOISE

To check that noise specifications are met, an agreed measuring procedure must be used. If, for single channel systems (telephone, sound broadcast or video broadcast), noise measurement appears to be relatively simple, the same is not true for multiplex systems where results are heavily dependent on system load or, to state it differently, on the influence of other channels on the channel under measurement.

The most frequently used noise measurement procedure is the noise power ratio, also known as *NPR* where multiplex signals are simulated with Gaussian noise, with the appropriate bandwidth.

To measure system noise, at first transmitter and receiver are brought side by side and interconnected through the antenna terminals, using an attenuator to simulate global attenuation. With a generator of white, Gaussian noise, followed by a band limiting filter, we simulate the multiplex signal with the required power (see Section 4.2). The transmitter and the receiver are adjusted so that the received signal power, at the base band, is equal to the transmitted signal at an equivalent point.

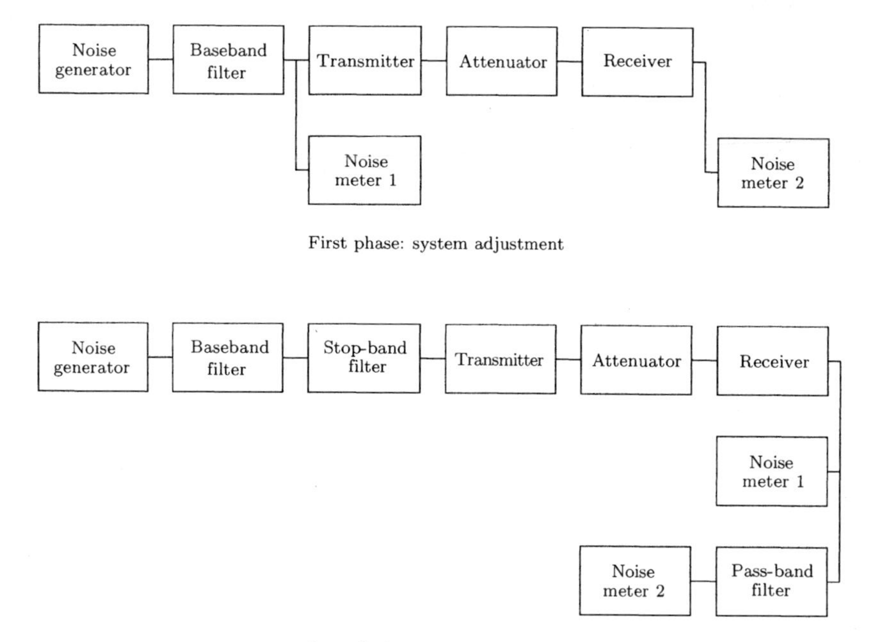

Fig. 4.12 *NPR* measurement.

With the same system load N_1 (in dBm0), a high quality stop band filter that removes the signal in the telephone channel under test is inserted, in series with the multiplex signal. On the receiver side a bandpass filter, tuned to the same telephone channel, enables the measurement the output noise power N_2 (in dBm0).

Noise power ratio *NPR* is defined as the ratio of the system load to the output noise power of the bandpass filter. In logarithmic units we have

$$NPR = N_1 - N_2. \tag{4.80}$$

Noise power N_2 is both due to thermal noise and to intermodulation noise (from modulators and demodulators), but it does not include intermodulation noise due to:

- impedance mismatches between at the interfaces between transmitter, waveguide and antenna nor receiver, waveguide and antenna;
- varying group delay due to propagation conditions.

Noting that the reference test signal, for the signal to noise ratio in a telephone channel, has a 0 dBm level, the unweighted signal to noise ratio, for a 4000 Hz channel, S/N may be derived from the measured *NPR* value as

$$S/N = NPR + 10\log_{10}\left(\frac{b_b}{4000}\right) - N_1, \tag{4.81}$$

where b_b is the multiplex signal bandwidth.

The weighted signal to noise ratio S/N_p is

$$S/N_p = S/N - 3.6. \tag{4.82}$$

Noise power may be derived from S/N_p using (4.64).

REFERENCES

1. Bell Telephone Laboratories Inc., *Transmission Systems for Communications*, 4th edition, USA, 1971.

2. Carl, H., *Radio Relay Systems*, translated from the original in German *Richtfunkverbindungen*, Macdonald and Co., London, 1966.

3. CCIR, *Recommendations et Rapports du CCIR, 1986*, Volume I, *Utization du spectre et contrôle des émissions*, UIT, Genève, 1986.

4. CCIR, *Recommendations et Rapports du CCIR, 1986*, Volume IX, Partie 1, *Service fixe utilisant les faisceaux hertziens*, UIT, Genève, 1986.

5. CCIR, *Recommendations of the CCIR, 1990*, Volume IX, Part 1, *Fixed Service Using Radio-Relay Systems*, UIT, Geneva, 1990

6. CCIR, *Recommendations et Rapports du CCIR, 1986*, Volume X-1, *Service de radiodiffusion (sonore)*, UIT, Genève, 1986.

7. CCIR, *Recommendations et Rapports du CCIR, 1986*, Volume XI-1, *Service de radiodiffusion (télévision)*, UIT, Genève, 1986.

8. CCIR, *Avis et Rapports du CCIR, 1986*, Volume XII, *Transmission de signaux de radiodiffusion sonore et de télévision sur une grande distance (CMTT)*, UIT, Genève, 1986.

9. CCIR, *Reports of the CCIR, 1990, Fixed Service Using Radio-Relay Systems*, UIT, Geneva, 1990.

10. Freeman, R. L. (1975). *Telecommunication Transmission Handbook*, John Wiley & Sons, New York, 1975.

11. Holbrook, B. D. and Dixon, J. T. , Load rating theory for multi-channel amplifiers, *Bell Systems Technical Journal, 18*, October, pp. 624–644, 1939.

12. ITU, *ITU-R Recommendations on CD-ROM*, UIT, Geneva, 2000.

13. ITU, *ITU-T Recommendations on CD-ROM*, UIT, Geneva, 1997.

14. Medhurst, R. G. , Explicit Form of FM Distortion Products with White Noise Modulation. *Proceedings of the Institute of Electrical Engineers*, Vol. 107, Part C, pp. 120–126, 1960.

15. Müller, M. , Die Zeigermethoe, ein anschauliches Verfahren zur Behandlung von Verzerrungen der Kleinhub-FM mit Anwendug auf FM-Richtfunk, *Archiv der elektrichen Ubertrag*, Vol. 16, pp. 25 and pp. 93, 1962.

16. Schwartz, M., *Information Transmission, Modulation, and Noise*, McGraw-Hill, New York, 1980.

17. White, R., *Engineering considerations for microwave communication systems*, 3rd edition, GTE Lenkurt Inc., San Carlos, CA., 1975.

Problems

4:1 Compare the multiplex peak power level for 960 analog telephone channels according to the ITU-T Recommendation G.223 and the Bell Telephone Laboratories.

4:2 Calculate Carson's bandwidth for a 600 channels frequency division multiplex signal, according to ITU-R Recommendations.

4:3 Calculate the noise power (in dBW) at the demodulator input, referred to the antenna terminals, for a 960 channels frequency division multiplex receiver with a 5 dB noise factor.

4:4 Calculate the unweighted channel output signal-to-noise ratio for the receiver of the previous problem, assuming the recommended ITU-T emphasis and a receiver input signal of −80 dBW.

4:5 Calculate the weighted output signal-to-noise ratio in worst channel of a 600 channel frequency division multiplex receiver for a carrier to noise ratio of 35 dB at the input of the demodulator.

4:6 Calculate the weighted output signal-to-noise ratio of a frequency modulated 525 lines TV receiver with a 6 dB noise factor, according to the relevant ITU Recommendations.

4:7 Calculate the intermodulation noise due to antenna feeder mismatch causing a voltage standing wave ratio of 1.10 on a 960 channel link with 50 m long feeders at both ends each with an attenuation of 4 dB.

4:8 Consider two microwave radio links which provide a 55 and 58 dB output signal-to-noise ratio. Calculate the output signal-to-noise ratio of the chain connection of these two links.

4:9 Calculate the equivalent noise factor of a receiver which provides a weighted 80 dB signal-to-noise ratio for the worst 3.1 kHz telephone channel at the zero point level (where the test signal has 1 mW power level)

4:10 Consider a 960 channel frequency division multiplex microwave link equipment, at 8 GHz, with transmitter output power of 1 W and a receiver noise figure of 5 dB. Neglecting atmospheric excess attenuation and assuming that fading follows Morita's law and that thermal noise is 50 percent of the maximum noise allowed under ITU-R Recommendation F.395-2 calculate the maximum hop length for transmitter and receiver antennas limited to 3.0 m diameter parabolic reflectors with 50 percent aperture efficiency

4:11 Using data from the previous problem and assuming a 0.01 percent rainfall of 50 mm/h, for a 60 km hop, calculate the worst channel unweighted noise power not exceeded for more than 0.01.

5
Digital Links

5.1 INTRODUCTION

Availability of high bit rates and competitive costs make digital microwave radio links more and more common. Currently all new microwave radio links are digital, particularly for multiplex telephony. For television signals, analog links were still the rule until the beginning of the 1990s, since studio quality digitalization of video signals, according to ITU-R [16] Recommendation BT.601-5, leads to a bit rate of 216 Mbit/s, higher than what was available in current radio links. The use of compression techniques for television signals reduces the transmission speed to 34 Mbit/s at most[1] and, in the market, there are already codecs able to fulfill specifications for contribution[2] links. With the recent standardization of compression techniques, new digital links for television are bound to become standard.

From the early 1980s onwards digital microwave radio links have been invariably preferred over analog links for transmission speeds up to 34 Mbit/s or 45 Mbit/s. For higher transmission speeds (140, 155, 560 and 620 megabits per second) the limited bandwidth available, varying attenuation and propagation delay due to multipath fading (usually known as selective fading) forces the use of complex modulation procedures, space diversity, adaptive equalizers,

[1]MPEG2 codecs achieve a picture quality similar to ITU-R BT.601-5 with a bit rate in the range from 8 to 10 Mbit/s.

[2]Contribution links transmit signals which may still undergo studio processing before broadcasting. On the other hand, distribution links are used to transmit signals from the studio to the transmitter.

and even path link shortening, all of which initially made digital microwave radio links of higher capacity less attractive. Nowadays, with the advances in high-speed electronics, complex modulation schemes, and improved equalizers, digital links are almost exclusively used.

Digital microwave radio links compete with optical fiber communications systems which achieve higher bit rates (of the order of some tens of Gbit/s per wavelength) but require a physical link between terminals. Microwave radio links can be installed in far less time and are more independent of accesses and path features. They also tend to be more robust and less affected by haphazards (natural or human made) .

Figures 5.1 and 5.2 represent simplified schematic block diagrams of a microwave radio links transmitter and a receiver, respectively.

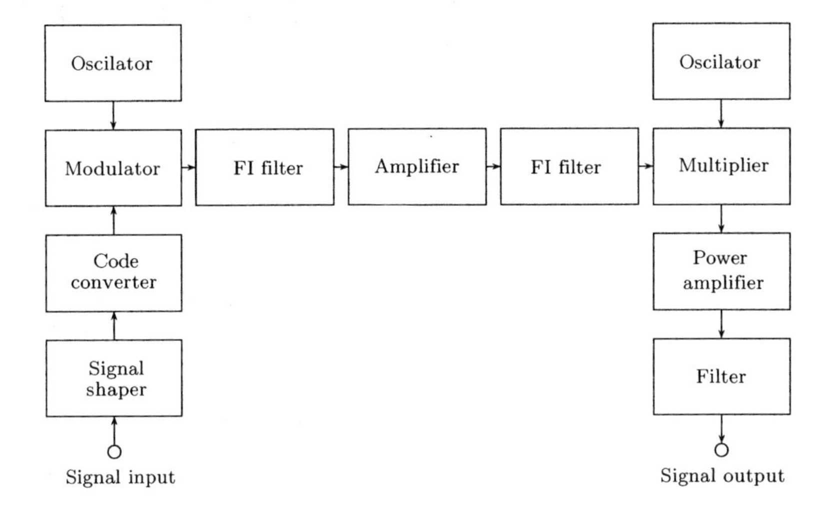

Fig. 5.1 Simplified block diagram of a digital transmitter used in a microwave radio link.

In the transmitter, the base band signal, after filtering or shape modification, is converted from the code used for transmission into another code, more appropriate for modulation. The occupied bandwidth may be limited with a filter before final amplification.

In the receiver, the base band is recovered by a demodulator that makes use of a detector and a decision circuit. The detector may require a signal with the same frequency and phase of the carrier – coherent detection – , or it may be based on the characteristics of the transmitted signal such as its energy or its frequency – incoherent detection – independently of any phase reference. Normally, after detection there is a decision process which converts the recovered signal into a bit stream. This procedure requires the knowledge

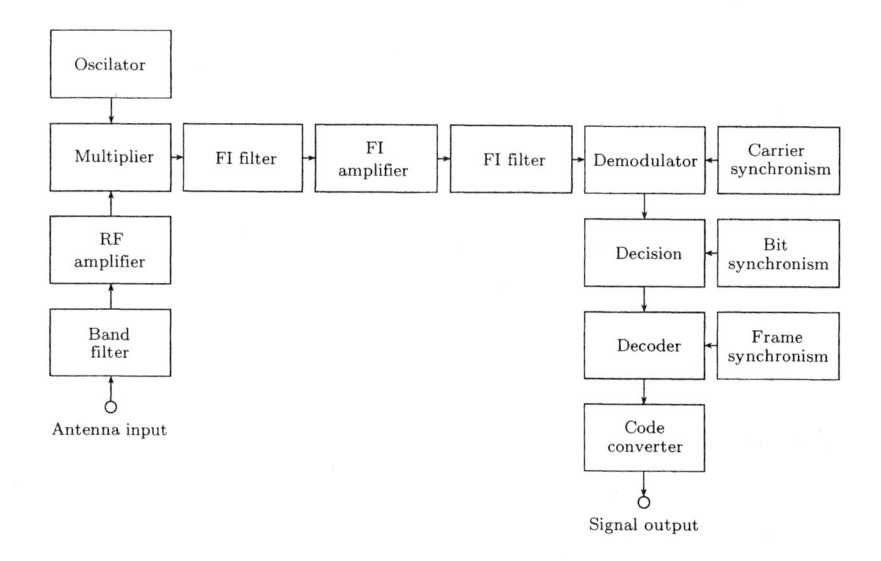

Fig. 5.2 Simplified block diagram of a digital receiver used in a microwave radio link.

of the best time in which to identify the transmitted bit. This is achieved by synchronizing the decision process with the received signal.

In most modulation methods decisions may be taken bit by bit, without any inconvenience. However, for some modulation techniques, it is possible to improve results, by analyzing the signal during an observation time corresponding to a number of bits.

Once the receiver has reconstructed the bit sequence it must identify signals therein, that is, it must synchronize the frame. After frame synchronism the code may be changed in order to adapt it to the transmission system, from the receiver to the final user.

Error detection correction codes are almost always used in the transmitted bit stream to enable coutinuous monitoring of link quality. In some cases error correcting codes may be used to improve link quality, by decreasing the bit error ratio.

In telecommunication systems most transmitted signals have their origin and destination as analog signals, independently from the fact that their transmission and switching may be carried-out digitally. Thus, in the link from the origin to the destination, conversion from analog to digital and from digital to analog assumes an important (and sometimes) decisive role.

This chapter starts with an analysis of the analog to digital and the digital to analog conversion procedures in order to establish the parameters – bit rate and bit error ratio – required for a given quality, defined in terms of bandwidth and signal-to-noise ratio of the final analog signal. Following digital codes used both in cable and wireless systems together with the ITU-R recommended digital hierarchies are described briefly. After reviewing the spectral properties of digital signals the most frequently used modulation techniques are recalled and, for each one, bandwidth and bit error probability are derived. The effects of selective fading are described and the chapter ends with a discussion of quality standards applicable to international links.

We assume that the reader is familiar with the theory of modulation and demodulation of digital signals in the presence of noise and in many cases results are presented with no demonstration.

5.2 ANALOG TO DIGITAL CONVERSION

5.2.1 Introduction

There are basically two procedures for analog to digital conversion:

1. sampling, quantization and codification, also known as pulse coded modulation or simply PCM, even if this strictly refers to the coding of pulses;

2. delta modulation, or DM.

Besides these two basic procedures, there quite a few others, which are here simply listed. The interested reader is referred to [18, 20]:

- sampling, quantization and pulse differential coding, also known as differential pulse coded modulation, or DPCM, where coding applies not to each pulse after quantization but rather to the difference between the quantized pulse and a prediction of this pulse based on the previous pulse;

- sampling, quantization and adaptive pulse coding, or APCM;

- sampling, quantization and adaptive differential pulse coding, or AD-PCM;

- adaptive delta modulation, or ADM;

- continuous delta modulation, or CDM.

5.2.2 Sampling

Consider an analog signal $u(t)$, with amplitude limited between a minimum u_m and a maximum u_M, and a maximum frequency f_M. The sampling theorem ensures recovering of the original (finite energy) signal from a sequence of samples, obtained by multiplying $u(t)$ by a sequence of Dirac pulses, with unity moment and repetition rate greater or equal to $2f_M$. This frequency is often called the Nyquist frequency or the Nyquist rate. The original signal is

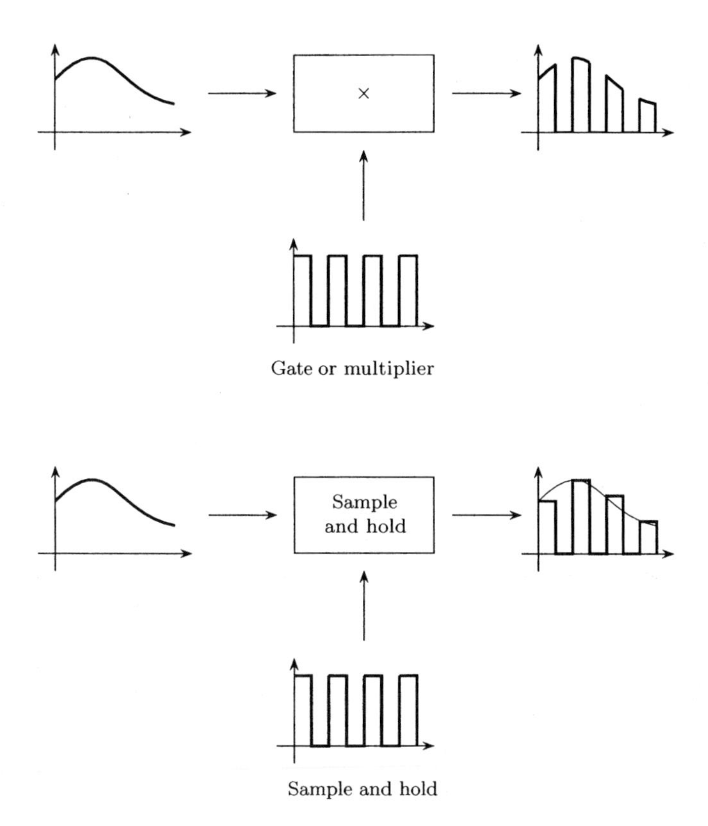

Fig. 5.3 Sampling performed by a gate (or a multiplier) and by sample and hold.

recovered by filtering the pulse sequence with an ideal low pass filter, having a cut-off frequency equal to f_M.

In practice, the sequence of Dirac pulses is approximated by a sequence of pulses with duration d and amplitude a. While this fact does influence the spectrum of the sequence of pulses it does not affect the capacity to retrieve the original signal using a low pass filter.

The product of signal $u(t)$ by the pulse sequence may be performed by an analog multiplier or by a gate. Alternatively, sampling may be obtained by a sample and hold circuit. In this case samples have constant amplitude, corresponding to the value of $u(t)$ at the time of sampling, while in the previous case samples have a varying amplitude, due to the changes in $u(t)$ during the sampling pulse, as shown in Figure 5.3.

As before and in spite of change in pulse shape, it is still possible to recover the original signal using a low pass filter, even if a equalizing circuit may be required when the pulse duration becomes non-negligible when compared to its period.

In real life it is not possible to recover the original signal from the sequence of samples because:

- the spectrum of $u(t)$ is not rigorously upper bounded,
- the ideal low pass filter cannot be implemented.

The resulting imperfections are known as aliasing and interpolation.

5.2.3 Quantization

Once the original signal has been transformed in a sequence of samples, or after amplitude modulating a sequence of pulses with the signal, we must quantize and code these pulses.

Quantization is a procedure whereby the amplitude of a pulse, which may have any value between u_m and u_M, is transformed into another value chosen from a discrete set of values $u_1, u_2, \ldots, u_n$. Quantization is called uniform or linear, if the difference between consecutive values of the set u_i is constant. This constant difference is known as the quantization step.

After quantization it is no longer possible to recover the original signal. The difference between the signal recovered from the quantized samples and the original signal is known as the quantization noise.

Consider a uniform quantization, with step a and m levels. Provided the signal obeys

$$u_M - u_m \leq ma,$$

the quantization error (noise) $n(t)$ is bounded

$$|n(t)| \leq \frac{a}{2}.$$

If amplitudes to be quantized follow a uniform distribution between u_m and u_M, then the distribution of $n(t)$ is also uniform between $-a/2$ and $+a/2$, that is,

$$f(n) = \begin{cases} \frac{1}{a} & \text{for} \quad -\frac{a}{2} \leq n(t) \leq +\frac{a}{2}, \\ 0 & \text{for} \quad -\frac{a}{2} > n(t) > +\frac{a}{2}. \end{cases}$$

The average value of the quantization noise is zero

$$\begin{aligned} \overline{n(t)} &= \int_{-a/2}^{+a/2} n f(n)\, dn \\ &= \frac{1}{a} \int_{-a/2}^{+a/2} n\, dn \\ &= 0. \end{aligned}$$

The quantization noise power n_q, the result of applying $n(t)$ to a unity resistor is

$$\begin{aligned} n_q &= \overline{n^2(t)} \\ &= \int_{-a/2}^{+a/2} n^2 f(n)\, dn \\ &= \frac{1}{a}\int_{-a/2}^{+a/2} n^2 dn \\ &= \frac{a^2}{12}. \end{aligned} \tag{5.1}$$

If we take a unipolar signal, with amplitude u and quantize it with m levels, these correspond to the amplitudes 0, a, $2a$, ... , $(m-1)a$. If, instead, we use a bipolar signal with dynamic range u the quantizing levels correspond to

- $\pm a/2$, $\pm 3a/2$, $\pm 5a/2$, ... , $\pm(m-1)a/2$ for m even;
- 0, $\pm a$, $\pm 2a$, $\pm 3a$, ... , $\pm(m-1)a/2$ for m odd.

In any case we have $a = \frac{u}{m-1}$.

The power s of the quantized signal may be computed from

$$s = \sum_{i=1}^{n} u_i^2\, \mathcal{P}(u_i), \tag{5.2}$$

where u_i is the voltage level corresponding to the quantized level i and $\mathcal{P}$ its probability of occurrence.

If signal $u(t)$ has a uniform amplitude distribution within its dynamic range u then all quantizing levels occur with the same probability

$$\mathcal{P}(u_i) = \frac{1}{m}. \tag{5.3}$$

Under these conditions, recalling that

$$\sum_{i=1}^{m} i^2 = \frac{m(m+1)(2m+1)}{6} \tag{5.4}$$

and that for m even

$$\begin{aligned} \sum_{i=1,3,5,\ldots}^{m-1} i^2 &= \sum_{i=1}^{m-1} i^2 - 2^2 \left(\sum_{i=1}^{\frac{m}{2}-1} i^2 \right) \\ &= \frac{m(m-1)(m+1)}{6}, \end{aligned} \tag{5.5}$$

we have:

- for a unipolar signal:

$$\begin{aligned} s &= \frac{1}{m}\sum_{i=1}^{m} u_i^2 \\ &= \frac{1}{m}\sum_{i=0}^{m-1} (ia)^2 \\ &= \frac{a^2(m-1)(2m-1)}{6}; \end{aligned} \tag{5.6}$$

- for a bipolar signal and an even number of levels

$$\begin{aligned} s &= \frac{1}{m}\sum_{i=1}^{m} u_i^2 \\ &= \frac{2}{m}\sum_{i=1,3,5,\ldots}^{m-1} \left(\frac{ia}{2}\right)^2 \\ &= \frac{a^2(m-1)(m+1)}{12}; \end{aligned} \tag{5.7}$$

- for a bipolar signal and an odd number of levels

$$\begin{aligned} s &= \frac{1}{m}\sum_{i=1}^{m} u_i^2 \\ &= \frac{2}{m}\sum_{i=1}^{\frac{m-1}{2}} (ia)^2 \\ &= \frac{a^2(m-1)(m+1)}{12}. \end{aligned} \tag{5.8}$$

For uniform quantization and a signal with uniform distribution of amplitudes within the dynamic range, the signal to quantization noise s/n_q becomes

- for a unipolar signal

$$s/n_q = 2(m-1)(2m-1); \tag{5.9}$$

- for a bipolar signal:

$$s/n_q = (m-1)(m+1). \tag{5.10}$$

In both cases, the signal to quantization noise is proportional to the square power of the number of quantization levels. According to Gregg [15] this situation occurs for many types of signals, such as sinewaves, triangular waves, rectangular waves and even for the Gaussian noise used to simulate frequency multiplexed voice .

If signal $u(t)$ is a sinewave with amplitude u_M the signal to quantization noise s/n_q is

$$s/n_q = \frac{6u_M^2}{a^2}. \tag{5.11}$$

Since the minimum number of quantization noise levels m is given as a function of the peak-to-peak signal amplitude $2u_M$ and the amplitude of the quantization levels a as

$$m = \frac{2u_M}{a},$$

the signal to quantization noise ratio may be written in the more usual form as

$$s/n_q = \frac{3m^2}{2}. \tag{5.12}$$

If we now recall that, for binary coding, the m quantization levels require a word with b bits where

$$b = \log_2 m,$$

then the signal to quantization noise, in logarithmic units, becomes

$$S/N_q = 10\log_{10}\left(\frac{3}{2}\right) + 20\log_{10}(m), \tag{5.13}$$

or:

$$S/N_q = 1.76 + 6.02b. \tag{5.14}$$

We must stress that this value of the signal to quantization noise corresponds to signals whose amplitude is the highest possible for the number of quantization levels and the quantization step adopted. For signals with arbitrary amplitude u_a, lower than the maximum u_M, signal to quantization noise is reduced accordingly

$$S/N_q = 1.76 + 6.02b + 20\log_{10}\left(\frac{u_a}{u_M}\right). \tag{5.15}$$

Voice signals in switched telephone networks, exhibit not only a reasonably high dynamic range (of the order of 20 to 30 dB) but also a maximum level, at the zero point level, which may vary within a 30 dB range. On the other hand, the amplitude distribution, even for signals with a standard maximum amplitude, is far from uniform. Under these conditions uniform quantization is not an adequate solution.

Non uniform quantization may be seen as the chain association of a non-linear amplifier (compressor-expander) and a uniform quantizer. Let $y(x)$ be the curve that represents the response of the compressor to the input signals. We plot $y(x)$ in Figure 5.4, after normalizing the input and output signal

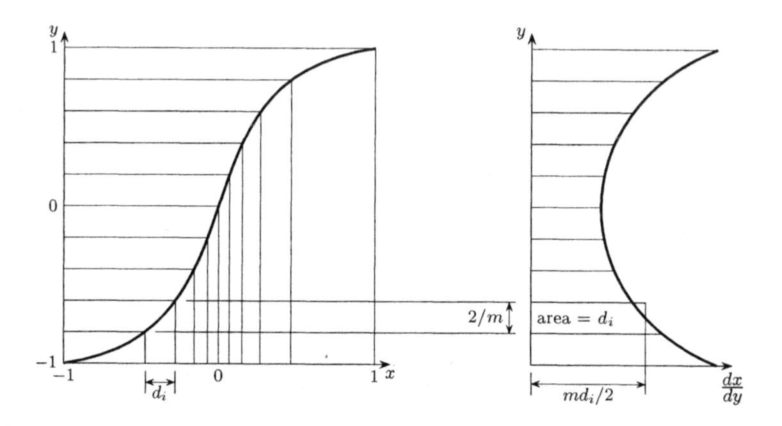

Fig. 5.4 Non-uniform quantization. Varying quantization step as a function of the amplitude of the output signal.

levels between -1 and $+1$. We also plot in the same figure the varying size of the quantizing steps as a function of the output signal amplitude

Quantization noise n_q may be calculated as a sum of m terms, each corresponding to the contribution of signals with amplitudes between x_{m_i} and x_{M_i}

$$n_q = \sum_{i=1}^{m} \left[\int_{x_{m_i}}^{x_{M_i}} n^2 f(n) dn \right]. \tag{5.16}$$

If the number of quantizing steps is large, it is reasonable to assume that the noise amplitude is constant within each quantizing step and that $f(n)$ does not vary much within the integration interval. Hence

$$n_q = \frac{1}{12} \sum_{i=1}^{m} p_i a_i^2, \tag{5.17}$$

where

$$\begin{aligned} a_i &= x_{M_i} - x_{m_i}, \\ p_i &= f(n_i) a_i, \\ n_i &= \frac{x_{M_i} - x_{m_i}}{2}. \end{aligned}$$

As may be seen from Figure 5.4, for m quantization levels we have:

$$a_i = \frac{2}{m} \left(\frac{dx}{dy} \right)_{x=x_i}, \tag{5.18}$$

and thus quantization noise is

$$
\begin{aligned}
n_q &= \frac{1}{12}\sum_{i=1}^{m} p_i a_i^2 \\
&= \frac{1}{3m^2}\sum_{i=1}^{m} p_i \left(\frac{dx}{dy}\right)^2_{x=x_i}.
\end{aligned} \tag{5.19}
$$

For a large number of quantization levels, the summations may be changed into integrals

$$
\begin{aligned}
s &= \sum_{i=1}^{m} u_i^2\, \mathcal{P}(u_i) \\
s &= \int_{-1}^{+1} x^2 f(x)\, dx,
\end{aligned} \tag{5.20}
$$

$$
\begin{aligned}
n_q &= \frac{1}{3m^2}\sum_{i=1}^{m} p_i \left(\frac{dx}{dy}\right)^2_{x=x_i} \\
&= \frac{1}{3m^2}\int_{-1}^{+1} \left(\frac{dx}{dy}\right)^2 f(x)\, dx.
\end{aligned} \tag{5.21}
$$

Signal to quantization noise is therefore

$$
s/n_q = 3m^2 \frac{\int_{-1}^{+1} x^2 f(x) dx}{\int_{-1}^{+1} (\frac{dx}{dy})^2 f(x) dx}, \tag{5.22}
$$

which shows that the signal to quantization noise depends not only on the compression curve but also on the signal amplitude distribution, even after normalizing signal amplitude between -1 and $+1$.

Since signal amplitude distribution is not known in advance it is wise to adopt a compression curve that (as much as possible) makes signal to quantization noise independent of signal statistics. Taking

$$
\frac{dx}{dy} = \begin{cases} kx & \text{for} \quad x > 0, \\ -kx & \text{for} \quad x < 0, \end{cases} \tag{5.23}
$$

and recalling Figure 5.4

$$
y = \begin{cases} 1 + \frac{\log(x)}{k} & \text{for} \quad x > 0, \\ -1 - \frac{\log(-x)}{k} & \text{for} \quad x < 0, \end{cases} \tag{5.24}
$$

we get a constant signal to quantization noise

$$
s/n_q = \frac{3m^2}{k^2}. \tag{5.25}
$$

With a finite number of quantization levels it is not possible to get a constant signal to quantization noise because, for values of x near zero, the compression curve varies abruptly and some approximations no longer hold.

Two compression curves are commonly used:

- The μ law adopted in the United States of America and Japan

$$y = \text{sign}(x)\frac{\log(1+\mu|x|)}{\log(1+\mu)}; \tag{5.26}$$

- The A law [3], adopted in Europe

$$y = \frac{ax}{1+\log(a)} \quad \text{for} \quad 0 \leq |x| \leq 1/a, \tag{5.27}$$

$$y = \text{sign}(x)\frac{1+\log(a|x|)}{1+\log(a)} \quad \text{for} \quad 1/a < |x| \leq 1, \tag{5.28}$$

  where $\text{sign}(x)$ is a function defined by

$$\text{sign}(x) = \begin{cases} +1 & \text{for} \quad x > 0, \\ -1 & \text{for} \quad x < 0, \\ 0 & \text{for} \quad x = 0. \end{cases}$$

For the A law

$$k_1 = 1 + \log(a), \tag{5.29}$$

where k_1 is the compression gain. For 8 bit PCM it is usual to take $a = 87.6$, and thus $k_1 = 5.47$.

As an example, let us calculate the signal to quantization noise for the μ law. From (5.21), substituting the derivative by the value obtained from (5.26) we get

$$n_q = \frac{\log_2^2(1+\mu)}{\mu^2}(1 + 2\mu|\bar{x}| + \mu^2 s). \tag{5.30}$$

To test the efficiency of the μ law let us take the simple case of a signal with a uniform distribution amplitudes in the range $[-a, +a]$. Then

$$s = \frac{a^2}{3}$$

and

$$f(x) = \frac{1}{2a},$$

[3]In the expressions related to curve A we represent by a the parameter of the curve which elsewhere is usually represented with A.

$$\begin{aligned} a &= \sqrt{3s}, \\ \bar{|x|} &= 2\int_0^a xf(x)dx \\ &= \frac{\sqrt{3s}}{2}. \end{aligned}$$

Substituting we get

$$s/n_q = 3m^2 \frac{s}{\frac{log_e^2(1+\mu)}{\mu^2}\left(1 + 2\mu\frac{\sqrt{3s}}{2} + \mu^2 s\right)}. \tag{5.31}$$

Taking the usual values $\mu = 255$ and $m = 256$

$$s/n_q = \frac{4.16 * 10^8}{1 + 441.7\sqrt{s} + 65025s},$$

whereas with no compression we would have simply

$$s/n_q = 3m^2 s.$$

Figure 5.5 shows the signal to quantization noise with a μ law compression and with no compression as a function of the input signal power. Although at high signal levels (that is, up to -15 dB) the signal to quantization noise is lower with compression at lower signal levels there is a very substantial improvement.

In order to evaluate the importance of the signal amplitude distribution we will now consider the voice signal whose amplitude distribution may be modeled by a Laplace distribution

$$f(x) = \frac{\alpha}{2}e^{-\alpha|x|}. \tag{5.32}$$

From (5.32) we get

$$\begin{aligned} s &= 2\int_0^\infty x^2 f(x)dx \\ &= \frac{2}{\alpha^2}, \\ \bar{|x|} &= 2\int_0^\infty xf(x)dx \\ &= \frac{1}{\alpha}. \end{aligned}$$

This distribution creates a problem because it cannot be limited to the interval $[-1, +1]$. However, the probability that $|x| > 1$ is equal to e^{α} which is less than 0.002 when $s < 0.1$. So with a small error we may use the values of s and $\bar{|x|}$ as given above in (5.30) and get

$$s/n_q = 3m^2 \frac{s}{\frac{log_e^2(1+\mu)}{\mu^2}\left(1 + 2\mu\frac{\sqrt{s}}{\sqrt{2}} + \mu^2 s\right)}. \tag{5.33}$$

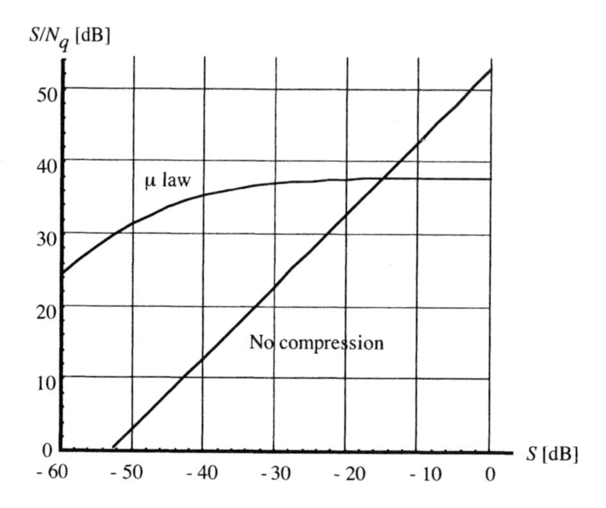

Fig. 5.5 Signal to quantization noise with a μ law compander ($\mu = 255$) and with no compander, for $m = 256$, as a function of signal power. Signal amplitude distribution is taken as uniform.

Simple comparison of equations (5.31) and (5.33) shows that the signal amplitude distribution plays a very limited role in compression performance.

Compression gain for small signals may be derived from (5.26), expanding the log function in series for $x << \mu$ and keeping only the first term

$$y = \frac{\mu}{\log_e(1+\mu)} x,$$

from which we get the compression gain in dB with $\mu = 255$

$$20 \log_{10} \left(\frac{\mu}{\log_e(1+\mu)} \right) = 33.25 \text{ dB}.$$

As a further example, let us calculate the signal to quantization noise for the A law. From (5.21), substituting the derivative by the value obtained from (5.27) and (5.28) and making use of the compression gain k_1 given in (5.29) we get

$$n_q = \frac{k_1^2}{3m^2} \left[\frac{1}{a^2} \int_{-1/a}^{+1/a} f(x)dx + \int_{-1}^{-1/a} x^2 f(x)dx + \int_{+1/a}^{+1} x^2 f(x)dx \right],$$

or

$$n_q = \frac{k_1^2}{3m^2} \left[\frac{1}{a^2} \int_{-1/a}^{+1/a} f(x)dx + \int_{-1}^{+1} x^2 f(x)dx - \int_{-1/a}^{+1/a} x^2 f(x)dx \right].$$

Recalling that the signal power is given by

$$s = \int_{-1}^{+1} x^2 f(x)dx,$$

we may write

$$n_q = \frac{k_1^2}{3m^2}\left[s + \int_{-1/a}^{+1/a} f(x)(\frac{1}{a^2} - x^2)dx\right],$$

and, finally,

$$s/n_q = \frac{3m^2}{k_1^2} \times \frac{1}{1 + \frac{1}{sa^2}\int_{-1/a}^{+1/a} f(x)(1 - a^2x^2)dx}. \tag{5.34}$$

Assume now that large amplitudes dominate, that is, ($|x| > 1/a$). Then

$$\int_{-1/a}^{+1/a} f(x)(1 - a^2x^2)dx \ll sa^2,$$

and

$$s/n_q < \frac{3m^2}{k_1^2}. \tag{5.35}$$

When small amplitudes prevail, that is when $|x| \ll 1/a$, noting that $1 - a^2x^2$ decreases with x and becomes negative for $|x| > 1/a$

$$\int_{-1/a}^{+1/a} f(x)(1 - a^2x2)dx \;\geq\; \int_{-1}^{+1} f(x)(1 - a^2x2)dx, \tag{5.36}$$

$$\int_{-1/a}^{+1/a} f(x)(1 - a^2x2)dx \;\geq\; 1 - sa^2,$$

and then

$$s/n_q \leq \frac{3m^2}{k_1^2} sa^2. \tag{5.37}$$

Looking to the denominator of the fraction in equation (5.34) and recalling equation (5.35) we can write

$$\int_{-1/a}^{+1/a} f(x)(1 - a^2x^2)dx \leq 1,$$

and so

$$s/n_q \geq \frac{3m^2}{k_1^2}\frac{1}{1 + \frac{1}{sa^2}}. \tag{5.38}$$

In Figure 5.6 we plot the upper and the lower limits of the signal to quantization noise as a function of the normalized signal power sa^2, for the A law

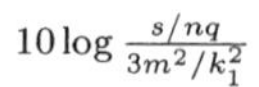

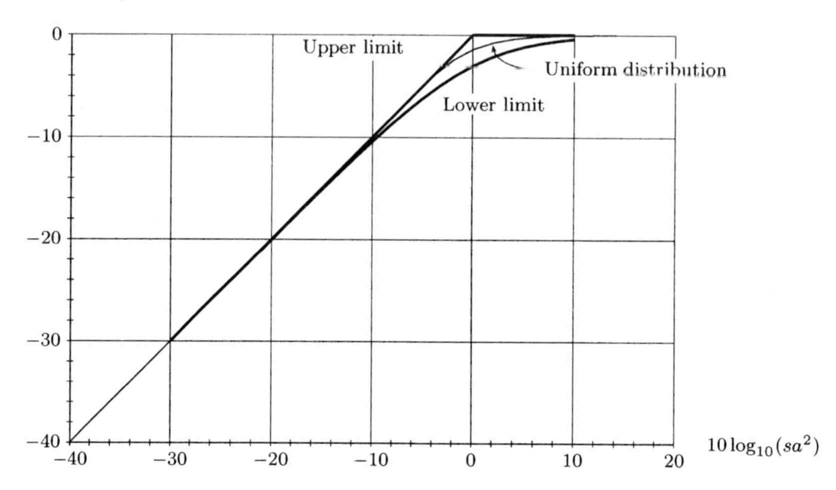

Fig. 5.6 Upper and lower limits of the signal to quantization noise with logarithmic compression (A law with $a = 87.6$) and change in this ratio with a uniform distribution of amplitudes.

(with $a = 87.6$), and the change in this ratio for a uniform amplitude distribution. It is obvious from this figure that the A law achieves a signal to quantization noise ratio, with compression, almost independent from signal statistics.

In a word: for the higher amplitudes all is as if the quantization was perfectly logarithmic (constant signal to quantization noise), while for the lower amplitudes the behavior is typical of the uniform quantization (signal to quantization noise proportional to the signal power). For $a = 87.6$ the compression gain is, from (5.27)

$$20 \log_{10} \left(\frac{a}{1 + \log a} \right) = 24 \text{ (dB)}.$$

Recalling that for ideal amplitude sinewaves, uniform quantization leads to a signal to quantization noise $s/n_q = 3m^2/2$, while logarithmic quantization (A law) yields a maximum value if $s/n_q = 3m^2/k_1^2$, it is easy to find out the superiority of uniform quantization, for large amplitudes, when $k_1 > \sqrt{2}$ (for the A law, when $a = 1.513$). Logarithmic quantization leads to a signal to quantization noise almost independent of the signal amplitude, for a large range of amplitudes, while uniform quantization leads to a signal to quantization noise that decreases when signal power decreases.

Compression laws are usually implemented, approximating the exact curve by an open polygon with a finite number of sides. The additional noise caused by this approximation is low (less than 3 dB), for the usual number of segments (from 13 to 15).

5.2.4 Coding

Coding is a procedure by which a sequence of pulses, also known as a code word, is assigned to each quantization level. Segmentation, referred to in quantization, is implemented in the coding procedure in one of two ways:

1. direct coding;

2. uniform coding, followed by digital compression.

In direct coding a bit is used for the sign, a few bits (usually 3) to identify the curve segment, and the remaining bits (4 in case of an 8-bit code word) for the level inside the segment.

In uniform coding followed by digital compression the signal is uniformly quantized, using a higher number of levels than ultimately required (e.g., 12), followed by binary coding of the quantization levels and finally compression of the output code. Table 5.1 shows the coding table for the 8 segment (per polarity) A-law.

Uniform code	Compressed code
0 0 0 0 0 0 0 w x y z a	0 0 0 w x y z
0 0 0 0 0 0 1 w x y z a	0 0 1 w x y z
0 0 0 0 0 1 w x y z a b	0 1 0 w x y z
0 0 0 0 1 w x y z a b c	0 1 1 w x y z
0 0 0 1 w x y z a b c d	1 0 0 w x y z
0 0 1 w x y z a b c d e	1 0 1 w x y z
0 1 w x y z a b c d e f	1 1 0 w x y z
1 w x y z a b c d e f g	1 1 1 w x y z

Table 5.1 A-law coding table.

We note that with a 12-bit AD converter the last bit is always neglected, which really means that we only require 11 bits to code the 8 segment A-law.

5.2.5 Delta modulation

Delta modulation is a French patent dating from 1946. It corresponds to quantization of samples of the difference between the signal and an estimate obtained by integrating the output signal.

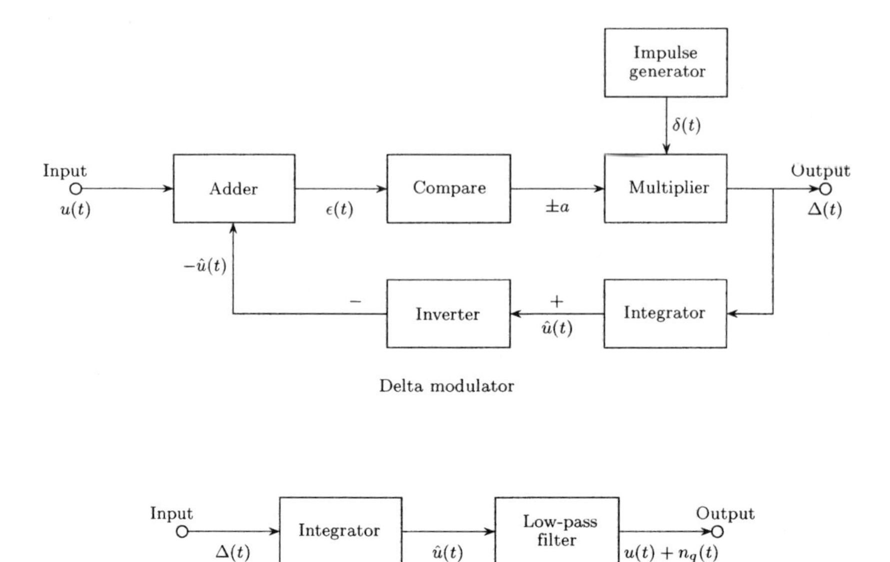

Fig. 5.7 Block scheme of a delta modulator and demodulator.

Figure 5.7 represents a block scheme for a delta modulator and demodulator and Figure 5.8 the wave forms corresponding to the input signal, its estimate, the difference (or delta) signal and the output pulse sequence.

With regard to Figure 5.8, we assume that, at the origin of time, the input signal estimate $\hat{u}$ is zero. At the first sampling , the difference $\epsilon(t)$ between $u_i(t)$ and $\hat{u}(t)$ is positive, thus the output is a positive pulse, with amplitude a. At the second sampling the difference between $u_i(t)$ and $\hat{u}(t)$ is still positive and thus the output is a positive pulse with amplitude a. From then onwards $\hat{u}(t)$ has an amplitude $2a$. At the third sampling $u_i(t)-\hat{u}(t)$ is already negative and, consequently, the output is a negative pulse with amplitude $-a$.

This procedure may become unstable or overloaded when the signal varies rapidly and the estimate does not follow it, either because the sampling rate is too low or because the output pulse amplitude is too low.

5.3 DIGITAL TO ANALOG CONVERSION

If the transmission is error free, the digital to analog conversion does not introduce further signal degradation and thus, at the output we can recover

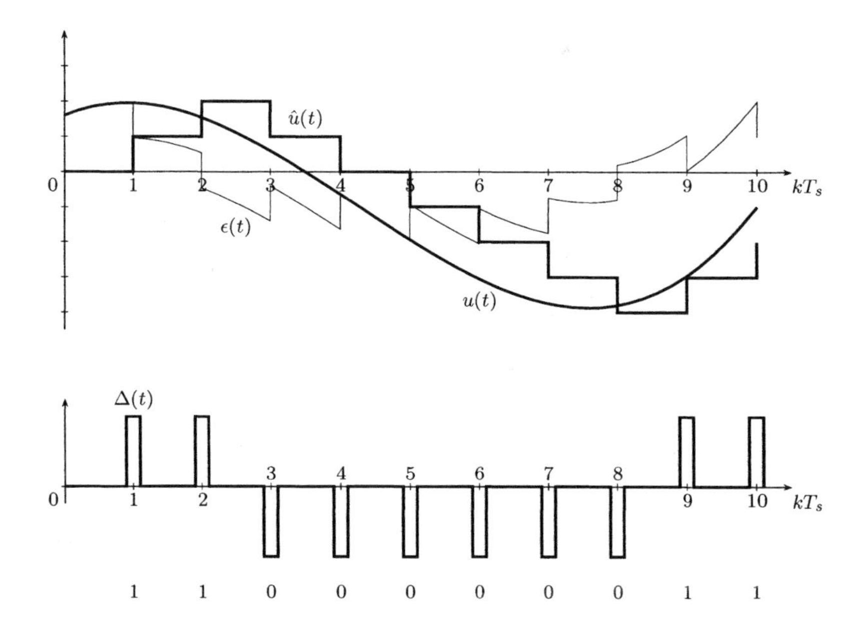

Fig. 5.8 Waveforms in a delta modulator.

the analog signal with a quality equivalent to the initial one, that is, with a signal-to-noise ratio equal to the quantization signal-to-noise ratio.

Since real links always introduce transmission errors, we must evaluate the influence of these errors in the output signal-to-noise ratio. In the next derivation we closely follow Gregg [15].

Let $u(t)$ be the original analog signal and $u_s(t)$ the recovered signal. Transmission noise power n on a unity resistor, due to transmission errors is

$$n = \mathcal{E}\{[u_s(t) - u_i(t)]^2\}, \tag{5.39}$$

where $\mathcal{E}$ is the operator expected value.

With ideal sampling, at the instants $t = kT_s$, with $k = 1, 2, ...$ where T_s is the sampling interval, we have

$$n = \mathcal{E}\{[u_s(kT_s) - u_i(kT_s)]^2\}. \tag{5.40}$$

Let $u_q(kT_s)$ be the values corresponding to samples u_i, after quantization. Adding and subtracting these values in (5.40) we highlight the contribution of transmission and quantization noise in the total noise

$$n = \mathcal{E}\{[u_s(kT_s) - u_q(kT_s) + u_q(kT_s) - u_i(kT_s)]^2\}. \tag{5.41}$$

Assuming quantization and transmission procedures to be independent:

$$n = n_t + n_q, \tag{5.42}$$

where transmission noise n_t and quantization noise n_q are given, respectively, by

$$n_t = \mathcal{E}\{[u_s(kT_s) - u_q(kT_s)]^2\}, \quad (5.43)$$
$$n_q = \mathcal{E}\{[u_q(kT_s) - u_i(kT_s)]^2\}. \quad (5.44)$$

Since quantization has already been dealt with, the new objective will be the influence of transmission errors, described by the bit error ratio, in short *ber*, on transmission noise.

Consider the sample of order k of signal $u_i(kT_s)$, quantized with m levels as $u_q(kT_s)$ and coded as a_{k_i} with b binary digits that will be represented as a vector of dimension b where the elements a_{k_1}, a_{k_2}, ... ,a_{k_b} are binary digits

$$a_k = (a_{k_1}, a_{k_2}, ..., a_{k_b}).$$

Let $\mathcal{P}_t(0)$ and $\mathcal{P}_t(1)$, respectively, be the probabilities that symbols 0 and 1 are transmitted correctly. Assume that the original signal is bipolar, with amplitude between $-U/2$ and $+U/2$, and that uniform quantization is applied. The relation between sample of order k after quantization and code word a_k is

$$u_q(kT_s) = \frac{U}{2}\sum_{i=1}^{b} \alpha_{k_i} 2^{-i}, \quad (5.45)$$

where

$$\alpha_{k_i} = \begin{cases} +1 & \text{if} \quad a_{k_i} = 1, \\ -1 & \text{if} \quad a_{k_i} = 0. \end{cases}$$

Take, for example, $m = 4$ $(b = 2)$, which leads to a quantization step equal to $U/4$. In this case, the quantization levels are $3U/8$, $U/4$, $-U/4$ e $-3U/8$, which correspond the following values of vector a_k: [11], [10], [01] and [00].

The code word represented by vector a_k is received as the word b_k, $b_k = (b_{k_1}, b_{k_2}, ..., b_{k_b})$, which corresponds to the received sample $u_s(kT_s)$

$$u_s(kT_s) = \frac{U}{2}\sum_{i=1}^{b} \beta_{k_i} 2^{-i}, \quad (5.46)$$

where

$$\beta_{k_i} = +1 \quad \text{if} \quad b_{k_i} = 1,$$
$$\beta_{k_i} = -1 \quad \text{if} \quad b_{k_i} = 0.$$

From (5.43) we get the transmission noise:

$$n_t = \frac{U^2}{4}\mathcal{E}\left\{\left[\sum_{i=1}^{b}(\alpha_{k_i} - \beta_{k_i})2^{-i}\right]^2\right\}. \quad (5.47)$$

Expanding the square of the sum and noting that the expected value of a sum is equal to the sum of the expected values

$$
\begin{aligned}
n_t &= \frac{U^2}{4}\sum_{i=1}^{b} \mathcal{E}\{[(\alpha_{k_i} - \beta_{k_i})^2]2^{-2i}\} + \cdots \\
&\cdots + \frac{U^2}{4}\sum_{i=1}^{b}\sum_{j=1, i\neq 1}^{b} \mathcal{E}\{[(\alpha_{k_i} - \beta_{k_i})(\alpha_{k_j} - \beta_{k_j})]2^{-(i+j)}\}. \quad (5.48)
\end{aligned}
$$

Before going any further, it is useful to calculate the expected value of the product of the elements of vectors α_k and β_k

$$
\begin{aligned}
\mathcal{E}(\alpha_{k_i}\alpha_{k_j}) &= \mathcal{P}(a_{k_i} = 1)\, \mathcal{P}(a_{k_j} = 0 \text{ if } a_{k_i} = 1)\,(+1)\,(+1) \\
&\quad + \mathcal{P}(a_{k_i} = 1)\, \mathcal{P}(a_{k_j} = 0 \text{ if } a_{k_i} = 1)\,(+1)\,(-1) \\
&\quad + \mathcal{P}(a_{k_i} = 0)\, \mathcal{P}(a_{k_j} = 1 \text{ if } a_{k_i} = 0)\,(-1)\,(+1) \\
&\quad + \mathcal{P}(a_{k_i} = 0)\, \mathcal{P}(a_{k_j} = 0 \text{ if } a_{k_i} = 0)\,(-1)\,(-1). \quad (5.49)
\end{aligned}
$$

Assuming that digits in a_k take the values 0 and 1, independently, we have for $i \neq j$

$$
\begin{aligned}
\mathcal{P}(a_{k_i} = 1) &= \mathcal{P}(1), \\
\mathcal{P}(a_{k_i} = 0) &= \mathcal{P}(0), \\
\mathcal{P}(a_{k_i} = 1 \text{ if } a_{k_j} = 0) &= \mathcal{P}(1), \\
\mathcal{P}(a_{k_i} = 1 \text{ if } a_{k_j} = 1) &= \mathcal{P}(1), \\
\mathcal{P}(a_{k_i} = 0 \text{ if } a_{k_j} = 1) &= \mathcal{P}(0), \\
\mathcal{P}(a_{k_i} = 0 \text{ if } a_{k_j} = 0) &= \mathcal{P}(0),
\end{aligned}
$$

and substituting in (5.49), we get for $i \neq j$

$$
\mathcal{E}(\alpha_{k_i}\alpha_{k_j}) = \mathcal{P}(1)^2 + \mathcal{P}(0)^2 - 2\mathcal{P}(0)\mathcal{P}(1). \quad (5.50)
$$

If digits 0 and 1 are equiprobable at the source, that is, if $\mathcal{P}(0) = \mathcal{P}(1) = 1/2$, then for $i \neq j$

$$
\mathcal{E}(\alpha_{k_i}\alpha_{k_j}) = 0. \quad (5.51)
$$

For $i = j$ we get:

$$
\begin{aligned}
\mathcal{P}(a_{k_i} = 1 \text{ if } a_{k_j} = 0) &= 0, \\
\mathcal{P}(a_{k_i} = 1 \text{ if } a_{k_j} = 1) &= \mathcal{P}(1), \\
\mathcal{P}(a_{k_i} = 0 \text{ if } a_{k_j} = 1) &= 0, \\
\mathcal{P}(a_{k_i} = 0 \text{ if } a_{k_j} = 0) &= \mathcal{P}(0),
\end{aligned}
$$

and, substituting in (5.49), we get for $i = j$

$$
\begin{aligned}
\mathcal{E}(\alpha_{k_i}\alpha_{k_j}) &= \mathcal{P}(0) + \mathcal{P}(1) \\
&= 1. \quad (5.52)
\end{aligned}
$$

Similarly, for the various elements of vector β_k we get

$$\mathcal{E}(\beta_{k_i}\beta_{k_j}) = \begin{cases} 1 & \text{for} \quad i = j, \\ 0 & \text{for} \quad i \neq j. \end{cases} \tag{5.53}$$

For the cross products of α_k and β_k, we get, for $i \neq j$

$$\begin{aligned} \mathcal{E}(\alpha_{k_i}\beta_{k_j}) \quad = \quad & \mathcal{P}(a_{k_i} = 1)\mathcal{P}(b_{k_j} = 1 \text{ if } a_{k_i} = 1)(+1)(+1) \\ & + \mathcal{P}(a_{k_i} = 1)\mathcal{P}(b_{k_j} = 0 \text{ if } a_{k_i} = 1)(+1)(-1) \\ & + \mathcal{P}(a_{k_i} = 0)\mathcal{P}(b_{k_j} = 1 \text{ if } a_{k_i} = 0)(-1)(+1) \\ & + \mathcal{P}(a_{k_i} = 0)\mathcal{P}(b_{k_j} = 0 \text{ if } a_{k_i} = 0)(-1)(-1). \end{aligned} \tag{5.54}$$

Assuming that:

- digits in a_k are independent,
- there is no intersymbol interference in the transmission,
- the probability of error in the transmission of digit 1 is $\mathcal{P}_t(1)$,
- the probability of error in the transmission of digit 1 is $\mathcal{P}_t(0)$,

we get

$$\begin{aligned} \mathcal{P}(a_{k_i} = 1) \quad &= \quad \mathcal{P}(1), \\ \mathcal{P}(a_{k_i} = 0) \quad &= \quad \mathcal{P}(0), \\ \mathcal{P}(b_{k_j} = 1 \text{ if } a_{k_i} = 1) \quad &= \quad \mathcal{P}(b_{k_j} = 1), \\ \mathcal{P}(b_{k_j} = 0 \text{ if } a_{k_j} = 1) \quad &= \quad \mathcal{P}(b_{k_j} = 0), \\ \mathcal{P}(b_{k_j} = 1 \text{ if } a_{k_i} = 0) \quad &= \quad \mathcal{P}(b_{k_j} = 1), \\ \mathcal{P}(b_{k_j} = 0 \text{ if } a_{k_j} = 0) \quad &= \quad \mathcal{P}(b_{k_j} = 0). \end{aligned}$$

Noting that the probability of receiving digit 1 is the sum of the probabilities of transmitting and correctly receiving digit 1 plus the probability of transmitting digit 0 and receiving it (incorrectly) as digit 1

$$\mathcal{P}(b_{k_j} = 1) = \mathcal{P}(1)[1 - \mathcal{P}_t(1)] + \mathcal{P}(0)\mathcal{P}_t(0), \tag{5.55}$$

and similarly

$$\mathcal{P}(b_{k_j} = 0) = \mathcal{P}(0)[1 - \mathcal{P}_t(0)] + \mathcal{P}(1)\mathcal{P}_t(1). \tag{5.56}$$

Substituting (5.55) and (5.56) in (5.54), we get, for $i \neq j$

$$\begin{aligned} \mathcal{E}(\alpha_{k_i}\beta_{k_j}) \quad = \quad & [\mathcal{P}(0) - \mathcal{P}(1)][\mathcal{P}(0)(1 - \mathcal{P}_t(0)) + \mathcal{P}(1)\mathcal{P}_t(0)] \\ & + [\mathcal{P}(1) - \mathcal{P}(0)][\mathcal{P}(1)(1 - \mathcal{P}_t(1)) + \mathcal{P}(0)\mathcal{P}_t(0)], \end{aligned} \tag{5.57}$$

which, for equiprobable digits at the source ($\mathcal{P}(0) = \mathcal{P}(1)$), is equal to zero.

For $i = j$ we get

$$\mathcal{E}(\alpha_{k_i}\beta_{k_j}) = \mathcal{P}(1)[1 - 2\mathcal{P}_t(1)] + \mathcal{P}(0)[1 - 2\mathcal{P}_t(0)]. \tag{5.58}$$

After obtaining the expected values of the products of the elements of vectors α_k and β_k we only have to expand the expression of the transmission noise (5.48) and substitute those values. For equiprobable digits we find out that the terms that correspond to the double sum vanish and thus we get

$$n_t = \frac{U^2}{2}[\mathcal{P}_t(0) + \mathcal{P}_t(1)] \sum_{i=1}^{b} 2^{-2i}. \tag{5.59}$$

Recalling that the sum of the first n elements of a geometric series whose first element is a_1 and whose rate r is given by

$$a_1 + \sum_{i=2}^{n} a_i = a_1 \frac{r^n - 1}{r - 1}, \tag{5.60}$$

the transmission noise becomes

$$n_t = \frac{U^2}{2}[\mathcal{P}_t(0) + \mathcal{P}_t(1)] \times \frac{1}{3}, \times \frac{2^{2b} - 1}{2^{2b}}$$

or:

$$n_t = \frac{U^2}{6}[\mathcal{P}_t(0) + \mathcal{P}_t(1)] \frac{m^2 - 1}{m^2}, \tag{5.61}$$

where $m = 2^b$ is the number of quantization levels.

For binary symmetric channels, that is, for channels where $\mathcal{P}_t(0) = \mathcal{P}_t(1) = ber$, the previous expression may be simplified:

$$n_t = \frac{U^2}{3}\, ber\, \frac{m^2 - 1}{m^2}, \tag{5.62}$$

which, in terms of the quantization step, $a = U/m$, may be written as

$$n_t = \frac{a^2}{3}\, ber\, (m^2 - 1). \tag{5.63}$$

As will be shown later, channels that use amplitude, frequency or phase modulation with coherent detection, and frequency and phase differential modulation with incoherent detection are binary symmetric channels.

Adding transmission noise and quantization noise we get the total noise

$$\begin{aligned} n &= \frac{U^2}{12m^2}\left[1 + 4\, ber\, (m^2 - 1)\right] \\ &= \frac{a^2}{12}\left[1 + 4\, ber\, (m^2 - 1)\right]. \end{aligned} \tag{5.64}$$

If the signal is a sinewave, with peak value $U/2$, the overall signal-to-noise ratio becomes, from (5.64)

$$s/n = \frac{\frac{3\,m^2}{2}}{1 + 4\,ber\,(m^2 - 1)}, \tag{5.65}$$

where m is the number of quantization levels.

In Figure 5.9 we plot the overall signal-to-noise ratio s/n, for a binary symmetric channel, as a function of the bit error ratio, taking as a parameter the number of digits b of code words ($m = 2^b$).

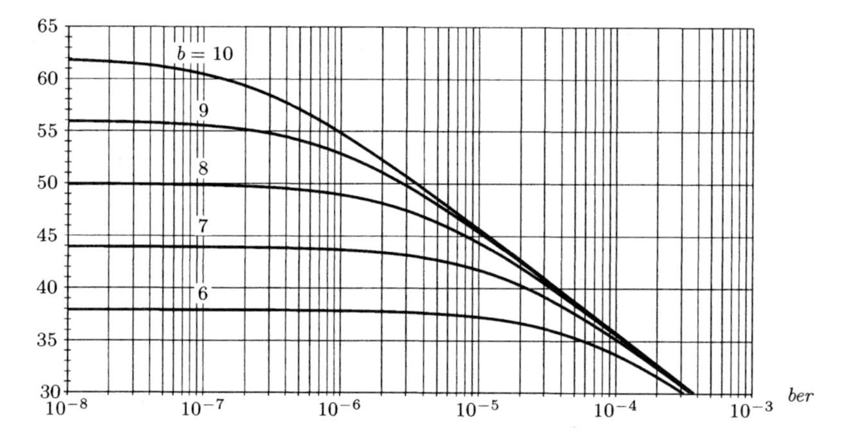

Fig. 5.9 Signal to noise for binary symmetric channels as a function of the bit error ratio for various code word lengths.

5.4 LINE CODES

5.4.1 Introduction

In many cases, digital signals, as supplied by sources, do not have the most adequate characteristics for transmission, either by cable or by radio. To overcome this difficulty, it is necessary to transform the source supplied digital signals into other signals more appropriate for direct transmission or for carrier modulation. This operation, known as coding, may be defined as an adaptation of the original signal to the requirements of the transmission, but it may have broader objectives, namely, to enable:

- bit synchronization, that is, to supply the receiver with information that enables it to decide on the best instant to evaluate the transmitted pulse;

- error detection and (or) error correction;
- word synchronization whenever the code uses groups or more than one pulse.

This section starts with a brief reference to the spectral power density (or power spectrum) of a synchronous binary signal and then proceeds to describe some of the most commonly used line codes. It ends with a brief reference to the ITU - T standard digital hierarchies.

5.4.2 Spectral power density of a binary synchronous signal

A binary synchronous signal is a sequence of two symbols, with the same duration T, different shapes, represented by functions $f_1(t)$ and $f_2(t)$, and probability of occurrence p and $1-p$, respectively.

Let $g_1(f)$ and $g_2(f)$ be the Fourier spectra corresponding to signals $f_1(t)$ and $f_2(t)$, assumed centered on the origin of time. It is possible to demonstrate [2] that the power spectrum $W(f)$ of a binary synchronous signal is given by the sum of a line spectrum $W_d(f)$, with lines on integer multiples of the inverse of the symbol duration, plus a continuous spectrum $W_c(f)$

$$W(f) = W_c(f) + W_d(f), \tag{5.66}$$

where

$$W_c(f) = \frac{p(1-p)}{T}|g_1(f) - g_2(f)|^2, \tag{5.67}$$

and

$$\begin{aligned} W_d(f) &= \frac{1}{T^2}|pg_1(0) + (1-p)g_2(0)|^2\delta(f) + \cdots \\ &\cdots + \frac{2}{T^2}\left[\sum_{k=1}^{\infty}\left|pg_1\left(\frac{k}{T}\right) + (1-p)g_2\left(\frac{k}{T}\right)\right|^2 \delta(f-\frac{k}{T})\right] \end{aligned} \tag{5.68}$$

5.4.3 Unipolar code

One of the simplest binary codes is the unipolar synchronous code , represented in Figure 5.10. The code name derives from the fact that all pulses have the same polarity and the same duration with no separation between pulses.

This code is not adequate for baseband transmission because it contains a continuous component, with no information, which leads to a power loss in transmission. The fact that the signal is synchronous implies a coordination (synchronization) between the transmitter and the receiver.

The unipolar power spectrum may be derived from (5.66), (5.67) and (5.68), with

$$f_1(t) = \begin{cases} 1 & \text{for} \quad 0 \le t \le T, \\ 0 & \text{for} \quad 0 < t > T, \end{cases}$$

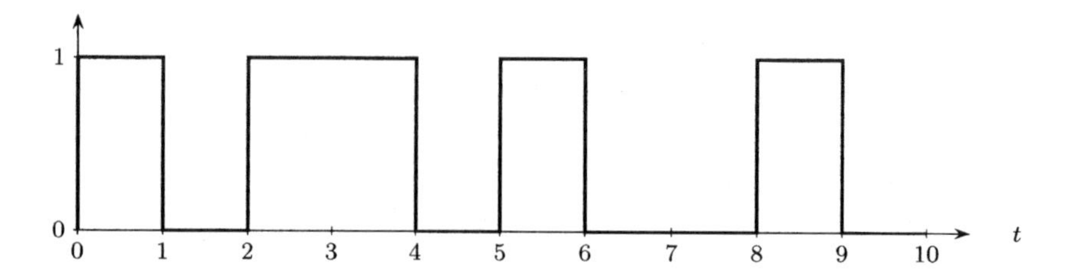

Fig. 5.10 Unipolar synchronous signal corresponding to the sequence 1011010010.

and, for any value of t

$$f_2(t) = 0.$$

We have

$$\begin{aligned} g_1(f) &= T\,\frac{\sin(\pi f T)}{\pi f T}, \\ g_2(f) &= 0, \end{aligned}$$

and, thus

$$W(f) = T\,\frac{\sin^2(\pi f T)}{(\pi f T)^2}p(1-p) + p^2\delta(f). \tag{5.69}$$

As shown in Figure 5.11, this code has a line at $f = 0$ and a continuous spectrum with a maximum at $f = 0$ and nulls at $f = n/T$ (where n is an integer).

5.4.4 Polar code

Synchronous polar code, also known as NRZ[4] (Figure 5.12) is another binary code that makes use of pulses with two polarities in order to remove the zero frequency component of the power spectrum.

In this case we have

$$f_1(t) = \begin{cases} 1 & \text{for} \quad 0 \le t \le T, \\ 0 & \text{for} \quad T < t < 0, \end{cases}$$

and

$$f_2(t) = -f_1(t),$$

thus

$$\begin{aligned} g_1(f) &= T\,\frac{\sin(\pi f T)}{\pi f T}, \\ g_2(f) &= -g_1(f), \end{aligned}$$

[4] *Non-Return to Zero.*

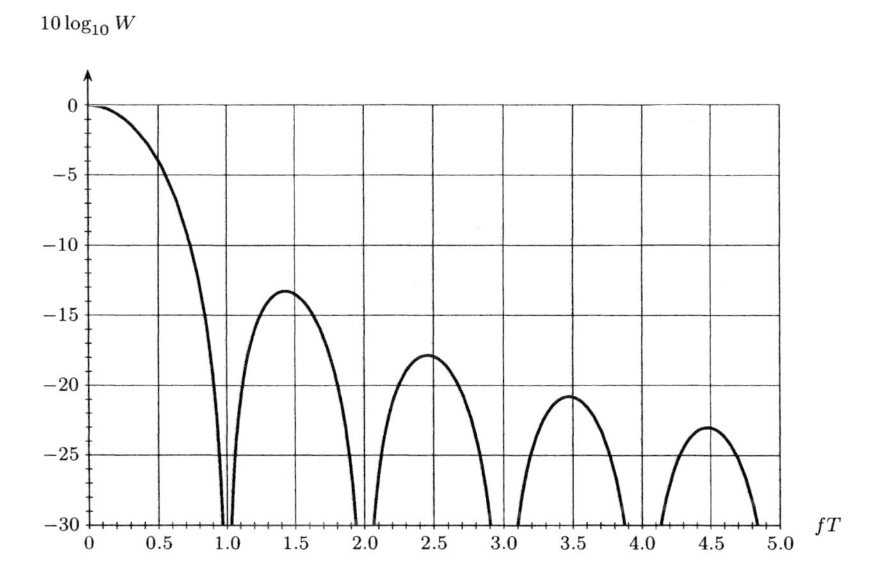

Fig. 5.11 Unipolar power spectrum for equiprobable symbols.

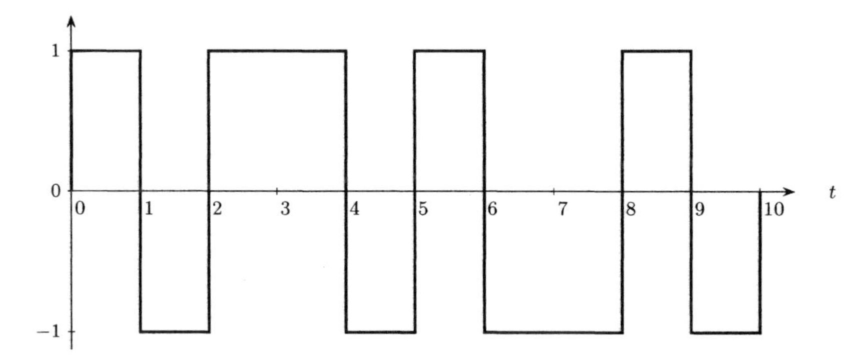

Fig. 5.12 Polar synchronous (NRZ) signal corresponding to the sequence 1011010010.

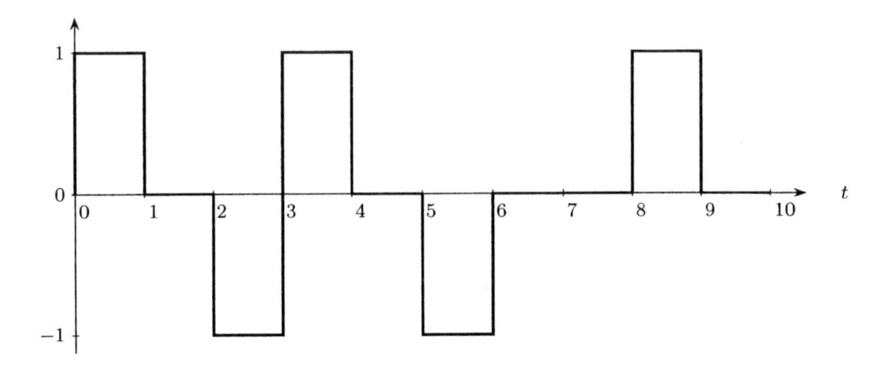

Fig. 5.13 Bipolar or AMI signal corresponding to the sequence 1011010010.

hence:

$$W(f) = 4T\,\frac{\sin^2(\pi f T)}{(\pi f T)^2}\,p(1-p) + \frac{(2p-1)^2}{2\pi}\,\delta(f). \tag{5.70}$$

For equiprobable symbols the polar code power spectrum has no lines. Apart from this fact, its spectral power density is similar to the unipolar code with a maximum at $f = 0$ and nulls at $f = n/T$ (where n is an integer).

Sometimes the acronyms NRZ and RZ [5] are used with a meaning different from the one here. In that case, both the polar and the unipolar code may be NRZ codes, when they follow the previous description, or RZ codes when each pulse is subdivided in two (usually with the same duration), of which the first holds the information and the second has always a null amplitude. For the same transmission speed RZ codes require pulses with half the duration of the equivalent NRZ codes.

5.4.5 Bipolar or AMI code

Bipolar or AMI[6] code, represented in Figure 5.13, solves the zero frequency component and the bit synchronization problems simultaneously. Bipolar code is generated with the following rules:

- binary zeros are represented by spaces (signals with zero amplitude);
- binary ones are represented, alternately, by positive and negative pulses.

The circuit in Figure 5.14 may be used to convert unipolar code into bipolar code. In this figure we plot the voltage waveforms in the circuit for the input sequence 1011010010.

[5] *Return to Zero.*
[6] *Alternate Mark Inversion.*

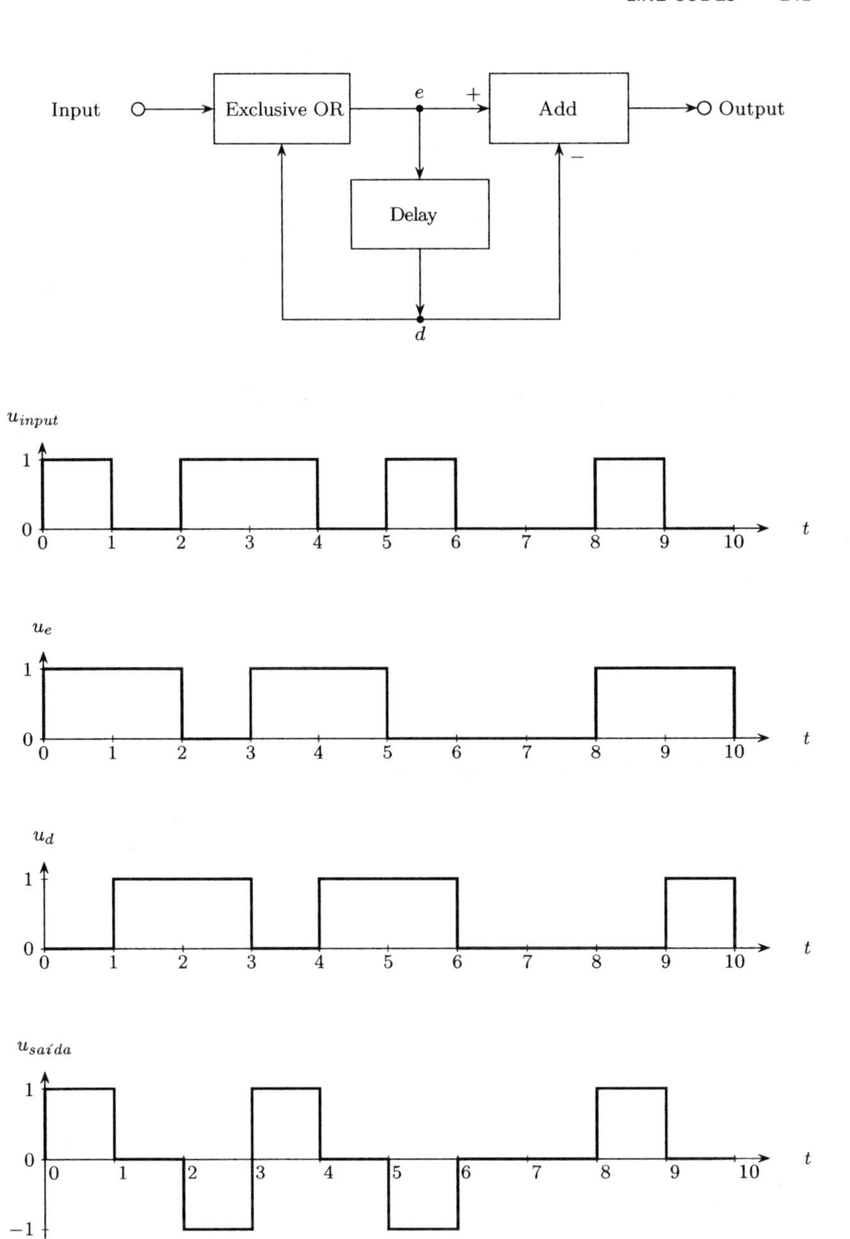

Fig. 5.14 Schematic diagram for a unipolar to bipolar code converter and voltage waveforms for the input sequence 1011010010.

A rectifier converts bipolar code into unipolar code, as shown in Figures 5.13 and 5.10.

When the same pulse duration is used for unipolar and bipolar code the required bandwidth for transmission is also the same, but since coding space is increased from binary to ternary, there is a possibility of error detection (two or more pulses with the same polarity), a feature that will be referred to later on.

It may be demonstrated [2] that the bipolar code spectral power density has no lines and is given by

$$W(f) = \frac{2p(1-p)}{T}|g_1(f)|^2 \frac{1-\cos(2\pi fT)}{1-2(2p-1)\cos(2\pi fT)+(2p-1)^2}, \tag{5.71}$$

where $g_1(f)$ is the pulse power spectrum. This spectral power density has nulls at $f = n/T$, where $n = 0, 1, 2, \ldots$ due to the zeros of function $1 - \cos(2\pi fT)$.

In Figure 5.15 we plot the spectral power density as a function of frequency f for elementary rectangular pulses, taking p as a parameter. As shown, the spectral power density maximum depends on the value of p but it is always near $1/2T$.

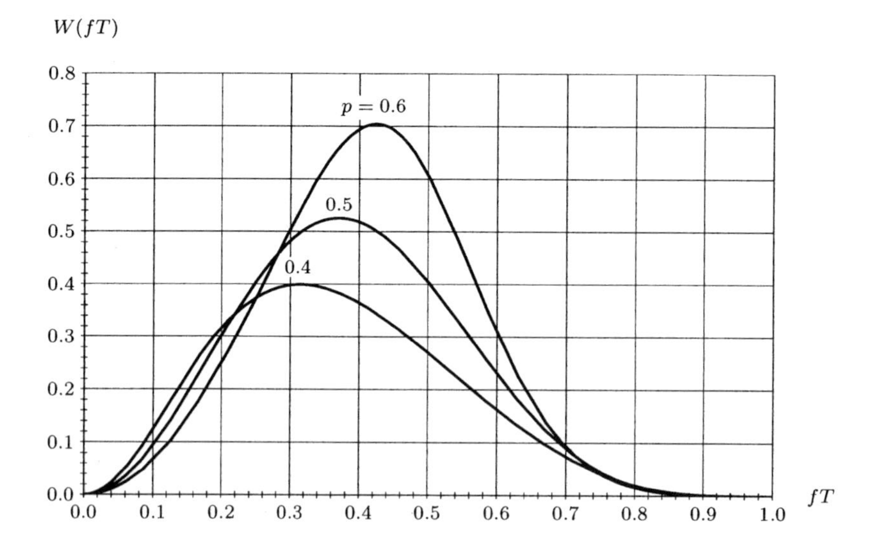

Fig. 5.15 Spectral power density for a bipolar signal using rectangular pulses, taking p as the parameter.

5.4.6 BNZS codes

Since polar code uses alternating positive and negative pulses to code sequences of ones, these sequences have a strong spectral component at $f = 1/2T$, where T is the pulse duration. The situation is quite different for sequences of zeros, which must be avoided in order not to loose bit synchronism. If long sequences of zeros cannot be avoided, then we must modify the bipolar signal.

In the bipolar signal with substituted sequences of n zeros, in short BNZS codes, for the transmission all sequences of n zeros are changed into n symbols that purposely contain a bipolar violation. In the reception, bipolar violations are detected and the original bipolar signal is recovered.

HDB3 code[7], recommended by ITU-T (Recommendation G.703 [17]) for the first, second and third digital hierarchies (except in North America and Japan) is a B4ZS code, that is, a bipolar code where all 4 zero sequences are substituted by a 4-bit sequence with a bipolar violation. The digit 3 in the code acronym HDB3, indicates that in this code there are no sequences of more than 3 consecutive zeros. The coding rules for the HDB3 code, referred to in Annex A to ITU-T Recommendation G.703 [17], are equal to the bipolar code, except when there are sequences with 4 consecutive zeros. In this case the fourth zero and, sometimes, also the first one, are modified so that successive polarity violations have an alternating parity and do not introduce a zero frequence component in the power spectrum.

5.4.7 PST code

The PST[8] code is another variation of the bipolar code, where words with two binary symbols are converted in words with two ternary symbols. Since there are 2^2 words with two binary symbols and 3^2 words with two ternary symbols, many conversion rules may be defined. One of the most usual ones is shown in Table 5.2.

Binary input	PST output	
	Mode +	Mode -
0 0	$- +$	$- +$
0 1	$0 +$	$0 -$
1 0	$+ 0$	$- 0$
1 1	$+ -$	$+ -$

Table 5.2 Binary to PST conversion rules.

[7] *High Density Bipolar.*
[8] *Pair Selected Ternary.*

This rule has two conversion modes + and −, where binary words 11 and 00 lead to the same result. To translate from binary into ternary, we choose one mode and stick to it until the word 11 appears at the input. Then we switch mode until the same word appears again, in which case we switch mode again. This cycle is repeated indefinitely.

PST code imposes framing of binary sequence into pairs of symbols which means that the decoder must recognize and keep track of borders between pairs. For random data the border is not difficult to identify since some of the possible misframes give rise to forbidden words (00, ++ and −−).

Spectral power densities for B6ZS and PST are similar, with nulls for $f = n/T$, with $n = 0, 1, 2, \ldots$ and a maximum close to $f = 1/2T$.

5.4.8 4B-3T code

Previously mentioned bipolar codes do not use coding space to decrease signaling speed but rather to facilitate synchronization and to achieve certain spectral properties.

4B-3T code translates four digit binary words into three ternary digit words achieving a 25 percent decrease in signaling speed. The translation rules from the 16 four binary digit words into the 27 three ternary digit words are shown in Table 5.3, where ternary words are grouped in −, 0 and + according to the polarity of the sum of their digits.

When more than one translation is possible, ternary words are alternately chosen among those grouped in − and +, so as to keep a zero dc component for the resulting code. Since the word 000 does not exist the 4B-3T code has a high timing content.

As previously mentioned for the PST code, for 4B-3T decoding we must ensure the right framing between three digit words.

4B-3T code spectral power density [21] is similar to other ternary codes (AMI, BNZS, PST) with a maximum near $f = 1/2T$.

5.4.9 Digital diphase or Manchester code

Bipolar and other codes derived from it, such as BNZS and PST, maintain and increase coding space to ease synchronism, to eliminate the zero frequency component or to allow for error detection. Similar results may be achieved while keeping the coding space and increasing the signaling speed. One of the most popular codes of the latter type is the digital diphase or Manchester code that makes use of a +− sequence to code a binary one and a −+ sequence to code a binary zero, as shown in Figure 5.16.

Manchester coding may be considered as resulting from the modulation of a square pulse (with period T equal to one half of the duration of the original

Binary input	4B-3T output		
	−	0	+
0 0 0 0	− − −		+ + +
0 0 0 1	− − 0		+ + 0
0 0 1 0	− 0 −		+ 0 +
0 0 1 1	0 − −		0 + +
0 1 0 0	− − +		+ + −
0 1 0 1	− + −		+ − +
0 1 1 0	+ − −		− + +
0 1 1 1	− 0 0		+ 0 0
1 0 0 0	0 − 0		0 + 0
1 0 0 1	0 0 −		0 0 +
1 0 1 0		0 + −	
1 0 1 1		0 − +	
1 1 0 0		+ 0 −	
1 1 0 1		− 0 +	
1 1 1 0		+ − 0	
1 1 1 1		− + 0	

Table 5.3 Translation rules from binary code into 4B-3T code.

polar signal) by the polar signal. The resulting power spectrum $W(f)$ is

$$W(f) = 4p(1-p)\left[\frac{\sin\left(\frac{\pi f T}{2}\right)}{\frac{\pi f T}{2}}\right]^2 \sin^2\left(\frac{\pi f T}{2}\right).$$

This power spectrum, plotted in Figure 5.17, corresponds to the polar signal spectrum shifted toward frequencies which are integer multiples of $1/T$.

Manchester code is mainly used for short links where cost of terminal equipment is more important than a non-optimal use of the available bandwidth. An example is Ethernet, a local area network (in short, a LAN) for data transmission.

5.4.10 CMI code

The coded mark inversion code, in short, CMI code, is another example of an increase in bandwidth in order to meet the objectives achieved in other codes by an increase in coding space. In CMI code, binary ones are coded as polar signals alternating with positive and negative polarity, and binary zeros are coded as a sequence $-+$ with a duration equal to the binary symbol, as shown in Figure 5.18. CMI coding is specified by ITU-T [17] (Recommendation G.703) for the fourth plesiochronous digital hierarchy (140 Mbit/s).

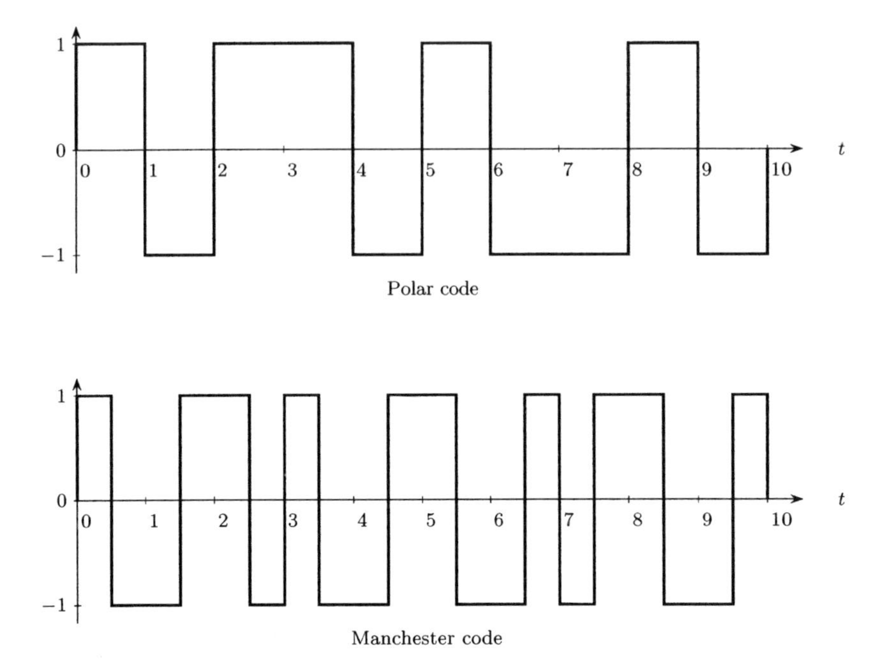

Fig. 5.16 Synchronous polar and digital diphase (or Manchester) code signals corresponding to the binary sequence 1011010010.

CMI power spectrum, computed from the expression in Poo [21] for rectangular shaped elementary pulses, has a null at $f = 0$ and a maximum at $f \approx 0.2/T$

5.4.11 Partial response codes

Partial response codes are an extension of the bipolar code principle. In Figure 5.19 we represent the block diagram of a circuit that produces a partial response code from a unipolar synchronous input. In the bipolar code the delay is T (the input signal period) and the signal $*$ in the addition is negative. This is why the bipolar code is also known as the $1 - \mathrm{D}$ code where D stands for the delay.

Partial response code class 1 or duo-binary code, denoted as the $1 + \mathrm{D}$ code is yet another partial response code where the delay is T and the sign $*$ in the addition is positive. The duo-binary code power spectrum, as given in Bic, Duponteil and Imbeaux [7], shows a maximum at the zero frequency and nulls at integer multiples of $1/2T$.

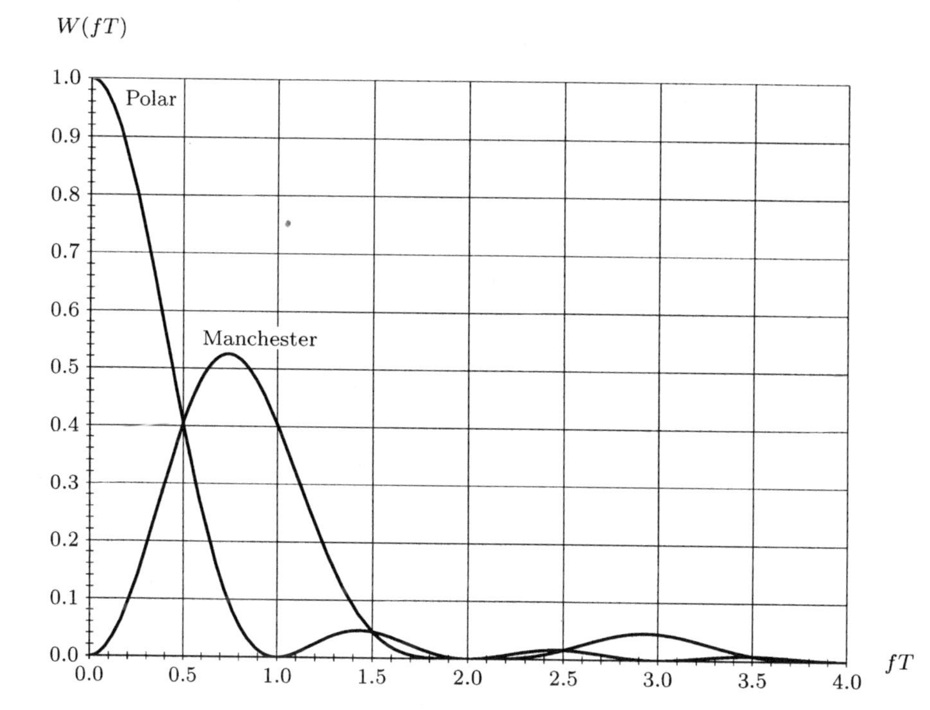

Fig. 5.17 Power spectrum for the digital diphase (or Manchester) and the polar codes.

It is possible to build partial response codes where the power spectrum, while similar to the duo-binary code with zeros at integer multiples of $1/2T$, also has a zero at the zero frequency. One of these codes is the modified duo-binary code, or $1 - \mathrm{D}^2$ code, where the delay is $2T$ and the signal $*$ in the addition is negative. The modified duo-binary code power spectrum is also given in Bic, Duponteil and Imbeaux [7].

5.4.12 Multilevel codes

To minimize bandwidth, multilevel codes may be used. These codes, with m signal levels, are usually polar and are derived from the unipolar code by grouping binary digits in blocks of $\log_2(m)$. Figure 5.20 represents the signal corresponding to the sequence 1011010010 in polar synchronous code and in quaternary polar code.

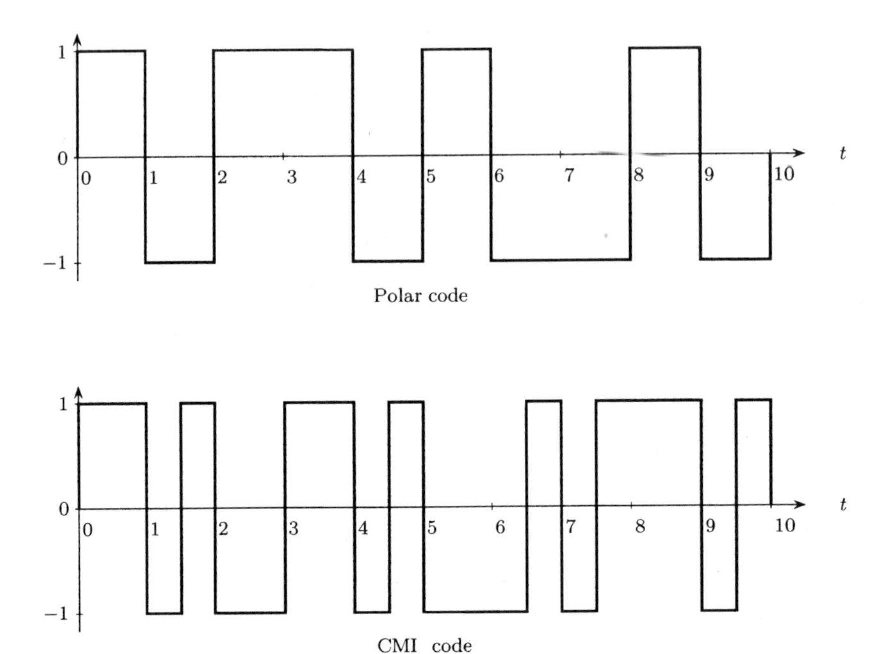

Fig. 5.18 Signals corresponding to the sequence 1011010010 in polar synchronous and in CMI codes.

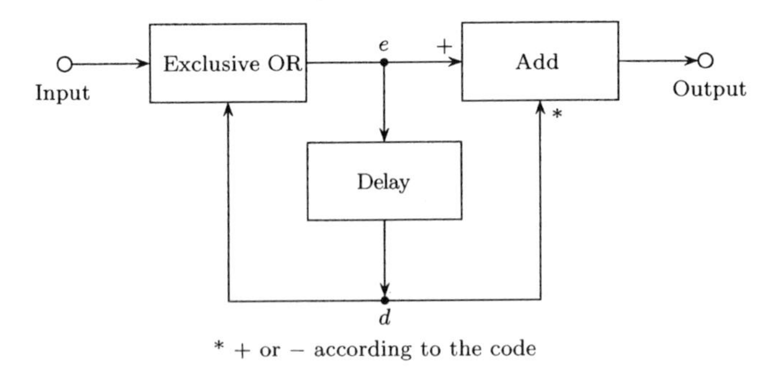

Fig. 5.19 Block diagram of a circuit that produces partial response codes from the unipolar code.

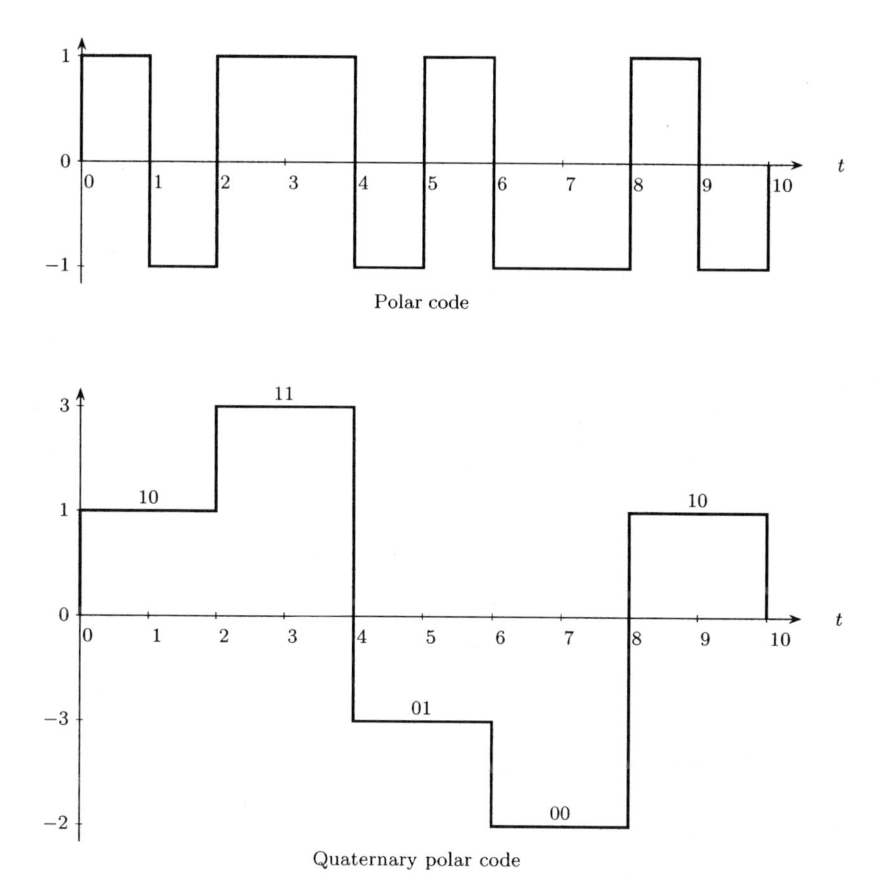

Fig. 5.20 Signals corresponding to the sequence 1011010010 in unipolar synchronous code and in quaternary polar code.

5.5 DIGITAL HIERARCHIES

5.5.1 Plesiochronous digital hierarchies

After reviewing the main codes used for digital signals we will briefly review the plesiochronous digital hierarchies, or PDH, recommended by ITU-T for use in time division multiplex telephony. These hierarchies, defined in ITU-T [17] Recommendation G.702, built on the basis of the 64 kbit/s used for voice coding, are shown in Table 5.4.

Hierarchies based on a first hierarchy at 1 544 kbit/s, denoted as T1, are used in North America and Japan. Elsewhere hierarchies are based on a 2048

Digital hierarchy	Binary rate (in kbit/s) for a first hierarchy with: 1544 kbit/s		2048 kbit/s
0	64		64
1	1 544		2 048
2	6 312		8 448
3	32 064	44 736	34 368
4	97 728		139 264

Table 5.4 Binary rates for the plesiochronous digital hierarchies defined in ITU-T [17] Recommendation G.702.

kbit/s first hierarchy, sometimes referred to as E1. The number of usable 64 kbit/s channels in each hierarchy is given in Table 5.5.

Digital hierarchy	Number of usable 64 kbit/s channels for a first hierarchy with: 1544 kbit/s		2048 kbit/s
1	24		30
2	96		120
3	480	672	480
4	1 440		1 920

Table 5.5 Number of usable 64 kbit/s channels for the ITU-T recommended plesiochronous digital hierarchies.

5.5.2 Synchronous digital hierarchies

Synchronous digital hierarchies, in short SDH, harmonize the different plesiochronous digital hierarchies based in 1544 and 2048 kbit/s. They make use of level 1 of the synchronous transport mode (STM-1) at 155.52 Mbit/s that supports $261 \times 9 = 2349$ usable channels at 64 kbit/s. ITU-T [17] Recommendation G.707 has already standardized the hierarchies shown in Table 5.6.

In order to accommodate both hierarchies based on 1544 and on 2048 kbit/s, mapping of plesiochronous hierarchies into synchronous hierarchies is not always very efficient. According to ITU-R[16] Recommendation F.750-2, one STM-1 module supports the following plesiochronous hierarchies, in alternative:

- one fourth hierarchy;

- three third hierarchies;
- twenty-one second hierarchies (excluding the 8 Mbit/s ones);
- sixty-three first hierarchies at 2 Mbit/s;
- eighty-four first hierarchies at 1.5 Mbit/s.

Codes used for baseband transmission in each hierarchy, identified by its binary rate, as per ITU-T [17] Recommendation G.703 are listed in Table 5.7

For radio transmission baseband code is converted into another code, usually unipolar, polar or multilevel, according to the type of modulation.

5.5.3 Frame structure for the T1 and the E1 hierarchies

The frame structure for the various hierarchies is specified in detail in ITU-T [17] Recommendation G.704 for plesiochronous hierarchies and in Recommendation G.707 for synchronous hierarchies. As an example in the next

Digital hierarchy	Binary rate (in kbit/s)
1	155 520
4	622 080
16	2 488 320
64	9 953 280

Table 5.6 Binary rates for synchronous digital hierarchies, according to ITU-T Recommendation G.707 [17].

Binary rate (kbit/s)	Code
1 544	AMI or B8ZS
6 312	B6ZS or B8ZS
44 736	B3ZS
97 728	AMI
2 048	HDB3
8 448	HDB3
34 368	HDB3
139 264	CMI
155 520	CMI

Table 5.7 Recommended codes for baseband transmission of ITU-T hierarchies (ITU-T Recommendation G.703 [17]).

few paragraphs we describe the frame structure for the first plesiochronous hierarchy at 1544 kbit/s (T1) and 2048 kbit/s (E1).

The T1 hierarchy accommodates 24 telephone channels each sampled at 8 kHz. Samples are quantized in 256 levels and coded with 8-bit words. The multiplex bit stream is obtained by interleaving the 24 8-bit words and adding an extra timing bit, preceding the information bits, in a 193-bit frame. The timing bits of 12 successive frames make up the 12-bit word [110111001000] used for frame synchronization.

Since there are 8000 frames per second, each with 193 bits, the bit rate is $8 \times 193 = 1544$ kbit/s.

Telephone signaling is transmitted within the voice bit stream as follows: the eight (least significant) bit of each sample is used for voice for 5 consecutive samples (and frames) and is used for signaling in the sixth frame. Thus in average samples are coded not with 8 bits but rather with $(5 \times 8 + 7)/6 = 7.83$ bits. Signaling uses a $8000/6 = 1333$ bit/s bit rate.

In the E1 hierarchy the multiplex bit stream is obtained by interleaving 32 8-bit words (one per channel) in a 256-bit frame. The frame repetition rate in 8000 Hz. A set of 16 consecutive frames, numbered from 0 to 15, is known as a multiframe. The 32 channels, numbered from 0 to 31, are used for:

- 30 PCM voice or 64 kbit/s digital signals (channels 1 to 15 and 17 to 31):
- 1 signalling channel (channel 16) which, when signalling is not required may be used as a 64 kbit/s channels in the same way as channels 1 to 15 and 17 to 31;
- 1 channel (channel 0) used for framing and other bits.

The multiframe alignment is indicated with 4 bits – 0000 – in the first bits of the signalling channel (channel 16), in the first frame (frame 0) of the multiframe.

In the even frames (including frame 0) the signal in channel 0 is composed of a bit, reserved for international use (that may be employed to implement an error detecting code, a cyclic redundancy code, or CRC), plus 7 bits – 0011011 – for frame synchronism. For the CRC code, each multiframe is divided into 2 blocks, each with 8 consecutive frames (or 2 048 bits), which correspond to 4 bits of the CRC code.

In the odd frames the signal in channel 0 is as follows:

- one bit reserved for international use which, when a CRC code is employed, is used to implement the 6-bit alignment code – 001011 –:
- one fixed 1 bit to differentiate the even from the odd frames;
- one bit for remote alarm signalling;
- five bits reserved for national use.

5.6 BASEBAND TRANSMISSION

5.6.1 Introduction

Although in microwave radio links digital signal transmission uses well-defined frequency bands, it is useful to approach this subject from the baseband point of view, that is, with the transmission of digital signals that extend down to the zero frequency.

Here we reduce baseband transmission to the derivation of expressions for the bit error ratio as a function of the main signal and noise characteristics. We will always assume that transmission channels are ideal, linear, with no attenuation nor intersymbol distortion, that simply add white Gaussian noise to the signal. These channels are usually known as AWGN channels

In general, for a binary signal, where the probabilities of occurrence of symbols are $\mathcal{P}(0)$ and $\mathcal{P}(1) = 1 - \mathcal{P}(0)$, the bit error ratio, in short *ber*, is equal to the sum of the probabilities that a 0 is transmitted and a 1 is received, plus that a 1 is transmitted and a 0 is received

$$ber = \mathcal{P}(0) \cdot \mathcal{P}(t = 0, r = 1) + \mathcal{P}(1) \cdot \mathcal{P}(t = 1, r = 0). \tag{5.72}$$

In this section we will derive expressions for the bit error ratio in terms of e_b/n_0 – ratio between the average bit power and noise power per unit bandwidth – in order to compare code performances.

5.6.2 Polar code (NRZ)

Assume that we apply a digital signal, in polar code (NRZ) with an amplitude $\pm u_s$ to one end – the transmitter – of an AWGN channel. At the other end of the channel – the receiver – we install a decision circuit that outputs a 1 when the received voltage is above zero and a 0 when it is below zero.

The probability that a 1 is received when a 0 is transmitted is

$$\begin{aligned} \mathcal{P}(t = 0, r = 1) &= \mathcal{P}(-u_s + n > 0) \\ &= \mathcal{P}(n > u_s), \end{aligned} \tag{5.73}$$

and the probability that a 0 is received when a 1 is transmitted is

$$\begin{aligned} \mathcal{P}(t = 1, r = 0) &= \mathcal{P}(u_s + n < 0) \\ &= \mathcal{P}(n < -u_s). \end{aligned} \tag{5.74}$$

Before the decision circuit, the signal with added noise, is filtered by a Nyquist lowpass filter, with cut-off frequency b (Hz). At the filter output, the signal $u_o(t)$ corresponding to a pulse with amplitude u_s is [9]

$$u_o(t) = u_s \frac{\sin(t/T)}{(t/T)} \frac{\cos(\beta t/T)}{1 - (2\beta t/T)^2}, \tag{5.75}$$

where β, between 0 and 1, is the excess band factor of the Nyquist filter and b, T and β are related by:

$$b = \frac{1+\beta}{T}. \tag{5.76}$$

The Nyquist filter ensures that, in the absence of noise, intersymbol interference is zero, if the sampling is performed at the right moment, that is, when $u_o(t)$ reaches its maximum value. Since

$$\begin{aligned} \max[u_o(t)] &= u_o(t=0) \\ &= u_s, \end{aligned} \tag{5.77}$$

due to the properties of the noise voltage and the symmetry of the Gaussian probability density function, at sampling time we will have

$$\mathcal{P}(n > u_s) = \mathcal{P}(n < -u_s), \tag{5.78}$$

where

$$\mathcal{P}(n > u_s) = \frac{1}{\sqrt{2\pi}} \int_{u_s/\sigma}^{\infty} \exp\left(-\frac{t^2}{2}\right) dt. \tag{5.79}$$

From

$$\mathcal{P}(n > u_s) + \mathcal{P}(n < u_s) = 1, \tag{5.80}$$

and from the properties of the Gaussian distribution it follows that

$$\mathcal{P}(n > u_s) = 1 - \frac{1}{\sqrt{2\pi}} \int_{-\infty}^{u_s/\sigma} \exp\left(-\frac{t^2}{2}\right) dt, \tag{5.81}$$

or, using the symbols in Abramowitz and Stegun [1]:

$$\begin{aligned} \mathcal{P}(n > u_s) &= 1 - P\left(\frac{u_s}{\sigma}\right) \\ &= Q\left(\frac{u_s}{\sigma}\right). \end{aligned} \tag{5.82}$$

For equiprobable transmitted symbols, that is, for $\mathcal{P}(0) = 1/2$ and$\mathcal{P}(1) = 1/2$, the bit error ratio becomes

$$\begin{aligned} ber &= \mathcal{P}(n > u_s) \\ &= 1 - P\left(\frac{u_s}{\sigma}\right) \\ &= Q\left(\frac{u_s}{\sigma}\right). \end{aligned} \tag{5.83}$$

Considering the complementary error function *erfc* , defined as

$$erfc(x) = \frac{2}{\sqrt{\pi}} \int_{x}^{\infty} \exp(-t^2) dt, \tag{5.84}$$

it is possible to show that

$$Q(x) = \frac{1}{2} erfc\left(\frac{x}{\sqrt{2}}\right). \tag{5.85}$$

Substituting the value of P in terms of *erfc* in the expression of the bit error ratio

$$ber = \frac{1}{2}\, erfc\left(\frac{u_s}{\sqrt{2}\sigma}\right). \tag{5.86}$$

If the filter is adapted to the digital signal, with period T, its noise bandwidth will be $b = 1/2T$, and the signal, after filtering, will have an average power per bit e_b

$$e_b = \lim_{T\to\infty}\left\{\int_{-T/2}^{+T/2} \frac{u_s^2}{R}\,\frac{\sin^2(\pi t/T)}{(\pi t/T)^2}\,\frac{\cos^2(\beta\pi t/T)}{[1-(\beta\pi t/T)^2]^2}\,dt\right\}. \tag{5.87}$$

This expression may be integrated if we recall that the power spectrum of a function is the Fourier transform of the autocorrelation of that function (Wiener-Khintchine theorem). Thus, if $f(t)$ is the given function, $G(\omega)$ its power spectrum and $\mathcal{R}(t)$ its autocorrelation function, we have

$$\mathcal{R}(t) = \lim_{T\to\infty}\left[\frac{1}{T}\int_{-T/2}^{+T/2} f(x)f(t-x)dx\right] \tag{5.88}$$

$$G(\omega) = \int_{-\infty}^{+\infty} R(t)\exp(-j\omega t)dt. \tag{5.89}$$

Applying the inverse Fourier transform to $G(\omega)$

$$\mathcal{R}(t) = \frac{1}{2\pi}\int_{-\infty}^{+\infty} G(\omega)\exp(j\omega t)d\omega, \tag{5.90}$$

equating the two expressions for the autocorrelation function, (5.88) and (5.90) for $t = 0$

$$\lim_{T\to\infty}\left[\frac{1}{T}\int_{-T/2}^{+T/2} f(x)f(-x)dx\right] = \frac{1}{2\pi}\int_{-\infty}^{+\infty} G(\omega)d\omega,$$

or:

$$\lim_{T\to\infty}\left[\frac{1}{T}\int_{-T/2}^{+T/2} f(x)f(t-x)dx\right] = \int_{-\infty}^{+\infty} G(f)df. \tag{5.91}$$

Bellamy [5] uses the following expression for the Fourier transform of the pulse at the output of a Nyquist filter:

$$g(f) = \begin{cases} 1 & \text{for } |f| \le \frac{1-\beta}{2T}, \\ \frac{1}{2}[1+\cos(\frac{\pi|f|T}{\beta} - \frac{\pi(1-\beta)}{2\beta})] & \text{for } \frac{1-\beta}{2T} \le |f| \le \frac{1+\beta}{2T}, \\ 0 & \text{for } |f| > \frac{1+\beta}{2T}. \end{cases}$$

Since

$$\int_{-\infty}^{+\infty} G(f)df = \int_{\infty}^{+\infty} |g(f)|^2 df$$
$$= \frac{1}{T}, \tag{5.92}$$

we have

$$e_b = \frac{u_s^2}{R}T. \tag{5.93}$$

As the noise power per unit bandwidth is

$$n_0 = \frac{\sigma^2}{R}\frac{1}{b}, \tag{5.94}$$

the bit error ratio may be rewritten as

$$ber = \frac{1}{2} erfc\left(\sqrt{\frac{e_b}{n_0}}\right). \tag{5.95}$$

In Figure 5.21 we plot the bit error ratio (*ber*) as a function of E_b/N_0, expressed in dB.

The complementary error function may be computed using the polynomial approximation given in [1] or, if its argument is larger than 2, by the following approximated formula (relative error less than about 0.1, and decreasing with increasing values of the argument)

$$erfc(x) \approx \frac{\exp(-x^2)}{x\sqrt{\pi}}. \tag{5.96}$$

5.6.3 Unipolar code

Unipolar code performance in additive white Gaussian noise may be derived from bipolar code performance. Since we only transmit pulses with an amplitude u_s (corresponding to binary ones) the detector level must be adjusted for $u_s/2$. This situation is similar to a polar code with pulses with an amplitude $u_s/2$ instead of u_s, that is, with an average energy per pulse that is 1/4 of the original polar code. If 0 and 1 symbols are equiprobable

$$ber = \frac{1}{2} erfc\left(\frac{u_s}{2\sqrt{2}\sigma}\right). \tag{5.97}$$

Since in polar code pulses corresponding to binary zeros and ones have the same average energy, whereas in the unipolar code only binary ones have non-zero energy, if the probability of binary ones is $\mathcal{P}(1) = 1/2$, the average energy

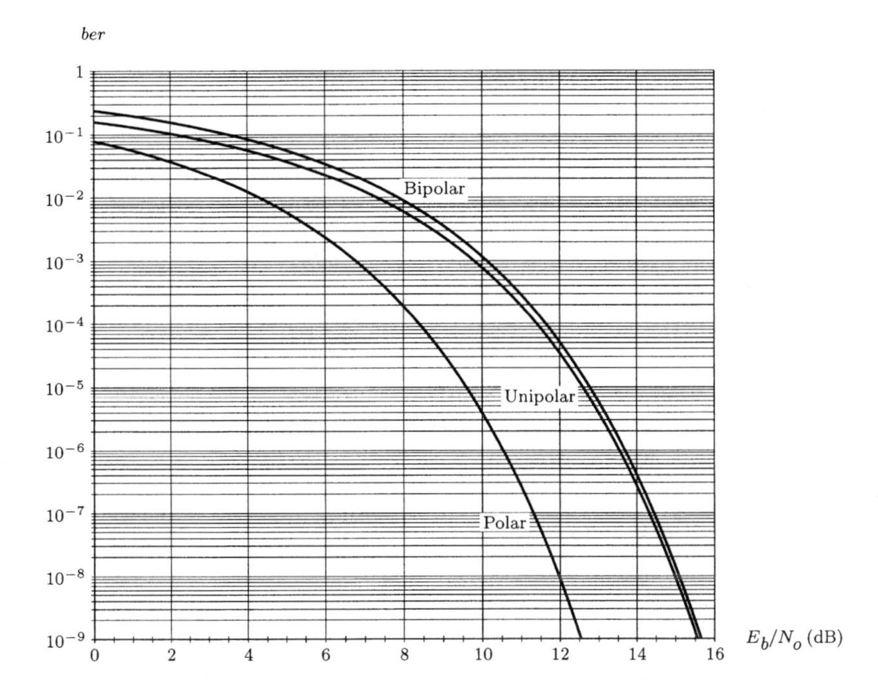

Fig. 5.21 Bit error ratio (*ber*) as a function of E_b/N_0, in dB, for the polar, unipolar and bipolar codes.

per bit of a polar code is 1/2 of the average energy per bit of the equivalent unipolar code. Under these conditions the bit error ratio may be written as

$$ber = \frac{1}{2} erfc\left(\sqrt{\frac{e_b}{2n_0}}\right). \tag{5.98}$$

In Figure 5.21 we plot the bit error ratio for unipolar code as a function of E_b/N_0, expressed in dB. Please note that it is possible to obtain the bit error ratio for unipolar code by shifting 3 dB to the right the polar code bit error ratio curve.

The previous expressions were derived using a detector adjusted for one half of the pulse height. In many practical cases the detector is adjusted for a constant level and thus the bit error ratio curves as a function of e_b/n_0 are no longer the same. Let u_d be the detection level, u_s the pulse amplitude after the filter and u_n the noise voltage at the same point. Then

$$ber = \mathcal{P}(0) \cdot \mathcal{P}(u_n > u_d) + \mathcal{P}(1) \cdot \mathcal{P}(u_s + u_n \leq u_d). \tag{5.99}$$

But, from (5.82)

$$\mathcal{P}(u_n > u_d) = Q\left(\frac{u_d}{\sigma}\right) \tag{5.100}$$

hence

$$\begin{aligned}\mathcal{P}(u_s + u_n \leq u_d) &= \mathcal{P}(u_n > u_s - u_d) \\ &= Q\left(\frac{u_s - u_d}{\sigma}\right).\end{aligned} \tag{5.101}$$

Substituting the probabilities given in (5.100) and (5.101) in (5.99) and making use of the relations between Q and *erfc* given in (5.85) we have:

$$ber = \frac{\mathcal{P}(0)}{2} erfc\left(\frac{u_d}{\sqrt{2}\sigma}\right) + \frac{\mathcal{P}(1)}{2} erfc\left(\frac{u_s - u_d}{\sqrt{2}\sigma}\right). \tag{5.102}$$

The average bit energy is now

$$e_b = \mathcal{P}(1)\frac{u_s^2}{R}T. \tag{5.103}$$

Introducing the concept of bit power equivalent to a detector voltage e_d, defined as the power of a pulse with the same shape as those used for symbol 1 but with the amplitude u_d we get, from (5.102)

$$ber = \frac{\mathcal{P}(0)}{2} erfc\left(\sqrt{\frac{e_d}{\mathcal{P}(1)no}}\right) + \frac{\mathcal{P}(1)}{2} erfc\left(\sqrt{\frac{e_b}{\mathcal{P}(1)n_0}} - \sqrt{\frac{e_d}{\mathcal{P}(1)n_0}}\right). \tag{5.104}$$

When the detector level is constant, the bit error ratio has a constant factor and a variable one, which is a function of the pulse height after the filter. This feature may be easily seen in Figure 5.22, where we plot the bit error ratio as a function of E_b/N_0, for different values of the detector level. To help comparison we also plot in the same Figure 5.22 the performance of an optimum detector, adjusted for a threshold voltage equal to 1/2 of the incoming pulses, that is,

$$\left(\frac{e_d}{n_0}\right)_{opt} = \frac{1}{4}\frac{e_b}{n_0}. \tag{5.105}$$

5.6.4 Bipolar code (AMI)

Bipolar code performance in the presence of additive white Gaussian noise is similar to that of unipolar code. In fact, for each transmitted signal, the decision circuit has to determine if it is a zero or an impulse, and in this case, which is its polarity. Thus we have to consider two decision thresholds, equal to half height of the pulse amplitude, one for each polarity.

The difference in performance between the unipolar and the bipolar codes derives from the fact that a zero in the bipolar code may be affected both by negative and positive noise, whereas in the unipolar code it can only be affected by positive noise.

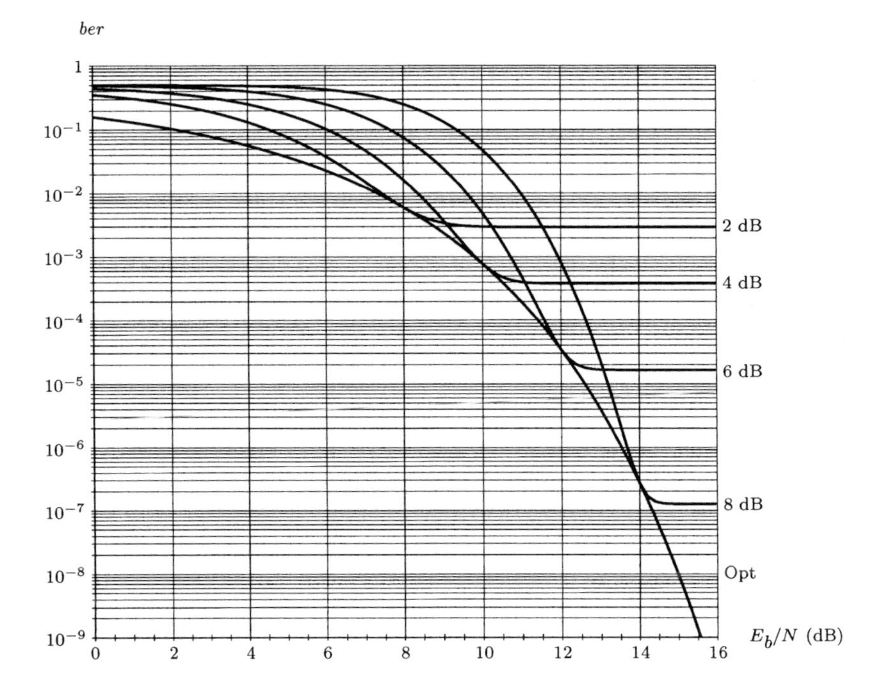

Fig. 5.22 Bit error ratio as a function of E_b/N_0, for the unipolar code, with equiprobable symbols and varying detector levels E_d/N_0.

Let $ber(0)$ and $ber(1)$ be the probabilities of error and $\mathcal{P}(0)$ and $\mathcal{P}(1)$ the probabilities of occurrence of symbols 0 and 1, respectively. Then, the probability of error will be

$$ber = ber(0) \cdot \mathcal{P}(0) + ber(1) \cdot \mathcal{P}(1). \tag{5.106}$$

Considering noise properties we have, for unipolar code

$$ber(0) = ber(1), \tag{5.107}$$

hence, if $\mathcal{P}(0) = 1/2$ and $\mathcal{P}(1) = 1/2$

$$ber = ber(0) = ber(1). \tag{5.108}$$

For unipolar code the general expression is the same but, as we have just shown, the zero bit error ratio is double the unipolar code bit error ratio

$$ber(0) = 2\,ber(0) \cdot \mathcal{P}(0) + ber(1) \cdot \mathcal{P}(1). \tag{5.109}$$

If $\mathcal{P}(0) = 1/2$ and $\mathcal{P}(1) = 1/2$, we get, for bipolar code

$$ber = 1.5\,ber(0), \tag{5.110}$$

which means that, for the same value of E_b/N_0, bipolar code has a bit error ratio 50 percent higher than unipolar code. In Figure 5.21 we plot the bit error ratio for bipolar code as a function of E_b/N_0, expressed in dB.

5.6.5 Multilevel codes

Consider a multilevel code, where the signal has a peak-to-peak amplitude $2u$, quantized into m equally spaced levels. To detect such a signal we require m equally spaced threshold levels, separated by a

$$a = \frac{2u}{m-1}. \tag{5.111}$$

The i level threshold will be

$$u_{d_i} = -u_s + ai. \tag{5.112}$$

We will assume that the received signal $u_s + u_n$ corresponds to level i when

$$u_{d_{i-1}} \leq u_s + u_n \leq u_{d_i}, \tag{5.113}$$

where u_n is the noise voltage that coexists with signal u_s.

At the intermediate values ($i = 2, 3, ..., m-1$) an error occurs when the absolute value of the noise voltage exceeds $a/2$. Thus the probability of error in identifying level i (with $i \neq 1$ and $i \neq m$) $ber(i)$ is

$$ber(i) = \mathcal{P}(i)\left[\mathcal{P}(u_n \leq -a/2) + \mathcal{P}(u_n \geq a/2)\right]. \tag{5.114}$$

We have previously shown that, for additive white Gaussian noise, we had

$$\mathcal{P}(u_n \leq -a/2) = \mathcal{P}(u_n \geq a/2) = \frac{1}{2}\, erfc\left(\frac{a/2}{\sqrt{2}\sigma}\right), \tag{5.115}$$

hence

$$ber(i) = erfc\left(\frac{a/2}{\sqrt{2}\sigma}\right). \tag{5.116}$$

At the first threshold level an error may only occur when the noise voltage is positive and exceeds $a/2$, thus

$$ber(1) = \mathcal{P}(1) \cdot \mathcal{P}(u_n \geq a/2),$$

or

$$ber(1) = \frac{\mathcal{P}(1)}{2} erfc\left(\frac{a/2}{\sqrt{2}\sigma}\right). \tag{5.117}$$

Similarly, at the last threshold level, an error may only occur when the noise voltage is negative and exceeds $-a/2$, hence

$$ber(m) = \mathcal{P}(m) \cdot \mathcal{P}(u_n \leq -a/2),$$

or

$$ber(m) = \frac{\mathcal{P}(m)}{2} erfc\left(\frac{a/2}{\sqrt{2}\sigma}\right). \quad (5.118)$$

The probability of symbol error, or symbol error ratio, *ser* will be

$$ser = \sum_{i=1}^{m} \mathcal{P}(i) ber(i). \quad (5.119)$$

If all levels are equiprobable

$$\mathcal{P}(i) = \frac{1}{m},$$

and

$$ser = \frac{m-1}{m} erfc\left(\frac{a/2}{\sqrt{2}\sigma}\right). \quad (5.120)$$

Recalling that a symbol of a multilevel code with m levels has an information I

$$I = \log_2(m),$$

we may write the bit error ratio as

$$ber = \frac{1}{\log_2(m)} \frac{m-1}{m} erfc\left(\frac{a/2}{\sqrt{2}\sigma}\right), \quad (5.121)$$

or, in terms of the peak signal level u

$$ber = \frac{1}{\log_2(m)} \frac{m-1}{m} erfc\left[\frac{u}{(m-1)\sqrt{2}\sigma}\right]. \quad (5.122)$$

Assuming that the multilevel signal has a period T and a rectangular shape, with a peak level u at the receiver Nyquist filter output, and that this signal is applied to a real impedance R, the average power e_b per received level will be

$$e_b = \mathcal{P}(1)\frac{u^2}{R}T + \mathcal{P}(m)\frac{u^2}{R}T + \sum_{i=2}^{m-1} \mathcal{P}(i)\frac{u_i^2}{R}T. \quad (5.123)$$

Assuming that all levels are equiprobable and recalling that the voltages corresponding to the various quantization levels are symmetrical in relation to zero, we may write the average pulse power e_b as twice the sum of the power of pulses with amplitudes $1, 3, 5, \ldots, m-1$ times the elementary pulse amplitude $u/(m-1)$, divided by m, that is,

$$e_b = \frac{u^2}{(m-1)^2}\, T\, \frac{2}{R\, m}\, [1^2 + 3^2 + 5^2 + \ldots + (m-1)^2].$$

Since the sum of the squares of odd numbers from 1 to $m-1$ may be calculated from (5.4) we may rewrite e_b as

$$e_b = \frac{u^2}{(m-1)^2} T \frac{1}{Rm} \frac{m(m-1)(m+1)}{3},$$

and manipulating

$$e_b = \frac{u^2}{m-1} \frac{T}{R} \frac{m+1}{3}. \tag{5.124}$$

Using the maximum power per pulse

$$e_{b_{max}} = u^2 \frac{T}{R},$$

the ratio between the average and the maximum power per pulse becomes

$$\frac{e_b}{e_{b_{max}}} = \frac{1}{3}\left(\frac{m+1}{m-1}\right). \tag{5.125}$$

As the noise power per unit bandwidth is given by

$$\begin{aligned} n_0 &= \frac{\sigma^2}{R}\frac{1}{b} \\ &= \frac{\sigma^2}{R} 2T, \end{aligned} \tag{5.126}$$

the bit error ratio may be expressed in terms of $e_{b_{max}}/n_0$

$$ber = \frac{1}{\log_2(m)} \frac{m-1}{m} \, erfc\left(\frac{1}{m-1}\sqrt{\frac{e_{b_{max}}}{n_0}}\right), \tag{5.127}$$

or e_b/n_0

$$ber = \frac{1}{\log_2(m)} \frac{m-1}{m} \, erfc\left[\sqrt{\frac{3e_b}{(m^2-1)n_0}}\right]. \tag{5.128}$$

For $m = 2$, the previous expression reduces to the polar (NRZ) code bit error ratio (5.95).

A quick look at (5.128) may induce the reader in error, since for the same bit error ratio the value of e_b/n_0 increases with m, which seems to make multilevel codes ($m > 2$) unattractive. However one should not forget that, for the same bit rate (in bits/s), multilevel codes increase pulse duration and thus decrease the occupied bandwidth. Here lies the importance of multilevel codes: they offer the possibility of exchanging a decrease in bandwidth for an increase in bit error ratio, for the same value of e_b/n_0.

In Figures 5.23 and 5.24 we plot the bit error ratio (ber) as a function of E_b/N_0 and $E_{b_{max}}/N_0$, both expressed in dB.

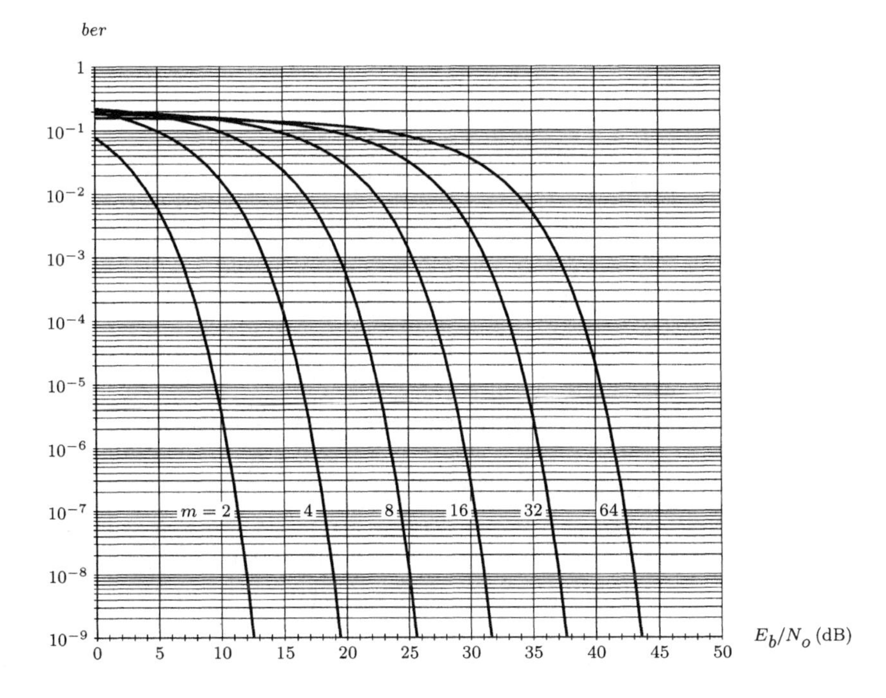

Fig. 5.23 Multilevel code bit error ratio as a function of E_b/N_0.

5.7 MODULATED SIGNAL TRANSMISSION

5.7.1 Introduction

Modulation is a technique used to transform baseband signals into signals with an appropriate bandwidth around a chosen center frequency. Since radio frequency spectrum is a scarce resource one should:

- adopt modulations which are efficient in spectrum usage;
- code digital signals to reduce bandwidth and to avoid spectral lines.

Polar code (NRZ) is the most frequently used code although, in some cases, multilevel codes or partial response codes $1 + \mathrm{D}$ are preferred.

Besides bit rate, the bit error ratio is the main quality indicator of a digital link. To compute its value we will assume, to start with, that the transmission channel is linear and additive, and that noise is spectrum limited, white and Gaussian.

Comparing different types of modulation is made easier if bit error ratios are always expressed in terms of the same link parameter. In a radio link

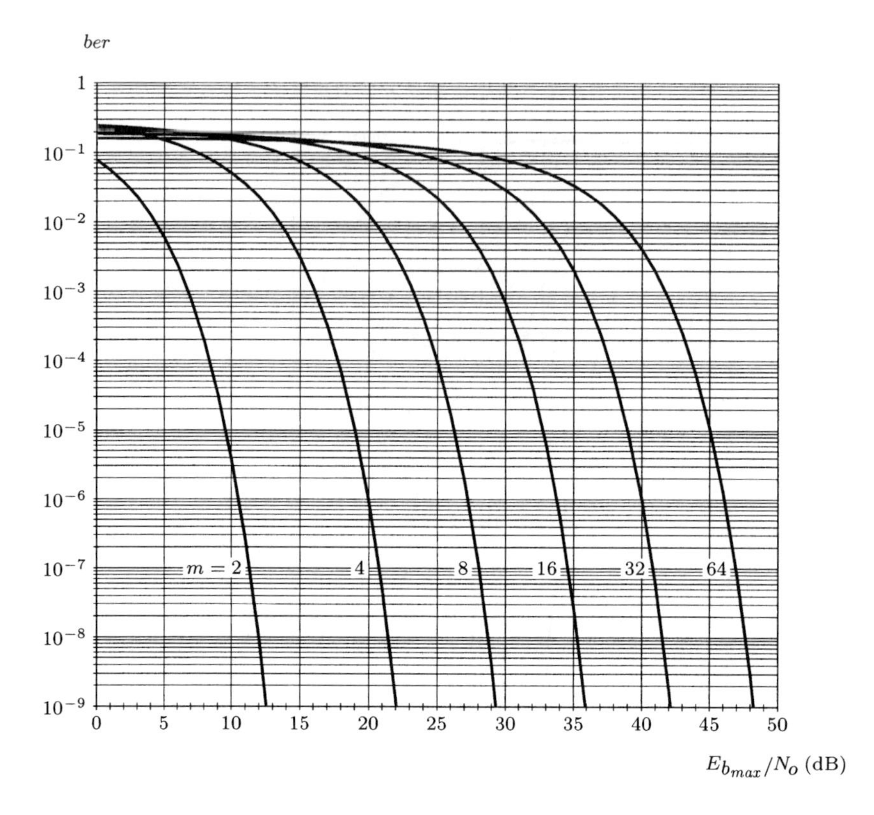

Fig. 5.24 Multilevel code bit error ratio as a function of $E_{b_{max}}/N_0$.

the most natural choice for this parameter is the carrier-to-noise ratio c/n. However since the ratio e_b/n_0, used in the baseband transmission, has also become a popular choice also the transmission of modulated signals, we will start by deriving a relation between c/n and e_b/n_0.

Let us assume that c is the average carrier power, n the noise power in the bandwidth b_{rf}, just before the demodulator, and f_b the bit frequency. The average power e_b associated with one bit is

$$e_b = \frac{c}{f_b}. \tag{5.129}$$

The noise power density no is

$$n_0 = \frac{n}{b_{rf}}. \tag{5.130}$$

Dividing equations (5.129) and (5.130) we get

$$\frac{e_b}{n_0} = \frac{c}{n}\frac{b_{rf}}{f_b}, \tag{5.131}$$

or, in logarithmic units

$$E_b/N_0 = C/N + 10\log_{10}\left(\frac{b_{rf}}{f_b}\right), \tag{5.132}$$

where, as usual

$$\begin{aligned} E_b/N_0 &= 10\log_{10}\left(\frac{e_b}{n_0}\right), \\ C/N &= 10\log_{10}\left(\frac{c}{n}\right). \end{aligned}$$

Introducing the concept of modulation efficiency, η_m, as the ratio between the bit frequency and the equivalent noise bandwidth

$$\eta_m = \frac{f_b}{b_{rf}}, \tag{5.133}$$

equation (5.131) may be rewritten as

$$\frac{e_b}{n_0} = \frac{c}{n}\frac{1}{\eta_m}, \tag{5.134}$$

or, in logarithmic units

$$E_b/N_0 = C/N - 10\log_{10}(\eta_m). \tag{5.135}$$

5.7.2 Modulation techniques

There are basically three modulation techniques, each with many variations:

1. amplitude modulation, or AM,
2. frequency modulation or FM,
3. phase modulation, or or PM,

plus quite a few hybrid techniques such as amplitude and phase modulation which are quite interesting because of potential savings in bandwidth.

When the modulating signal is digital, modulation techniques change their names. Hence, amplitude modulation (AM), frequency modulation (FM) and phase modulation (PM) become amplitude shift keying (ASK), frequency shift keying (FSK) and phase shift keying (PSK), respectively.

All modulation techniques may be conceived in terms of sinewaves (carriers) modulated by digital signals (baseband) with the information in coded form.

The next description of the modulating techniques follows closely Oetting [19].

5.7.3 Amplitude modulation

Let $s_m(t)$ be a polar (NRZ) code signal. Using this signal to modulate a sincwave carrier, with angular frequency ω_c and unit amplitude amplitude , we get

$$r_{AM}(t) = [1 + k \cdot s_m(t)] \cos(\omega_c t). \tag{5.136}$$

When the modulation depth is unity ($k = 1$), the resulting signal is known as on-off keying, or OOK. When the carrier is suppressed we have double side band suppressed carrier or DSB-SC

$$r_{DSB}(t) = s_m m(t) \cos(\omega_c t). \tag{5.137}$$

If $s_m(t)$ only takes values 0 and 1, as in the unipolar code, we get the OOK modulation again. If, on the other hand, $s_m(t)$ takes values -1 and $+1$, as in the polar code, the resulting modulation, which is still an amplitude modulation, may also be considered as a binary phase modulation or 2-PSK, since the carrier sign change may also be achieved with the addition and subtraction of π radians to its phase.

Suppressing one sideband of a DSB signal results in a single sideband signal, or SSB, which is spectrally more efficient than the DSB signal. To facilitate the demodulation of an SSB signal, at the cost of a slightly lower spectral efficiency, we may start from an AM signal, to which we remove one of the sidebands (and possibly attenuate the carrier) producing yet another type of amplitude modulation: vestigial sideband modulation, or VSB.

The use of SSB and VSB in digital radio is rather infrequent since it is possible to devise other kind of amplitude modulation, which is simpler to implement and yet provides the same spectral efficiency. We are referring to quadrature amplitude modulation, QAM or 4-QAM, the result of adding two DSC-SC signals with carriers out of phase by $\pi/2$.

Let $si_m(t)$ and $sq_m(t)$ be two independent, polar coded, digital signals and ω_c the carrier angular frequency. The general form of the QAM signal is

$$r_{QAM}(t) = si_m(t) \cos(\omega_c t) + sq_m(t) \sin(\omega_c t). \tag{5.138}$$

The carrier modulated by $si_m(t)$ is known as the in-phase carrier or I carrier, while the carrier modulated by $sq_m(t)$ is known as the quadrature carrier, or Q carrier. Just as a DSB signal, where $m(t)$ only takes the values -1 and $+1$, may be known as a 2-PSK signal, the QAM signal when $si_m(t)$ and $sq_m(t)$ are restricted to $+1$ and -1, may be known as a quaternary phase modulation or 4-PSK.

If modulating signals are coded with multilevel codes m_i and m_q, the result is known as m-QAM, where $m = m_i \times m_q$. Often $m_i = m_q$, and 16-QAM, 64-QMA and even 256-QAM modulations are nowadays quite common. Laboratory prototypes may use higher values of m_i and m_q .

If modulating signals $si_m(t)$ and $sq_m(t)$ are coded with a ternary code (with levels -1, 0 and $+1$) that minimize the intersymbol interference caused by filtering, the result is quadrature partial response amplitude modulation, or QPR.

Amplitude modulations based in DSB-SC, such as 2-PSK and 4-PSK (or 4-QAM), can only be demodulated using coherent detection.

5.7.4 Frequency modulation

The simplest form of frequency modulation, known as frequency shift keying, or FSK, is obtained by shifting the frequency of a sinewave oscillator between two frequencies ω_1 and ω_2 separated by $\Delta\omega$, such that

$$\Delta\omega \ll \omega_1, \omega_2.$$

Let ω_c be the average angular frequency of ω_1 and ω_2 and let $s_m(t)$ be the digital signal in polar code. The general form of FSK modulation is

$$r_{FSK}(t) = \cos\{[\omega_c + s_m(t)\Delta\omega]t\}. \tag{5.139}$$

It is usual to define the FSK modulation index k_f as the ratio between the frequency difference and the symbol duration (which for binary codes is equal to the inverse of the bit rate):

$$\begin{aligned} k_f &= \frac{\Delta\omega}{2\pi}T \\ &= \Delta f\, T. \end{aligned} \tag{5.140}$$

FSK signals may be demodulated either with coherent or with incoherent detection. For incoherent detection it is sufficient to use two bandpass filters, centered on angular frequencies ω_1 and ω_2, followed, in each case by envelope detectors. To avoid the superposition of the filters pass bands the difference between center frequencies must be larger than $1/T$ (or $k_f \geq 1$). Alternatively, we may use a discriminator (which converts frequency changes into amplitude changes) followed by an envelope detector. In the alternative solution the previous requirement no longer applies.

Some variants of the basic frequency modulation technique are based on the idea of continuous phase and hence are known as continuous phase frequency shift keying, or CPFSK. Abrupt phase transitions are avoided and a significant reduction of off-band spectral components is achieved. It is possible to show that ,with coherent detection, k_f values around 0.7 lead to an optimal performance for any observation interval.

Yet another variant of the basic frequency modulation is known as minimum shift keying, or MSK, or fast frequency modulation. MSK is a special case of CPFSK, that makes use of coherent detection with $k_f = 1/2$. Its performance is equivalent to phase shift keying, with the advantages of CPFSK and a simple self-synchronizing method (which does not exist for CPFSK with $k_f = 0.7$).

5.7.5 Phase modulation

In phase modulation, or phase shift keying, or PSK, the digital signal modulates in phase a sinewave carrier, with $2\pi/m$ phase shifts, where m is the number of levels of the digital signal. Although m-PSK systems, for any integer value of m are conceivable, in practice m is restricted to 2, 4, and, possibly, 8.

Let $s_m(t)$ be the modulating signal, in unipolar multilevel code of order m and ω_c the carrier angular frequency. The general form of a m-PSK signal is

$$r_{PSK}(t) = \cos\left[\omega_c t + s_m(t)\frac{2\pi}{m}\right]. \tag{5.141}$$

The simplest form of PSK, is known as 2-PSK or binary phase shift keying, or BPSK, where the carrier is phase shifted 0 or π following the modulating signal values of 0 or 1, respectively.

The demodulation of phase modulated signals requires coherent detection, where the reference phase is usually produced using a non-linear operation. To resolve the π ambiguities, inherent to some phase reference capturing techniques, we may use differential encoding phase shift keying, or DE-PSK and convey information by way of phase transitions rather than by the absolute phase. For instance, no phase transition could correspond to a 0 and a phase transition to a 1. DE-PSK performance is somewhat lower than PSK because an error in one bit causes errors in the following bits.

Another variant of 2-PSK is differential phase shift keying, or DPSK, where just as in DE-PSK information is differentially encoded. The difference between DPSK and DE-PSK lies in the detector. In DPSK there is no attempt to extract a phase reference. Instead, the previous signal phase is used for the current bit. Since the phase reference is not obtained using the previous bits, but only the previous one, DPSK performance is slightly worse than DE-PSK.

A very popular form of m-PSK corresponds to $m = 4$ and is known as 4-PSK, quadrature phase shift keying or QPSK. As before (with $m = 2$) input data may be differentially encoded, yielding DQPSK or 4-DPSK, with the inherent performance degradation.

A modification of QPSK is known as offset-keyed quadrature phase shift keying, OQPSK, staggered quadrature phase shift keying or SQPSK. The OQPSK signal may be visualized starting from the QPSK (or QAM) signal. In the latter during the time interval corresponding to two bits of the modulating signal the I carrier is modulated by one bit and the Q carrier by the other bit. The resulting signal may take any if the following phase values (0, $\pm\pi/2$, π) and, consequently, sudden phase shifts of $\pm\pi/2$ and $\pm\pi$ radians may occur. In the OQPSK the signal that modulates the Q carrier is shifted $T/2$ in relation to the signal that modulates the I carrier (T being the bit duration). The phase transition rules are such that when we add the two modulated carriers the resulting signal exhibits sudden phase shifts limited to $\pm\pi/2$ separated by $T/2$, whereas in QPSK these shifts may reach $\pm\pi$ and are separated by T.

5.7.6 Modulation constellation

To visualize the various modulation techniques, be it amplitude, phase or quadrature, we may simply plot the points corresponding to the amplitude and the (relative) phase of each possible modulation state. This plot is usually known as the constellation of states of modulation or simply the modulation constellation. In Figure 5.34 we plot the constellations for DSB (or 2-PSK), 4-PSK (or 4-QAM), 8-PSK and 16 QAM.

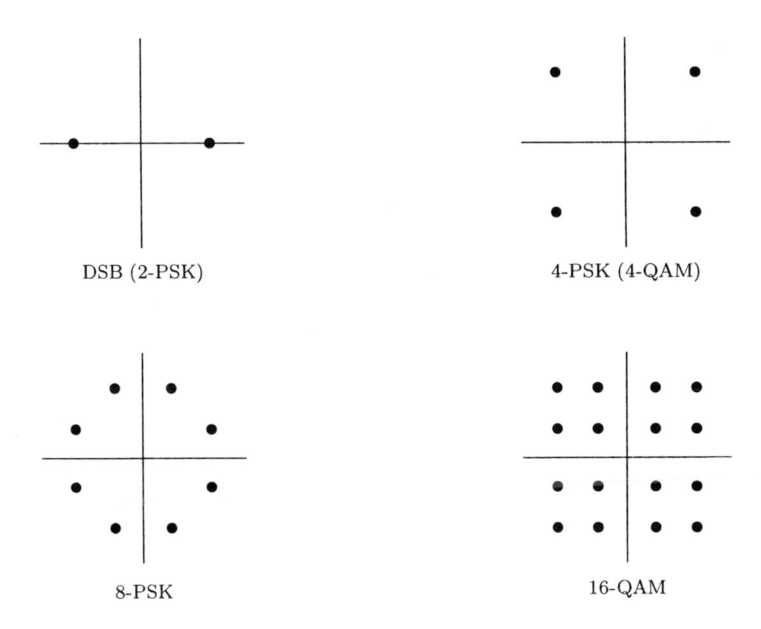

Fig. 5.25 Constellations for DSB (2-PSK), 4-PSK (4-QAM), 8-PSK and 16-QAM.

5.8 NOISE PERFORMANCE OF MODULATION TECHNIQUES

5.8.1 Amplitude modulation with coherent detection

When we multiply a signal by a sinewave carrier, we get a double sideband amplitude modulated digital signal. If the digital code has a non-zero direct current component, such as for the unipolar code, the result is known as on-off keying, or OOK. Otherwise the modulated signal is known as double sideband suppressed carrier, or DSB-SC. In any case, the modulated signal may be detected coherently (or synchronously) by multiplying it by a sinewave with the same frequency and phase of the original carrier and applying the result to a lowpass filter.

Let $u_d(t)$ be the digital signal and ω_c the carrier angular frequency. If we represent narrow band white Gaussian noise, as usual, by the sum of two carriers, with the same frequency as the signal carrier, modulated in quadrature in double sideband suppressed carrier by independent, Gaussian distributed, random variables $x(t)$ and $y(t)$, the signal and noise at the receiver input may be written as

$$u_e(t) = u_d(t)\cos(\omega_c t) + x(t)\cos(\omega_c t) + y(t)\sin(\omega_c t). \tag{5.142}$$

In the receiver, we generate a local carrier, with unity amplitude, synchronized with the transmitter carrier. Multiplying (mixing) the incoming signal and noise with the local carrier and applying the output to a lowpass filter we get

$$u_s(t) = \frac{u_d(t) + x(t)}{2}. \tag{5.143}$$

This result shows that the bit error ratio in DSB-SC with coherent detection is equal to bit error ratio in the non-modulated (baseband) digital signal.

In radio frequency, instead of the ratio $e_b/{n_0}_{rf}$ between the average bit power and the noise power per unit bandwidth, it is often preferable to use the ratio c/n between the carrier power and the noise power. The carrier power c is defined as the average carrier power during one radio frequency cycle, during the modulating signal peak amplitude. The noise power n is the thermal noise power in the occupied radio frequency band.

In the present case, whichever the code, we have

$$c = \frac{u_d^2}{2R}, \tag{5.144}$$

and

$$\begin{aligned} n &= KTb_{rf} \\ &= \frac{\sigma^2}{R}, \end{aligned}$$

where σ is the standard deviation of $x(t)$.

For the polar (NRZ) code

$$ber = \frac{1}{2} erfc\left(\frac{u_d}{\sqrt{2}\sigma}\right), \tag{5.145}$$

or, in terms of c/n

$$ber = \frac{1}{2} erfc\left(\sqrt{\frac{c}{n}}\right). \tag{5.146}$$

Similarly, we have:

- Unipolar code

$$ber = \frac{1}{2} erfc\left(\sqrt{\frac{c}{2n}}\right); \tag{5.147}$$

- Bipolar code

$$ber = \frac{3}{4} erfc\left(\sqrt{\frac{c}{2n}}\right); \tag{5.148}$$

- Multilevel code

$$ber = \frac{1}{\log_2(m)} \frac{m-1}{m} erfc\left(\frac{1}{m-1}\sqrt{\frac{c}{n}}\right). \tag{5.149}$$

Please note that these bit error ratio expressions are identical to those for the non-modulated (baseband) signals, if we substitute e_b/n_0 for c/n. The exception being the multilevel code where $e_{b_{max}}/n_0$ should be substituted for c/n.

Not all authors use the same expression for the binary error ratio as a function of the carrier to noise ratio for multilevel code, because they define carrier power as average carrier power and not, as we did, as the average carrier power during a radio frequency cycle when the modulating signal has its maximum level. If we represent the average carrier to noise ratio as $\bar{c}/n$ we get for the multilevel code

$$ber = \frac{1}{\log_2(m)} \frac{m-1}{m} erfc\left(\sqrt{\frac{3}{m^2-1}\frac{\bar{c}}{n}}\right). \tag{5.150}$$

Recalling that, according to the sampling theorem, it is possible to recover a spectrum limited signal, with maximum frequency f_M , from a set of samples taken at a frequency $2f_M$, and that a amplitude modulated signal has a bandwidth which is double that of the modulating signal, we may easily reach the conclusion that the theoretical maximum modulation efficiency η_m of amplitude modulated binary signals is 1 bit/s/Hz. For an m level multilevel code, since a symbol has information of $I = \log_2(m)$ bits, the theoretical maximum modulation efficiency is, from (5.133)

$$\begin{aligned} \eta_m &= \frac{f_b}{b_{rf}} \\ &= \log_2(m). \end{aligned} \tag{5.151}$$

5.8.2 Amplitude modulation with incoherent detection

With on-off keying (OOK) incoherent detection is obtained when we apply the modulated signal to a peak or envelope detector (a rectifier followed by a lowpass filter) followed by a decision circuit that outputs a 1 when the input voltage is above a given threshold (u_d) and a 0 otherwise. As before, the bit error ratio is given by the sum of the probabilities that a 0 is received as a 1 and a 1 is received as a 0

$$ber = \mathcal{P}(0) \cdot \mathcal{P}(t = 0, r = 1) + \mathcal{P}(1) \cdot \mathcal{P}(t = 1, r = 0). \qquad (5.152)$$

The probability that we receive a 1 when a 0 is transmitted is

$$\mathcal{P}(t = 0, r = 1) = \mathcal{P}(|u_n| > u_d), \qquad (5.153)$$

and the probability that a 0 is received when a 1 is transmitted is

$$\begin{aligned}\mathcal{P}(t = 1, r = 0) &= \mathcal{P}(u_s + u_n < u_d) \\ &= \mathcal{P}(u_n < u_d - u_s).\end{aligned} \qquad (5.154)$$

When we transmit a 0, that is, in the absence of a signal, the peak detector input is thermal noise: narrow band, white, Gaussian noise $u_n(t)$. We will represent noise voltage as the sum of two noise components, $x(t)$ and $y(t)$, independent and in quadrature

$$u_n(t) = x(t)\cos(\omega_c t) + y(t)\sin(\omega_c t). \qquad (5.155)$$

Noise voltage may be represented as a vector in a rectangular coordinate system which rotates at an angular velocity ω_c around the origin, as shown in Figure 5.26.

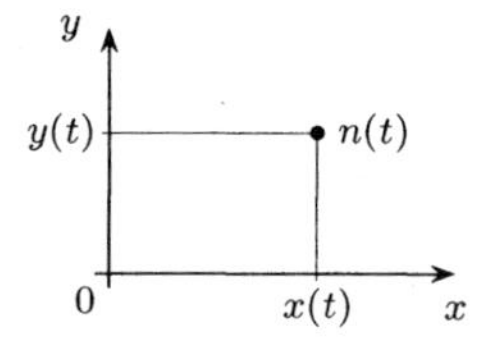

Fig. 5.26 Graphic representation of narrow band, white, Gaussian, noise voltage.

Since the amplitudes of $x(t)$ and $y(t)$ follow statistically independent Gaussian distributions, with zero average and standard deviation σ, the probability $d\mathcal{P}$ that the tip of vector $n(t)$ lies in elementary area $dx\,dy$ centered on a point with coordinates (x,y) is

$$d\mathcal{P}(x, y) = \frac{1}{2\pi\sigma^2} \exp\left(-\frac{x^2 + y^2}{2\sigma^2}\right) dx\,dy. \qquad (5.156)$$

Changing from rectangular to polar coordinates, we get the probability $d\mathcal{P}$ that the tip of vector $u_n(t)$ lies in the annular area with inner radius r, thickness dr, delimited by the angles θ and $\theta + d\theta$

$$d\mathcal{P}(r, \theta) = \frac{1}{2\pi\sigma^2} \exp\left(-\frac{r^2}{2\sigma^2}\right) r\,dr\,d\theta. \qquad (5.157)$$

The probability density function given in (5.157) is the probability density function of the Rayleigh distribution.

Integrating (5.157) between 0 and 2π we get the probability that the tip of $u_n(t)$ lies within r and $r + dr$

$$d\mathcal{P}(r) = \frac{r}{\sigma^2} \exp\left(-\frac{r^2}{2\sigma^2}\right) dr. \tag{5.158}$$

The probability that we receive a 1 when a 0 is transmitted is the integral of (5.158) between u_d and $+\infty$

$$\begin{aligned} \mathcal{P}(t=0, r=1) &= \int_{u_d}^{+\infty} \frac{r}{\sigma^2} \exp\left(-\frac{r^2}{2\sigma^2}\right) dr \\ &= \exp\left(-\frac{u_d^2}{2\sigma^2}\right). \end{aligned} \tag{5.159}$$

When a 1 is transmitted, the input of the peak detector is a sinewave with amplitude u_s, added to white Gaussian narrow band noise (see Figure 5.27)

$$u_e(t) = u_s \cos(\omega_c t) + x(t)\cos(\omega_c t) + y(t)\sin(\omega_c t). \tag{5.160}$$

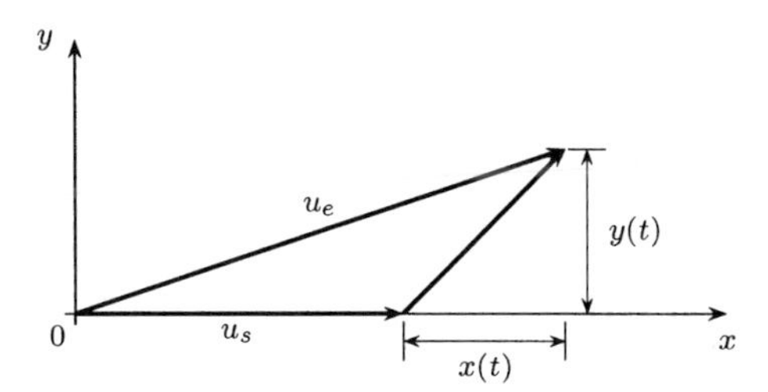

Fig. 5.27 Graphic representation of the signal added to narrow band, white, Gaussian noise.

For large signal-to-noise ratios, that is, when $u_s >> \sigma$, we may neglect the quadrature component and approximate $u_e(t)$ by:

$$|u_e(t)| = u_s + x(t). \tag{5.161}$$

Under these conditions

$$\begin{aligned} \mathcal{P}(t=1, r=0) &= \mathcal{P}(u_s + x < u_d) \\ &= \mathcal{P}(x < u_d - u_s) \\ &= \frac{1}{2} erfc\left(\frac{u_s - u_d}{\sqrt{2}\sigma}\right). \end{aligned} \tag{5.162}$$

Substituting (5.159) and(5.162) in the probability of error (5.72), we get

$$ber = \mathcal{P}(0)\exp\left(-\frac{u_d^2}{2\sigma^2}\right) + \frac{\mathcal{P}(1)}{2} erfc\left(\frac{u_s - u_d}{\sqrt{2}\sigma}\right), \tag{5.163}$$

which is equivalent to the expression obtained for the unipolar code.

We may rewrite the bit error ratio as a function of the carrier to noise ratio c/n and the detector threshold to noise[9] c_d/n

$$ber = \mathcal{P}(0)\exp\left(-\frac{c_d}{n}\right) + \frac{\mathcal{P}(1)}{2} erfc\left(\sqrt{\frac{c}{n}} - \sqrt{\frac{c_d}{n}}\right). \tag{5.164}$$

Substituting the complementary error function by its approximate value given in (5.96), we have, for equiprobable symbols ($\mathcal{P}(0) = \mathcal{P}(1) = 1/2$)

$$ber = \frac{1}{2}\exp\left(-\frac{c_d}{n}\right)\left[1 + \frac{\exp(-c/n + 2\sqrt{c/n}\,\sqrt{c_d/n})}{2\sqrt{\pi}(\sqrt{c/n} - \sqrt{c_d/n})}\right], \tag{5.165}$$

which for large values of the signal to noise ratio may again be approximated as

$$ber = \frac{1}{2}\exp\left(-\frac{c_d}{n}\right). \tag{5.166}$$

Under the above conditions the optimum detection level is equal to half the peak value of the received signal, hence

$$ber = \frac{1}{2}\exp\left(-\frac{c}{4n}\right). \tag{5.167}$$

In the general case, the probability of receiving a 0 when a 1 was transmitted is more complex, as we will show next. To begin with, we start by defining a new variable u:

$$u = u_e + x(t). \tag{5.168}$$

Since u_e is constant and $x(t)$ is a Gaussian variable, with zero average and standard deviation σ, u is also a Gaussian variable with average u_e and standard deviation σ. The probability that the tip of u_e lies in the area $dx\,dy$ centered in $(u_e + x, y)$ is

$$\mathcal{P}(x,y) = \frac{1}{2\pi\sigma^2}\exp\left[-\frac{(x-u_e)^2 + y^2}{2\sigma^2}\right] dx\,dy. \tag{5.169}$$

Changing into polar coordinates

$$\mathcal{P}(r,\theta) = \frac{r}{2\pi\sigma^2}\exp\left[-\frac{r^2 + u_e^2 - ru_e\cos(\theta)}{2\sigma^2}\right] r\,dr\,d\theta, \tag{5.170}$$

[9] Ratio of the average power of a sinewave with a peak value equal to the detection threshold voltage c_d and the noise power

and integrating between 0 and 2π

$$\mathcal{P}(r) = \frac{r}{\sigma^2} \exp\left(-\frac{r^2 + u_e^2}{2\sigma^2}\right) \left\{\frac{1}{2\pi} \int_0^{2\pi} \exp\left[\frac{ru_e \cos(\theta)}{2\sigma^2}\right] d\theta\right\} dr. \tag{5.171}$$

Recalling the integral definition of the modified Bessel function of the first kind and order zero given in Abramowitz and Stegun[1]):

$$I_0(x) = \frac{1}{\pi} \int_0^{\pi} \exp[x \cos(\theta)] d\theta, \tag{5.172}$$

and noting that the cosine function is periodic, with period π, the probability density function becomes

$$\mathcal{P}(r) = \frac{r}{\sigma^2} \exp\left(-\frac{r^2 + u_e^2}{2\sigma^2}\right) I_0\left(\frac{ru_e}{\sigma^2}\right) dr, \tag{5.173}$$

hence

$$\mathcal{P}(t = 1, r = 0) = 1 - \int_{u_d}^{+\infty} \frac{r}{\sigma^2} \exp\left(-\frac{r^2 + u_s^2}{\sigma^2}\right) I_0\left(\frac{ru_s}{\sigma^2}\right) dr. \tag{5.174}$$

Using the Marcum Q function, defined in equation (3.31), it is possible to rewrite (5.174) as

$$\mathcal{P}(t = 1, r = 0) = 1 - Q\left(\frac{u_s}{\sigma}, \frac{u_d}{\sigma}\right). \tag{5.175}$$

Introducing the carrier-to-noise and the threshold-to-noise ratios c/n and c_d/n, respectively, corresponding to sinewaves of amplitudes u_s and u_d

$$\frac{c}{n} = \frac{u_s^2}{2\sigma^2}, \tag{5.176}$$

$$\frac{c_d}{n} = \frac{u_d^2}{2\sigma^2}, \tag{5.177}$$

expression (5.175) is modified into

$$\mathcal{P}(t = 1, r = 0) = 1 - Q(\sqrt{2c/n}, \sqrt{2c_d/n}), \tag{5.178}$$

and the bit error ratio becomes

$$ber = \mathcal{P}(1)[1 - Q(\sqrt{2c/n}, \sqrt{2c_d/n})] + \mathcal{P}(0) \exp(-c_d/n). \tag{5.179}$$

The optimum detection level, that is the detection level that, for a given signal level, leads to the minimum bit error ratio is given by [15]:

$$\exp\left(-\frac{u_d^2}{2\sigma^2)}\right) I_0\left(\frac{u_d u_s}{\sigma^2}\right) = \frac{\mathcal{P}(0)}{\mathcal{P}(1)}, \tag{5.180}$$

or, in terms of c/n and c_d/n

$$\exp(-c_d/n)I_0(2\sqrt{c_d/n}\,\sqrt{c/n}) = \frac{\mathcal{P}(0)}{\mathcal{P}(1)}. \qquad (5.181)$$

According to Schwartz and Bennett [23] for $\mathcal{P}(0) = \mathcal{P}(1) = 1/2$, a solution for the previous expression, for values of c/n between 0.1 and 1000 is approximately given by

$$\frac{c_d}{n} = 1 + \frac{c}{4n}. \qquad (5.182)$$

Note that the optimum detection level is dependent on the received signal level as for the unipolar code. In Figure 5.28 we plot the bit error ratio as a function of the carrier-to-noise ratio, for different values of the threshold-to-noise ratio.

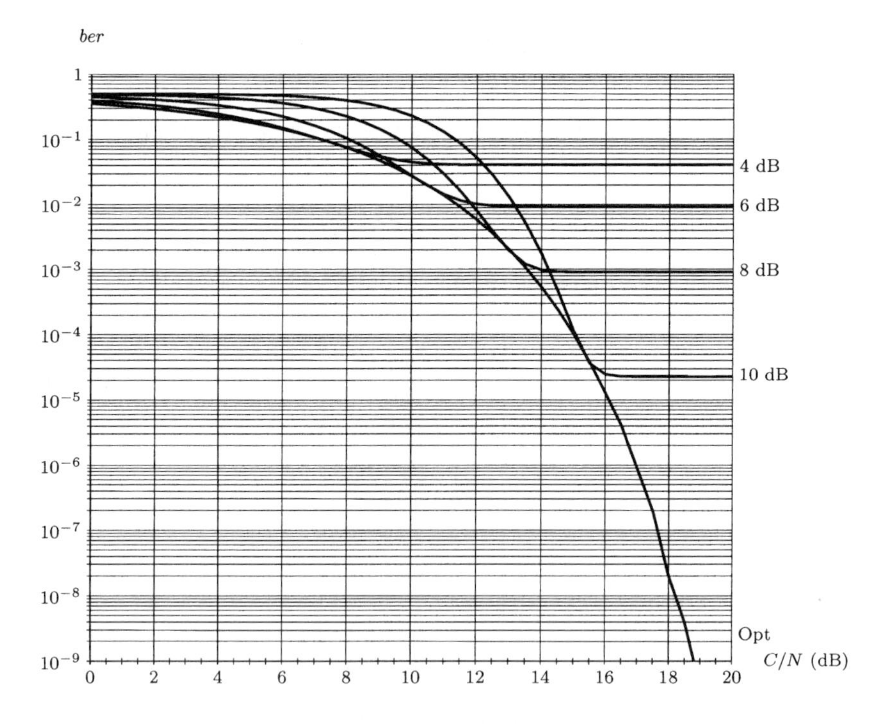

Fig. 5.28 Bit error ratio as a function of the carrier-to-noise ratio, for an amplitude modulated digital unipolar signal for different values of the threshold-to-noise ratio.

5.8.3 Frequency shift keying with coherent detection

Frequency shift keying, or FSK, corresponds to the change of frequency of a sinewave carrier according to a digital signal, so that symbol 0 corresponds to frequency ω_{c0} and symbol 1 to frequency ω_{c1}.

Modulated FSK signals may be demodulated both coherently and incoherently. As for the ASK modulation, we will start with coherent demodulation.

FSK signals may be considered as the sum of two carriers, with angular frequencies ω_{c1} and ω_{c0}, OOK modulated by two digital signals, of which the first is the modulating signal and the second is its complement, that is, the signal obtained from the first one, substituting all the 1s by 0s and all the 0s by 1s.

The FSK receiver may be though of as two independent receivers, one tuned for ω_{c1} and another one tuned for ω_{c0}, each with a synchronous demodulator.

To compute the bit error ratio, we will assume that the carrier spread is such that, the output of one of the receivers, after the adapted filter, is not practically affected by the presence of signal in the other receiver. Thus, when symbol 1 is transmitted, voltages u_{s1} and u_{s2} at the demodulators output are

$$\begin{aligned} u_{s1} &= u_s + u_{n1}(t), \\ u_{s2} &= u_{n2}(t). \end{aligned}$$

The decision circuit considers that a 1 was transmitted, when the output voltage of receiver tuned to ω_{c1} is equal to, or higher than, the output voltage of the receiver tuned to ω_{c0}. A decision that a 0 was transmitted is taken when the opposite is true. The bit error ratio is, therefore

$$ber = \mathcal{P}(1)\mathcal{P}(t=1, r=0) + \mathcal{P}(0)\mathcal{P}(t=0, r=1), \tag{5.183}$$

where

$$\begin{aligned} \mathcal{P}(t=1, r=0) &= \mathcal{P}(u_s + u_{n1} < u_{n2}) \\ &= \mathcal{P}(u_{n1} - u_{n2} < -u_s), \qquad (5.184) \\ \mathcal{P}(t=0, r=1) &= \mathcal{P}(u_{n1} \geq u_s + u_{n2}) \\ &= \mathcal{P}(u_{n1} - u_{n2} \geq -u_s). \qquad (5.185) \end{aligned}$$

Assuming that the noise voltages in both receivers follows two independent Gaussian distributions, with zero average and standard deviation σ, the sum (and the difference) of these noise voltages is also Gaussian, with zero average and standard deviation σ_2

$$\sigma_2^2 = 2\sigma^2. \tag{5.186}$$

Under these conditions

$$\begin{aligned} \mathcal{P}(t=1, r=0) &= \frac{1}{2}\, erfc\left(\frac{u_s}{\sqrt{2}\sigma_2}\right) \\ &= \frac{1}{2}\, erfc\left(\frac{u_s}{2\sigma}\right), \qquad (5.187) \\ & \qquad (5.188) \end{aligned}$$

$$\mathcal{P}(t=0, r=1) = \frac{1}{2}\, erfc\left(\frac{u_s}{\sqrt{2}\sigma_2}\right) = \frac{1}{2}\, erfc\left(\frac{u_s}{2\sigma}\right). \tag{5.189}$$

Whatever the probabilities of the transmitted symbols, the bit error ratio becomes, from (5.72), (5.187) and (5.189)

$$ber = \frac{1}{2}\, erfc\left(\frac{u_s}{2\sigma}\right). \tag{5.190}$$

Recalling that the carrier-to-noise ratio c/n in each of the receivers that makes up the FSK receiver is given by

$$\frac{c}{n} = \frac{u_s^2}{2\sigma^2}, \tag{5.191}$$

the bit error ratio (5.190) may be written as

$$ber = \frac{1}{2}\, erfc\left(\sqrt{\frac{c}{2n}}\right). \tag{5.192}$$

We note that this bit error ratio is identical to the bit error ratio for ASK with coherent demodulation, for unipolar coded digital signals with optimum detection level. Although the FSK receiver is more complex than the ASK receiver (since it requires two independent ASK receivers) the detection level is not dependent on the received signal level, an important advantage in real systems.

5.8.4 Frequency shift keying with incoherent detection

Just as for the coherent detection, we may consider an incoherent FSK receiver as two separate receivers, one tuned for the angular frequency ω_{c1} and the other for ω_{c0}. Each of these receivers uses a simple peak detector. The decision circuit compares the output voltages of each of the peak detectors and decides that a 1 was transmitted when the output voltage of the first detector is higher than the that of the second detector and a 0 otherwise.

To derive the bit error ratio, we will assume, as before, that the frequency difference between the two carriers is such that the output (after the adapted filter) of one of the receivers is not affected by the presence of signal in the other receiver. Thus, when symbol 1 is transmitted, output voltages u_{s1} and u_{s2} at the peak detectors will be

$$u_{s1} = \sqrt{[u_s + x_1(t)]^2 + y_1^2(t)}, \tag{5.193}$$

$$u_{s2} = \sqrt{x_1^2(t) + y_1^2(t)}, \tag{5.194}$$

where x_1, x_2, y_1 and y_2 represent the amplitude of noise voltages in the first and the second detector, in phase and in quadrature, respectively, and are independent, random, Gaussian variables, with zero average and standard deviation σ.

The decision circuit considers that a 1 was transmitted when the output voltage of the receiver tuned to ω_{c1} is equal to or higher than that of the receiver tuned to ω_{c0}. In the opposite case the decision circuit considers that a 0 was transmitted. The bit error ratio is

$$ber = \mathcal{P}(1) \cdot \mathcal{P}(t = 1, r = 0) + \mathcal{P}(0) \cdot \mathcal{P}(t = 0, r = 1), \tag{5.195}$$

where

$$\begin{aligned}\mathcal{P}(t = 1, r = 0) &= \mathcal{P}(u_{s2} > u_{s1}) \\ &= \mathcal{P}\{[x_2^2(t) + y_2^2(t)] > [(u_s + x_1(t))^2 + y_1^2(t)]\}, \qquad (5.196)\\ \mathcal{P}(t = 0, r = 1) &= \mathcal{P}(u_{s1} > u_{s2}) \\ &= \mathcal{P}\{[x_1^2(t) + y_1^2(t)] > [(u_s + x_2(t))^2 + y_2^2(t)]\}. \qquad (5.197)\end{aligned}$$

Given the continuity of the distributions of noise voltages and the fact that their average and standard deviation are identical, we have

$$\mathcal{P}(t = 1, r = 0) = \mathcal{P}(t = 0, r = 1), \tag{5.198}$$

and, as $\mathcal{P}(0) + \mathcal{P}(1) = 1$, we get from (5.72), (5.196), (5.197) and (5.198)

$$ber = \mathcal{P}\{[x_2^2(t) + y_2^2(t)] > [(u_s + x_1(t))^2 + y_1^2(t)]\}, \tag{5.199}$$

which may be written as

$$ber = \mathcal{P}(0 \leq u_{s1}\ ,\ u_{s1} < u_{s2} < \infty). \tag{5.200}$$

Recalling that the distribution of u_{s1} is given by (5.173) and that u_{s2} follows a Rayleigh distribution, whose probability density function is given by (5.157), we get

$$ber = \int_0^\infty \frac{x}{\sigma^2} \exp\left(-\frac{x^2 + u_s^2}{2\sigma^2}\right) I_0\left(\frac{xu_s}{\sigma^2}\right) \int_x^\infty \frac{r}{\sigma^2} \exp\left(-\frac{r^2}{2\sigma^2}\right) drdx,$$

or

$$ber = \int_0^\infty \frac{x}{\sigma^2} \exp\left(-\frac{x^2 + u_s^2}{2\sigma^2}\right) I_0\left(\frac{xu_s}{\sigma^2}\right) \exp\left(-\frac{x^2}{2\sigma^2}\right) dx,$$

from where, with the following change of variable

$$y = \frac{\sqrt{2}x}{\sigma},$$

we get

$$ber = \frac{1}{2} \exp\left(-\frac{u_s^2}{4\sigma^2}\right) \int_0^\infty y \exp\left[-\frac{y^2 + (\frac{u_s}{2\sigma})^2}{2}\right] I_0\left(\frac{yu_s}{\sqrt{2}\sigma}\right) dy. \tag{5.201}$$

Using again the Marcum Q function defined in equation (3.31) and recalling that the carrier-to-noise ratio c/n, corresponding to a sinewave of amplitude u_s, is

$$\frac{c}{n} = \frac{u_s^2}{2\sigma^2},$$

we may write the bit error ratio as

$$ber = \frac{1}{2}\exp(-c/2n)Q\left(\sqrt{\frac{2c}{n}}, 0\right). \tag{5.202}$$

Since it is possible to prove that $Q(x, 0) = 1$, (see Schwartz and Bennett [23]) we finally get for incoherent FSK demodulation

$$ber = \frac{1}{2}\exp\left(-\frac{c}{2n}\right). \tag{5.203}$$

Note that this expression for the bit error ratio is identical to the approximate expression for incoherent OOK demodulation (with large values of the carrier-to-noise ratio). In Figure 5.29 we plot the bit error ratio as a function of the carrier-to-noise for coherent and incoherent FSK demodulation.

5.8.5 Phase shift keying

Phase shift keying, also known as m-PSK, is the phase change of a sinewave carrier of angular frequency ω_c according to a digital signal, where each of the m signal states corresponds to a phase value of the set $0, \frac{2\pi}{m}, ..., \frac{2\pi(m-1)}{m}$. For a binary digital signal, that is, for $m = 2$, we get 2-PSK, or binary phase shift keying, or BPSK, or simply PSK.

Since

$$\cos(\omega_c t + \pi) = -\cos(\omega_c t),$$

a 2-PSK signal may also considered to be an ASK signal, with a polar coded modulating signal. As we have seen, this signal may be demodulated synchronously with a bit error ratio given by

$$ber = \frac{1}{2}erfc\left(\sqrt{\frac{c}{n}}\right). \tag{5.204}$$

If we compare bit error ratio expressions we find out that 2-PSK has a 3-dB advantage over FSK, that is, for the same bit error ratio it requires half the signal to noise ratio.

In the general case, that is for m-PSK with $m > 2$, the approximate bit error rate is given by Bic, Duponteil and Imbeaux [7]

$$ber = \frac{1}{\log_2(m)}erfc(z), \tag{5.205}$$

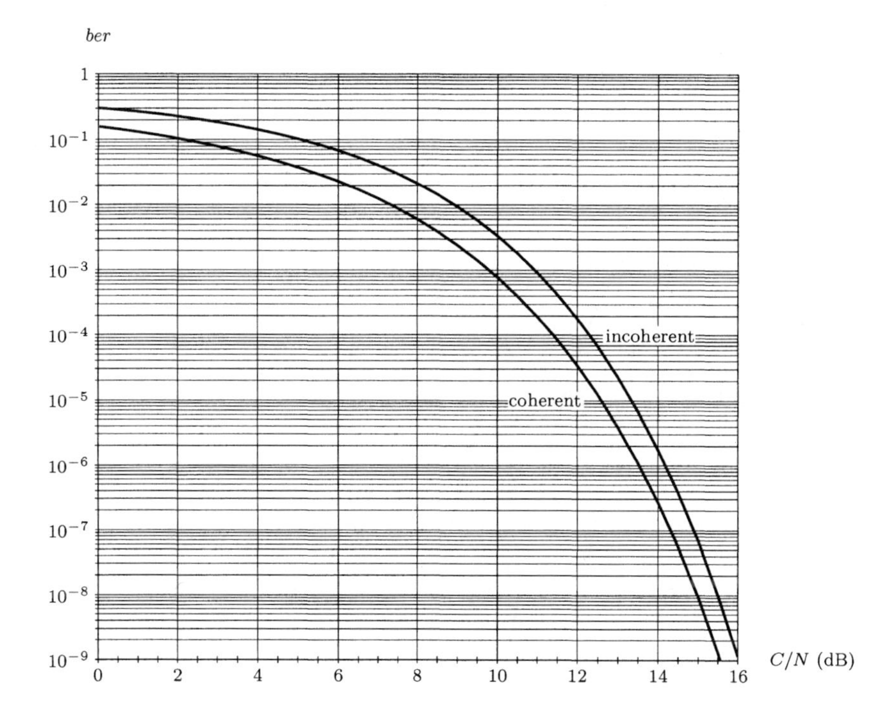

Fig. 5.29 Bit error ratio as a function of the carrier-to-noise ratio for coherent and incoherent FSK demodulation.

where

$$z = \sin\left(\frac{\pi}{m}\right)\sqrt{\frac{c}{n}}, \tag{5.206}$$

and $erfc(x)$ is given in (5.84).

Substituting in (5.206) the value of c/n in terms of e_b/n_0 we get

$$z = \sin\left(\frac{\pi}{m}\right)\sqrt{\log_2(m)\,\frac{e_b}{n_0}}. \tag{5.207}$$

The bit error ratio as a function of E_b/N_0 and C/N is plotted, in Figures 5.30 and 5.31, for m between 2 and 32. We recall that, for $m = 2$ the bit error ratio is not computed by (5.205) and (5.207), but rather by (5.204).

5.8.6 Quadrature amplitude modulation

Quadrature amplitude modulation of order m, or m-QAM, is a type of multilevel amplitude modulation. It corresponds to the superposition, in the

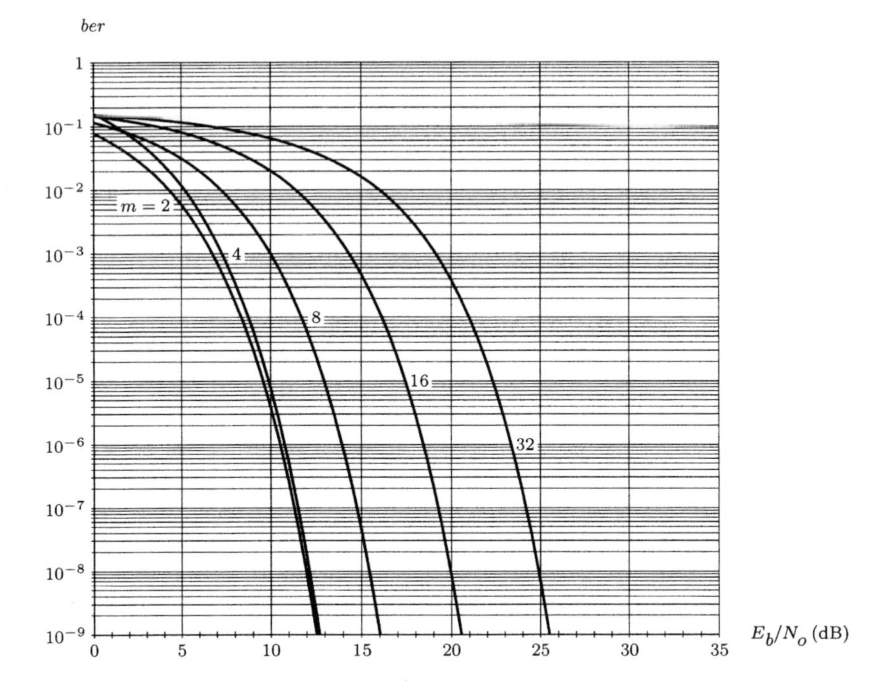

Fig. 5.30 Bit error ratio as a function of E_b/N_0 for m-PSK, with m varying between 2 and 32.

same radio channel, of an m_1-ASK carrier and another m_2-ASK carrier, in quadrature with the first one (i.e., phase shifted $\pi/2$ radians), where

$$m = m_1 \times m_2. \tag{5.208}$$

Although m-QAM systems, where m is larger than 1 and not prime, may be conceived, in practice m is usually an even perfect square (4, 16, 64 or 256). For these systems the occupied bandwidth corresponds to the $\sqrt{m}$-ASK and the m-QAM modulation efficiency is

$$\begin{aligned} \eta_m &= 2\log_2(\sqrt{m}) \\ &= \log_2(m). \end{aligned} \tag{5.209}$$

The m-QAM bit error ratio may be written directly from the bit error ratio expressions for the equivalent $\sqrt{m}$-ASK

$$ber = \frac{2}{\log_2(m)} \frac{\sqrt{m}-1}{\sqrt{m}} \, erfc(z), \tag{5.210}$$

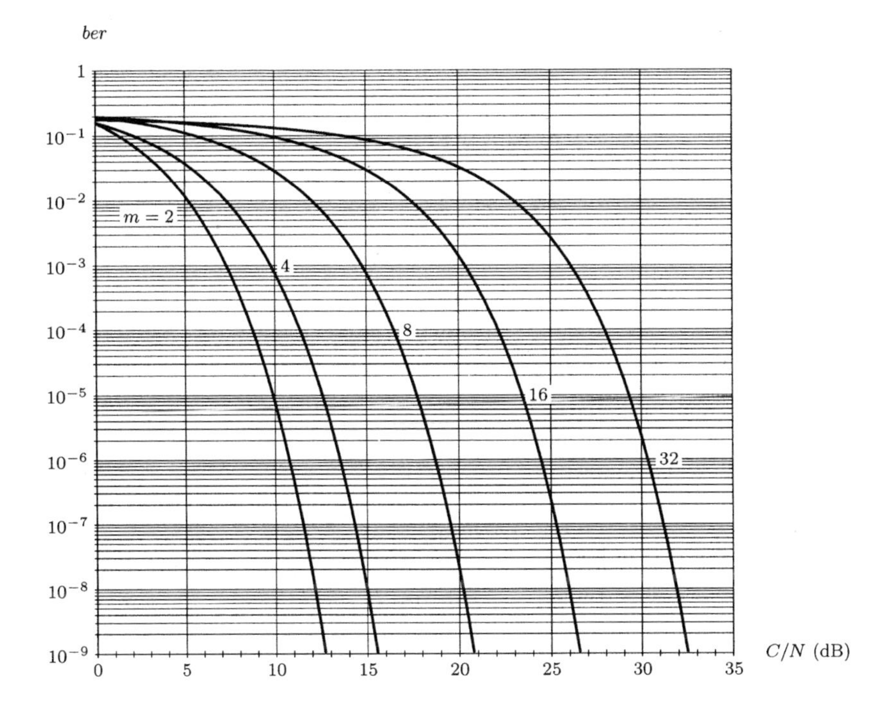

Fig. 5.31 Bit error ratio as a function of C/N for m-PSK, with m varying between 2 and 32.

where

$$z = \frac{\sqrt{c_{\sqrt{m}}/n}}{\sqrt{m} - 1}, \tag{5.211}$$

and

- *erfc* is the complementary error function defined in (5.84),
- $c_{\sqrt{m}}/n$ is the ratio between the carrier power for the $\sqrt{m}$-ASK case and the noise power in radio frequency.

Since radio frequency power of the m-QAM signal is double that of the two $\sqrt{m}$-ASK, it is possible to rewrite (5.211) in terms of the c/n for the m-QAM

$$z = \frac{\sqrt{\frac{c}{2n}}}{\sqrt{m} - 1}. \tag{5.212}$$

Had we used the average carrier-to-noise ratio $\bar{c}/n$, instead of the carrier-to-noise ratio c/n, we would have had

$$z = \sqrt{\frac{3}{m-1}\frac{\bar{c}}{2n}}. \tag{5.213}$$

Recalling that for the m-QAM we have

$$\frac{e_b}{n_0} = \frac{c}{n}\frac{1}{\log_2(m)}, \tag{5.214}$$

we may write z in terms of e_b/n_0

$$z = \sqrt{\frac{3}{2}\frac{\log_2(m)}{m-1}\frac{e_b}{n_0}}, \tag{5.215}$$

or, noting (5.125), in terms of $e_{b_{max}}/n_0$

$$z = \frac{1}{\sqrt{m}-1}\sqrt{\frac{\log_2(m)}{2}\frac{e_{b_{max}}}{n_0}}. \tag{5.216}$$

The bit error ratio as a function of E_b/N_0 and C/N is plotted in Figures 5.32 and5.33, for values of m between 4 and 256, in the first case, and 4 and 4096 in the second case.

5.8.7 Radio frequency bandwidth

Although the concept of modulation efficiency has already been referred to in the analysis of modulation procedures, here we present the calculation of radio frequency bandwidth in a more condensed format.

In microwave radio links, particularly in high-capacity ones, the radio frequency bandwidth b_{rf} is defined by a filter at the modulator or at the output of one of the radio frequency amplifier stages. Sometimes this filter is simplified, and its main role is the reduction of spurious products. In these cases, and approximately also in all others, the modulated signal bandwidth is the result of a Nyquist filter that limits the spectrum of the modulating signal. Thus for 2-ASK and 2-PSK the radio frequency bandwidth is equal to the bit rate.

For the case of m-ASK and m-PSK, the same is true, the only difference being that the modulation uses a multilevel rather than a binary code. Since a symbol of a m level multilevel code has an information $I = \log_2(m)$, we have

$$b_{rf} = \frac{f_b}{\log_2(m)}. \tag{5.217}$$

As far as m-QAM is concerned, it suffices to consider it as the superposition of two $\sqrt{m}$-ASK modulations, with carriers in quadrature, each one modulated

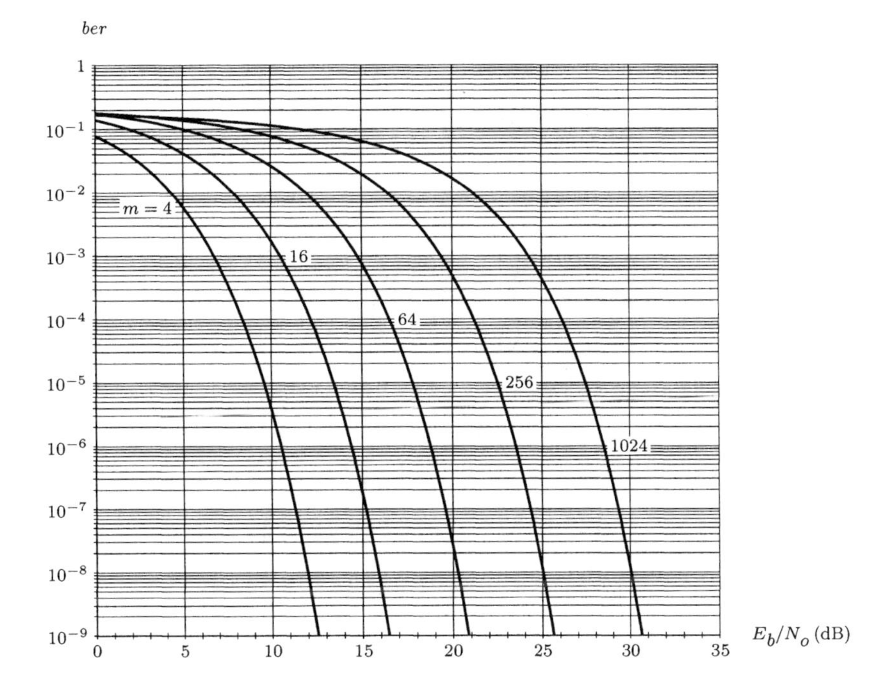

Fig. 5.32 Bit error ratio as a function of E_b/N_0 for m-QAM, with m between 4 and 256.

by a signal with a bit rate equal to one half of the original bit rate, to write

$$
\begin{aligned}
b_{rf} &= \frac{f_b}{2\log_2(\sqrt{m})} \\
&= \frac{f_b}{\log_2(m)}.
\end{aligned} \tag{5.218}
$$

As was stated before, the bandwidth occupied by a modulated digital signal is defined by the Nyquist filter. This value should be used to compute the noise power. The occupied radio frequency spectrum the bandwidth b_{orf} is given by

$$b_{orf} = (1+\beta)b_{rf}, \tag{5.219}$$

where β is the Nyquist filter excess band factor, usually between 0.1 and 0.5.

The ratio f_b/b_{orf} is often known as spectral efficiency. It differs from the modulation efficiency η_m by the factor $1+\beta$.

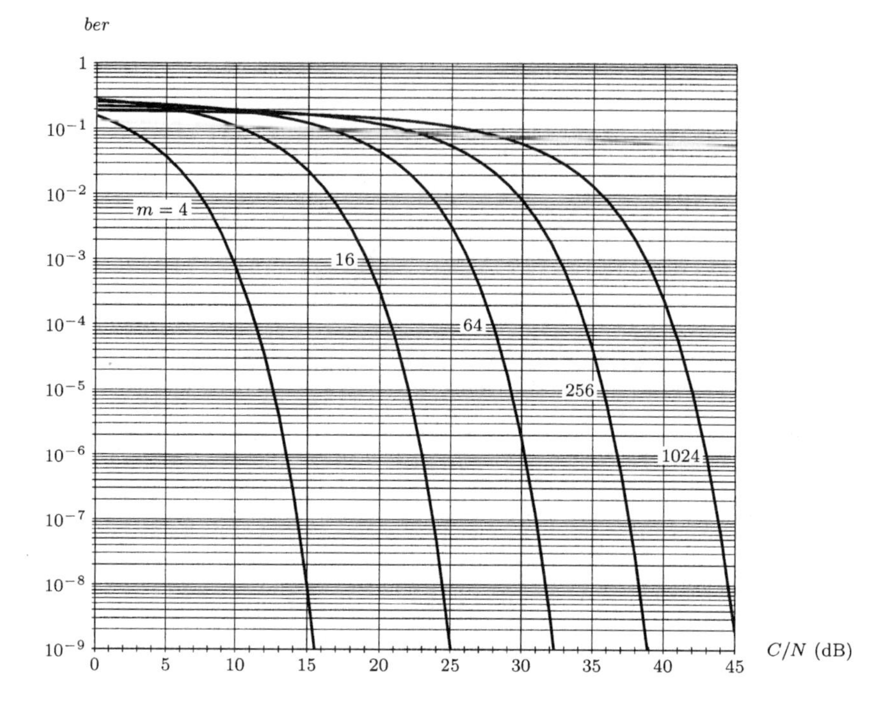

Fig. 5.33 Bit error ratio as a function of C/N for m-QAM, with m between 4 and 4096.

5.8.8 Receiver sensitivity

The power required at the receiver input (antenna) terminals for a given output bit error ratio (usually 10^{-3} or 10^{-6}) is known as receiver sensitivity and can be obtained from the receiver specifications.

Receiver sensitivity in logarithmic units Rs may be estimated by adding the input noise power, N_0, as given in (4.13) or (4.14), the receiver noise factor Nf and the carrier-to-noise ratio C/N for the modulation and required bit error ratio

$$Rs = N_0 + Nf + C/N. \tag{5.220}$$

5.9 SELECTIVE FADING

5.9.1 Introduction

Besides reducing signal level and increasing bit error ratio, due to a reduction of the carrier to noise ratio, fading in digital radio links may significantly increase bit error ratio due to intersymbol distortion. This effect, known as selective fading, is due to the fact that the transmission channel is no longer linear. In this context, the simple decrease in signal level, is known as uniform fading.

Due to multipath, radio channel transfer function exhibits significant amplitude fluctuations and its phase characteristic ceases to be linear with frequency, which may be even worse. This behavior is more pronounced as the channel bandwidth increases, and becomes a particularly serious concern in large capacity links.

Since the channel transfer function varies rapidly with time, it is close to impossible to avoid the effects of selective fading with simple equalizing circuits. It is thus imperative to consider this effect in the design of real circuits. If and when perfect equalizers, able to adapt instantly to the channel, become available, it will be possible to neglect the effects of selective fading and to consider only uniform fading .

In analog radio links selective fading is hardly noticeable due to the insensitivity of frequency modulation to its effects.

We will define link margin as the ratio, usually expressed in dB, between the received carrier power, in stable propagation conditions, that is, without fading, and the carrier power at the receiver input that leads to a given bit error ratio. In general, this bit error ratio is rather high, typically 10^{-3}. Higher bit error ratios lead to frequent link interruptions, due to frame synchronism loss. The link margin corresponds to fading that can be tolerated without significant link interruptions. In most cases link margin for uniform fading, or uniform margin, is higher than $30 - 35$ dB.

Selective fading reduces link margin to values which are much lower than those to be expected with uniform fading alone. On the other hand, increasing the uniform margin, by increasing the transmitter power or the antenna gain, does not lead to a corresponding increase in the link margin, as shown in Figure 5.34. For the higher bit rates, it may even happen that the link margin tends asymptotically to a maximum.

The decrease in link margin corresponds to the increase of the fraction of time during which the link exhibits a higher than specified bit error ratio. This fraction of time may be estimated, as shown later.

5.9.2 Transmission channel characteristics

According to Rummler [22] transmission channel behavior between transmitter and receiver antennas during selective fading may be explained as the result of the existence of three independent paths. Hence the field at the receiver antenna is expressed as the sum of three field components, whose amplitudes and delays may be written as a function of the field received in non-fading conditions as:

1. the direct ray, with unity amplitude and zero delay;

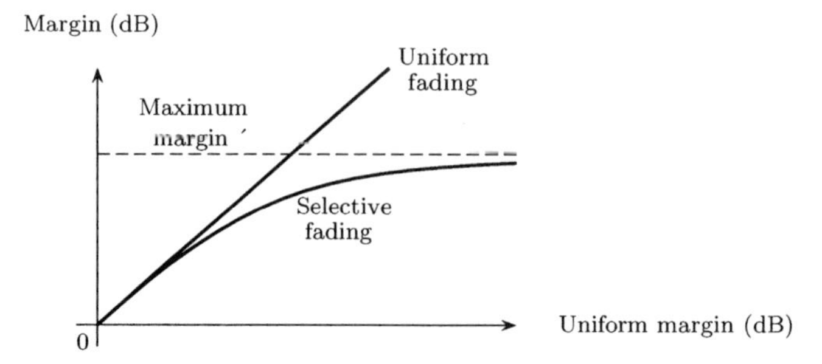

Fig. 5.34 Link margin as a function of the received carrier level under stable propagation conditions.

2. the ray responsible for uniform fading, with amplitude a_1 and delay τ_1;
3. the ray responsible for selective fading, with amplitude a_2 and delay τ_2, such that $\tau_2 \gg \tau_1$.

Under these conditions, the channel transfer function $h(\omega)$, referred to the non-fading propagation conditions, at the angular frequency ω_0, may be written as

$$h(\omega) = 1 + a_1 \exp[-j(\omega - \omega_0)\tau_1] + a_2 \exp[-j(\omega - \omega_0)\tau_2]. \tag{5.221}$$

Noting that $\tau_1 \ll \tau_2$, $h(\omega)$ may be approximated to

$$h(\omega) = a\{1 + b \exp[-j(\omega - \omega_0)\tau]\}, \tag{5.222}$$

where

$$\tau = \tau_2, \tag{5.223}$$

$$ab = a_2. \tag{5.224}$$

The transfer function in amplitude and phase, as a function of the frequency, is plotted in Figure 5.35. We note in this figure the periodicity of the amplitude and phase changes as well as the importance of selective fading whenever the occupied radio frequency bandwidth is a significative fraction of $1/\tau$. Since $\tau \approx 6$ ns and $1/\tau \approx 167$ MHz it follows that selective fading is negligible for 2 Mbit/s, not very important for 8 Mbit/s, significant for 34 Mbit/s and decisive for higher capacity links.

Since the transfer function varies periodically with frequency, in a narrow bandwidth (less than about $1/6\tau$), it is possible to choose different sets of values for a, b, ω_0 e τ that suit equally well the experimental values. Baccetti and Tartara [4] have demonstrated that this choice is indifferent provided that

$$a(1 - b) = \text{constant} \tag{5.225}$$

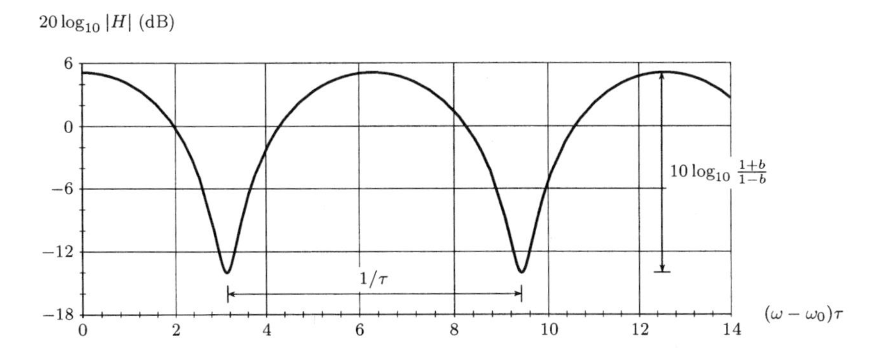

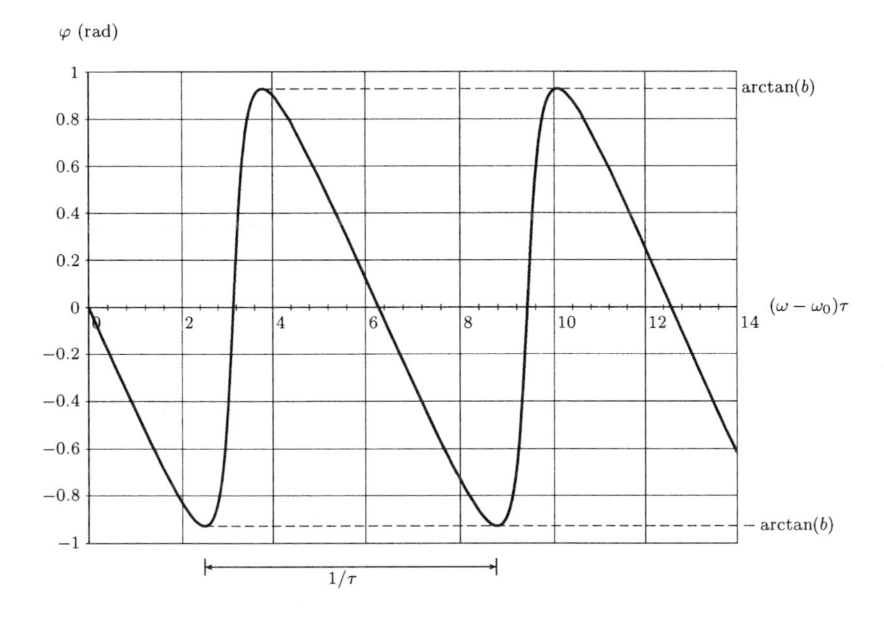

Fig. 5.35 Transfer function amplitude H and phase φ as a function of frequency f.

$$\frac{\sqrt{b}\tau}{1-b} = \text{constant.} \tag{5.226}$$

According to these authors in the three-ray model, Rummler invariably took $b < 1$. However it is also possible to take $b > 1$, which corresponds to distinct fading characteristics. Note that from (5.226) for $b < 1$, we have $\tau > 0$ and, for $b > 1$, we have $\tau < 0$. The first case is known as minimal phase fading and the second as non-minimal phase fading.

Occasionally we may use the differential channel attenuation ΔA instead of the channel transfer function to characterize selective fading.

5.9.3 Sensitivity to selective fading

The behavior of a link in the presence of selective fading may be derived from the fading parameters and the link margin for uniform fading M_u, by measuring, or computing, the surfaces made up by all points of the space (a, b, f_0) that lead to the same (given) value of the bit error ratio. As it was referred to before, the values of a, b e f_0 completely define selective fading.

The plot of such surfaces is rather impractical and so one usually prefers to project the intersect curves of these surfaces with planes corresponding to constant values of a in the plane b, f_0. These curves, also known as sensitivity curves, are sketched in Figure 5.36.

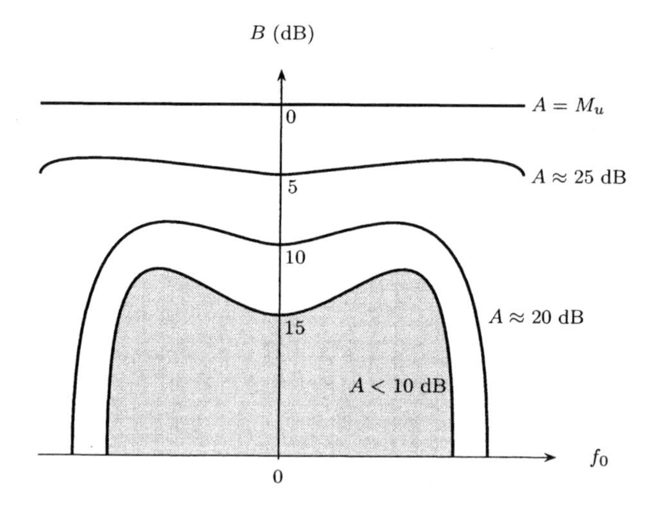

Fig. 5.36 Sensitivity curves.

Please examine Figure 5.36 carefully and start by noting that parameters a and b are represented in logarithmic units (A and B) defined as

$$A = -20\log_{10}(a), \tag{5.227}$$

$$B = -20\log_{10}(1-b). \tag{5.228}$$

For $B = 0$ dB, that is, for uniform fading, the curve corresponding to the system performance is horizontal, or independent of f_0, and the value of A corresponds to the link margin for uniform fading ($A = M_u$). When the value of A decreases, at first performance curves lower, which correspond to higher values of B, but after a given value of A all performance curves coincide. In this case the responsible for transmission errors is the intersymbol distortion rather than the increase in link attenuation. For usual values of the uniform link margin (higher than 35 dB) the sensitivity curves coincide for $A \leq 10$ dB.

The critical sensitivity curve $B_c(f_0)$ represents the effect of intersymbol distortion and system sensitivity to selective fading. It is commonly known as system signature and it may be measured or derived from system parameters.

5.9.4 Fading margin

We have seen, in the Chapter 3, that for line-of-sight links with no appreciable reflections, the probability $\mathcal{P}$ that the received power p is less or equal to p_0, in the worst month, may be estimated using an expression of the form

$$\mathcal{P}(p \leq p_0) = \mathcal{F}\,\frac{p_0}{p_n},$$

where:

- $\mathcal{F}$ is the deep fade ocurrence factor, a constant that includes link dependence on the following factors: geoclimatic, terrain, direct ray slope, length and carrier frequency;
- p_n is the received power without fading.

Introducing the concept of link margin m

$$m = \frac{p_n}{p_0}, \tag{5.229}$$

the previous expression becomes

$$\mathcal{P}(p \leq p_0) = \frac{\mathcal{F}}{m}. \tag{5.230}$$

Identifying p_0 as the receiver sensitivity, that is, the carrier power at the receiver antenna terminals that corresponds to a given (maximum) bit error ratio, the probability that the received carrier power is less that p_0 is equivalent to the probability $\mathcal{P}_c$ that the maximum bit error ratio is exceeded.

Assume, now, that the probability that the bit error ratio is exceeded may be divided into two parts: one $\mathcal{P}_u$, due to uniform fading, and the other

$\mathcal{P}_s$, due to distortion caused by selective fading. ITU-R[16] Recommendation F.1093-1 suggests that:

$$\mathcal{P}_c = \left(\mathcal{P}_u^{\alpha/2} + \mathcal{P}_s^{\alpha/2}\right)^{2/\alpha}, \tag{5.231}$$

with α between 1.5 and 2. Note that with $\alpha = 2$ (5.231) may be simplified to

$$\mathcal{P}_c = \mathcal{P}_u + \mathcal{P}_s. \tag{5.232}$$

Simple inspection of (5.231) shows that, for the same values of $\mathcal{P}_u$ and $\mathcal{P}_s$, the probability that the maximum bit error ratio is exceeded $\mathcal{P}_c$ increases when α decreases. Hence the simplified expression may be rated as optimistic.

The division of the probability that the bit error ratio is exceeded in two parts corresponds to an equivalent division of the link margin m

$$\frac{1}{m} = \left[\left(\frac{1}{m_u}\right)^{\alpha/2} + \left(\frac{1}{m_s}\right)^{\alpha/2}\right]^{2/\alpha}, \tag{5.233}$$

where m_u is the link margin for uniform fading, or uniform margin, m_s the link margin for selective fading, or selective fading margin. For $\alpha = 2$ we have

$$\frac{1}{m} = \frac{1}{m_u} + \frac{1}{m_s}. \tag{5.234}$$

To avoid confusion between the concepts of link margin and uniform link margin, some authors prefer to use effective or real link margin instead of link margin, and theoretical link margin instead of uniform link margin.

Take p_{ber} as the carrier power required by the receiver for a given bit error ratio. The uniform link margin is

$$m_u = \frac{p_n}{p_{ber}}. \tag{5.235}$$

Some experimental data [3] indicates that the selective fading margin may be approximated by

$$m_s = 1000\,\frac{s_{ref}}{s}, \tag{5.236}$$

where s is the area under the receiver signature in MHz, measured for a given value of τ, and $s_{ref} = 15$ MHz. Substituting we get

$$m_s = \frac{15000}{s}. \tag{5.237}$$

Comparing these values with other proposed by Baccetti and Tartara [4] we suggest that the selective fading margin should be reduced to

$$m_s = \frac{8000}{s}. \tag{5.238}$$

Typical values of s for non-equalized systems are as follows:

- 0.5 to 0.8 MHz, for 4-PSK, 34 Mbit/s receivers ;
- 25 a 30 MHz, for 16-QAM, 140 Mbit/s receivers.

Sometimes instead of the area under the receiver signature this curve is approximated by a rectangle with width b (in MHz) and height $B < 0$ (in dB). In this case the area under the signature is

$$s = b \times 10^{\frac{B}{20}}. \tag{5.239}$$

ITU-R Recommendation F.1093-1 [16] presents four methods, known as A, B, C and D, to predict selective fading effects. In the following we describe method B, the only one for which the above-mentioned Recommendation includes all data required for its use. For this method one should take $\alpha = 2$.

As before, here we will also assume that the probability that the bit error rate is exceeded may be divided into two parts, one $\mathcal{P}_u$, due to uniform fading, and the other $\mathcal{P}_s$ due to intersymbol distortion caused by selective fading (5.232). Now, however the value of $\mathcal{P}_s$ is given by

$$\mathcal{P}_s = \eta \times \mathcal{P}_{s/mp}, \tag{5.240}$$

where:

- η is a parameter related to the propagation conditions calculated from the deep fade occurrence factor $\mathcal{F}$ by the following empirical relation:

$$\eta = 1 - \exp\left(-0.2\mathcal{F}^{3/4}\right); \tag{5.241}$$

- $\mathcal{P}_{s/mp}$ is the probability that the bit error ratio is exceeded due to intersymbol interference during multipath fading, computed from an adimensional normalized parameter K_n which depends on the type of modulation, the average echo delay τ_m and the symbol period T_s:

$$\mathcal{P}_{s/mp} = 2.16 \cdot K_n \cdot \frac{(2\tau_m^2)}{T_s^2}. \tag{5.242}$$

The following values of K_n are typical for non-equalized receivers:

- 1.0 for 4-PSK;
- 7.0 for 8-PSK;
- 5.5 for 16-QAM;
- 15.4 for 64-QAM.

Baseband transverse adaptive equalizers may reduce to about 1/10 the above values of K_n.

The average echo delay is related with link length. For paths with no significant reflections we have:

$$\tau_m = 0.7 \cdot \left(\frac{d}{50}\right)^n \quad \text{[ns]}, \tag{5.243}$$

where d is expressed in km and n is in range 1.3 – 1.5.

To compare the results of both methods take a typical 140 Mbit/s link, at $f = 4$ GHz, using 16-QAM modulation, in a $d = 50$-km path with no significant reflections and employing non-equalized receivers. For a 30-dB fade, we have, from (5.230) and (eq:3.3.1) $\mathcal{P} = 4.95 \cdot 10^{-5}$ and from (5.230) $\mathcal{F} = 0.0495$. Taking for the receiver signature $s = 30$ MHz we have, from (5.238), $m_s = 266.7$, that is, from (5.230) $\mathcal{P}_s = 1.85 \cdot 10^{-4}$.

Using method B in ITU-R Recommendation F.1093-1 we get $K_n = 5.5$, $\tau_m = 0.7$ ns, $T_s = 28.6$ ns, $\mathcal{P}_{s/mp} = 0.0142$, $\eta = 0.0208$ and finally $\mathcal{P}_s = 2.95 \cdot 10^{-4}$.

The ratio of the results provided by the two methods is less than 2, a result within the expected range of results from the various methods in ITU-R Recommendation F.1093-1.

5.9.5 Reducing the effects of selective fading

To reduce the effects of selective fading, which are particularly serious in high-capacity links, the following procedures, whose effectiveness may be judged from the reduction in the signature, may be used:

- adaptive equalizers in the frequency domain, whereby one tries to level the frequency response of the transmission channel, at the level of the receiver intermediate frequency;
- adaptive equalizers in the time domain, where one tries to equalize the amplitude and phase response of the transmission channel, at the intermediate frequency level, before carrier recover;
- an association of adaptive equalizers in the frequency and the time domain;
- space diversity;
- frequency diversity;
- an association of diversity with adaptive equalizers.

Table 5.8 displays estimates of the effectiveness of adaptive equalizers for a 16-QAM, 140 Mbit/s link.

Diversity is very effective in fighting fading effects both for uniform and selective fading. In addition to what was stated in Section 3.7, where we dealt

Device	Signature reduction factor	
	Minimal phase	Non-minimal phase
Adaptive equalizer in the frequency domain	4.9	4.9
Adaptive equalizer in the time domain	490	22
Association of adaptive equalizers in the frequency and the time domain	490	35

Table 5.8 Signature reduction factors for different types of equalizers for a 16-QAM, 140 Mbit/s [3].

with combiner performance, we note that in digital links optimal combiners are substituted by combiners aiming at minimal intersymbol distortion.

The improvement factor for double diversity i_d, calculated from (3.132), (3.133) or (3.134), may be written as

$$i_d = k_d \cdot \frac{p_n}{p_0}, \tag{5.244}$$

where

- k_d is a constant, which depends on the link characteristics, namely, link length, operating frequency, main and secondary antenna gains, distance between antenna centers or the difference between operating frequencies, as in (3.132), (3.133) or (3.134);
- p_n is the average received carrier power, in ideal propagation conditions, that is, with no fading;
- p_0 is the received carrier power for which we require diversity gain, in general, the carrier power corresponding to a minimum guaranteed quality of service.

Without diversity the probability that the received carrier power is less than p_0 may be written, (3.66), as

$$\mathcal{P} = \mathcal{F}\,\frac{p_0}{p_n}$$

or, defining the margin m_0

$$m_0 = \frac{p_n}{p_0}$$

as

$$\mathcal{P} = \frac{\mathcal{F}}{m_o}.$$

With diversity the previous expression changes into

$$\mathcal{P}_d = \frac{\mathcal{F}}{m_{o_d}} \frac{1}{i_d}, \tag{5.245}$$

where

$$i_d = k_d m_{o_d}, \tag{5.246}$$

and k_d is calculated from (3.132), (3.133) or (3.134), as appropriate.

The improvement factor for selective fading is given by

$$i_{sd} = k_d \cdot m_s, \tag{5.247}$$

where m_s is the link margin for selective fading, without diversity.

As before, one is led to the following expression for the link margin for selective fading and double diversity:

$$m_{sd} = i_{sd} \cdot m_s. \tag{5.248}$$

Uniform link margin m_u, is the ratio between the received carrier power without fading p_n and the received carrier power required to ensure a given quality of service p_{qs}

$$m_u = \frac{p_n}{p_{qs}}. \tag{5.249}$$

Hence link margin with double diversity m_d, as a function of uniform link margin m_u and link margin for selective fading with double diversity m_{sd}, becomes

$$\frac{1}{m_d} = \frac{1}{m_u} + \frac{1}{m_{sd}}. \tag{5.250}$$

Link margin m_l for the specified quality of service is now

$$m_l = \frac{m_d}{m_{o_d}}, \tag{5.251}$$

or, in logarithmic units

$$M_l = 10 \log_{10}(m_l). \tag{5.252}$$

The association of diversity and adaptive equalizers leads to a very significative reduction of the probability of link interruption.

5.9.6 Minimal phase and non-minimal phase fading

According to Alcatel Thomson [3], the proportion of minimal phase and non-minimal phase depends on fade depth. In a first approximation:

- short path lengths (up to 20 km) have a low proportion (20 percent) of non-minimal phase fades;

- long paths (over 40 km) have a high proportion (50 percent) of non-minimal phase fades;
- intermediate paths (between 20 and 40 km) have an intermediate proportion of non-minimal phase fades.

Let d be the path length in km, and i_{mp} and i_{nmp} the improvement factors for minimal phase and non-minimal phase fades. The resulting improvement factor will be:

$$i = \begin{cases} \left(\frac{0.5}{i_{mp}} + \frac{0.5}{i_{nmp}}\right)^{-1} & \text{for} \quad d \geq 40 \text{ km}, \\ \left(\frac{0.8}{i_{mp}} + \frac{0.2}{i_{nmp}}\right)^{-1} & \text{for} \quad d \leq 20 \text{ km}, \\ \left(\frac{k_1}{i_{mp}} + \frac{k_2}{i_{nmp}}\right)^{-1} & \text{for} \quad 40 \text{ km} > d > 20 \text{ km}, \end{cases} \tag{5.253}$$

where:

$$k_1 = 0.5 + 0.3\,\frac{40 - d}{20}, \tag{5.254}$$

$$k_2 = 0.5 - 0.3\,\frac{40 - d}{20}. \tag{5.255}$$

5.10 LINK QUALITY

5.10.1 Microwave radio link chains

Like analog links, most microwave digital radio links contain a (large) number of hops. It is thus necessary to derive the effect of link chains in the quality of real links.

The quality of a microwave radio link, taken as a quadrupole, is only dependent on the bit error ratio, ber. If we chain n digital links, with pulse regeneration, provided the bit error ratio of each link is low ($ber \ll 1$), the overall bit error ratio ber_r is given by

$$ber_r = \sum_{i=1}^{n} ber_i, \tag{5.256}$$

where ber_i is the bit error ratio of ith link.

This expression is of little use in the design of microwave radio links since, in general, only the statistics of the bit error ratio are known and not its value for a given instant of time. In the following paragraphs we present the hypothetical reference digital path and its required quality and describe procedures to determine quality for real links. A simple example is used to provide insight into the various quality criteria.

5.10.2 The hypothetical reference digital path

Since the bit error ratio in a digital path depends on the number of hops, it is to be expected that, before establishing quality standards, ITU-R defines a reference path, known as the hypothetical reference digital path, or HRDP.

For microwave radio links used for telephony, with a capacity above the second hierarchy (120 telephone channels, or 8 Mbit/s), ITU-R[16] Recommendation F.556-1 defines the hypothetical digital reference path with the following properties:

- length, 2500 km;
- capacity according to one of the ITU-T hierarchies, or an integer multiple thereof;
- nine sets of digital multiplexers, composed by a variable number of multiplexers and demultiplexers, in transmission sense;
- nine equal-length radio sections.

We should note that, due to code conversions, insertion of justification, parity and service bits, bit rate inside a radio electric section may differ from an ITU-T hierarchy bit rate (or an integer multiple thereof).

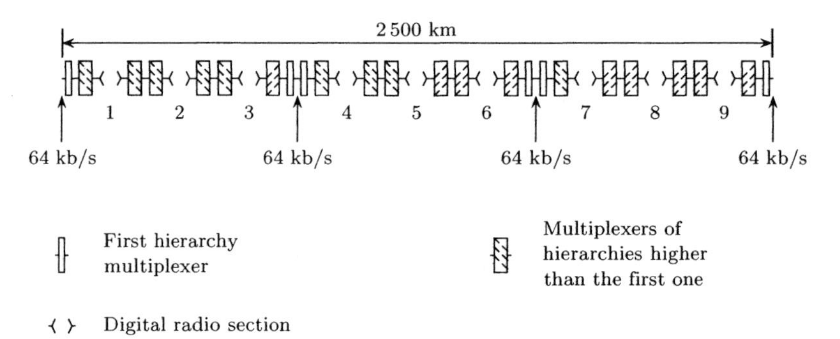

Fig. 5.37 Hypothetical reference digital path, for telephony microwave radio links, with a capacity above the second hierarchy [16].

5.10.3 Hypothetical reference configuration

Microwave digital links may be used as part of an integrated services digital network, or ISDN. ITU-T [17] Recommendation G.821 provides performance parameters and objectives for ISDN connections at 64 kbit/s for a hypothetical reference configuration, or HRX, of a integrated services digital network, operating at a bit rate below the first digital hierarchy (or primary rate). The

HRX has 27 500 km in length (between T points) and foresees three circuit grades: high, medium and local. Local and average grade circuits may make up to 1250 km at each end of the reference configuration. Since this was the first recommendation to be adopted (in 1980) it found wide applications even in areas where it was not developed for. One of these applications was error performance evaluation at bit rates higher than 64 kbit/s.

In 1993, ITU-T [17] adopted Recommendation G.826, which specifies performance objectives for constant bit rate digital paths at or above the primary rate, independent of the physical network that supports the path, based on an hypothetical reference path 27 500 km long between path end points , (or PEP), with one central international portion and two end national portions. Compliance with this recommendation will in most cases ensure that a 64 kbit/s connection will also meet the requirements of ITU-T Recommendation G.821.

Microwave radio links may be a part of any of the circuit grades foreseen in Recommendation G.821. For the high-quality grade the criteria in ITU-R [16] Recommendation F.634-4 applies, with a slight difference related to the apportionment of the overall objectives. In Recommendation F.634-4, the objectives are proportional to the path length (assumed to be higher than 280 km), whereas in Recommendation G.821 they vary by steps as shown later on.

Given the role played by microwave radio links in the hypothetical reference path, referred to in ITU-T Recommendation G.826, both in its international and national portions, Recommendations ITU-R F.1092-1 and F.1189 [16] define an apportionment of the overall objectives, based on a fixed part plus a variable part function of the radio path lengths.

5.10.4 Error performance objectives for the hypothetical reference path

Up to 1994 ITU-R quality criteria were set in terms of error performance objectives defined by the maximum fraction of time during which a given bit error ratio (usually 10^{-3} and (or) 10^{-6}) should not be exceeded. As will be shown in the next chapter, quality criteria based on bit error ratios are ideally suited for link design but are rather inconvenient for monitoring link performance, because to measure bit error ratios at system rates the link must be taken out of service and traffic substituted by a pseudo-random bit stream.

With the introduction and the widespread use of error control codes, dealt with in detail in the next chapter, it became easier to define error performance objectives based on the fraction of time during which errors were detected.

Note 2 of ITU-T [17] Recommendation G.821 provides the following definitions:

- errored seconds, or *es*: one-second periods in which one or more bits are in error;

- severely errored seconds, or *ses*, one-second periods in which the bit error ratio exceeds 10^{-3};
- errored second ratio, or *esr*, ratio of errored seconds to total seconds, during a given period of time;
- severely errored second ratio, or *sesr*, ratio of severely errored seconds to total seconds, during a given period of time.

Error performance objectives for the 64 kbit/s circuits, established on the hypothetical digital reference path, at below the first digital hierarchy, defined in ITU-R [16], Recommendation F.556-1, are specified in ITU-R [16] Recommendation F.594-4 . According to this recommendation and including not only the effects of fading but also all other possible causes of quality degradation, in an available, unidirectional, 64 kbit/s, circuit of the hypothetical digital reference path:

1. The errored second ratio, *esr*, should not exceed 0.0032 of any month;
2. The severely errored second ratio, *sesr* should not exceed 0.00054 in any month.

A previous version of the same recommendation noted that the objective of the errored second ratio is usually met when the objective of the severely errored second ratio is met.

ITU-R [16] Recommendation F.557-4 states that a microwave radio link is unavailable when, in any transmission sense, one or both of the following conditions are met during at least 10 consecutive seconds:

- loss of signal, with loss of synchronism or frame alignment;
- bit error ratio higher than 10^{-3}.

The same Recommendation imposes that the hypothetical reference circuit should be available 99.7 percent of the time (per year). This value does not include possible improvements due to traffic rerouting to other systems. The value of 99.7 percent is provisional. The real value should be in the range 99.5 to 99.9 percent.

For the ITU-T hierarchies and the recommended frame synchronism techniques, a bit error ratio of 10^{-3} corresponds to a high probability of losing frame synchronism. This value is a compromise between lower values that would lead to very low fractions of time (less than about 0.01 percent) and very high costs, and higher values, corresponding to higher fraction of time (about 0.1 percent) that would lead to systems with a quality significantly lower than that of analog links.

ITU-T Recommendation G.826 [17] specifies the end-to-end objectives for a 27 500 km hypothetical digital reference path at or above the first digital hierarchy given in Table 5.9. This recommendation introduces a new parameter in the error performance objectives: the background block error ratio, or *bber*, defined as the ratio of background block errors (or *bbe*) to the total blocks (excluding those during severely errored seconds), in a fixed measurement interval (usually a month), where a background block error is an errored block not occurring as part of an severely errored second.

Bit rate (Mbit/s)	Block size (bit)	esr	sesr	bber
1.5 – 5	800 – 5000	0.04	0.002	2×10^{-4}
>5 – 15	2000 – 8000	0.05	0.002	2×10^{-4}
>15 – 55	4000 – 20000	0.075	0.002	2×10^{-4}
>55 – 160	6000 – 20000	0.16	0.002	2×10^{-4}
> 160 – 3500	15000 – 30000	—	0.002	1×10^{-4}

Table 5.9 End-to-end error performance objectives for a 27 500 km hypothetical reference digital path at or above the first digital hierarchy, according to ITU-T Recommendation G.826 [17].

It is obvious that there are quite a few inconsistencies between ITU-T Recommendation G.821 and G.826 (which, in fact, are listed in Appendix I to Recommendation G.821), among which the criteria to define a severely errored second and the allocation principle. These inconsistencies are likely to be resolved in future revisions of these recommendations.

In the next sections, whenever possible, we adhere to the bit error ratio based definition of the severely errored second as this follows quite naturally from the basic performance criteria set for digital links: bit rate and bit error ratio.

5.10.5 Error performance objectives for real circuits

When microwave radio links are chained, the overall bit error ratio is equal to the sum of the individual link bit error ratios. Thus it seems that the apportionment of the overall bit error ratio by the various links could be performed according to the length of each link. In fact, the situation is far more complex because fading acts independently in each of the links in the chain.

Assume that the bit error ratio of the ith link in a chain is a random variable with a probability density function p_i. The probability density function p_r of the overall bit error ratio in a chain with n links is given by

$$\begin{aligned} p_r(b) &= \int_0^b dx_1 \int_0^{b-x_1} dx_2 \ldots \\ &\ldots \int_0^{b-\sum_{i=1}^{n-2} x_i} \left[\prod_{i=1}^{n-1} p_i(x_i)\right] p_n\left(b - \sum_{i=1}^{n-1} x_i\right) dx_n. \quad (5.257) \end{aligned}$$

The cumulative probability that $\mathcal{P}(b \leq b_0)$ is

$$\begin{aligned}\mathcal{P}(b \leq b_0) &= \int_0^{b_0} db \int_0^{b} dx1 \int_0^{b-x_1} dx_2 \ldots \\ &\ldots \int_0^{b-\sum_{i=1}^{n-2} x_i} \left[\prod_{i=1}^{n-1} p_i(x_i)\right] \\ &\times p_n\left(b - \sum_{i=1}^{n-1} x_i\right) dx_n. \end{aligned} \tag{5.258}$$

Although it is quite possible to formulate educated guesses about fading in each individual link and it is acceptable to assume that fading in different links is uncorrelated, the fact that the bit error ratio of a given link is not related in a simple way with fading, makes the use of (5.257) and (5.258) rather difficult.

In addition, quality criteria based on bit error ratios are difficult to monitor in real link operation. Thus, current ITU-R recommendations are based on the errored seconds ratio (esr) and the severely errored seconds ratio ($sesr$).

For an available link with length d (in km), between 280 and 2500, at a bit rate below the first digital hierarchy, ITU-R[16] Recommendation F.634-4 states that:

1. The errored second ratio esr should not exceed $\frac{0.0032d}{2500}$ during any month;

2. The severely errored second ratio $sesr$ should not exceed $\frac{0.00054d}{2500}$ during any month.

If we compare ITU-R Recommendations F.594-4 and F.634-4, we find out that both require for each link the bit error ratio of the hypothetical digital reference circuit, during a period of time which is reduced proportionally to the link length.

The reason for this is the same mentioned for analog links for short periods of time. That is, we assume that causes of an increase in the error ratio in the different hops are uncorrelated and, since we are only interested in short periods of time, we neglect simultaneous occurrences in two or more hops.

It is interesting to note that a severely errored second is defined here as a one-second period where the bit error ratio exceeds 10^{-3}. Thus the second item of the current recommendation is the same (with different wording) as the previous recommendation.

ITU-R Recommendation F.1189-1 [16], divides the national portion of a 27 500 hypothetical digital path in three basic sections:

- long haul, from the international gateway up to the primary, secondary or tertiary center, depending on country network architecture;

- short haul, from the primary, secondary or tertiary center, depending on country network architecture, up to the local exchange;

- access, from the local exchange to the path end point;

and sets error performance objectives for each direction of constant bit rate digital paths, given in Table 5.10. Following ITU-T Recommendation G.826 performance objectives in this case are block based rather that bit based as in the previous case. Block size to monitor path performance varies, and usually increases with bit rate, according to the digital hierarchy, from about 2 000 bits up to 20 000 bits.

Now the severely errored second is defined as one-second period which contains 30 percent or more errored blocks, or at least one defect. Defects are system dependent. They include among others the following: loss of signal, loss of frame alignment, alarm indications.

In Table 5.10 the value of X is given by:

- for the long haul network section $X = A$:

$$A_1 = 0.01 \quad \text{to} \quad 0.02;$$
$$A = A_1 + 0.01\frac{l_{lh}}{500};$$

  where l_{lh}, in km, is the actual route length of the long-haul section;

- for the short-haul network section $X = B$:

$$B = 0.075 \quad \text{to} \quad 0.085;$$

- for the access network section $X = C$:

$$C = 0.075 \quad \text{to} \quad 0.085.$$

In should be noted that:

- If the actual route length of the long-haul section is not known, then the air route ar multiplied by the appropriate routing factor rf should be used, according to:
  - $ar < 1000$ km, then $rf = 1.5$;
  - $1000 \geq ar < 1200$ km, then $l_{lh} = 1500$;
  - $ar \geq 1200$ km, then $rf = 1.25$.
- the sum $A_1 + B + C$ should not exceed 0.175;
- the sum $A_1 + B + C$ should be in the range 0.155 to 0.165.

Bit rate (Mbit/s)	esr	sesr	bber
1.5 – 5	0.04 × X	$0.002 \times X$	$2 \times X \times 10^{-4}$
> 5 – 15	0.05 × X	$0.002 \times X$	$2 \times X \times 10^{-4}$
>15 – 55	0.075 × X	$0.002 \times X$	$2 \times X \times 10^{-4}$
>55 – 160	0.16 × X	$0.002 \times X$	$2 \times X \times 10^{-4}$
>160 – 3500	to be defined	$0.002 \times X$	$1 \times X \times 10^{-4}$

Table 5.10 Error performance objectives for radio relay paths of the national portion of the hypothetical reference digital path at or above the first digital hierarchy, (ITU-R Recommendation F.1189-1 [16]).

5.11 CONVERTING QUALITY PARAMETERS INTO BIT ERROR RATIOS

Since the severely errored second is no longer defined as a second with a bit error ratio over 10^{-3}, for link design and planning a relation between these two parameters is required. Similar relations are also required for the other quality parameters: the errored seconds ratio and the background block error ratio.

5.11.1 Severely errored seconds ratio

Assume uniformly distributed errors along time and a low bit error rate, so that a single bit per block will be in error. Recalling that a severely errored second (ses) is a second during which 30% of the blocks transmitted have one (or more) bits in error, the bit error ratio (ber_{ses}) is given by

$$ber_{ses} = \frac{0.3 * n_{block}}{f_b}, \tag{5.259}$$

where n_{block} stands for the number of transmitted blocks per second and f_b for the bit rate.

The number of transmitted blocks per second, as per ITU-R [16] Recommendation P.530-8, is 2000 for the first and second digital hierarchy and 8000 for the higher hierarchies.

The initial assumption that each errored bit causes an errored block is rather crude, since in microwave radio links errored bits tend to follow a Poisson distribution, where the time difference τ between errored bits obeys has the following cumulative distribution function:

$$\mathcal{P}(\tau \geq \tau_0) = e^{-\frac{\tau_0}{\tau_m}}, \tag{5.260}$$

where τ_m is the average distance between errored bits.

We note that ber_{ses} is the ratio of errored bits and transmitted bits, that is:

$$ber_{ses} = \frac{1}{f_b \tau_m}. \tag{5.261}$$

Since errors are not uniformly distributed in time, some of the blocks accounted for as errored blocks will have more than one errored bit, while others will have no errored bit. Thus the value of ber calculated in (5.259) may be slightly increased. If the time occupied by a block in the bit stream is τ_{block} the probability that a block has k bits in error is:

$$\mathcal{P}(k) = \left(\frac{\tau_{block}}{\tau_m}\right)^k \frac{e^{-\frac{\tau_{block}}{\tau_m}}}{k!}. \tag{5.262}$$

For a single errored bit we have:

$$\mathcal{P}(1) = \frac{\tau_{block}}{\tau_m} e^{-\frac{\tau_{block}}{\tau_m}}. \tag{5.263}$$

Imposing the condition for a severely errored second, that is, $\mathcal{P} = 0.3$ in (5.263) we get:

$$\frac{\tau_{block}}{\tau_m} = 0.4894.$$

Recalling (5.261) and taking $\tau_{block} = b_{block} \times \tau_1$ where b_{block} is the number of bits per block and $\tau_1 = 1/f_b$ is the duration of a bit we get

$$ber_{ses} = \frac{0.4894 f_b}{b_{block}}. \tag{5.264}$$

The number of bits per block b_{block} may be obtained from the relevant ITU-R Recommendation or may be estimated from the bit rate divided by the number of transmitted blocks per second n_{block}.

The previous analysis assumes a single errored bit per block. If there are more errored bits per block either the error control code flags the block as an errored block, as it should, or the error control code is unable to detect the error and the block is accepted as correct.

Let us consider a simple case: the error control code detects all odd number of errored bits but is unable to detect an even number of errored bits. Then, from (5.262) we have

$$\mathcal{P}(k = odd) = \sum_{i=0}^{\infty} \left(\frac{\tau_{block}}{\tau_m}\right)^{2i+1} \frac{e^{-\frac{\tau_{block}}{\tau_m}}}{(2i+1)!}. \tag{5.265}$$

Since $\frac{\tau_{block}}{\tau_m} << 1$ the upper limit of the summation in (5.265) may be reduced to 9 (or even less) without incurring into significant errors.

Taking $\mathcal{P}(k = odd) = 0.3$ we get

$$ber_{ses} = \frac{0.4581}{b_{block}}. \tag{5.266}$$

In Table 5.11 we compare the values given by (5.266) with those provided in ITU-R [16] Recommendation P.530-8. As shown, these two values are sufficiently close for design and planning. To improve on the accuracy provided by (5.266) would require a detailed analysis of the error control performance used in each case.

Path type	Bit rate (kbit/s)	Bit rate supported (kbit/s)	ber_{ses} P.530-8	ber_{ses} (5.266)
VC-11	1 664	1 500	5.4×10^{-4}	5.5×10^{-4}
VC-12	2 240	2 000	4.0×10^{-4}	4.1×10^{-4}
VC-2	6 848	6 000	1.3×10^{-4}	1.3×10^{-4}
VC-3	48 960	34 000	6.5×10^{-5}	7.5×10^{-5}
VC-4	150 336	140 000	2.1×10^{-5}	2.4×10^{-5}
STM-1	159 520	155 000	2.3×10^{-5}	2.3×10^{-5}

Table 5.11 Bit error ratio ber_{ses} corresponding to the severely errored seconds ratio $sesr$.

The number of transmitted blocks per second and block size in bits/block for different path types and bit rates are shown in Table 5.12.

Path type	Bit rate (kbit/s)	Bit rate supported (kbit/s)	Block/s	Bit/block
VC-11	1 664	1 500	2 000	832
VC-12	2 240	2 000	2 000	1 120
VC-2	6 848	6 000	2 000	3 424
VC-3	48 960	34 000	8 000	6 120
VC-4	150 336	140 000	8 000	18 792
STM-1	159 520	155 000	8 000	19 940

Table 5.12 Number of transmitted blocks per second n_{block} and number of bits per block b_{block} for different path types and bit rates.

With $\mathcal{P}(k = odd) = 0.3$ a simple bit parity error would fail to detect 0.4581 of the errored blocks, far in excess of the 0.1 failure rate set by the ITU-T [17] Recommendation G.826.

In order to improve the detection rate we may divide the block into m bytes with n bits each and provide a parity bit for each bit of the m bytes. Such a parity bit code is known as a Bit Interleaved Parity code, BIP (n, m), or simply as BIP$-n$ (see Section 6.4.7).

The probability that an errored block is detected is $\mathcal{P} = 1 - \mathcal{Q}$, where $\mathcal{Q}$ is the probability that the errored block fails to be detected. Since each of the parity bits will only detect an odd number of errored bits the probability that there are an odd number of errored bits in the ith position of each of the m bytes is given by (5.265) where τ_{block} is now τ_{block}/n. The probability that the BIP$-n$ code fails to detect an error is the probability that none of the parity bits indicates an error

$$\mathcal{Q} = [1 - \mathcal{P}(k = odd)]^n. \tag{5.267}$$

For the BIP-2 code when we equate $\mathcal{P} = 0.3$ we get

$$\frac{\tau_{block}}{n\tau_m} = 0.198,$$

which means that now only 0.198 of the errored blocks fail detection.

Increasing n to 4 or 8 reduces the fraction of the errored blocks that fail to be detected to 0.0935 and 0.0456, respectively, thus meeting the ITU-T recommendations.

According to ITU-R [16] Recommendation P.530-8 complex modulations and error correcting codes, such as 128 trellis coded modulation tend to produce error burst with 10 to 20 errors per burst. In these cases the bit error ratio values in Table 5.12 should be multiplied by the burst length (in bits).

Even if Table 5.11 strictly applies only to SDH, ITU-R Recommendation P.530-8 suggests its application to PDH using the closest transmission rate.

5.11.2 Background block error ratio

Conversion of the background block error ratio ($bber$) in the corresponding bit error ratio ber_{bber} may be carried out according to the following simplified step-by-step method described in ITU-R [16] Recommendation P.530-8:

1. Calculate the probability $\mathcal{P}_{rber}$ that the residual bit error rate $rber$[10] (varying from about 10^{-10} for 2 Mbit/s to 10^{-13} for 155 Mbit/s.) is exceeded;

2. Calculate the severely errored seconds ratio $sesr$ as the probability that the bit error ratio for the severely errored seconds ber_{ses} given in Table 5.11 is exceeded;

3. Calculate the background block error ratio $bber$ as

$$bber = sesr\frac{\alpha_1}{2.8\alpha_2(sl_{bber} - 1)} + \frac{b_{block}rber}{\alpha_3}, \tag{5.268}$$

where:

[10]The residual bit error ratio is the minimum bit error ratio provided by the transmission system

- α_1 is the number of errors per burst for $10^{-3} > ber > ber_{ses}$, typically from 10 to 30;
- α_2 is the number of errors per burst for $ber_{ses} > ber > rber$, typically from 1 to 10;
- $\alpha_3 = 1$ is the number of errors per burst for $ber < rber$;
- b_{block} is the number of bits per block;
- sl_{bber} is the absolute value of the slope of the ber distribution curve on a log-log scale for $ber_{ses} > ber > rber$

$$sl_{bber} = \left| \frac{\log_{10}(rber) - \log_{10}(ber_{ses})}{\log_{10}[\mathcal{P}(rber)] - \log_{10}[\mathcal{P}(ber_{ses})]} \right| . \quad (5.269)$$

The second term in (5.268) corresponds to the errored blocks due to the residual error rate, when errors tend to appear isolated (hence $\alpha_3 = 1$), whereas the first represents the fraction of the errored blocks due to values of the bit error rate from the residual bit error rate up to (but excluding) the bit error rate corresponding to the severely errored seconds.

The background block error ratio, to be met during the "most unfavorable month" may be due to fading and to rain. For the former the fading distribution applicable to the link should be used while for the latter the rain attenuation is appropriate. In case the severely errored seconds ratio due to rain, or the background block error ratio due to rain, fall below the validity limits of the expressions used (2.111) we may neglect the corresponding values in the calculations.

Rain affects the performance of microwave radio links in two ways. Heavy rain may cause link unavailability, whereas lighter rains increases the bit error ratio of an available link. Since unavailability is defined by periods where the severely errored seconds ratio is exceeded for more than 10 successive seconds, part of the severely errored seconds ratio due to rain would be included in unavailability. At present ITU-R Recommendation[16] P.530-8 suggests that (heavy) rain should only cause unavailability, that is, there should be no rain contribution to the severely errored seconds ratio. This assumption simplifies the use of (5.268) since only the second term needs to be evaluated.

5.11.3 Errored seconds ratio

Similarly to the background block errored ratio, conversion of the errored seconds ratio (esr) in the corresponding bit error ratio ber_{es} may be carried out according to the following simplified step-by-step method described in ITU-R [16] Recommendation P.530-8:

1. Calculate the probability $\mathcal{P}(rber)$ that the residual bit error rate $rber$ (varying from about 10^{-10} for 2 Mbit/s to 10^{-13} for 155 Mbit/s) is exceeded;

2. Calculate the severely errored seconds ratio $sesr$ as the probability that the bit error ratio for the severely errored seconds ber_{ses} given in Table 5.11 is exceeded;

3. Calculate the errored seconds ratio esr as

$$esr = sesr \sqrt[sl_{esr}]{n_{block}} + \frac{b_{block} b_{block} rber}{\alpha_3}, \tag{5.270}$$

where:

- $\alpha_3 = 1$ is the number of errors per burst for $ber < rber$,
- n_{block} is the number of transmited blocks per second,
- b_{block} is the number of bits per block,
- sl_{esr} is the absolute value of the slope of the ber distribution curve on a log-log scale for $ber_{sesr} > ber > rber$

$$sl_{esr} = \left| \frac{\log_{10}(rber) - \log_{10}(ber_{sesr})}{\log_{10}[\mathcal{P}(rber)] - \log_{10}[\mathcal{P}(ber_{sesr})]} \right|. \tag{5.271}$$

As noted for the background block error ratio, the second term in (5.270) also represents the fraction of the errored seconds due to the residual bit error rate when errors tend to appear isolated whereas the first term accounts for errored seconds due to values of the bit error rate from the residual bit error rate up to (but excluding) the bit error rate corresponding to the severely errored seconds.

As for the background block error ratio, the errored seconds ratio to be met during the "most unfavorable month" may be due to fading and to rain. Again, for the former, the fading distribution applicable to the link should be used while for the latter the rain attenuation is appropriate. In case the severely errored seconds ratio due to rain, or the background block error ratio due to rain, fall below the validity limits of the expressions used (2.111) we may neglect the corresponding values in the calculations.

As noted in Section 5.11.2 in the calculation of the errored seconds ratio due to rain the first term in (5.270) may be neglected.

5.12 APPLYING QUALITY CRITERIA: AN EXAMPLE

In order to illustrate the application of the various quality criteria consider a 30-km-long, 4-PSK, 34 Mbit/s microwave radio link, operating at 13 GHz with horizontal polarization. The residual bit error rate $rber$ will be taken as 10^{-12}. Since this is a PDH hierarchy, following ITU-R[16] Recommendation P.530-8, we should use the VC-3 path type (48 960 kbit/s) in Table 5.11.

To simplify calculations we will assume that link fading follow Morita's law (3.47) and that atmospheric attenuation due to oxygen and water vapor may be neglected. Rain intensity not exceeded for more than 0.01 percent of the time will be taken as 42 mm/h.

Assuming that the link is available (unavailability will be dealt with in the last chapter), the ITU-R previous quality criteria imposed bit error ratios of 10^{-6} and 10^{-3}, not to be exceeded for more than $0.004d/2500$ and $0.00054d/2500$ of the time, respectively, where d is the link length in km, with a minimum of 280 (km). For our case these time fractions correspond to 0.00045 and 0.00006, respectively.

Using equation (3.47), with $d = 30$ and $f = 13$, we get $m = 60.1$ for $\mathcal{P} = 0.00045$ and $m = 445$ for $\mathcal{P} = 0.00006$. Using (5.230) together with (3.47) we get $\mathcal{F} = 0.0269$ which, from (5.241), yields $\eta = 0.0132$. From (5.243) with $d = 30$ and $n = 1.4$ we get $\tau_m = 0.342$ ns. Using (5.242) with $K_n = 1$, and $T_s = 58.8$ ns we get $\mathcal{P}_{s/mp} = 1.46 \cdot 10^{-4}$ and using (5.240) $\mathcal{P}_s = 1.93 \cdot 10^{-6}$.

The probability of selective fading may be converted into the selective margin using (5.230) and we obtain $m_s = 13930$. Using (5.238) we get the area under the receiver signature $s = 0.57$ MHz, well within the range of typical values for 34 Mbit/s receivers using 4-PSK modulation.

Once the selective margin m_s has been obtained the uniform margin m_u may be calculated from the margin m using (5.234) and we get $m_{u_1} = 60.3$ and $m_{u_2} = 459.7$ or $M_{u_1} = 17.8$ dB and $M_{u_2} = 26.6$ dB.

From (5.205) and (5.206) we find out that the carrier-to-noise ratio at the demodulator input is equal to 13.5 dB for $ber = 10^{-6}$ and 9.8 dB for $ber = 10^{-3}$.

Adding the uniform margins (in dB) to the carrier-to-noise for the specified bit error rates we finally get the required carrier-to-noise ratios under no fading conditions: $C/N_0 = 31.3$ dB for $\mathcal{P} = 0.00045$ and $C/N_0 = 36.4$ dB for $\mathcal{P} = 0.00006$. To meet both conditions we must have $C/N_0 = 36.4$ dB.

We will now consider the severely errored seconds ratio $sesr$ which, for the short-haul section and for the access network should obey (Table 5.10) $sesr \leq 0.002 \times X$ with X in the range $[0.075, 0.085]$. Taking $X = 0.08$ we get $sesr = 1.6 \cdot 10^{-4}$. From Table 5.11 we get the required bit error ratio $ber_{ses} = 6.5 \cdot 10^{-5}$.

As stated before, we will assume that the severely errored seconds ratio is only due to fading. Hence from (3.47), with $d = 30$, $f = 13$, and $\mathcal{P}_{ses} = sesr = 1.6 \cdot 10^{-4}$ we get $m_{u_{ses}} = 171.6$. Using the selective margin previously calculated $m_s = 13930$ and (5.234) we get the fading margin $m_{ses} = 169.5$ or $M_{ses} = 22.3$ dB.

From (5.205) and (5.206) with $ber_{ses} = 6.5 \cdot 10^{-5}$ we calculate carrier-to-noise ratio at the demodulator input as $C/N_{ses} = 11.7$ dB. Adding this value to the uniform fading margin (in dB) we get the carrier-to-noise ratio at the demodulator input required to meet the required severely errored seconds ratio $C/N_{sesr} = 34.0$ dB.

For the background block error ratio we will calculate separately the effects of fading and rain.

Following the procedure laid out in Section 5.11.2 we should calculate the probability $\mathcal{P}_{rber}$ that the residual bit error rate is exceed due to fading. But for this calculation we require the value of the carrier-to-noise ratio C/N at the demodulator input. As a starting point we will take $C/N = C/N_{sesr} = 34$ dB.

From (5.205) and (5.206) we get the carrier-to-noise ratio required for the residual bit error rate $C/N_{rber} = 16.9$ dB. The uniform margin for the residual bit error rate is simply the difference between the values of C/N and C/N_{rber}, that is, 17.1 dB or 50.8. Introducing the selective margin $m_s = 13930$ we get the corresponding margin $m_{rber} = 50.6$, or $M_{rber} = 17$ dB and, from (3.47), $\mathcal{P}_{rber} = 5.3 \cdot 10^{-4}$.

Since we used the carrier-to-noise ratio required to meet the severely errored seconds we already know that $\mathcal{P}_{ses} = sesr = 1.6 \cdot 10^{-4}$ and $ber_{ses} = 6.5 \cdot 10^{-5}$. Introducing these values together with $rber$ and $\mathcal{P}_{rber}$ in (5.269) we get $sl_{bber} = 14.9$. Finally from (5.268), with $\alpha_1 = 10$, $\alpha_2 = 1$, $\alpha_3 = 1$ and $b_{block} = 6120$ (from Table 5.11) we get $bber = 4.1 \cdot 10^{-5}$ which is rather more than the recommended value $1.6 \cdot 10^{-5}$.

To find out the minimum value of the carrier-to-noise ratio at the demodulator input that meets the specified background block error ratio we must proceed by trial and error, remembering that for each value of the carrier-to-noise we must recalculate the severely errored seconds ratio (i.e., the probability that ber_{ses} is exceeded).

Taking $C/N = 38.1$ dB we get a uniform margin $M_u = 38.1 - 11.7 = 26.4$ dB, or $m_u = 441$ which, from (5.234) with $m_s = 13930$, yields $m = 427.4$. Introducing this margin in (3.47) we get $\mathcal{P}_{ses} = 6.3 \cdot 10^{-5}$. Proceeding as before for the residual bit error ratio we get $M_u = 38.1 - 16.9 = 21.2$ dB, or $m_u = 130.5$, $m = 129.3$ and $\mathcal{P}_{rber} = 2.1 \cdot 10^{-4}$. Using (5.269) we get $sl_{bber} = 15$ and finally from (5.268) $bber = 1.6 \cdot 10^{-5}$, the maximum recommended value.

For the errored seconds ratio we require again a starting value of carrier-to-noise ratio at the demodulator input. We may use again the value that meets the severely errored seconds ratio, 34 dB, for which we have already obtained $sesr = \mathcal{P}_{ses}$, ber_{ses}, $rber$, $\mathcal{P}_{rber}$ and sl_{sesr}. Noting that in this case $sl_{esr} = sl_{sesr}$ and using (5.270) with $b_{block} = 8000$ we get $esr = 3.4 \cdot 10^{-4}$, considerably less than the maximum recommended value $6 \cdot 10^{-3}$.

It is left to the reader to verify that the minimum value of the carrier-to-signal ratio that meets the recommended errored seconds ratio is $C/N_{esr} = 20.9$ dB.

The procedure to calculate the background block error ratio due to the rain is similar to one used for fading, with the difference that the excess attenuation now is due to rain and is calculated according to the method described in Section 2.5.3 for the worst month.

As it was stated before, at present ITU-R Recommendation P.530-8 assumes that the severely errored seconds ratio due to rain is totally accounted for in the link unavailability. Thus we need only to calculate the second term in (5.268), which, in fact, is the same used for the background block error ratio due to fading. Thus $bber = 6.1 \cdot 10^{-9}$, a much lower value than the background block error ratio due to fading.

The same reasoning applies to the errored seconds ratio due to rain. Since the severely errored seconds ratio due to rain is totally accounted for in the link unavailability, the second term in (5.270) yields $esr = 4.9 \cdot 10^{-5}$, again a much lower value than the one due to fading.

From the above we conclude that the most difficult criterion to meet in this link is the background block error ratio, which requires a carrier-to-noise ratio at the demodulator input equal to 38.1 dB. We note that the previous quality criteria required a slightly lower value (36.4 dB).

5.13 QUALITY OBJECTIVES FOR ISDN LINKS

ITU-R [16] Recommendation F.696-2 defines error performance objectives in terms of errored second ratio (esr) and severely error second ratio ($sesr$) for radio electric sections of an Integrated Services Digital Network (ISDN), which are shown in Table 5.13. These values refer to available, unidirectional 64 kbit/s links.

Parameter	**Fraction of the worst month**			
	Class 1 280 km	**Class 2 280 km**	**Class 3 50 km**	**Class 4 50 km**
esr	0.00036	0.0016	0.0016	0.004
$sesr$	0.00006	0.000075	0.00002	0.00005

Table 5.13 Error performance objectives for radio electric sections which are part of an Integrated Services Digital Network, (ITU-R [16] Recommendation F.696-2).

Recommendation F.696-2 also puts forward the following values for the bidirectional unavailability:

- Class 1: 0.033 percent;
- Class 2: 0.05 percent;
- Class 3: 0.05 percent;
- Class 4: 0.1 percent.

Links shorter than those in Table 5.13 must comply with the values in this table. For longer links the following applies:

- Class 1: the values of the bit error ratio should be calculated as per Recommendation 634-3 proportionally to the link length:

- Classes 2,3 and 4: the overall bit error ratio should correspond to an integer number of sections of the same class, whose total length, at least, is equal to the link length.

When microwave radio links make up all the 1250 km local grade link, at each end of the hypothetical reference connection, specified in ITU-T [17] Recommendation G.821 error performance objectives applicable to each 64 kbit/s channel, in each direction, are stated in ITU-R [16] Recommendation F.697-2:

- The severely errored second ratio (*sesr*) should not exceed 0.000 15 in any month;
- The errored second ratio (*esr*) should not exceed 0.001 2 in any month.

Since the use of microwave radio links in the local grade section, particularly in urban areas, frequently leads to hops with frequencies above 10 GHz and lengths less than 10 km, for which it is not economic to provide reserve channels, we must define reasonable unavailability values, according to current technologies.

For these frequencies and hop lengths, the availability is usually determined by the rainfall and the equipment. CCIR [13] Report 1053-1 puts forward the following values for the unavailability due to the rainfall:

- 0.004 to 0.0004 percent, per year, in Japan;
- 0.001 to 0.01 percent, per year, in the United Kingdom.

REFERENCES

1. Abramowitz, M. and Stegun, I., *Handbook of Mathematical Functions with Formulas, Graphs and Mathematical Tables*, National Bureau of Standards, Washington, DC, 1964.
2. Albuquerque, A. A., *Sistemas de Telecomunicações. Notas das aulas teóricas*, Lisboa, Associação de Estudantes do Instituto Superior Técnico, 1987.
3. Alcatel Thomson, Note on High Capacity Digital Radio Link Calculation, *Faisceaux Hertziens*, 1985.
4. Baccetti B. and Tartara G. , *Equalization and Quality Prediction in Digital Radio Systems*, GTE Telecomunicazioni, Milan, 1983.
5. Bellamy, J. C. , *Digital Telephony*, John Wiley & Sons, New York, 1982.
6. Benedetto,S., Biglieri, E. and Castellani, V., *Digital Transmission Theory*, Prentice-Hall International Editions, Englewood Cliffs, NJ, 1987.

7. Bic, J. C., Duponteil D. and Imbeaux J. C. , *Élements de communications numériques - Transmission sur fréquence porteuse*, Dunod, Paris, 1986.

8. Biglieri, E., Divsalar, D., McLane, P. and Simon, M., *Introduction to Trellis-Coded Modulation with Applications*, Macmillan Publishing Company, New York, 1991.

9. Carlson, A. B., *Communication Systems. An Introduction to Signals and Noise in Electrical Communication*, 2nd edition, McGraw-Hill International Book Company, Tokyo, 1981.

10. CCIR, *Recommendations et rapports du CCIR, 1982*, Volume IX - 1, *Service fixe utilisant les faisceaux hertziens*, UIT, Genève, 1982.

11. CCIR, *Recommendations and rapports du CCIR, 1986*, Volume XI -1, *Service de radiodiffusion (télévision)*, UIT, Genève, 1986.

12. CCIR, *Recommendations of the CCIR, 1990*, Volume IX - Part 1, *Fixed Service Using Radio-Relay Systems*, UIT, Geneva, 1990.

13. CCIR, *Reports of the CCIR, 1990*, Annex to Volume IX, Part 1, *Fixed Service Using Radio-Relay Systems*, UIT, Geneva, 1990.

14. Clark, G. and Cain, J., *Error-Correction Coding for Digital Communications*, Plenum Press, New York, 1982.

15. Gregg, W. D., *Analog and Digital Communication*, John Wiley & Sons, New York, 1997

16. ITU, *ITU-R Recommendations on CD-ROM*, UIT, Geneva, 2000.

17. ITU, *ITU-T Recommendations on CD-ROM*, UIT, Geneva, 2000.

18. Jayant, S. N., Digital Coding of Speech Waveforms: PCM, DPCM and DM Quantizers. *Proceedings of the IEEE*, May pp. 611–632, 1974.

19. Oetting, J. D., A Comparison of Modulation Techniques for Digital Radio. *IEEE Transactions on Communications*, Vol. COM-27, No. 12, December, pp. 1752–1762, 1979.

20. Oppenheim, A. V. and Schafer R. W., *Digital Signal Processing*, Prentice-Hall, New York, 1975.

21. Poo, G. S., Computer aids for code spectra calculations, *Analytical Foundations for Digital Line Codecs. Digest of Colloquium Held on 4th June 1981*, Department of Electrical Engineering Science, University of Essex, Colchester, 1981.

22. Rummler, W. D., A New Selective Fading Model: Application to Propagation Data. *Bell Systems Technical Journal*, Vol. 58, May-June, pp. 1037–1071, 1979.

23. Schwartz, M., Bennett, W. and Stein, S., *Communication systems and techniques*, McGraw-Hill Book Company, New York, 1966.

24. Sklar, B. *Digital Communications. Fundamentals and applications*, Prentice-Hall International Editions, Englewood Cliffs, NJ, 1988.

Problems

5:1 Calculate the maximum signal-to-quantization noise (in dB) for a uniformly quantized sinusoidal signal coded with 8 bits.

5:2 Calculate the signal-to-quantization noise for the μ-law assuming a uniform distribution of amplitude.

5:3 Calculate the signal-to-noise ratio for a sinewave signal, uniformly coded with 8 bits in a binary symmetric channel with $ber = 10^{-5}$.

5:4 Sketch the bipolar (or AMI), the Manchester, the CMI and the 4 level multicode corresponding to the sequence 11011100010.

5:5 Compute the *ber* for $E_b/N_0 = 12$ dB for polar, unipolar (optimum detection level) and bipolar codes.

5:6 Compute the *ber* for $E_b/N_0 = 15$ dB for 4 and 8 level multilevel codes.

5:7 Compute the *ber* for $C/N = 15$ dB for amplitude modulation and coherent detection of polar, unipolar, bipolar and 4 level multipolar codes.

5:8 Compute the *ber* for $C/N = 15$ dB and $C_d/N = 9.5$ dB for amplitude modulation and incoherent detection of a unipolar code.

5:9 Compute the *ber* for $C/N = 13$ dB for FSK, with coherent and incoherent detection, and for 2-PSK and 4-PSK.

5:10 Compute the *ber* for $C/N = 15$ dB for 4-QAM and 16-QAM.

5:11 Calculate the occupied radio frequency spectrum for a 4-PSK 1.5 Mbit/s signal, with $\beta = 0.25$.

5:12 Estimate the sensitivity of a 16-QAM 155 Mbit/s receiver with a noise factor of 6 dB for $ber = 10^{-3}$.

5:13 Calculate the link margin for a link with a uniform margin of 35 dB and a signature of 25 MHz.

5:14 Calculate the probability that the bit error rate is exceeded due to inter-symbol distortion, using method B in ITU-R Recommendation F.1093-1, for a typical non-equalized 155 Mbit/s, 16-QAM receiver in a 35-km-long link and a 30-dB fade.

5:15 Calculate the double space diversity improvement factor for selective fading in a 40-km-long link, at 6 GHz, with a receiver signature of 25 MHz, and equal receiver antennas whose centers are 10 m apart. Use the ITU-R Recommendation P.530.8 expression.

5:16 Consider a 155 Mbit/s microwave radio link equipment, at 8 GHz, with transmitter output power of 1 W and a receiver noise figure of 5 dB. Neglecting atmospheric excess attenuation and assuming that fading follows Morita's law calculate the maximum hop length for transmitter and receiver antennas limited to 3.0-m-diameter parabolic reflectors with 50 percent aperture efficiency.

5:17 Consider an 8-GHz, 50-km-long, 140 Mbit/s link, with a transmitter output power of 31 dBm, a receiver noise factor of 5.5 dB and equal antenna gains of 39.1 dBi at each terminal. The feeder losses at each terminal are equal to 3 dB and the excess atmospheric attenuation is equal to 0.35 dB. Check if this links meets all the relevant ITU-R quality criteria.

5:18 Consider again the link and the equipment of the previous problem, using either equalizers in the time and frequency domain or space diversity and check if this link meets all the relevant ITU-R quality criteria.

6

Error Control Codes

6.1 INTRODUCTION

This chapter deals with codes capable to detect and (or) correct transmission errors. After introducing the importance of control codes we deal in some detail with block and cyclic codes. Given the difficulty of block codes to deal with burst of errors, so frequent in microwave radio links, code interleaving is introduced as a possible solution for this problem. Next we approach convolutional codes and Viterbi decoding. Code concatenation is presented as another way to deal with error bursts. Following, there is an introduction to turbo codes and signal space codes, currently hot research topics. Finally we derive performance bounds for transmission channels based on Shannon's work and compare the performance of a few codes based on these bounds. To simplify the subject, most of the presentation is restricted to binary codes, thus avoiding exposure of the reader to Galois field theory.

In the early days of digital links, performance was mostly defined by the bit error ratio, *ber*, as in Chapter 5. Nowadays it is mostly the occurrence of errors in the blocks (from about 800 to 20 000 bits), in which the bit stream is organized. These blocks include some form of error control coding to detect errors. This simplifies link supervision and enables a continuous check of link performance, which was not previously possible.

Error correction codes are still rather uncommon in digital links, with the very important exception of deep space links, basically due to the lack of coders and decoders for medium and high bit rates with very low latency (few milliseconds), as required for multihop links, with a large number of relay stations. Spectrum shortage has also pushed for bandwidth efficiency, rather than power efficiency, thus making most error correction coding less atractive. The increasing performance of coders and decoders and the recent appearance of signal space codes may well change the current situation and make error correcting codes a common feature in digital links.

6.2 THE IMPORTANCE OF ERROR-CORRECTING CODES

In order to introduce the importance of error correcting codes let us consider a channel where the probability of receiving an errored symbol is $ser = 0.01$. If we consider each symbol in turn it is very difficult, if not impossible, to find out if this symbol is correct or not, although on average it would be correct 99 out of 100 times.

Now let us take a block of n symbols and compute the probability that, in this block, the fraction of errored symbols exceeds ser_{max}. If errors in symbols are independent, the probability $ser(n, j)$ that in a block of n symbols there are j errors is given by the binomial distribution

$$ser(n, j) = C_j^n p^j (1-p)^{n-j}, \tag{6.1}$$

where

$$C_j^n = \frac{n!}{j!(n-j)!} \tag{6.2}$$

represents the number of combinations of n elements j to j.

A plot of $ser(n, j)$ as a function of j/n is shown in Figure 6.1 for $n = 10$, 50 and 200. This plot requires some careful thinking. Take, for instance, $n = 10$. We can only have $j = 1, 2, \ldots, 10$ errors in a block, and thus j/n can only assume the values $1/10, 2/10, \ldots, 10/10$ and no intermediate values. Thus the curves are not simply obtained by joining with straight lines the dots that represent $ser(j/n)$ but rather by drawing a stepwise curve.

As block size n increases the plot of $ser(j/n)$ will tend to a vertical line at $j/n = 0.01$.

Before going any further let us try go get a deeper insight into the meaning of Figure 6.1, bearing in mind that is was calculated for $ser = 0.01$. Take again $n = 10$. The probability that a block of 10 symbols has zero symbols in error is 0.90, that it has 1 errored symbol is 0.091 and 2 errored symbols 0.0041. If we can tolerate a small percentage of errors in a block without destroying the message within, then large blocks will provide us a way of "ensuring" (to a prescribed probability of being wrong) that the number of errors in each block will not be more than the block size times the average error probability. That is, long blocks, by the averaging effect of large numbers, "ensure" that the fraction of errors will be limited, a feat that small blocks cannot match. In effect, 10% errored symbols will occur with a probability of 0.091 in blocks with 10 symbols, a probability 0.000135 in 50 symbol blocks and an almost

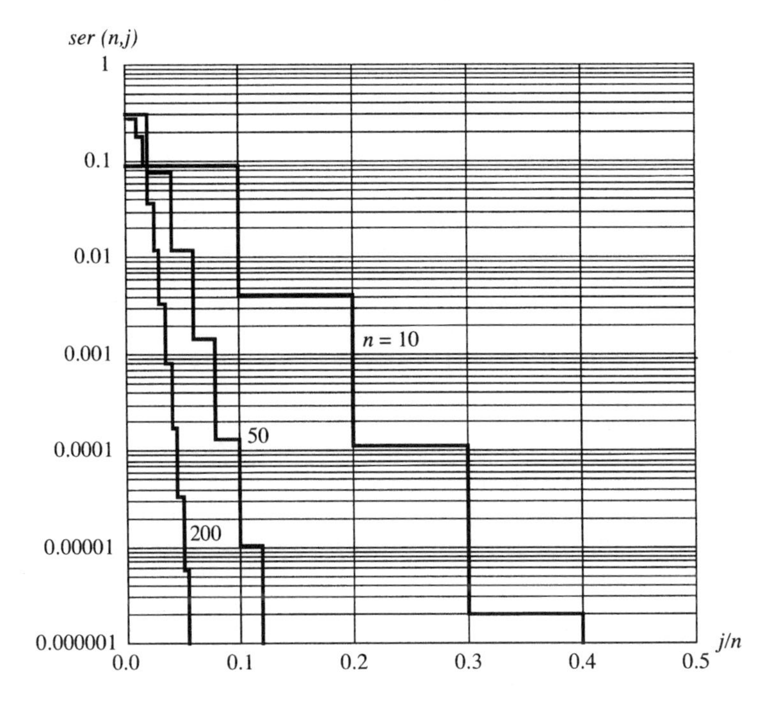

Fig. 6.1 Symbol error probability in a block of n symbols as a function of the fraction j/n of errored symbols for a symbol error probability of $ser = 0.01$, with $n = 10$, 50 and 200.

negligible probability in 200 symbol blocks. For infinite size blocks we could be sure that there would be a zero probability to find more than 1% of the symbols in error.

The important question now is: how can errors in a block be accepted without destroying the message within? The answer is redundancy, that is, introducing symbols in addition to those strictly needed to convey the message.

6.3 A FIRST APPROACH TO ERROR-CORRECTING CODES

As was stated stated in the previous chapter, coding is a procedure to transform the source information digits in coded digits more appropriate for transmission. Here, we will be mainly concerned with detecting and correcting errors that may occur in the transmission, by deliberately introducing redundancy in the source signal.

Codes may be classified in two main groups: block and tree. The first ones do not require memory and, given a set of k input digits, always produce a set of n output digits. Tree codes have a finite memory and thus output depends not only on current input but also on the past input.

Another way of classifying codes is: linear and nonlinear. Linear codes have two very interesting properties. One is that two code words may be added (modulo-2) to form another code word. The other is that the distance between any two code words is the same as the distance between the all-zero code word and any other word, that is, it equals the number of non-zero symbols, or the Hamming weight of that word.

Linear block codes are usually referred to as group codes because code words form a mathematical structure known as a group. Linear tree codes are commonly known as convolutional codes since the encoding operation can be thought of as the discrete time convolution of the input (data) sequence with the impulse response of the encoder.

Yet another way of classifying codes is: random-error-correcting codes and burst-error-correcting codes. Most codes are designed to deal with random (or independent) errors. Concatenation, a technique that will described later, may be used to counteract, rather effectively, error bursts.

In the following we will be dealing mainly with linear block codes, and convolution codes. We will be paying particular attention to cyclic codes, a special class of linear block codes, due to their importance and to the techniques used to describe and analyze their properties.

A block with n binary symbols will allow for 2^n different messages, or code words. In order to tolerate errored symbols, not all possible code words may be used. The number of symbols by which two code words differ is known as the Hamming distance d_H. The smallest value of d_H for all pairs of code words is known as the minimum distance d_{min} of the code. The number of errors t that a given code is capable of correcting and its minimum distance d_{min} are related by

$$t = (d_{min} - 1) \div 2, \tag{6.3}$$

where $\div$ stands for integer division.

To correct an error, a maximum likelihood decoder chooses, among the legitimate code words, the one that is closest, that is, the one that has the smallest Hamming distance, to the received code word.

Codes are usually denoted by a set of two integers, (n, k), meaning that for every k source digits the code produces n coded digits. The code rate r_c is the ratio between k and n

$$r_c = \frac{k}{n}, \tag{6.4}$$

and thus is an appropriate measure of code (transmission) efficiency.

The following example, taken from [14], we will help to understand the relation between the amount of redundancy in a code and the number of errors that are correctable.

Assume a (5,2) code that is a 5-digit code which, out of the possible $2^5 = 32$ code words (or n-tuples), conveys only the following $2^2 = 4$ legitimate messages: 00000, 00111, 11100 and 11011. According to our earlier statements, this code has $d_{min} = 3$ and hence $t = 1$ and so it should be able to correct all single errors. The question is: how should we build the decoder?

From equation 6.1 it is easy to prove that for a symbol error ratio $ser < 1/2$ in any received code word it is more likely that there are no errors than that there is a single error on any position, which, in turn, is more likely than any double error and so on, that is,

$$(1-ser)^5 > ser(1-ser)^4 > \ldots > ser^4(1-ser) > ser^5.$$

We may then build a look-up table, the first line of which contains the 4 legitimate words. The following lines contain all single error words (5 per legitimate code word) positioned below the closest legitimate word. We are then left with $32 - 6 * 4 = 8$ words, to which we cannot assign a unique position in the table, because they differ at least two positions from legitimate code words and can thus be placed in any two columns.

The decoder uses the look-up table as follows. It receives a word and finds its column in the table. It then outputs the legitimate code word in the first row and the same column, obviously correcting all single errors.

Since look-up table size grows exponentially with block size this technique is only practical for small block sizes.

The process of building a look-up table divides the original set of n-tuples into k subsets – the columns –. For a t-error-correcting code the number of n-tuples n_t in each subset obeys the inequality

$$n_t \geq 1 + n + C_2^n + \ldots + C_t^n. \tag{6.5}$$

This inequality may be easily understood if we note that, for each legitimate code word, there are n code words that differ from it in one position, C_2^n in two positions, and so on. The inequality is justified because after affecting all the code words up to t errors there may be some code words left (as it happened in our earlier example).

Equation (6.5) may be turned into a useful bound. Since out of the 2^n possible code words, there are only 2^k legitimate ones and as many columns in the look-up table, and we have a lower limit to the number of code words per column, we may write

$$2^k \leq \frac{2^n}{1 + n + C_2^n + \ldots + C_t^n}. \tag{6.6}$$

This inequality is known as the Hamming bound or sphere-packing bound. The equality is only satisfied for the so-called perfect codes (to be defined later). The Hamming bound may be written as

$$n - k \geq \log_2 \left(\sum_{j=0}^{t} C_j^n \right). \tag{6.7}$$

The code described above has an important property which is outlined in Table 6.1. Input digit sequences are uniquely found in code words and redundant digits are easily identified. Such codes are known as systematic codes. Codes that do not exhibit this property are referred to as non-systematic.

Input sequence	Code word
0 0	0: 0 0 : 0 0
0 1	0: 0 1 : 1 1
1 0	1: 1 0 : 1 1
1 1	1: 1 1 : 0 0

Table 6.1 Input sequence and code words for a (5,2) code.

Now that we have shown that it is possible to correct errors in a block of symbols, it is appropriate to return to Figure 6.1. If there is a block code of length 50, that with a 50% redundancy ($r_c = 0.5$) is capable of correcting up to 5 random errors, then a channel with a symbol error ratio of 10^{-2} would result in a block error ratio of about 10^{-5}, since all errors up to 5 would be corrected by the code.

6.4 BLOCK CODES

6.4.1 Introduction

A (n, k) block code produces a set of n digits – a code word – for each set of k source digits. According to our previous definition, the code rate is $r_c = k/n$.

With few exceptions, the only block codes of practical importance, are linear block (or group) codes. Values for k lie in the range from 3 to several hundred and values of r_c from 1/4 to 7/8.

Parity check (n, k) codes are a particular class of block codes in which code word digits are a set of $n - k$ parity checks performed on the k information digits. When information digits are binary, as will be the case in all following examples, parity checks may be seen as the remainder of the division by 2.

A block code is said to be systematic when the first k digits are a replica of the information digits and the remaining $n - k$ digits are parity checks on the k information digits.

Before proceeding let us consider a few simple codes and the respective definition equations, where x_i is the ith digit of the code word and u_i the ith input digit:

- Repetition code $(3, 1)$

$$x_1 = u_1,$$

$$\begin{aligned} x_2 &= u_1, \\ x_3 &= u_1. \end{aligned}$$

- Parity check code $(3,2)$

$$\begin{aligned} x_1 &= u_1, \\ x_2 &= u_2, \\ x_3 &= u_1 + u_2. \end{aligned}$$

- Hamming code $(7,4)$

$$\begin{aligned} x_i &= u_i \ (i = 1,2,3,4), \\ x_5 &= u_1 + u_2 + u_3, \\ x_6 &= u_2 + u_3 + u_4, \\ x_7 &= u_1 + u_2 + u_4. \end{aligned}$$

All the operations used in the previous examples use modulo-2 arithmetic. It should be obvious from these examples that standard matrix notation is the most appropriate tool to describe these codes. The code word $\mathbf{x}$ (as a column vector) may be derived from the input data $\mathbf{u}$ (another column vector) by multiplication by the generator matrix $\mathbf{G}$

$$\mathbf{x} = \mathbf{G}\,\mathbf{u}. \tag{6.8}$$

For the $(7,4)$ Hamming code the generator matrix is

$$\mathbf{G} = \begin{bmatrix} 1 & 0 & 0 & 0 \\ 0 & 1 & 0 & 0 \\ 0 & 0 & 1 & 0 \\ 0 & 0 & 0 & 1 \\ 1 & 1 & 1 & 0 \\ 0 & 1 & 1 & 1 \\ 1 & 1 & 0 & 1 \end{bmatrix}.$$

The reader is invited to check that the input $\mathbf{u} = [1100]$ produces the code word $\mathbf{x} = [1100010]$.

When the code is systematic the generator matrix is

$$\mathbf{G} = \begin{bmatrix} \mathbf{I_k} \\ \mathbf{P} \end{bmatrix}, \tag{6.9}$$

where $\mathbf{I}_k$ is the identity matrix of order k and $\mathbf{P}$ is the parity matrix, an $(n-k,k)$ matrix with the parity check information, which defines the coding rules for a systematic code.

The following properties of parity check codes may be easily proved:

1. Each code word is a sum of columns of the generator matrix.
2. The block code words consist of all possible sums of the columns of the generator matrix.
3. The sum of two code words is also a code word.
4. A sequence of n zeros is a code word in any parity check code.

Because of these properties, parity check codes are also known as linear codes.

An important property of a code word is its Hamming weight, that is, the number of ones it contains. The set of weights, together with the number of code words with a given weight, is known as the weight distribution of the code. The weight distribution of the Hamming code $(7,4)$ is given in Table 6.2

Weight	**Number of code words**
0	1
3	7
3	7
7	1

Table 6.2 Weight distribution of the Hamming code $(7,4)$.

The measure of the difference between two code words **a** and **b** is known as the Hamming distance $d(a,b)$, defined as the number of positions in which these code words differ. The smallest of these distances is the code minimum distance, denoted as d_{min}. The distance $d(a,b)$ is equal to the weight of their sum $w(a+b)$.

Since the all zero sequence is a code word and the modulo-2 sum of any pair of code words is also a code word, in any linear block code the minimum distance of a linear block code is the minimum weight of its non-zero code words. To design a linear block code with high minimum distance it is sufficient to make sure that the minimum weight of its non-zero words is as large as possible.

6.4.2 Error detection and error correction with block codes

Let us assume that a transmitted code word $\mathbf{x}$ is received as $\mathbf{y}$

$$\mathbf{y} = \mathbf{x} + \mathbf{e}, \tag{6.10}$$

where $\mathbf{e}$ is the error vector. Each component of the error vector is 1 when this component of the received word is in error and 0 otherwise.

The decoder compares the received vector and computes the $n-k$ check digits from the k information digits. Then it compares these digits with the received ones and flags each difference.

Let us now define the syndrome $\mathbf{s}$ of the received word, a $n-k$ dimension vector whose elements are 0 when the corresponding check digit is satisfied and 1 otherwise. It may be shown that

$$\mathbf{s} = \mathbf{H}\,\mathbf{y}, \tag{6.11}$$

where $\mathbf{H}$ is the parity check matrix, defined as

$$\mathbf{H} = [\mathbf{P}|\mathbf{I_{n-k}}], \tag{6.12}$$

where $\mathbf{P}$ is the parity matrix and $\mathbf{I}_{n-k}$ is the identity matrix of order $n-k$.

For a code word $\mathbf{x}$ we have

$$\mathbf{H}\,\mathbf{x} = 0.$$

Introducing the definition of generator matrix (6.8) we have

$$\mathbf{H}\,\mathbf{G}\,\mathbf{u} = 0$$

for any $\mathbf{u}$. Hence

$$\mathbf{H}\,\mathbf{G} = 0. \tag{6.13}$$

We will take again the Hamming code $(7,4)$ as an example. From the definition equations we see that the decoder must compute the following syndrome elements:

$$s_1 = (y_1 + y_2 + y_3) + y_5, \tag{6.14}$$
$$s_2 = (y_2 + y_3 + y_4) + y_6, \tag{6.15}$$
$$s_3 = (y_1 + y_2 + y_4) + y_7, \tag{6.16}$$

and thus

$$\mathbf{H} = \begin{bmatrix} 1 & 1 & 1 & 0 & 1 & 0 & 0 \\ 0 & 1 & 1 & 1 & 0 & 1 & 0 \\ 1 & 1 & 0 & 1 & 0 & 0 & 1 \end{bmatrix}. \tag{6.17}$$

This form of the parity check matrix is known as the echelon canonical form.

Since the syndrome is a null vector if, and only if, $\mathbf{y}$ is a code word, the decoder can detect all received words which are not code words.

The fact that the product of the parity check matrix $\mathbf{H}$ with a code word is zero may be interpreted as requiring that a given subset of columns of $\mathbf{H}$ should sum to zero. Since this relation holds for all code words it must also hold for the minimum weight words. Assume that d_{min} is the minimum distance of the code, that is, the minimum weight of its non-zero code words. Then all subsets of $d_{min}-1$ columns should not sum to zero, that is, they must be linearly independent. This property may be used to find the minimum distance d_{min} of a code given its parity check matrix.

Simple inspection of the parity check matrix for the Hamming code $(7,4)$ shows that $d_{min} = 3$.

With a (n,k) code we can generate 2^n words of which only 2^k are valid, because they produce a zero syndrome. However, since there are only $n-k$ parity digits we can only identify 2^{n-k} error vectors out of a possible total of $2^n - 2^k$. Another way to state the same fact is that with $n-k$ digits we can only produce 2^{n-k} different syndromes for the $2^n - 2^k$ possible errors.

Take again the Hamming code $(7,4)$. The syndrome is a vector with dimension 3, that is, it may take 2^3 different values. Assuming that there is only a single error, a careful look to the code definition shows the error vector and the errored digit for each value of the syndrome given in Table 6.3.

Syndrome	Error vector	Errored digit
0 0 0	0 0 0 0 0 0 0 0	none
0 0 1	0 0 0 0 0 0 0 1	7
0 1 0	0 0 0 0 0 0 1 0	6
0 1 1	0 0 0 0 1 0 0 0	4
1 0 0	0 0 0 0 0 1 0 0	5
1 0 1	1 0 0 0 0 0 0 0	1
1 1 0	0 0 1 0 0 0 0 0	3
1 1 1	0 1 0 0 0 0 0 0	2

Table 6.3 Syndrome, error vector and errored digit for single errors with the Hamming code $(7,4)$.

If we assume more than one digit in error, there are quite a few error patterns that yield the same syndrome, the set of which is known as a coset. The first error pattern in each coset is called the coset leader. The coset leader is the minimum weight word that generates a given syndrome. A table with all the 2^n patterns, arranged in 2^k rows, each yielding the same syndrome, and 2^{n-k} columns is known as the standard array of the code. For large values of n and k the table with all the possible values of the syndrome and the coset leaders becomes too large to be of practical use as a look-up table.

The following algorithm may be used to provide error correction:

1. Compute the syndrome for the received sequence.

2. Find the coset leader.

3. Compute the estimated code word by adding the coset leader to the received sequence.

A simple example using the Hamming code $(7,4)$ will help to clarify this subject. Assume, for instance, that we receive vector $\mathbf{y} = [1011011]$. The syndrome computed with the parity check matrix given in (6.12) according to equation (6.11) is $\mathbf{s} = [011]$. Using the look-up table in Table 6.3 we build the error vector as $\mathbf{e} = [00001000]$. Adding the error vector to the received vector we get the estimated original code word $\mathbf{x} = [1010011]$ which, as expected, has a null syndrome.

Take again the definition of syndrome and, as before, assume that the received vector $\mathbf{y}$ is the sum of the transmitted code word $\mathbf{x}$ and an error vector $\mathbf{e}$. Then we may write:

$$\begin{aligned} \mathbf{y} &= \mathbf{x} + \mathbf{e}, \\ \mathbf{s} &= \mathbf{H}\,\mathbf{y} \\ &= \mathbf{H}\,\mathbf{x} + \mathbf{H}\,\mathbf{e}. \end{aligned}$$

But, since $\mathbf{x}$ is a code word, $\mathbf{H}\,\mathbf{x} = \mathbf{0}$, hence:

$$\mathbf{s} = \mathbf{H}\,\mathbf{e}. \tag{6.18}$$

Equation (6.18) indicates that each syndrome element is the sum of the columns of $\mathbf{H}$ corresponding to the positions of the ones in the error vector. Thus, if a column of $\mathbf{H}$ is zero an error in that position cannot be detected. Also, if two columns of $\mathbf{H}$ are equal, an error in these positions cannot be corrected since it yields the same syndrome. We may then conclude that a block code can correct all errors if the columns of $\mathbf{H}$ are all non-zero and can only correct as many possibilities of errors as there are different columns in $\mathbf{H}$.

Applying the previous conclusion to the Hamming code $(7, 4)$ we realize that, since its parity check matrix has 7 different and non-zero columns it can correct all single-digit errors.

In fact, it may be proved that a linear block code (n, k) with minimum distance d_{min} can detect all error vectors of weight not greater than $d_{min} - 1$ and can correct all error vectors of weight not greater than $t = (d_{min} - 1) \div 2$, where $\div$ stands for the integer division.

Thus the Hamming code $(7, 4)$, for which $d_{min} = 3$, can detect all single and double errors and can correct all single errors.

6.4.3 Hamming codes

Hamming code $(7, 4)$ has a code rate $r_c = 4/7$. An interesting question which we will try to answer is the following: is it possible to build codes with the same value of $d_{min} = 3$ but with a higher code rate, approaching unity?

We already known that with 3 parity digits there are $2^3 - 1 = 7$ different and non-zero n-tuples. With 4 parity digits we would have $2^4 - 1 = 15$ such n-tuples. Thus it is possible to build a family of Hamming codes of the form $(2^p - 1, 2^p - 1 - p)$, where $p = n - k$, with $d_{min} = 3$. These are the Hamming codes $(3, 1)$, $(7, 4)$, $(15, 11)$, $(31, 26)$, $\ldots$, with code rates $1/3$, $4/7$, $11/15$, $26/31$, $\ldots$.

Hamming codes have parity check matrixes whose columns are all the possible sequences of $n - k$ binary digits with the exception of the the all-zero sequence.

Let us define the code weight enumerating function as the polynomial

$$p(x) = \sum_{i=0}^{n} n_i x^i, \tag{6.19}$$

where w is the word weight and n_w is the number of words with weight w. For Hamming codes the weight enumerating function is [2]

$$p(x) = \frac{1}{n+1}\left[(1+x)^n + n(1+x)^{(n-1)/2}(1-x)^{(n+1)/2}\right]. \tag{6.20}$$

Using the weight enumerating function we have, for Hamming codes $(3,1)$, $(7,4)$, and $(15,11)$, respectively

$$\begin{aligned} p(x) &= 1 + x^3, \\ p(x) &= 1 + 7x^3 + 7x^4 + x^7, \\ p(x) &= 1 + 35x^3 + 105x^4 + 168x^5 + 280x^6 + 435x^7 \\ &\quad + 435x^8 + +280x^9 + 168x^{10} + 105x^{11} \\ &\quad + 35x^{12} + x^{15}. \end{aligned}$$

6.4.4 Extended Hamming codes

Hamming codes may be extended by appending an additional parity digit that checks all previous n digits. This is a new class of $(2^p, 2^p - 1 - p)$ block codes, known as the extended Hamming codes.

The parity check matrix $\mathbf{H}_e$ is formed by extending the parity check matrix $\mathbf{H}$ of the Hamming codes as shown below:

$$\mathbf{H}_e = \begin{bmatrix} \mathbf{H} & 0 \\ 1 & 1 \end{bmatrix}. \tag{6.21}$$

6.4.5 Dual codes

As we have seen in equation (6.13), the product of the parity check matrix $\mathbf{H}$ and the generator matrix $\mathbf{G}$ is zero. Then both matrixes may be interchanged and the parity check matrix may become the generator matrix of a new $(n, n-k)$ code which is known as the dual of the (n, k) code generated by $\mathbf{G}$.

There is a relation between the weight enumerating function $p(x)$ of the original (n, k) code and weight enumerating function $q(x)$ of its dual code [4]

$$q(x) = 2{-}k(1+x)^n p\left(\frac{1-x}{1+x}\right). \tag{6.22}$$

Maximal length codes are duals of Hamming codes. Thus for every $p = 2, 3, 4, \ldots$ there is a Hamming code $(2^p - 1, 2^p - 1 - p)$ and a maximal length code $(2^p - 1, p)$, whose weight enumerating function is, from (6.22) and (6.20)

$$q(x) = 1 + (2^p - 1)x^{2p-1}. \tag{6.23}$$

Maximal length codes have a minimum distance $d_{min} = 2^{p-1}$, which increases exponentially with p, but its code rate $r_c = \frac{p}{2^p-1}$ decreases also exponentially with p. An interesting property of these codes is that all their words, except the all-zeros word, have the same weight 2^{p-1}.

6.4.6 Reed-Muller codes

Reed-Muller codes are linear block codes equivalent to cyclic codes (defined in Section 6.5) with an added overall parity check. They provide a wide range of code rates and minimum distances. For every m and $r < m$ there is a Reed-Muller code (n, k) where

$$\begin{aligned} n &= 2^m, & (6.24)\\ k &= \sum_{i=0}^{r} C_i^m, & (6.25)\\ d_{min} &= 2^{m-r}. & (6.26) \end{aligned}$$

The Reed-Muller code generator matrix makes use of a set of vectors ν_0, ν_1, ..., ν_m and all their products first two at a time, then three at a time, and so on, up to r at a time. Vector ν_0 is the all ones vector. All other ν_1, ..., ν_m vectors are rows of a matrix with all the m-tuples as columns.

For $m = 3$ and $r = 2$ we have a Reed-Muller code $(8, 7)$ with $d_{min} = 2$, where vectors $\nu_1, \ldots, \nu_m$ are

$$\begin{aligned} \nu_0 &= [11111111],\\ \nu_1 &= [00001111],\\ \nu_2 &= [00110011],\\ \nu_3 &= [01010101],\\ \nu_1\nu_2 &= [00000011],\\ \nu_1\nu_3 &= [00000101],\\ \nu_2\nu_3 &= [00010001], \end{aligned}$$

and the generator matrix $\mathbf{G}$ is

$$\mathbf{G} = \begin{bmatrix} 1 & 1 & 1 & 1 & 1 & 1 & 1 & 1 \\ 0 & 0 & 0 & 0 & 1 & 1 & 1 & 1 \\ 0 & 0 & 1 & 1 & 0 & 0 & 1 & 1 \\ 0 & 1 & 0 & 1 & 0 & 1 & 0 & 1 \\ 0 & 0 & 0 & 0 & 0 & 0 & 1 & 1 \\ 0 & 0 & 0 & 0 & 0 & 1 & 0 & 1 \\ 0 & 0 & 0 & 1 & 0 & 0 & 0 & 1 \end{bmatrix}. \tag{6.27}$$

First-order Reed-Muller codes are closely related to maximal length codes. If we extend a maximal length code by adding an overall parity check digit we obtain an orthogonal code where the number of code words is 2^m: an all zero word and $2^m - 1$ words all with the same weight $d = 2^{m-1}$. In this code, all code words agree in $d = 2^{m-1}$ positions and disagree in $d = 2^{m-1}$ positions with all other code words. Using a polar line code it will result in a set of 2^m signals, hence the name orthogonal code.

The first Reed-Muller code is obtained from the orthogonal code by augmenting it with the all ones code word. For this reason it is often known as the bi-orthogonal code.

6.4.7 Bit interleaved parity codes

Bit Interleaved Parity codes, or simply BIP (n, m), are a simple form of parity codes used for error detection in synchronous digital transmission (SDH), where n stands for the number of bits used in error detection and m is the number of bytes (each with n bits) controlled.

In a BIP (n, m) the n-bit error detection word is generated as follows:

1. The $n \times m$ bits are organized in n-bit bytes.
2. The ith bit (with $i = 1, 2, \ldots n$) of every byte are added together (modulo-2).
3. The i bit of the error detection word, is set to 0 if the previous sum is zero, and 1 otherwise (even parity).

Bit Interleaved Parity codes share with interleaved codes (see Section 6.6) the ability to deal with error bursts. It is easy to show that a BIP (n, m) can detect all error bursts up to $2n - 1$ bits long.

6.4.8 Product codes

Product codes are a way of double encoding information. A very simple example will be used describe the concept. Assume that a message is divided into groups of 4 bits, arranged in a 2×2 matrix, which is filled from left to right and top to bottom

$$\mathbf{u} = \begin{bmatrix} u_{1,1} & u_{1,2} \\ u_{2,1} & u_{2,2} \end{bmatrix}.$$

Let us now add parity bits to the information bits, both horizontally, that is, by rows, and vertically, that is, by columns. Horizontally we have

$$p_{i,3} = u_{i,1} + u_{i,2} \qquad (j = 1, 2),$$

and vertically

$$p_{3,j} = u_{1,j} + u_{2,j} \qquad (j = 1, 2).$$

The output code word may now be represented as a 3×3 matrix

$$\mathbf{x} = \begin{bmatrix} u_{1,1} & u_{1,2} & p_{1,3} \\ u_{2,1} & u_{2,2} & p_{2,3} \\ p_{3,1} & p_{3,2} & \end{bmatrix},$$

and will be transmitted as a vector

$$\mathbf{x} = [u_{1,1}, u_{1,2}, u_{2,1}, u_{2,2}, p_{1,3}, p_{2,3}, p_{3,1}, p_{3,2}].$$

For each line (or column) we have a code rate $r_c = 2/3$ and the overall code rate will be $r_c = 1/2$.

What are product code advantages, as compared with the simple parity code? The simple parity code detects all single errors in each set of two information bits, but cannot correct them. The product code goes one step further. It can detect and correct all single errors from the combination of parity bits. It can also detect all double errors and although it can correct quite a few of these errors it cannot correct them all. The performance of this product code is similar to that of the (8, 4) extended Hamming code.

6.5 CYCLIC CODES

Cyclic codes are a special form of parity-check codes. An (n, k) linear block code is a cyclic code if any cyclic shift of a code word produces another code word.

The reader is invited to verify that the Hamming code $(7, 4)$ is a cyclic code.

To deal with cyclic codes it is useful to take the coefficients of a polynomial in an arbitrary d to represent a code word. Thus code word $\mathbf{x} = [x_{n-1}, x_{n-2}, \ldots, x_1, x_0]$ is represented as

$$x(\delta) = x_{n-1}\delta^{n-1} + x_{n-2}\delta^{n-2} + \ldots + x_1\delta + x_0. \tag{6.28}$$

As before, the binary coefficients of this polynomial are manipulated using modulo-2 arithmetic.

If $x(\delta)$ is a code polynomial of a cyclic code (n, k) then a cyclic shift (to the left) of i positions generates another code polynomial denoted by $x^i(\delta)$, which is the remainder of the division of $d^i x(\delta)$ by $\delta^n + 1$

$$\delta^i x(\delta) = q(\delta)(\delta^n + 1) + x^i(\delta), \tag{6.29}$$

where $q(\delta)$ is the quotient polynomial.

Take, as an example, a code word $\mathbf{c} = [0111010]$ from Hamming code $(7, 4)$ and shift it 4 positions to the left. The code polynomial is $x(\delta) = \delta^5 + \delta^4 + \delta^3 + \delta$ hence applying (6.29) we have

$$\begin{aligned} \delta^4 x(\delta) &= q(\delta)(\delta^7 + 1) + x^i(\delta), \\ q(\delta) &= \delta^2 + d + 1, \\ x^i(\delta) &= \delta^5 + \delta^2 + \delta + 1, \end{aligned}$$

and a new code word [0100111].

Let us look again at the generator matrix $\mathbf{G}$ of Hamming code $(7, 4)$ which, for convenience, we repeat here:

$$\mathbf{G} = \begin{bmatrix} 1 & 0 & 0 & 0 \\ 0 & 1 & 0 & 0 \\ 0 & 0 & 1 & 0 \\ 0 & 0 & 0 & 1 \\ 1 & 1 & 1 & 0 \\ 0 & 1 & 1 & 1 \\ 1 & 1 & 0 & 1 \end{bmatrix}.$$

The last column of this generator matrix may be written as the polynomial $\delta^3 + \delta + 1$. This polynomial must have a 1 in the last position, otherwise with 6 shifts to the left (or 1 to the right) we would get four zeros as the information digits with two ones as the parity digits, which is impossible. Thus, this is the only polynomial of degree $n - k = 3$ in the code. If there was another one it would also have to have a one in its last position, but since the sum of two code words (hence polynomials) must also be a legitimate code word, we would encounter the previous impossibility. This example may be generalized into the following theorem:

Given an (n, k) cyclic code there is a unique code polynomial of degree $n-k$ of the form $g(d) = d^{n-k} + \ldots + 1$. All other code polynomials are multiples of $g(d)$ and every polynomial of degree $n - 1$ or less that is divisible by $g(d)$ must be a code polynomial. Polynomial $g(d)$ is known as the cyclic code generator polynomial.

For the Hamming code $(7, 4)$ the generator polynomial is $g(d) = d^3 + d + 1$.

The generator polynomial $g(\delta)$ of a cyclic code divides $\delta^n + 1$. Conversely, every polynomial of degree $n - k$ that is a divisor of $\delta^n + 1$ generates a (n, k) cyclic code.

Finally, we should remark that if $g(\delta)$ divides both $\delta^m + 1$ and $\delta^n + 1$ with $m < n$ then $\delta^m + 1$ is a code word in the cyclic code (n, k) whose minimum distance is 2. To avoid this drawback n must be taken as the smallest integer n such that $\delta^n + 1$ is a multiple of $g(\delta)$.

Factorization of $\delta^n + 1$ is not obvious, except for a few values of n such as 3, 5, 11, 13, 19. Tables such as MacWilliams and Sloane's [2] provide factors for n from 7 to 127. From such tables we find that

$$\delta^9 + 1 = (\delta + 1)(\delta^2 + \delta + 1)(\delta^6 + \delta^3 + 1).$$

Polynomials may be used to generate code words. As an example we will take again the Hamming code $(7, 4)$, whose generator polynomial is $g(\delta) = \delta^3 + \delta + 1$ and calculate the code polynomial $x(\delta)$ for the information sequence [1101] (which corresponds to $u(\delta) = \delta^3 + \delta^2 + 1$). Applying $x(\delta) = u(\delta)g(\delta)$ we get $x(\delta) = \delta^6 + \delta^5 + \delta^4 + \delta^3 + \delta^2 + \delta + 1$ which corresponds to the code word [1111111].

The parity check polynomial is defined as

$$h(\delta) = \frac{\delta^n + 1}{g(\delta)}. \tag{6.30}$$

For the Hamming code $(7, 4)$, from (6.30), we have $h(\delta) = \delta^4 + \delta^2 + \delta + 1$.

6.5.1 Error detection and correction with cyclic codes

Assume that code polynomial $x(\delta)$ is transmitted over a noisy channel, that introduces errors, and is received as $y(\delta)$, where

$$y(\delta) = x(\delta) + e(\delta). \tag{6.31}$$

If we divide $y(\delta)$ by the code generator polynomial we obtain

$$y(\delta) = m(\delta)g(\delta) + s(\delta), \tag{6.32}$$

where $m(d)$ is the quotient and $s(d)$ is the remainder, or syndrome polynomial.

All code polynomials, and only code polynomials, are exact multiples of the code generator polynomial, thus $y(\delta)$ is a code polynomial if $s(\delta) = 0$.

Since we have assumed that $x(\delta)$ is a code polynomial

$$x(\delta) = q(\delta)g(\delta). \tag{6.33}$$

Substituting the value of $x(d)$ given in (6.33) into (6.31), equating the resulting value of $y(d)$ with (6.32) and noting that all operations are performed using modulo-2 arithmetic we get

$$e(\delta) = [m(\delta) + q(\delta)]g(\delta) + s(\delta), \tag{6.34}$$

which shows that the syndrome is also the syndrome of the error polynomial.

Let us take again the Hamming code $(7, 4)$ that we have shown to be a cyclic code, with code generator polynomial $g(\delta) = \delta^3 + \delta + 1$, and assume that we receive a polynomial $y(\delta) = \delta^6 + \delta^5 + 1$. Using (6.32) we get $s(\delta) = \delta + 1$ and hence the syndrome is [011]. We may therefore conclude that $y(\delta)$ is in error. To correct it we use the parity check matrix (6.17), which enables us to write the following three equations (6.14), (6.15) and (6.16):

$$\begin{aligned} y_1 + y_2 + y_3 + y_5 &= s_1, \\ y_2 + y_3 + y_4 + y_6 &= s_2, \\ y_1 + y_2 + y_4 + y_7 &= s_3. \end{aligned}$$

Obviously this system has more equations than unknowns. That is why the code cannot correct all possible errors. Assuming that we restrict the number of errors to one, what we would like to know is the position of the error. Trying all the possibilities we easily arrive at Table 6.3, from which we find out that the correct code word is $x = [1101001]$.

It is possible to prove that, when the errors are confined to the parity check positions of $y(\delta)$, the syndrome $s(\delta)$ is equal to $e(\delta)$.

As an example, consider a BCH code[1] with $n = 15$ and $d_{min} = 5$. As will be shown later, we have $m = 4$, $t = 2$ and $k = 7$. This code is able to detect and correct any 2-digit errors. Its generator polynomial is $g(\delta) = \delta^8 + \delta^7 + \delta^6 + \delta^4 + 1$. Its parity check matrix may be written as

$$\mathbf{H} = \begin{bmatrix} a_{1,1} & a_{1,2} & \cdots & a_{1,15} \\ \vdots & \vdots & \cdots & \vdots \\ a_{i,1} & a_{i,2} & \cdots & a_{i,15} \\ \vdots & \vdots & \cdots & \vdots \\ a_{j,1} & a_{j,2} & \cdots & a_{j,15} \\ \vdots & \vdots & \cdots & \vdots \\ a_{8,1} & a_{8,2} & \cdots & a_{8,15} \end{bmatrix}.$$

Let us assume that there are two errors, in positions i and j that is vector $\mathbf{e}$ is an all zero column vector with the exception of rows i and j which are 1

$$\mathbf{e} = \begin{bmatrix} 0 \\ \vdots \\ 1 \\ \vdots \\ 1 \\ \vdots \\ 0 \end{bmatrix}.$$

[1]BCH are dealt with in Section 6.5.4

The syndrome is

$$\mathbf{S} = \mathbf{He} = \begin{bmatrix} a_{1,i} + a_{1,j} \\ \vdots \\ a_{8,i} + a_{8,j} \end{bmatrix}.$$

We are thus left with 8 equations and 2 unknowns. As we know all values of $a_{i,j}$ by trying all possibilities we can find out the values of i and j for all possible values of the syndrome. Since the syndrome has 8 (binary digits), besides the all zero value (which corresponds to the all zero error vector) it may assume $2^8 - 1 = 255$ different values. It can thus indicate the position of all (15) possible 1-digit errors plus all ($15 \times 14/2 = 105$) 2-digit errors plus about 30 % of all 3-digit errors ($15 \times 14 \times 13/6 = 455$).

There are quite a few alternatives to trying all possibilities, as it was presented. The reader is referred to [7] for further details.

6.5.2 Cyclic redundancy check codes

Cyclic redundancy check in short CRC codes, provide a simple yet effective way to detect transmission errors. Together with bit interleaved parity (or BIP) codes they are the ITU specified codes for error detection.

The basic idea behind cyclic redundancy check codes is to add a few bits to the message before transmission, so that the resulting polynomial is divisible by the code generator polynomial $g(d)$. The receiver divides the message by the generator polynomial and, if the remainder is zero, concludes (rightly or wrongly) that the received message is correct.

The generator polynomial is choosen in such a way as to minimize the possibility that the receiver accepts a message with errors as being correct.

Let $x(\delta)$ be the transmitted message (with the added bits), $e(\delta)$ the error and $y(\delta)$ the received message

$$y(\delta) = x(\delta) + e(\delta).$$

From the way $x(\delta)$ was formed we have

$$x(\delta) = g(\delta)q(\delta),$$

thus

$$y(\delta) = g(\delta)q(\delta) + q_a(\delta) + r(\delta).$$

Single-bit errors imply $e(\delta) = \delta^i$ (where i is the errored bit). If $g(\delta)$ has two or more terms it will never divide $e(\delta)$, that is, all single-bit errors are detected.

For two bit errors we have

$$\begin{aligned} e(\delta) &= \delta^i + \delta^j \quad (i > j) \\ &= \delta^j(\delta^{i-j} + 1). \end{aligned}$$

To detect double errors δ^k+1 should not be divisible by $g(\delta)$ for $\max(i-j) \geq k > 0$ where $\max(i-j)$ corresponds to the block size. It may be proved [16] that $d^{15} + d^{14} + 1$ does not divide $\delta^k + 1$ for $k < 32768$.

It can be shown [15] that an odd number of errors is always detected if $g(\delta)$ contains the factor $\delta + 1$ and that all error bursts whose length is less than the number of error detection bits are detected as well as many longer ones.

ITU uses, among others, the following generator polynomials:

- CRC-4 — $\delta^4 + \delta + 1$;
- CRC-5 — $\delta^5 + \delta^4 + \delta^2 + 1$;
- CRC-6 — $\delta^6 + \delta + 1$;
- CRC-12 — $\delta^{12} + \delta^{11} + \delta^3 + \delta^2 + \delta + 1$;
- CRC-16 — $\delta^{16} + \delta^{15} + \delta^2 + 1$;
- CRC-CCITT — $\delta^{16} + \delta^{12} + \delta^5 + 1$.

The CRC-12 is used for transmission of 6-bit characters. Both CRC-16 and CRC-CCITT are used for streams of 8-bit characters, in the United States and Europe, respectively.

The algorithm to generate the check digits with a CRC is the following:

1. Let l be the order of the generator polynomial.
2. Append l zeros to the original messase x, and denote the result as x_0.
3. Compute the remainder (modulo-2) $rm(d)$ of the division of the polynomial $x_0(\delta)$ by the generator polynomial $g(\delta)$.
4. The check digits are the coefficients of the remainder polynomial $rm(\delta)$.

An example will make the use of the algorithm clear. Let $x = [1010001101]$ and $g(\delta) = \delta^5 + \delta^4 + \delta^2 + 1$ (the CRC-5 polynomial). We have successively

$$
\begin{aligned}
l &= 5, \\
\mathbf{x}_0 &= [101000110100000], \\
rm(\delta) &= \delta^3 + \delta^2 + \delta,
\end{aligned}
$$

and the checks digits are, from left to right, [01110].

Although the verification of check digits appears to be complicated, a simple hardware circuit comprising a few shift registers and exclusive-or's (i.e. modulo-2 additions), may be used to perform the task, as shown in Figure 6.2. In this figure the check digits, from left to right, are represented as rm_4, rm_3, rm_2, rm_1 and rm_0.

Table 6.4 reproduces the shift registers contents for each of the input bits, including the appended zero bits. Initially the shift registers are assumed to be all set to zero. When the last appended bit is input the contents of the shift registers are the check bits.

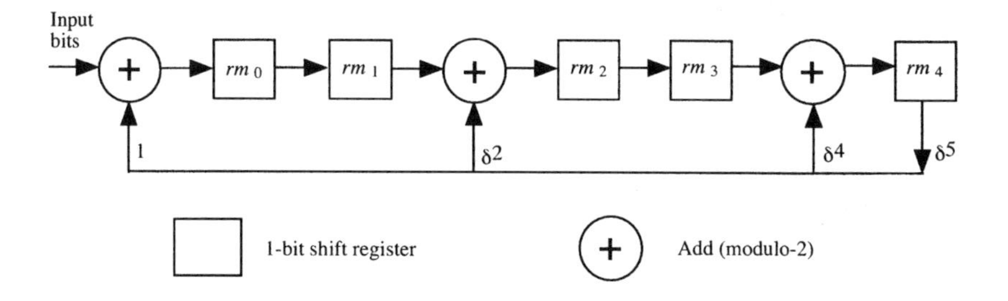

Fig. 6.2 Simple circuit to perform the check digit calculation for CRC-5.

Input Bits	Shift registers				
	rm_0	rm_1	rm_2	rm_3	rm_4
	0	0	0	0	0
1	1	0	0	0	0
0	0	1	0	0	0
1	1	0	1	0	0
0	0	1	0	1	0
0	0	0	1	0	1
0	1	0	1	1	1
1	0	1	1	1	0
1	1	0	1	1	1
0	1	1	1	1	0
1	1	1	1	1	1
0	1	1	0	1	0
0	0	1	1	0	1
0	1	0	0	1	1
0	1	1	1	0	0
0	0	1	1	1	0

Table 6.4 Input sequence and shift register contents for a CRC-5 coder with the input sequence [1010001101].

6.5.3 Golay codes

Golay discovered a (23, 12) cyclic code with the generator polynomial $g(\delta)$ given by

$$g(\delta) = \delta^{11} + \delta^9 + \delta^7 + \delta^6 + \delta^5 + \delta + 1, \tag{6.35}$$

which has a minimum distance $d_{min} = 7$, that is, that can correct all triple errors. This is the only non-trivial linear binary perfect code (i.e., a code where all possible sequences have distance one from one of the code words) with multiple error correcting capabilities. Besides the Hamming single error correcting codes, the repetition codes (with n odd) and the Golay code, no other linear binary perfect codes exist.

Another interesting Golay code is the (24, 12) extended Golay code which has $r_c = 1/2$ and $d_{min} = 8$.

6.5.4 Bose-Chaudhuri-Hocquenhem (BCH) codes

Bose-Chaudhuri-Hocquenhem (in short BCH) codes are particularly useful to correct random errors because decoding algorithms can be implemented with acceptable complexity. For any pair of positive integers m, t there is a BCH code with $n = 2^m - 1$, $n - k \leq mt$ and $d_{min} = 2t + 1$. This code can correct all combinations of t or less errors. The generator polynomial may be constructed from factors of $\delta^{2m} + 1$. A list of generator polynomials may be found in [2].

An example of a BCH code generator polynomial, for $m = 4$, $n = 15$, $k = 5$ and $t = 3$, is

$$g(\delta) = \delta^8 + \delta^7 + \delta^6 + \delta^4 + 1. \tag{6.36}$$

BCH code performance can be estimated from an upper bound where we take the pessimistic assumption that a t-error-correcting BCH code when receiving a word with $i > t$ errors will cause the decoder to output a word differing from the correct one in $i + t$ digits and thus a fraction $\frac{i+t}{n}$ of the binary digits will be in error. The bit error ratio will be

$$ber = \sum_{i=t+1}^{n} \frac{i+t}{n} C_i^n p^i (1-p)^{n-i}, \tag{6.37}$$

where p is the bit error probability in the uncoded message.

In Figure 6.3 we compare the performance of uncoded and BCH coded polar transmission, defined by the bit error ratio as a function of E_b/N_0, for BCH codes (15, 7), (31, 16), (63, 30), (127, 64) and (255, 123), all of which have a code rate r_c close to 0.5. In order for this comparison to be meaningful the value of e_b/n_0 is, in each case, multiplied by the code rate, which is equivalent to saying that the energy used to convey the message is the same in all cases.

Although this is the traditional way to represent code performance, it may not be the most useful one since it depends on the uncoded modulation chosen for reference. In Figure 6.4 we compare the same codes with no reference to any particular uncoded modulation. In order to use this figure one must however not forget that the apparent code gain is not the real code gain, because the reduction in bit error ratio due to the code must be offset by its increase due to the bandwidth increase (in the inverse proportion to the code rate). Thus the uncoded bit error ratio to be used should be the one corresponding to the signal to noise ratio after the bandwidth increase due to coding. In this way the benefits (if any) of coding will be obvious.

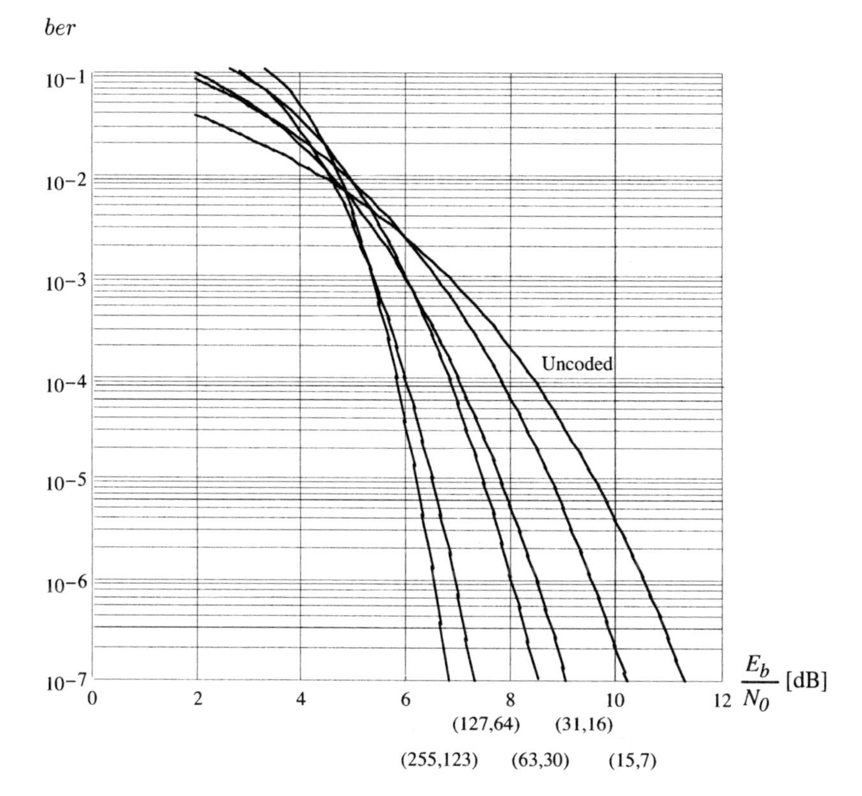

Fig. 6.3 Performance of the BCH coded and uncoded polar transmission for BCH codes $(15,7)$, $(31,16)$, $(63,30)$, $(127,64)$ and $(255,123)$.

6.5.5 Reed-Solomon codes

Reed-Solomon codes are a class of BCH codes generalized to the nonbinary case, that is, to a symbol (or byte) set of size $q = 2^m$. Reed-Solomon codes are capable of correcting all combinations of t or fewer symbol errors. Besides symbol size, BCH codes have the following parameters:

- block length – $n = 2^m - 1$ symbols, (or $m(2^m - 1)$ binary digits);
- information per block – $k = n - 2t$ symbols, (or $m(2^m - 1 - 2t)$ binary digits);
- parity check symbols – $n - k = 2t$, (or $n - k = 2mt$ binary digits).

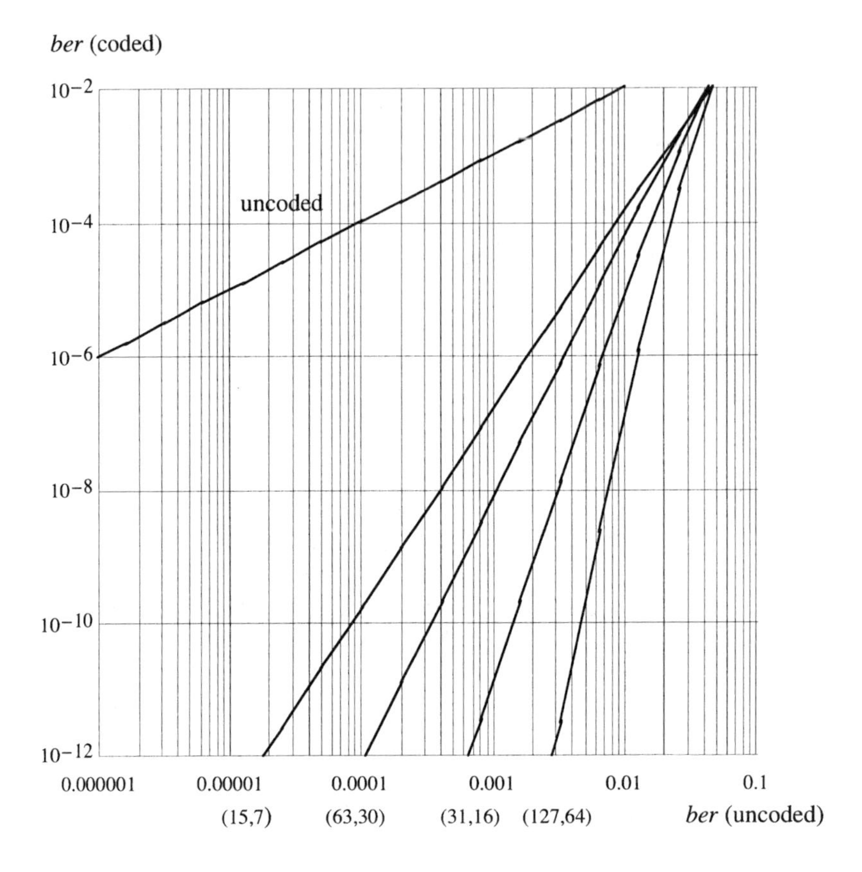

Fig. 6.4 Performance of BCH codes $(15, 7)$, $(31, 16)$, $(63, 30)$, and $(127, 64)$.

Taking $m = 3$ and $t = 1$ we get an $(21, 15)$ Reed-Solomon code, with a code rate $r_c = 5/7$ capable of correcting up to 8 errored bits out of 21 consecutive bits.

Since they are geared toward symbol correction, Reed-Solomon codes, when interpreted as binary codes, are well suited for error burst correction.

6.5.6 Majority logic decodable codes

Most majority logic decodable codes are cyclic codes whose decoding may be implemented using simple circuitry based on majority gates. These codes are slightly inferior to BCH codes in terms of error-correction capabilities and efficiency but decoding is easier to implement.

Reed-Muller codes are an example of majority logic codes. They are extended cyclic codes, that is, cyclic codes with an overall parity check.

For the Reed-Muller codes we have

$$\begin{aligned} n &= 2^m, \\ k &= \sum_{i=0}^{r} C_i^m, \\ d_{min} &= 2^{m-r}. \end{aligned}$$

Taking as an example the first-order $r = 1$ code, we have: $m = 3$, $n = 8$, $k = 4$ and $d_{min} = 4$.

As we have seen, the generator matrix $\mathbf{G}$ for this code is given in (6.27) and it transforms message $\mathbf{u} = [0101]$ into code vector $\mathbf{x} = \mathbf{G}\mathbf{u} = [01011010]$.

In order to understand how the decoder works let us take $\mathbf{u} = [u_0, u_1, u_2, u_3]$. The elements of code word $\mathbf{x}$ are

$$\begin{aligned} x_0 &= u_0, \\ x_1 &= u_0 + u_3, \\ x_1 &= u_0 + u_2, \\ x_3 &= u_0 + u_2 + u_3, \\ x_4 &= u_0 + u_1, \\ x_5 &= u_0 + u_1 + u_3, \\ x_6 &= u_0 + u_1 + u_2, \\ x_7 &= u_0 + u_1 + u_2 + u_3. \end{aligned}$$

If we recall that u_0, u_1, u_2 and u_3 are binary digits and that all arithmetic operations are performed using modulo-2 arithmetic, we may write:

$$\begin{aligned} u_3 &= x_0 + x_1, \\ u_2 &= x_0 + x_2, \\ u_1 &= x_0 + x_4, \\ u_3 &= x_6 + x_7, \\ u_2 &= x_4 + x_6, \\ u_1 &= x_2 + x_6, \\ u_3 &= x_2 + x_3, \\ u_2 &= x_1 + x_3, \\ u_1 &= x_1 + x_5, \\ u_3 &= x_4 + x_5, \\ u_2 &= x_5 + x_7, \\ u_1 &= x_3 + x_7. \end{aligned}$$

Take now $\mathbf{u} = [0101]$, which produces $\mathbf{x} = [01011010]$ and assume that the received vector is $\mathbf{y} = [11011010]$, with an error in the first position.

From the above we notice that there are four ways to compute u_1, u_2 and u_3. In this case we have:

$$u_1 = [0111],$$
$$u_2 = [1000],$$
$$u_3 = [0111].$$

By majority voting we may conclude that $u_1 = 1$, $u_2 = 0$ and $u_3 = 1$. It remains to obtain u_0. If we sum (modulo-2) all equations we get $\sum_{i=0}^{7} x_i = 0$ whereas $\sum_{i=0}^{7} y_i = 1$. This clearly indicates that $x_0 = u_0 = 0$.

Majority voting can be generalized for all Reed-Muller codes.

6.5.7 Codes for burst-error detection and correction

In some channels used in telecommunications, such as microwave radio links, errors tend to occur in bursts. In this situation codes designed to correct random errors are rather inefficient. Nevertheless, cyclic codes may become useful also in this case.

Let us define a burst error of length b as an error pattern where errors are confined to b consecutive positions. A burst error may be represented by the polynomial

$$e(d) = d^i e_b(d), \tag{6.38}$$

where d^i locates the error burst in the sequence of length n and $e_b(d)$ is a polynomial

$$e_b(d) = d^{b-1} + \ldots + 1. \tag{6.39}$$

Any $(n.k)$ cyclic code can detect all bursts whose lengths are not greater than $(n-k)$. The syndrome is the remaining of the division of $\delta^i e_b(\delta)$ by the generator polynomial $g(\delta)$. But the syndrome is always different from zero since neither δ^i nor $e_b(\delta)$ are multiples of $g(\delta)$ provided that $b < n-k$.

A burst error correcting code can correct all bursts of length b or less provided the number of check digits satisfies the Reiger bound

$$n - k \geq 2b. \tag{6.40}$$

A list of efficient cyclic codes is given in [2].

6.6 INTERLEAVED CODES

Random error correcting codes together with a suitable interleaver/deinterleaver pair are a practical way of dealing with error bursts. The interleaver rearranges the order of symbols to be transmitted in a deterministic way. The deinterleaver applies the inverse operation.

Given an (n, k) cyclic code, an (in, ik) interleaved code can be obtained by arranging i code words of the original code as rows of a rectangular array with i rows. The elements of this array are then transmitted by column. Since the deinterleaver rearranges the matrix in rows, error bursts are effectively broken in i parts, which at most have $b \div i$ digits (or $b \div i + 1$ digits if b is not a multiple of i). The cyclic code can then correct the digits in error.

The parameter i is called the interleaving degree of the code. If the original code is able to correct up to t errors, the interleaved code will have the same random-error-correction capability but, in addition, it will be able to correct all error bursts up to length $i \times t$.

An example, taken from [2], will help to clarify matters. Take a BCH $(15, 5)$ code, whose generator polynomial is

$$g(\delta) = \delta^{10} + \delta^8 + \delta^5 + \delta^4 + \delta^2 + \delta + 1,$$

capable of correcting all errors of 3 or less digits in blocks of 15 digits. Taking $i = 5$, we create a $(75, 25)$ interleaved code. Assume that an information sequence of 25 digits is to be transmitted. This sequence will result in five 15-digit words which will be positioned as the rows of a 5×15 matrix, as shown in Figure 6.5. The 75 digits are transmitted as a code word, by column, in the order they are numbered.

1	6	11	16	21	26	31	. . . . .	66	71
2	7	12	17	22	27	32	. . . . .	67	72
3	8	13	18	23	28	33	. . . . .	68	73
4	9	14	19	24	29	34	. . . . .	69	74
5	10	15	20	25	30	35	. . . . .	70	75

Fig. 6.5 Construction of a 75-bit code word out of $i = 5$ 15-bit words from a BCH $(15, 5)$ code in order to build a $(75, 25)$ interleaved code.

Assume further that, during transmission, digits from 13 to 27 (shaded in Figure 6.5) are corrupted by an error burst. In the reception, these digits will be deinterleaved into five 15-digit code words, none of which has more than 3 digits in error and thus the initial information can be fully recovered.

From Figure 6.5 it is easy to see that any error burst up to 15 digits will imply at most a 3-digit error in a word.

6.7 CONVOLUTIONAL CODES

6.7.1 Introduction

A binary convolutional encoder is a finite memory system that outputs n binary digits for every k information digits input. As before, the code rate is defined as $r_c = k/n$. Usually k and n are small numbers. Typically, $n \leq 8$ and r_c is in the range from 1/4 to 7/8.

Information digits are introduced in sets of k digits as input to a shift register, with Lk positions. As a block of k digits enters the register, n modulo-2 adders feed the output register with n digits and the input digits are shifted one position to the right, the eldest being lost.

In a convolutional encoder the n digits generated depend not only on the k input digits but also on the $(L-1)k$ previous digits.

Such a (n, k) convolutional encoder is known as a convolutional encoder of constraint L and the code it produces as a convolutional code of constraint L.

To describe the encoder we may use the same technique employed for block codes. Consider $\mathbf{u}$ as an semi-infinite message column vector and $\mathbf{x}$ the corresponding semi-infinite encoded vector. The encoding operation may be written as

$$\mathbf{x} = \mathbf{G}\mathbf{u},$$

where $\mathbf{G}$ is a semi-infinite matrix with the following structure

$$\mathbf{G} = \begin{bmatrix} \mathbf{G}_1 & 0 & 0 & 0 & 0 \\ \mathbf{G}_2 & \mathbf{G}_1 & 0 & 0 & 0 \\ \vdots & \mathbf{G}_2 & \mathbf{G}_1 & 0 & 0 \\ \mathbf{G}_n & \vdots & \mathbf{G}_2 & \mathbf{G}_1 & 0 \\ 0 & \mathbf{G}_n & \vdots & \mathbf{G}_2 & \mathbf{G}_1 \\ \vdots & \vdots & \vdots & \vdots & \vdots \end{bmatrix}. \tag{6.41}$$

In (6.41)

$$\mathbf{G}_1 = \begin{bmatrix} \mathbf{I} \\ \mathbf{P}_1 \end{bmatrix},$$

$$\mathbf{G}_i = \begin{bmatrix} \mathbf{Z} \\ \mathbf{P}_i \end{bmatrix},$$

where $\mathbf{I}$ is the identity matrix with dimension $[k, k]$, $\mathbf{P}_1$ is the parity matrix for the $(n-k)$ first digits (dimension $[n-k, k]$), $\mathbf{Z}$ is the all zeros matrix with dimension $[k, k]$ and $\mathbf{P}_i$ the parity matrix for the remaining digits.

A $(3, 1)$ convolutional code with constraint $L = 3$, taken from [2], will be used as an example. Consider Figure 6.6, where the output register is replaced by a commutator that reads sequentially the outputs of the adders.

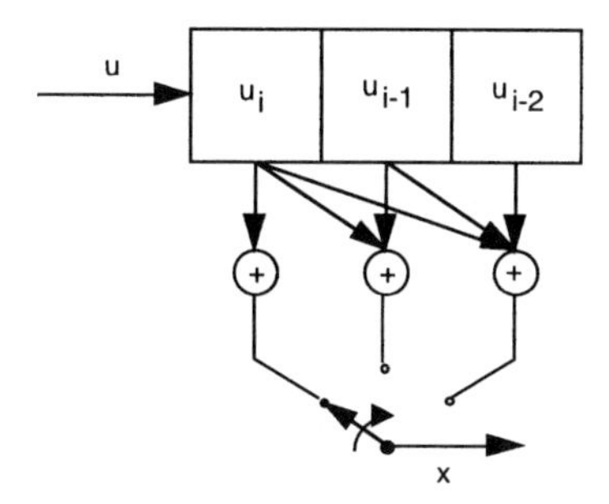

Fig. 6.6 Schematic circuit to implement a convolutional code $(3, 1)$ with constraint $L = 3$.

The commutator makes a full turn (i.e., provides the output of all the adders) for each input digit.

A simplified and more convenient scheme for the same encoder is shown in Figure 6.7.

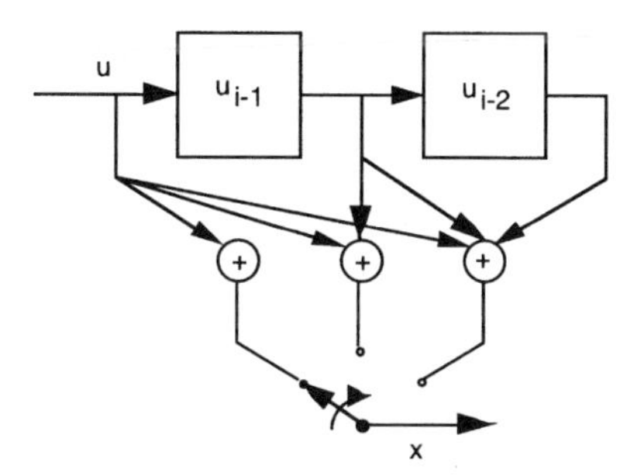

Fig. 6.7 Simplified schematic circuit to implement a convolutional code $(3, 1)$ with constraint $L = 3$.

Element $g_{i,j}$ of submatrix $\mathbf{G}_i$, (actually a column vector, since $k = 1$) represents the connections between the output of the jth shift register and the ith adder, with the usual convention that a 1 represents a connection and a 0 a non-existing connection. Hence, $\mathbf{G}_1 = [111]$, $\mathbf{G}_2 = [011]$, $\mathbf{G}_3 = [001]$

and

$$\mathbf{G} = \begin{bmatrix} 1 & 0 & 0 & 0 & 0 & \cdots \\ 1 & 0 & 0 & 0 & 0 & \cdots \\ 1 & 0 & 0 & 0 & 0 & \cdots \\ 0 & 1 & 0 & 0 & 0 & \cdots \\ 1 & 1 & 0 & 0 & 0 & \cdots \\ 1 & 1 & 0 & 0 & 0 & \cdots \\ 0 & 0 & 1 & 0 & 0 & \cdots \\ 0 & 1 & 1 & 0 & 0 & \cdots \\ 1 & 1 & 1 & 0 & 0 & \cdots \\ 0 & 0 & 0 & 1 & 0 & \cdots \\ 0 & 0 & 1 & 1 & 0 & \cdots \\ 0 & 1 & 1 & 1 & 0 & \cdots \\ 0 & 0 & 0 & 0 & 1 & \cdots \\ 0 & 0 & 0 & 1 & 1 & \cdots \\ 0 & 0 & 1 & 1 & 1 & \cdots \\ \vdots & \vdots & \vdots & \vdots & \vdots & \cdots \end{bmatrix}.$$

The reader is invited to check that vector $\mathbf{u} = [11011\ldots]$ is coded into $\mathbf{x} = [111100010110100\ldots]$.

As another example take the convolutional code $(3, 2)$ with constraint $L = 2$ whose encoder is shown in Figure 6.8.

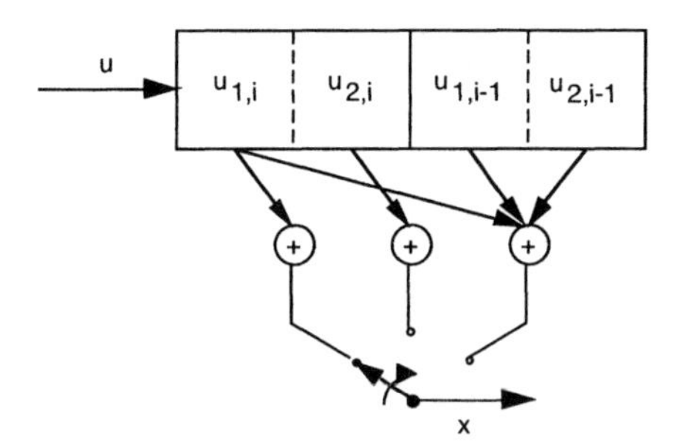

Fig. 6.8 Simplified schematic circuit to implement a convolutional code $(3, 2)$ with constraint $L = 2$.

Now input digits enter the encoder two at a time, and occupy positions 1 and 2 of the input shift register, on a first-in-first-out order, that is, the first digit goes to position 2 and the second digit to position 1. Recalling that the $\mathbf{G}_i$ matrixes represent the connections of input register i with each adder and noting that now we have a two-position shift register matrix, $\mathbf{G}_i$ must have two columns, one for each position of the shift register. Hence

$$\mathbf{G}_1 = \begin{bmatrix} 1 & 0 \\ 0 & 1 \\ 1 & 0 \end{bmatrix},$$

$$\mathbf{G}_2 = \begin{bmatrix} 0 & 0 \\ 0 & 0 \\ 1 & 1 \end{bmatrix},$$

$$\mathbf{G} = \begin{bmatrix} 1 & 0 & 0 & 0 & 0 & 0 & 0 & 0 \\ 0 & 1 & 0 & 0 & 0 & 0 & 0 & 0 \\ 1 & 0 & 0 & 0 & 0 & 0 & 0 & 0 \\ 0 & 0 & 1 & 0 & 0 & 0 & 0 & 0 \\ 0 & 0 & 0 & 1 & 0 & 0 & 0 & 0 \\ 1 & 1 & 1 & 0 & 0 & 0 & 0 & 0 \\ 0 & 0 & 0 & 0 & 1 & 0 & 0 & 0 \\ 0 & 0 & 0 & 0 & 0 & 1 & 0 & 0 \\ 0 & 0 & 1 & 1 & 1 & 0 & 0 & 0 \end{bmatrix}.$$

An alternative way to implement the convolutional code $(3, 2)$, in Figure 6.9, shows clearly that the input stream is divided in two, that is the first digit becomes the first digit in u_1 and the second digit the first in u_2.

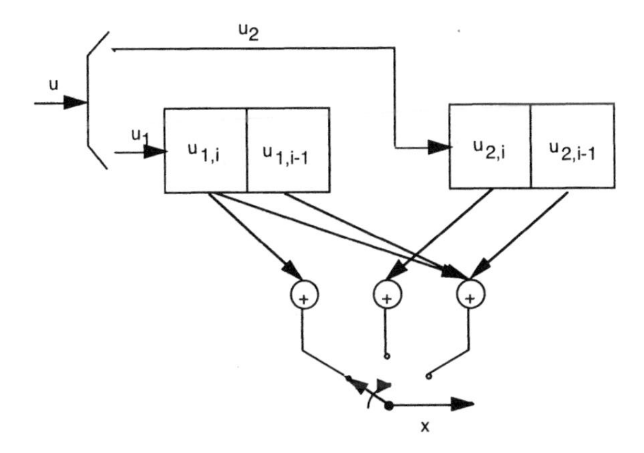

Fig. 6.9 Alternative schematic circuit to implement a convolutional code $(3, 2)$ with constraint $L = 2$.

6.7.2 State diagram representation of convolutional codes

The state diagram is a powerful alternative to the algebraic description of a convolutional code. As an example, let us consider again the convolutional code $(3,1)$, with constraint length $L = 3$. This means that any output digit depends on an input digit and its previous two digits, that is, this encoder has memory $M = L - 1 = 2$

The state of the encoder at a given time is defined by the contents of its memory at that time.

Since it has $M = 2$, the $(3,1)$ encoder has $2^2 = 4$ possible states: 00, 01, 10 and 11. Assume the encoder is at state $S_0 = (00)$. When a 1 is input the encoder outputs [100] and moves to state $S_2 = (10)$. This behavior is completely described by the state diagram in Figure 6.10.

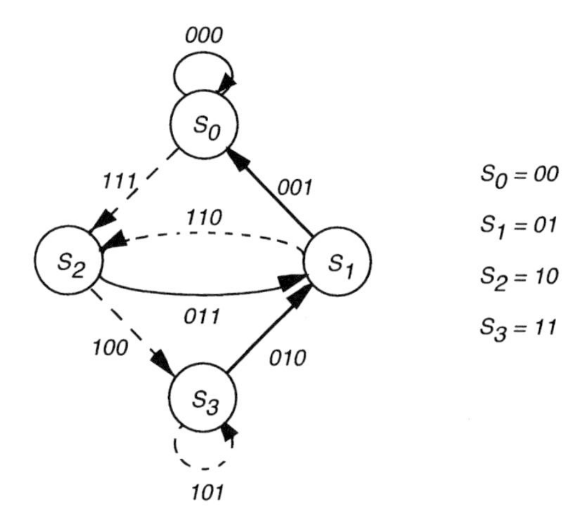

Fig. 6.10 State diagram for the convolutional code $(3,1)$.

The state diagram is easily read. From one state we proceed to the next following the solid lines when a zero is input and following the dashed lines otherwise. If we start from state $S_0 = (00)$ and take the first digit (1) we move from S_0 to S_2 and output [111]. Another 1 is input and we move to state S_3 outputting [100]. Following we proceed to S_1 then S_2 and S_3 and correspondingly we output [010], [110] and [100].

The concept of state diagram can be applied to any convolutional code (n,k) of constraint length L and memory $M = L-1$. The number of states is $2^k M$. There are 2^k edges entering and 2^k edges leaving each state. The labels on each edge are sequences of length n. As L or M increase, the number of states grows exponentially, and the state diagram becomes impractical.

In order to introduce some form of time dependence in the transition of states we use a trellis whose nodes, represented by dots, are arranged vertically, for each instant. The vertical size is the trellis depth. Dashed and solid lines have the same meaning as in the state diagram (see Figure 6.11).

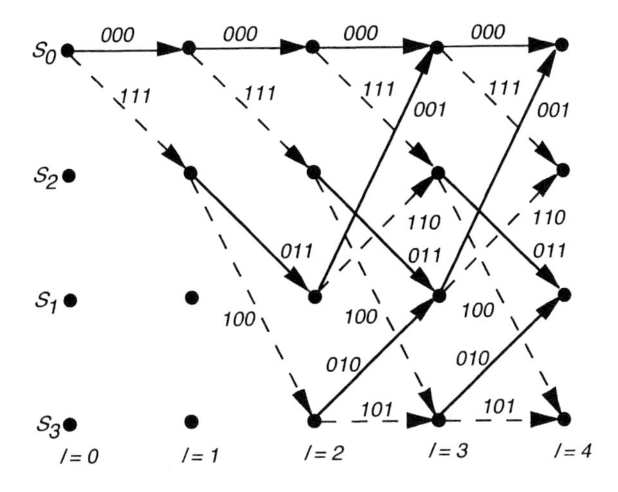

Fig. 6.11 Trellis representing the convolutional code $(3, 1)$.

Using the trellis in Figure 6.11 it is easy to follow the coding path of sequence [1101]. Starting from $S_0 = (00)$ we see that the first 1 input causes a [111] output, the subsequent 1 a [100] and the following 0 and 1 the triplets [010] and [110], respectively.

Error detection and correction capabilities of convolutional codes are directly related to the distance properties of the encoded sequences.

Consider a pair of encoded sequences up to the depth l in the trellis and assume they disagree at the first branch. We define the lth order column distance $d_c(l)$ as the minimum Hamming distance between all pairs of such sequences. For the computation of $d_c(l)$ we can use the all zero sequence.

There are two important values of l. The first is L the code constraint length $d_{min} = d_c(L)$. The second is the free distance of the convolutional code $d_f = \lim_{l \to \infty} d_c(l)$.

A simple algorithm to compute d_f is based on the following steps:

1. Compute $d_c(l)$ for $l = 1, 2, \ldots$.

2. If the sequence giving $d_c(l)$ merges into the all-zero sequence, keep its weight as d_f.

Take the trellis from the Figure 6.11 but instead of the output sequence associated with each transition take its weight (Figure 6.12).

Starting from S_0 at $l = 0$ a 1 input leads to an [111] output, that is, $d_c(1) = 3$. For $l = 2$ a 1 input causes a [100] output, and so $d_c(2) = 3 + 1 = 4$. Had we taken a 0 input to proceed from $l = 1$ to $l = 2$ we would have reached S_1 and the output would be [011]. From there another zero would

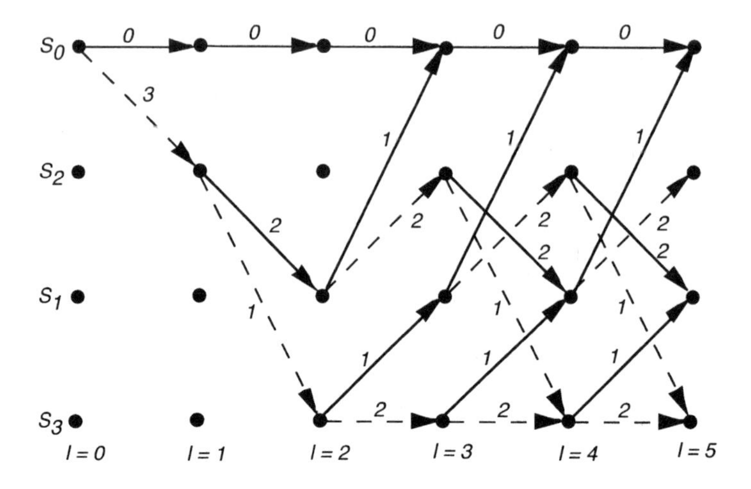

Fig. 6.12 Trellis for the convolutional code (3, 1) with the weight associated with each transition.

lead us back to S_0 (with a [001] output) and the Hamming distance would be $d_c(3) = 3 + 2 + 1 = 6$. Returning to S_3 at $l = 2$, a 0 input yields a [010] output and so a $d_c(3) = 4 + 1 = 5$. Another 0 input takes us back to the all-zero sequence and outputs a [001], hence $d_c(4) = 5 + 1 = 6$. Clearly we have $d_f = 6$, since all other paths lead to a higher value of the Hamming distance with regards to the all-zero sequence. Table 6.5 resumes our findings.

l	$d_c(l)$
1	3
2	4
3	$5 \leftarrow d_{min}$
4	$6 \leftarrow d_f$
5	6

Table 6.5 Path to compute d_{min} and d_f.

In the general case, computation of d_f may require very long sequences. The problem may however be solved in another way as we will show in the following.

First we redraw the state diagram and label the edges with an indeterminate d^n where n is the weight of the output sequence corresponding to the transition. The loop at S_0 is eliminated since it does not contribute to the weight of the output sequence. Finally the S_0 state is broken in two: one representing the input and the other the output of the state diagram (see Figure 6.13).

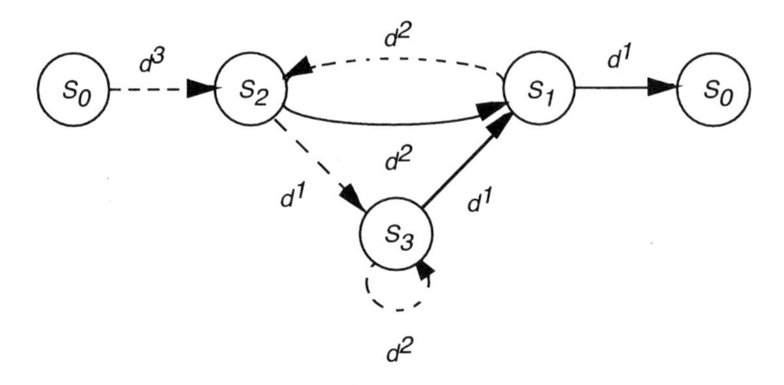

Fig. 6.13 Modified state diagram.

A path label is the product of the labels of all its edges. From the infinite number of paths that start at S_0 and return to S_0 we now look to the one whose label corresponds to the lowest d power. Following the modified state diagram it is very easy to find out that $S_0S_2S_1S_0$ (or $S_0S_2S_3S_1S_0$) lead to a path label of d^6 and thus to $d_f = 6$.

6.7.3 Transfer function

Consider a simple directed graph as given in Figure 6.14.

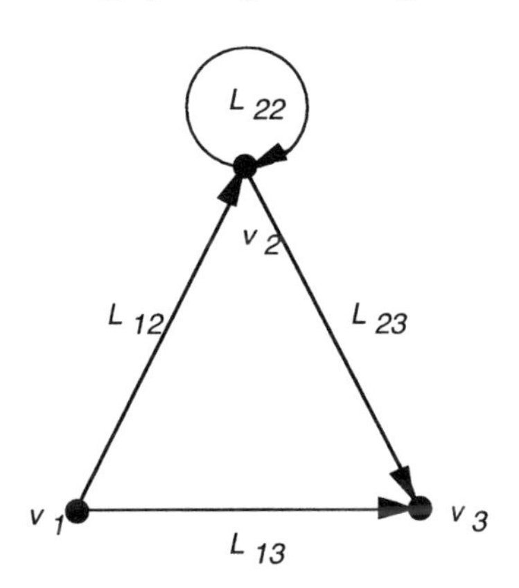

Fig. 6.14 A simple directed graph.

Transmission between ν_1 and ν_3 is given by:

$$\begin{aligned}
T(\nu_1, \nu_3) &= L_{13} + L_{12}L_{23} + L_{12}L_{22}L_{23} + L_{12}L_{22}^2L_{23} + \ldots \\
&= L_{13} + L_{12}L_{23}(1 + L_{22} + L_{22}^2 + \ldots) && (6.42) \\
&= L_{13} + \frac{L_{12}L_{23}}{1 - L_{22}}. && (6.43)
\end{aligned}$$

The directed graph in Figure 6.14 may be reduced to a simple two-point graph ν_1, ν_3 connected by L'_{13}

$$L'_{13} = L_{13} + \frac{L_{12}L_{23}}{1 - L_{22}}.$$

Given any directed graph it is possible to compute the transmission between a pair of vertices by removing, one by one, the intermediate vertices and redefining the new labels. As an exercise we will reduce the graph corresponding to the modified state diagram, step by step.

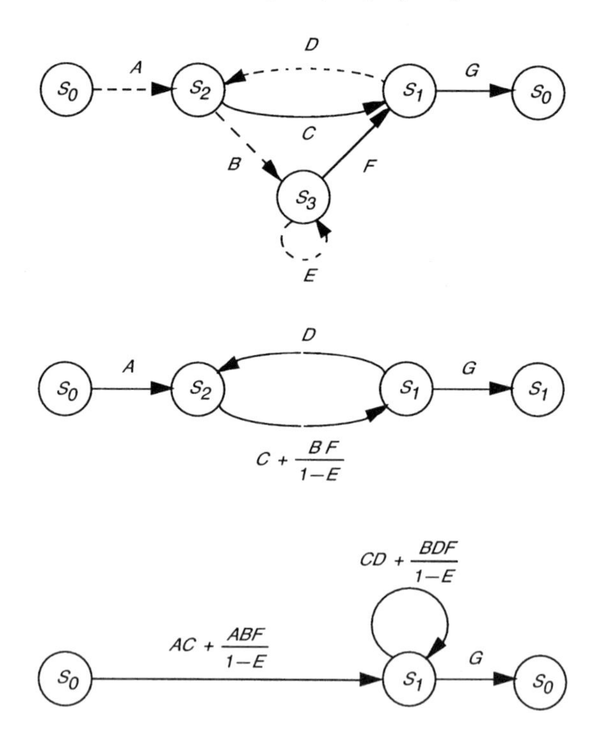

Fig. 6.15 Reducing the direct graph corresponding to the convolutional code $(3, 1)$ step by step.

Transmission between S_0 and S_0 is now

$$\begin{aligned} T(S_0, S_0) &= \frac{AC + \frac{ABF}{1-E}G}{1 - CD - \frac{BDF}{1-E}} \\ &= \frac{ACG(1-E) + ABFG}{1 - E - CD + CDE - BDF}. \end{aligned}$$

If we substitute edge labels $A, B, \ldots, G$ by those in the modified state diagram we get the transfer function between S_0 and S_0

$$\begin{aligned} T(S_0, S_0) &= \frac{d^6(1-d^2) + d^6}{1 - d^2 - d^4 + d^6 - d^4} \\ &= \frac{2d^6 - d^8}{1 - (d^2 + 2d^4 - d^6)} \\ &= 2d^6 + d^8 + 5d^{10} + \ldots. \end{aligned}$$

This result means that there are two paths with weight 6, one with weight 8, five with weight 10, and so on.

The transfer function can be used to provide additional information on code properties. This is achieved by considering the modified graph where the edge label, besides d^n, includes a label l that counts the path length and another label j if the corresponding transition is caused by a 1 input.

Now the transfer function becomes

$$\begin{aligned} T(d, j, l) &= \frac{d^6 l^3}{j}(1 + lj - d^2 lj)1 - d^2 lj(1 + d^2 l + d^2 l^2 j - d^4 l^2 j) \\ &= d^6 l^3 j(1 + lj) + d^8 l^5 j^3 \\ &\quad + d^{10} l^5 j^2 (1 + 2lj + lj^2 + l^2 j^2) + \ldots. \end{aligned} \tag{6.44}$$

The transfer function indicates that there are two paths with weight 6, one with length 3 and another with length 4. It also tells us that the first path was caused by an input with a one whereas for the second the input had two ones. The path with weight 8 has length 5 and its input sequence has three ones. There are 5 paths with weight 10, one with length 5, three with length 6 and one with length 7. And so on.

When deriving the generating function of a convolutional code is was (implicitly) assumed that $T(d)$ converges. Otherwise the expansions would not have been valid. When the convergence is not verified the code is said to be catastrophic. An example is the convolutional code $(3, 1)$ with constraint $L = 2$, described above. The state diagram for this code shows that the self loop at S_3 does not increase the distance from the all-zero sequence. Therefore, the sequence $S_0 S_2 S_3 \ldots S_3 S_1 S_0$ will be at a distance 6 from the all-zero path, hence the chance of having an arbitrary large number of decoding errors even for a finite number of channel errors. It is possible to prove that systematic convolutional codes cannot be catastrophic.

Computer search methods have been employed to find convolutional codes that, for a given rate and constraint length, have the largest possible free distance.

6.7.4 Maximum likelihood decoder for convolutional codes

The maximum likelihood decoder selects the code word whose distance from the received sequence is minimum. Unlike block codes, convolutional codes have no fixed length. Each possible sequence is a path in the code trellis. Therefore, the decoder must choose a path that is the closest to the received sequence. The distances to be used are, as before, the Hamming distance for hard decoders and the Euclidean distance for soft decoders.

Let us start with hard decoding, assuming polar modulation in a binary symmetric channel with uncoded bit error probability p. Consider the convolutional code $(3, 1)$ used before and assume an information sequence

$$\mathbf{u} = [0100000],$$

which is encoded into

$$\mathbf{x} = [000111011001000000000],$$

but received as

$$\mathbf{y} = [110111011001000000000],$$

that is, with two errors in the first triplet. Obviously, the received sequence does not correspond to any path in the coding trellis.

The Viterbi algorithm provides a way to find out the closest possible path to the received sequence.

The first two steps of the Viterbi algorithm are shown in Figure 6.16. The algorithm, at each step l in the trellis, stores, for each state, the surviving path, that is, the minimum distance path from the starting state S_0 and the corresponding accumulated metric.

Start from S_0 and consider that we may receive either a 0, which leaves us at S_0 or a 1, leading to S_2. In Figure 6.16 in each transition we mark the difference in weight of the received sequence and the sequence which should have been produced by the encoder, had it been fed with a 0 (solid line) or a 1 (dashed line). From the state diagram (Figure 6.10) we notice that if the encoder is at S_0 and it receives a 0 or a 1, it outputs 000 or 111, respectively. Since the weight of the first triplet of digits in the received sequence is 2, we have, on the transition from S_0 to S_0, a 2 and, on the transition from S_0 to S_1, a 1. Following, with the encoder at S_0, if we input a 0 or a 1, we change to S_0 or S_2, whereas if we had been at S_2 we would have moved into S_1 and S_3, respectively. At any stage in the procedure we may ignore all paths that, from a given state, lead into the same state, but the one with the minimum weight.

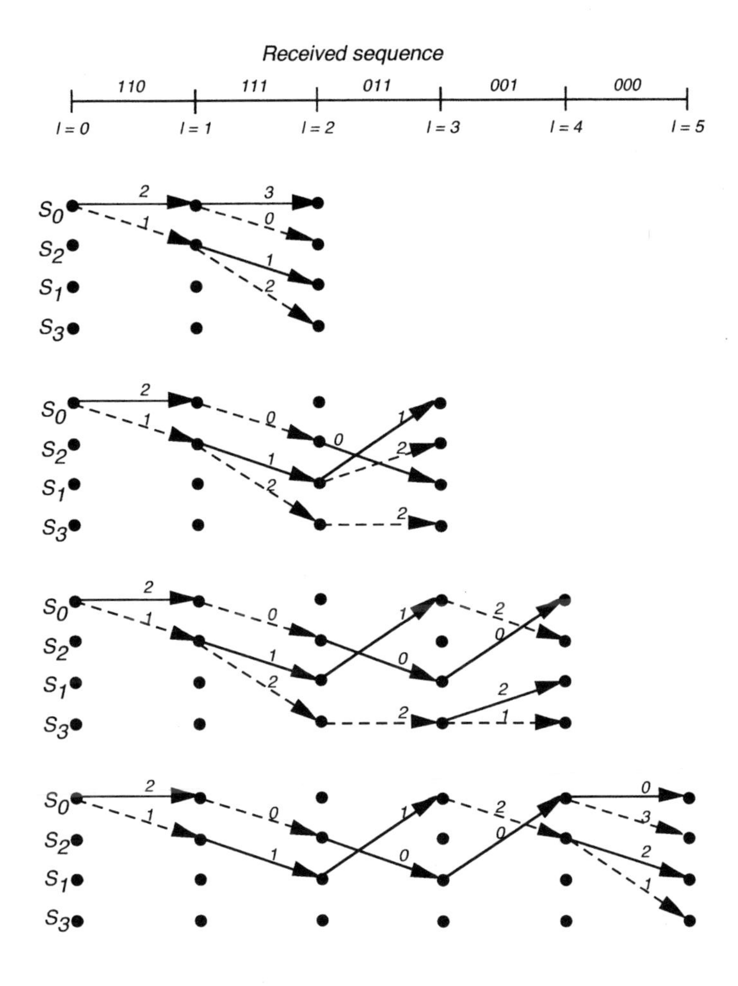

Fig. 6.16 Diagram showing the operation of the Viterbi algorithm.

Proceeding from $l = 1$ to $l = 2$ we neglect path $S_0S_0S_0$ because its weight would be 7 whereas other paths such as $S_0S_2S_1S_0$ have a weight of only 3. Similarly we may neglect the path $S_0S_2S_3S_1$ because its weight (4) is higher than that of path $S_0S_0S_2S_1$ which is only 2.

At $l = 4$ we already returned to S_0 with a minimum weight of 2, that is, we have recovered the original sequence [01000] in spite of the fact that the received sequence had two digits in error.

The trellis structure has enabled us to correct errors because wrong paths tend to merge into correct ones after some steps. Ideally one should output the result only after all input digits have been received. This is obviously not practical, because of the long delay and the excessive memory requirements. Simulation has shown that a delay of about $5L$ (where L is the constraint) provides (almost) the same performance.

The implementation of a soft decision Viterbi decoder is not significantly different from a hard decision one. In fact, it suffices to replace the Hamming metric by the Euclidean metric.

6.7.5 Performance of convolutional codes with Viterbi decoding

The probability of an error event eer using convolutional codes with hard decision Viterbi decoding may be upper bound as

$$eer < T(d)|_{d=\sqrt{4p(1-p)}}. \tag{6.45}$$

Turning now to the bit error ratio we have

$$ber < \frac{\partial T(d,j)}{\partial j}\bigg|_{j=1,d=\sqrt{4p(1-p)}}, \tag{6.46}$$

which may be approximated by

$$\begin{aligned} ber &\le A[4p(1-p)]^{d_f/2} & (6.47)\\ &\approx A\,2^{d_f}p^{d_f/2}, & (6.48)\end{aligned}$$

where A is a constant depending on the code and p the uncoded bit error probability.

Let us now derive the bit error ratio for the $(3,1)$ convolutional code previously described ($L=3$, $r_c=1/3$ and $d_f=6$) with hard decision Viterbi decoding. From (6.44) we have

$$\frac{\partial T(d,j,l=1)}{\partial j}\bigg|_{j=1} = 3d^6 + 3d^8 + \ldots,$$

whereby, making $d=\sqrt{4p(1-p)}$ we get

$$ber = 192p^3(1-p)^3.$$

The following tighter bound may be derived

$$ber \le 30p^3(1-p)^3. \tag{6.49}$$

For the same code with soft Viterbi decoding, we get from [2] the following approximation for the union bound

$$ber < \frac{1}{2}\frac{\partial T(d,j)}{\partial j}\bigg|_{j=1,d=e^{r_c\frac{e_b}{n_0}}}, \tag{6.50}$$

which, when $e_b/n_0 \to \infty$ may be approximated by

$$ber = \frac{A}{2} e^{d_f r_c \frac{e_b}{n_0}}, \tag{6.51}$$

where, as with hard decision, A is a constant depending on the code.

For the convolutional code $(3,1)$, we get from the transfer function applying (6.50)

$$ber < \frac{3}{2} e^{-2\frac{e_b}{n_0}}. \tag{6.52}$$

Figure 6.17 shows the performance of uncoded polar transmission, and coded polar transmission using the convolutional code $(3,1)$ with hard and soft decision Viterbi decoding.

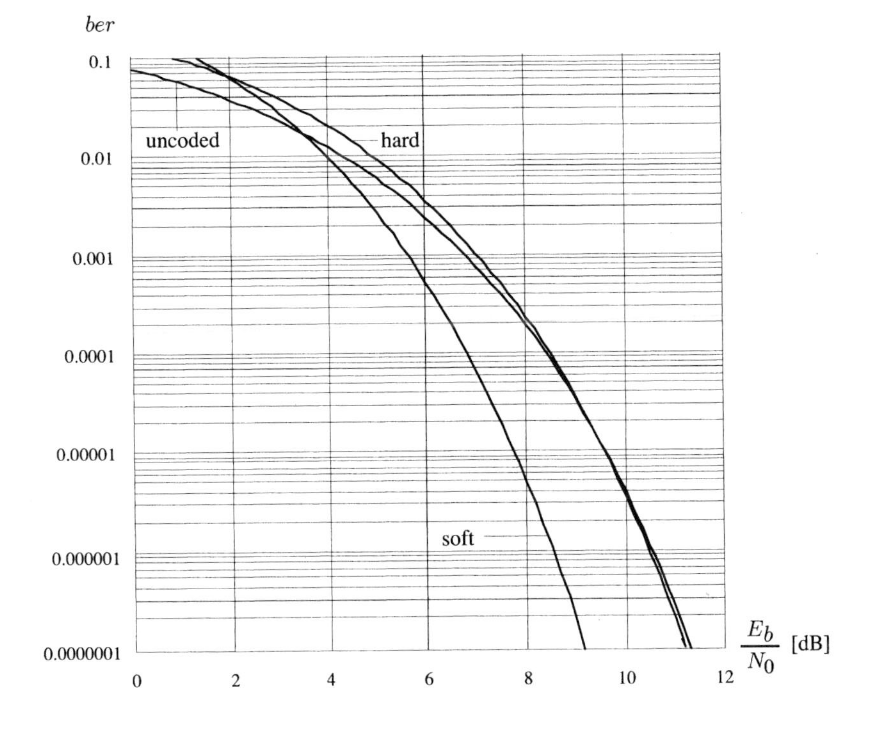

Fig. 6.17 Performance of uncoded polar transmission and coded polar transmission using the convolutional code $(3,1)$ with hard and soft decision Viterbi decoding.

The coding gain is obviously not very high: from 0 to 1 dB for hard decision and 2 dB for soft decision Viterbi decoding. Much better results may be achieved with the same code rate $r_c = 1/3$ by increasing the constraint length

L from 3 to 7 or 8. For a bit error ratio of 10^{-6} coding gains of about 6 dB can be obtained.

Unquantized soft decision decoding is hardly ever used. Practical circuits use eight level quantization, for which the degradation, as compared to the unquantized case is quite small, about 0.25 dB.

It may not be obvious why the $(3,1)$ convolutional code has such a small coding gain. In order to get a deeper understanding it is instructive to plot the performance (i.e., the *ber*) of the convolutional code $(3,1)$ with hard decision Viterbi decoding, as a function of the uncoded bit error probability p. The result computed with (6.49) is shown in Figure 6.18.

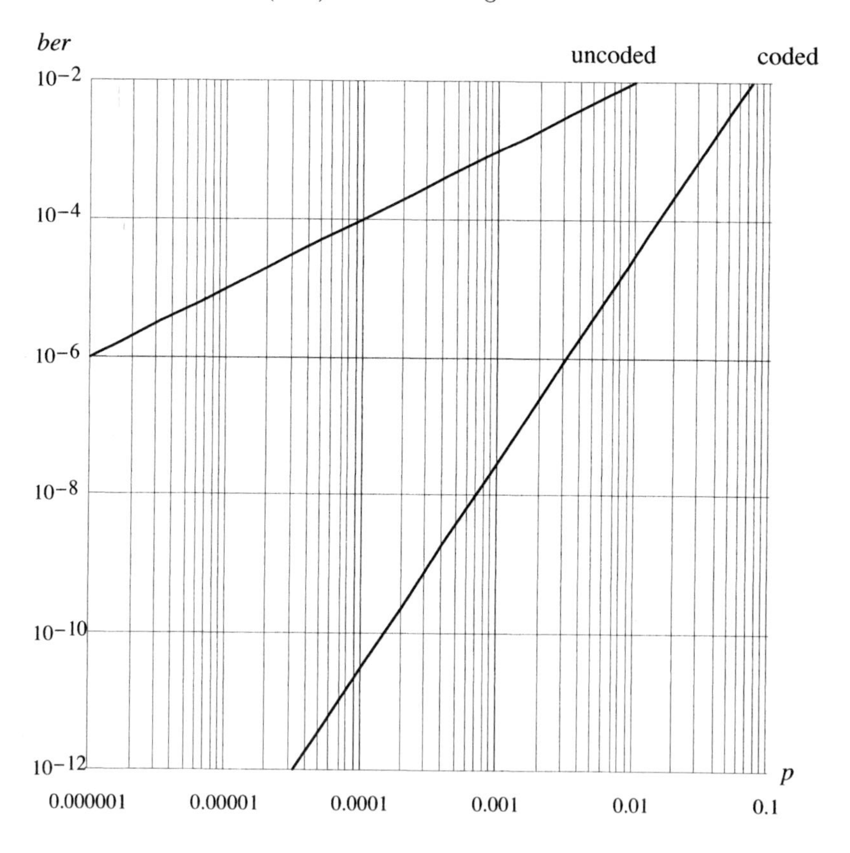

Fig. 6.18 Performance (*ber*) of the convolutional code $(3,1)$ with hard decision Viterbi decoding as a function of the uncoded bit error probability p.

From Figure 6.18 the performance gain of the convolutional code $(3,1)$ is apparently very high indeed. One should however not forget that this code implies a threefold increase in bit rate (or bandwidth) and thus a threefold decrease (−4.79 dB) in carrier-to-noise ratio. If we now consider 2-PSK and synchronous demodulation, a carrier-to-noise ratio of 6 dB yields a bit error ratio of 0.0024. Using the $(3,1)$ convolutional code (and three times the bandwidth) the same performance requires a 5.6 dB carrier-to-noise ratio, hence a 0.4 dB coding gain.

The previous example shows that a code performance must always be considered together with the bandwidth increase it entails. In practical radio links this often means code rates higher than 1/2.

An alternative to derive the generating function $T(d)$ involves writing the state equations, that is, the equations that represent the accumulated path gain x_{i+1} from state i to state $i+1$ as influenced by all other states. For the convolutional code $(3,1)$ we have

$$\begin{aligned} x_1 &= d^2 l x_2 + dl x_3, \\ x_2 &= d^2 l j x_1 + d^3 l j, \\ x_3 &= dl j x_2, \\ x_4 &= dl x_1. \end{aligned}$$

State equations may be written in matrix form as

$$\mathbf{x} = \mathbf{A}\mathbf{x} + \mathbf{x}_0, \tag{6.53}$$

where

$$\mathbf{x} = \begin{bmatrix} x_1 \\ x_2 \\ x_3 \\ x_4 \end{bmatrix},$$

$$\mathbf{x}_0 = \begin{bmatrix} 0 \\ d^3 l j \\ 0 \\ 0 \end{bmatrix},$$

and the state matrix $\mathbf{A}$ is

$$\mathbf{x} = \begin{bmatrix} 0 & d^2 l & dl & 0 \\ d^2 l j & 0 & 0 & 0 \\ 0 & dlj & 0 & 0 \\ dl & 0 & 0 & 0 \end{bmatrix}.$$

State equation (6.53) may be rewritten as

$$\mathbf{x} = (\mathbf{I} - A)^{-1}\mathbf{x}_0,$$

which may be expanded into

$$\mathbf{x} = (\mathbf{I} + \mathbf{A} + \mathbf{A}^2 + \mathbf{A}^3 + \ldots)\mathbf{x}_0. \tag{6.54}$$

Since all paths terminating in state S_0 are given by x_4, then

$$T(d,l,j) = x_4.$$

This procedure may be used for all codes, except catastrophic ones. A sufficient condition is

$$|\mathbf{I} - \mathbf{A}| \neq 0.$$

For a small number of states one may compute the matrix inversion in the standard way. For a large number of states, expression (6.54) provides an algorithm for the rapid computation of $T(d,l,j)$.

Applying this method to the convolutional code $(3,1)$ and using the algorithm in (6.54) we get

$$T(d,l,j) = d^6 l^3 j + d^6 l^4 j^2 + \ldots.$$

6.8 CONCATENATED CODES

Concatenated codes are a practical tool to implement very long codes with large error-correcting capabilities. Concatenated codes make use of two codes: an inner, (n_i, k_i) binary code, and an outer (n_o, k_o) nonbinary code, (often a Reed-Solomon code), as shown in Figure 6.19.

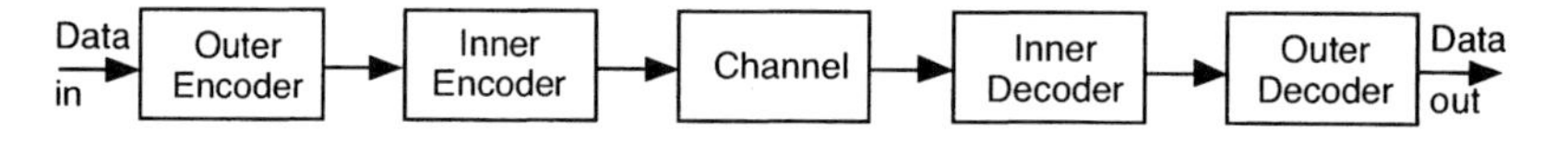

Fig. 6.19 Concatenated coder and decoder.

The outer encoder considers each set of k_i input bits as an input symbol and encodes each set of k_o symbols (corresponding to $k_i k_o$ input bits) into n_o k_i-bit symbols. The inner encode, in turn, encodes each of these k_i-bit symbols into a n_i-bit symbol. The overall code rate is thus $r_c = \frac{k_i k_o}{n_i n_o}$.

On the receiver side, the inner decoder provides an estimate of the n_o k_i-bit symbols for outer decoder with a modest symbol error rate ser and the outer decoder outputs a bit stream with $k_o k_i$ bits with a very low bit error rate ber.

Using an outer (n_O, k_O) t-symbol-error-correcting Reed-Solomon code and assuming that the inner decoder, which uses an (n_i, k_i) code, only produces symbol decisions, with a symbol error probability ser, the bit error rate at the outer decoder output may be estimated from [7]

$$ber < \frac{2^{k_i}}{2^{k_i} - 1} \sum_{j=t+1}^{n_O} C_j^{n_O} ser^j (1 - ser)^{n_O - 1}. \tag{6.55}$$

According to [9], the minimum distance of the concatenated code is equal to $d_{o_{min}} d_{i_{min}}$ where $d_{o_{min}}$ is the minimum distance of the outer code and

$d_{i_{min}}$ is the minimum distance of the inner encoder. Thus concatenation of codes tends to improve the error performance of individual codes.

Concatenated codes are effective against a mixture of random errors, corrected by the inner code, often a convolutional code and burst errors left to the outer code (usually a Reed Solomon code).

Suitable convolutional codes are short constraint, with soft decision Viterbi decoding. Interleaving may be used to counteract the tendency of Viterbi decoders to correlate errors. Reed Solomon codes use symbols from 6 to 9 bits. Coding gains of about 7.5 dB at a bit error ratio of 10^{-6} are achievable, and a small increase (1–2 dB) in the ratio E_b/N_0 leads to a very pronounced decrease in the bit error ratio.

It is also possible to represent each of the k_i-bit symbols of the Reed-Solomon code with one of the 2^k signals of an orthogonal signal set. The decoder complexity grows since a separate correlator is required for each of the 2^k signals of the inner code, thus only small values of k_i are practical. These codes require a large bandwidth but may provide large coding gains (about 10 dB at a bit error rate of 10^{-6}). Such codes are used in deep space probes.

6.9 TURBO CODES

So far we have formed concatenated codes by serially applying codes to a bit stream. Here we use parallel concatenation of component codes to make a turbo code.

Before starting on turbo codes let us look again at convolutional codes. Take a simple $r_c = 1/2$ convolutional encoder with constraint L and memory $L-1$. Assume that at instant k the encoder input is bit d_k and the corresponding encoder output is the bit pair (u_k, v_k)

$$
\begin{aligned}
u_k &= \sum_{i=0}^{L-1} g_{1,i} d_{k-i}, \\
v_k &= \sum_{i=0}^{L-1} g_{2,i} d_{k-i},
\end{aligned}
$$

where the arithmetic, as usual, is modulo-2, and $g_{1,i}$ and $g_{2,i}$ are the coefficients of the generator polynomials G_1 and G_2.

A simple example of such an encoder is given in Figure 6.20, where $L = 3$, $G_1 = [111]$ and $G_2 = [101]$.

This encoder has a finite impulse response and produces a non-systematic convolutional code. It is known [14] that non-systematic convolutional codes have a better performance at high E_b/N_0 than systematic codes with the same memory. At low E_b/N_0 the opposite is usually true [14].

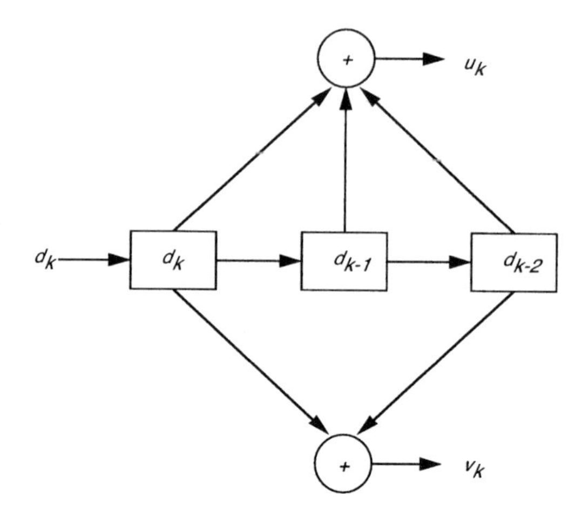

Fig. 6.20 A simple convolutional encoder with $L = 3$, $G_1 = [111]$ and $G_2 = [101]$.

A class of infinite impulse response codes, also known as recursive systematic convolutional, because previously encoded bits are continuously fed back to the encoder input, has been proposed as building blocks for a turbo code. For high code rates recursive systematic convolutional codes have better performance than the best non-systematic convolutional codes at any E_b/N_0 [14].

A binary $r_c = 1/2$ recursive systematic convolutional code is obtained from the non-systematic convolutional code in Figure 6.20 by using a feed-back loop and setting one of the two outputs equal to the input, as shown in Figure 6.21.

Now for input d_k the encoder outputs the bit pair (d_k, a_k), where

$$a_k = d_k + \sum_{i=1}^{L-1} g_i a_{k-i},$$

and $g_i = g_{1,i}$ if $u_k = d_k$ (as in the figure), or $g_i = g_{2,i}$ if $v_k = d_k$.

According to [14] a recursive systematic convolutional code does not modify the output code weight, as compared to the original non-systematic convolutional code, it only changes the mapping between input data and corresponding output data sequences.

Consider now the parallel concatenation of two encoders such as the one just described (Figure 6.22). In practice both component encoders are not necessarily identical.

This encoder alternatively produces codewords from each component encoder. The interleaver tries to avoid pairing of low weigh codewords from each encoder.

For a recursive systematic encoder it is possible to show that:

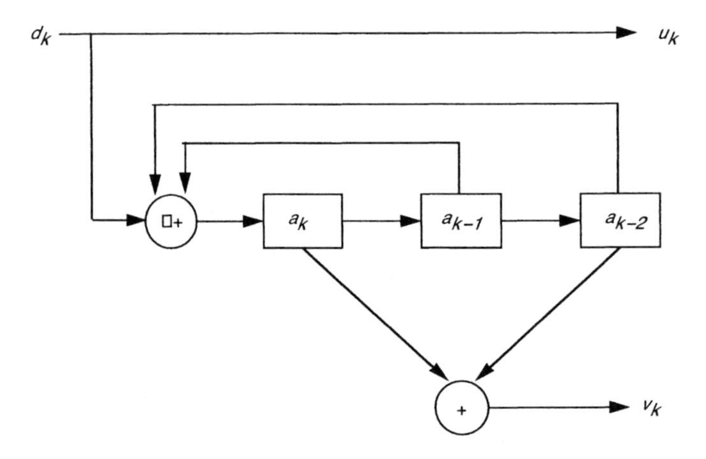

Fig. 6.21 A simple binary recursive systematic convolutional encoder, with $r_c = 1/2$.

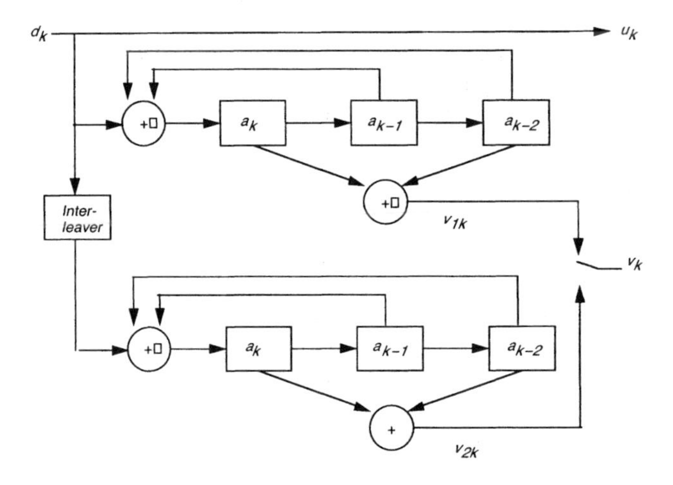

Fig. 6.22 Parallel concatenation of two recursive systematic convolutional encoders to produce a turbo encoder.

- an all-zero input sequence produces an all-zero output sequence;

- a single one in an otherwise all zero input sequence, produces a sequence which will never return to an all-zero sequence, or in other words, a weight 1 input produces an infinite weight output.

By dividing the input stream in two, the interleaver may break weight-2 input sequences into weight-1 input sequences (with produce infinite weight output) and it may break weight-3 (or longer) input sequence producing at least one weight-2 input sequence. Hence turbo code performance is largely dependent on minimum weight codewords produced by weight-2 input sequences because weight-1 input sequences give rise to infinite weight output sequences and weight-3 and larger input sequences can largely be prevented by a properly designed interleaver.

In Figure 6.22 the minimum weight code word is generated by the weight 3 input sequence [000...01110...0]. Another input sequence that produces a fairly low weight codeword is [00...010010...0]. After the interleaver, none of these sequences is likely to appear again at the other encoder, so that a minimum weight codeword from one encoder is unlikely to be combined with another minimum weight codeword from the other encoder. The important aspect of building blocks for turbo codes is that they are recursive thus protected against low weight codeword pairing that cannot be remedied by the interleaver.

The Viterbi algorithm is the optimal decoding method to minimize the probability of sequence error. However, it is not capable to provide the *a posteriori* probabilities required for iterative (turbo) decoding. An algorithm capable of such is the Bahl algorithm, that has been modified by Berrou for decoding recursive systematic convolutional codes. The reader is invited to look into references [1] and [3] for details.

6.10 SIGNAL SPACE CODES

6.10.1 Introduction

Up to now coding has been shown as a means of improving performance (*ber*) at the cost of bandwidth. Can performance be improved without bandwidth increase?

The answer may be found by considering coding and modulation simultaneously. Coding adds more digits to the information digits. If the modulation technique provides more signals than those strictly required to transmit the source digits, signal sequences may be chosen to increase the Euclidean distance among them.

To exploit signal space redundancy the modulator requires memory. The encoding process is known as a signal space code.

Assume that k bits per symbol must be transmitted and that 2^n symbols, $(n > k)$ are available. Then the rate of the signal space code is $r_c = k/n$. Provided the dimensionality of the signal space is the same for the 2^k and the 2^n signals, no bandwidth increase is needed.

Decoding is based on the Viterbi algorithm and the relevant performance parameter is the minimum Euclidean distance between coded sequences of signals.

Consider a given binary stream modulated in 4-PSK, which occupies a bandwidth b_{rf}. If a $r_c = 2/3$ code is applied to the binary stream and the transmission time is kept constant, then the bit rate and the bandwidth must be increased by a factor of $1/r_c = 3/2$. Bandwidth increase is the inevitable result of applying coding while keeping the transmission time constant. Since noise increases with bandwidth, if transmitter power is kept constant, the received coded stream will contain more errors than the uncoded one. So there will be a coding gain if, and only if, the code can correct more errors than those due to the bandwidth increase.

Let us now assume than we use an 8-PSK modulation. For an equal bit rate 8-PSK has 2/3 of the bandwidth of 4-PSK, thus we may apply a a 2/3 rate code, use an 8-PSK modulation and keep the same bandwidth as the uncoded bit stream and 4-PSK. Since, for the same transmitter power, 8-PSK provides a higher error ratio than 4-PSK we will have a positive coding gain if and only if the code offsets the higher error rate due to the adoption of 8-PSK instead of the original 4-PSK. If no coding was used, the adoption of 8-PSK would entail a 2/3 bandwidth and noise reduction (equivalent to -1.76 dB). Since 4-PSK and 8-PSK, for the same error ratio, differ by about 5 dB in carrier-to-noise ratio, the 2/3 rate code would provide an improvement if its coding gain is higher than about $5 - 1.76 = 3.24$ dB.

6.10.2 Ungerboeck codes

Let us consider that the k binary digits are to be assigned to a signal belonging to a set of 2^{k+1} signals. The code rate will be $r_c = \frac{k}{k+1}$. The problem is now to map each sequence of k information digits into the redundant set of channel signals. This mapping should be such that it maximizes the Euclidean distance between sequences of channel signals.

A possible approach is mapping by set partitioning. It is based on the division of the signal set into subsets of increasing minimum distance.

An example will be used to describe set partitioning. Assume we want to transmit 2 binary digits per symbol ($k = 2$) with 8-PSK. We have $r_c = 2/3$. The partitioning of the signal space into subsets is shown in Figure 6.25.

For unity symbol energy the original signal set A_0 has a minimum distance d_0 given by

$$\begin{aligned} d_0 &= 2\sin\left(\frac{\pi}{8}\right) \\ &= \sqrt{2-\sqrt{2}}. \end{aligned}$$

The first subsets B_0 and B_1 have a minimum distance d_1

$$d_1 = \sqrt{2},$$

whereas the second subsets C_0, C_1, C_2 and C_3 have a minimum distance $d_2 = 2$.

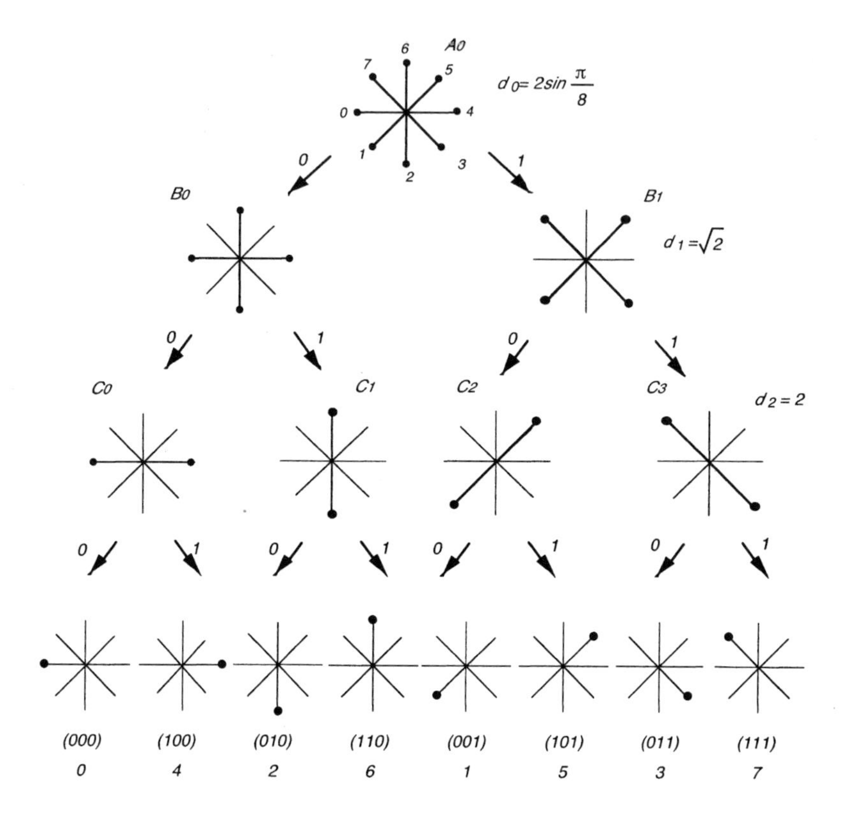

Fig. 6.23 Set partitioning for 8-PSK.

Let us start by using only the first subset, B_0 and B_1. The encoder will have two states S_1 and S_2 and its state diagram is shown in Figure 6.24. Signals from subset B_0 are assigned to the transitions from S_1 and those from B_1 to the transitions from S_2. Transition rules are chosen to correspond to pairs of signals with the largest possible Euclidean distance. That is, 2 and 6, on the one hand, and 1 and 5, on the other hand, correspond to transitions from S_1 to S_2 and from S_2 to S_1, respectively. The coding table derived from the state diagram is given in Table 6.6. Each path in the state diagram is labeled with the input (in binary digits) and the output stream (in octal).

Using the state diagram one finds that the input stream [00 01 01 10 10 00] is coded as [000 010 011 101 100 000].

To find out what we have gained let us consider the trellis in Figure 6.25.

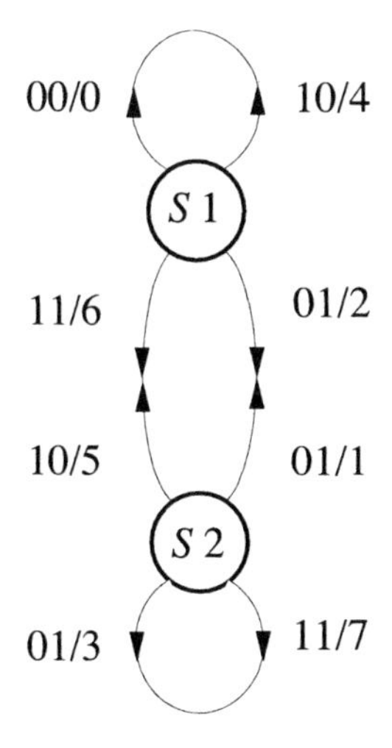

Fig. 6.24 State diagram for the first subset of the 8-PSK partitions.

Data	**State**	
	S_1	S_2
0 0	000 (0)	001 (1)
0 1	010 (2)	011 (3)
1 0	100 (4)	101 (5)
1 1	110 (6)	111 (7)

Table 6.6 Coding table for an Ungerboeck code using the first subset of the 8-PSK partitions.

The minimum distance between paths on the trellis is the distance between paths $S_1S_1S_1$ and $S_1S_2S_1$ which, from Figure 6.25 is the distance between 0 and 2 plus the distance between 1 and 0, that is, $d_{min} = \sqrt{d_0^2 + d_1^2} = 1.608$. For the uncoded reference system (4-PSK) $d_{ref} = \sqrt{2}$. Thus the coding gain is $10\log_{10}(d_{min}^2/d_{ref}^2)$ or 1.1 dB, which is rather low, for a code rate $r_c = 2/3$.

Increasing the number of states leads to higher coding gains as we will show next.

Assume now that we have four states S_1, S_2, S_3 and S_4, whose transitions correspond to the signals of subsets C_1, C_2, C_3 and C_4 (Figure 6.26).The coding table is given in Table 6.7.

As an exercice the reader may check out that the output sequence corresponding to the same input sequence as before ([00 01 01 10 10 00]) is now [000 010 011 111 110 000].

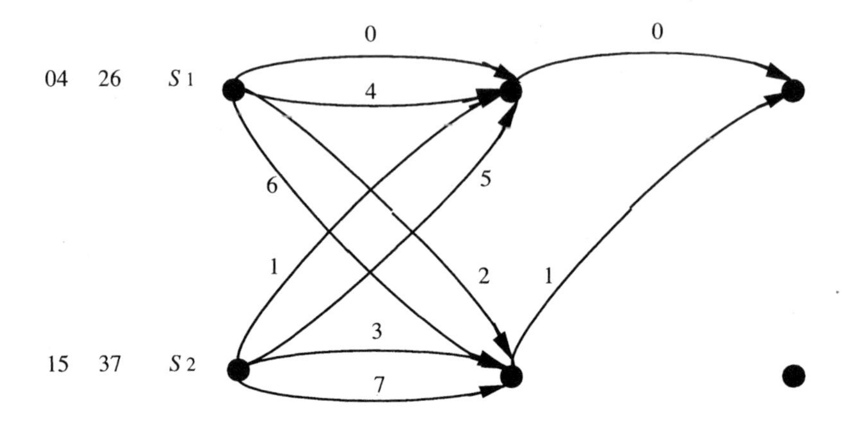

Fig. 6.25 Trelllis for the first subset of the 8-PSK partitions. Transitions are identified by the output produced (in octal).

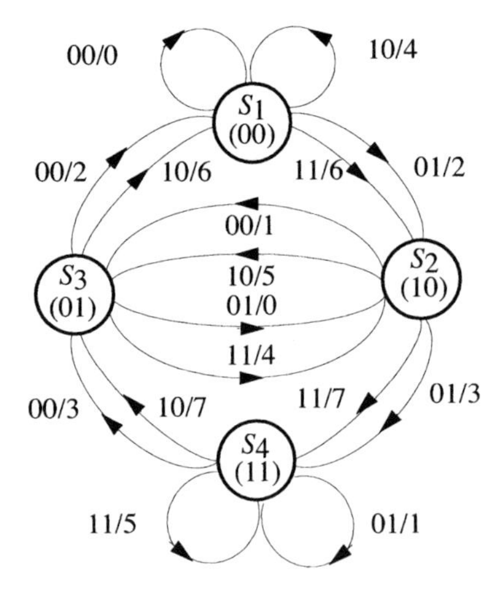

Fig. 6.26 State diagram for the second subset of the 8-PSK partitions.

The trellis corresponding to the four-state coder is shown in Figure 6.27.

Now the minimum distance between paths on the trellis is between paths $S_1S_1S_1S_1$ and $S_1S_2S_3S_1$, which from Figure 6.27, is the distance between 0 and 2 plus the distance between 2 and 1 plus the distance between 1 and 0, that is, $d_{min} = \sqrt{d_1^2 + d_0^2 + d_0^2}$ or $d_{min} = \sqrt{6 - 2\sqrt{2}} = 1.78$.

Data	State			
	S_1	S_2	S_3	S_4
0 0	000 (0)	001 (1)	100 (4)	101 (5)
0 1	010 (2)	011 (3)	000 (0)	001 (1)
1 0	100 (4)	101 (5)	110 (6)	111 (7)
1 1	110 (6)	111 (7)	010 (2)	011 (3)

Table 6.7 Coding table for an Ungerboeck code applied to the second subset of the 8-PSK partitions.

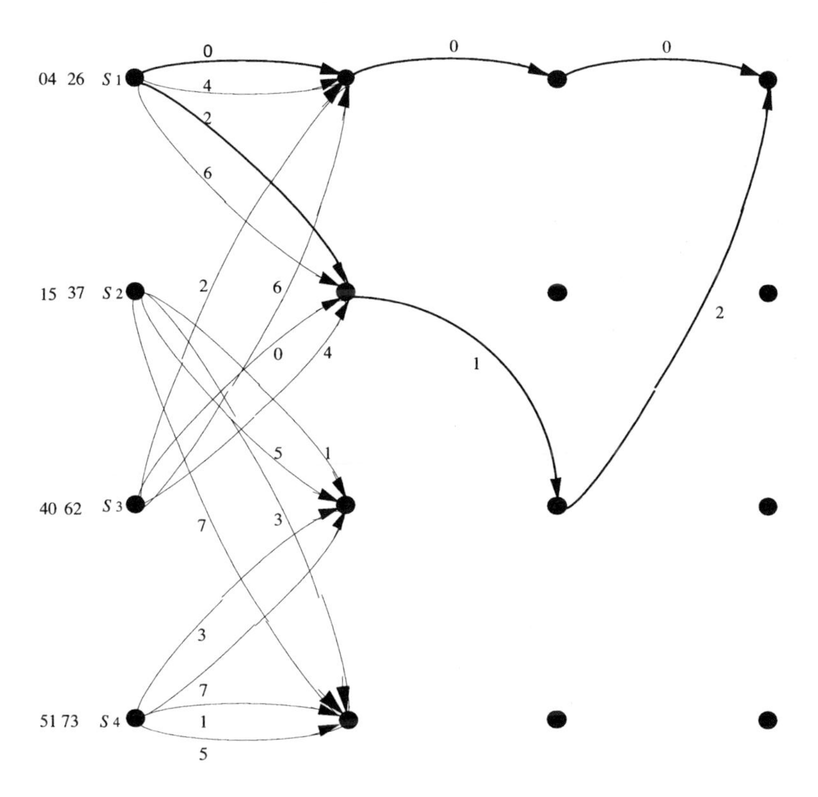

Fig. 6.27 Trellis for the second subsets of the 8-PSK partitions. Transitions are identified by the output produced (in octal).

Since for the uncoded reference system (4-PSK) $d_{ref} = \sqrt{2}$, the coding gain is $10 \log_{10}(d_{min}^2/d_{ref}^2)$ or 2 dB, for the same code rate $r_c = 2/3$.

A still higher coding gain could be achieved if we increased the number of states to the maximum possible, that is, 8. The reader is invited to find out that in this case the coding gain would be 3 dB, always for the same code rate $r_c = 2/3$.

6.11 ITERATIVE SOFT DECODING

6.11.1 Introduction

So far we have considered decoding of block codes as a "single pass" operation, that is, each set of n bits has produced a set of k bits regardless error probabilities. However, as will be shown on the following it is possible to improve on this performance by iterative soft decoding, also known as turbo decoding.

6.11.2 Bayes theorem

The joint probability of events A and B, $p(A,B)$, may be written in terms of the "conditional" probability of event A given event B, $p(A|B)$ or in terms of the "conditional" probability of event B given event A, $p(B|A)$

$$\begin{aligned} p(A,B) &= p(A|B)\,p(B) \\ &= p(B|A)\,p(A). \end{aligned}$$

Thus

$$p(A|B) = \frac{p(B|A)p(A)}{p(B)}. \tag{6.56}$$

If we have an additive white Gaussian noise (AWGN) channel whose output is a continuous random variable x (data and noise), the probability of receiving a data signal $d = i$ is given from Bayes theorem as

$$p(d = i|x) = \frac{p(x|d = i)p(d = i)}{p(x)} \qquad i = 1, \ldots, m. \tag{6.57}$$

6.11.3 Maximum likelihood decision

Assume that data signal d may take only the values -1 and $+1$. The probability density function of the channel when there is no signal is the usual $(\sigma = 1)$

$$p(x) = \frac{1}{\sqrt{2\pi}} e^{-x^2/2}. \tag{6.58}$$

The likelihood of x given $d = +1$, $p(x|d = +1)$, and given $d = -1$, $p(x|d = -1)$ are, respectively,

$$p(x) = \frac{1}{\sqrt{2\pi}} e^{-(x+1)^2/2},$$

and

$$p(x) = \frac{1}{\sqrt{2\pi}} e^{-(x-1)^2/2}.$$

Given the received signal x the maximum likelihood decision is to choose $d = +1$ if $p(x|d = +1) > p(x|d = -1)$ and $d = -1$ otherwise.

6.11.4 Maximum *a posteriori* decision

The maximum *a posteriori* decision takes into account the *a priori* probabilities that is, one should choose $d = +1$ if $p(d = +1|x) > p(d = -1|x)$ or $\frac{p(d=+1|x)}{p(d=-1|x)} > 1$ and $d = -1$ otherwise.

Using Bayes theorem, the maximum *a posteriori* decision rule to choose $d = +1$ may be written as

$$p(x|d = +1)p(d = 1) > p(x|d = -1)p(d = -1),$$

or, as a ratio,

$$\frac{p(x|d = +1)p(d = +1)}{p(x|d = -1)p(d = -1)} > 1.$$

This ratio is known as the *a posteriori* odds ratio. The ratio

$$\frac{p(x|d = +1)}{p(x|d = -1)}$$

is the likelihood ratio.

6.11.5 Log-likelihood ratio

Taking the logarithm of the likelihood ratio we get

$$\begin{aligned} L(d|x) &= \log\left[\frac{p(d = +1|x)}{p(d = -1|x)}\right] \\ &= \log\left[\frac{p(x|d = +1)p(d = +1)}{p(x|d = -1)p(d = -1)}\right] \\ &= \log\left[\frac{p(x|d = +1)}{p(x|d = -1)}\right] + \log\left[\frac{p(d = +1)}{p(d = -1)}\right] \\ &= L_c(x) + L(d), \end{aligned} \tag{6.59}$$

where $L_c(x)$ is the log-likelihood ratio measured at the detector under the alternate conditions that $d = +1$ or $d = -1$ may have been transmitted and $L(d)$ is the *a priori* log-likelihood ratio of data bit d.

If a systematic code is used, it may be proven that the log-likelihood ratio at the decoder output (soft decoding) is

$$L(d|x) = L_c(x) + L(d) + L_e(x), \tag{6.60}$$

where the new term $L_e(x)$ is the extrinsic log-likelihood ratio that represents the extra knowledge obtained from the decoding process. The soft decision $L(d|x)$ is s real number whose sign denotes the hard decision (if positive $d = +1$) and whose the magnitude is a measure of the confidence in the decision.

The following algorithm is used for decoding:

1. If no *a priori* knowledge is available, assume equally probable symbols, that is
$$L(d) = 0.$$
2. Decode by rows and get
$$L_{er}(x) = L(d|x) - L_c(x) - L(d). \tag{6.61}$$
3. Make
$$L(d) = L_{er}(x).$$
4. Decode by columns and get
$$L_{ec}(x) = L(d|x) - L_c(x) - L(d). \tag{6.62}$$
5. Make
$$L(d) = L_{ec}(x).$$
6. If a reliable estimation is not yet obtained go back to step 2 and repeat. Otherwise the output is
$$L(d|x) = L_c(x) + L_{er}(x) + L_{ec}(x).$$

Iterative soft decoding, as described, is also known as turbo decoding.

We will now apply this technique to the product code described in Section 6.4.8 and work out an example described in [14] for the product code detector, in an additive white Gaussian noise channel, where

$$\begin{aligned}
L_c(x|d) &= \log_e \left[\frac{p(x|d=+1)}{p(x|d=-1)}\right] \\
&= \log_e \left[\frac{\frac{1}{\sigma\sqrt{2\pi}} e^{-\frac{1}{2}(\frac{x-1}{\sigma})^2}}{\frac{1}{\sigma\sqrt{2\pi}} e^{-\frac{1}{2}(\frac{x+1}{\sigma})^2}}\right] \\
&= -\frac{1}{2}\left(\frac{x-1}{\sigma}\right)^2 + \frac{1}{2}\left(\frac{x+1}{\sigma}\right)^2 \\
&= \frac{2x}{\sigma}.
\end{aligned}$$

To simplify notation $L_c(x|d)$ will henceforth be simply denoted as $L_c(x)$. We will also assume that $\sigma = 1$.

Following [14] we will assume that the data bits are [1001], leading to the code word [10011111], which expressed in terms of a bipolar voltage means: $+1, -1, -1, +1, +1, +1, +1, +1$. Assume further the following channel values $\mathbf{x} = 0.75, 0.05, 0.10, 0.15, 1.25, 1.0, 3.0, 0.5$, which, for convenience, will be organized as a 2×2 data matrix:

$$\mathbf{d} = \begin{bmatrix} 0.75 & 0.05 \\ 0.1 & 0.15 \end{bmatrix},$$

and a 2×2 parity matrix

$$\mathbf{p} = \begin{bmatrix} 1.25 & 1.0 \\ 3.0 & 0.5 \end{bmatrix}.$$

Since $L_c(x) = 2x$ we get, organized as data and parity matrixes:

$$\mathbf{L}_c(d) = \begin{bmatrix} 1.5 & 0.1 \\ 0.2 & 0.3 \end{bmatrix},$$

$$\mathbf{L}_c(p) = \begin{bmatrix} 2.5 & 2.0 \\ 6.0 & 1.0 \end{bmatrix}.$$

If decisions were taken either on $\mathbf{d}$ or on $\mathbf{L}_c(d)$ the second or the third bit would be incorrectly decoded as 1 rather than as 0. The product code would then detect the errors but would not be able to correct them.

In the following we will see that soft iterative decoding can better this performance. However, before proceeding we need to be know how to compute the sum of two log-likelihood ratios.

6.11.6 Log-likelihood algebra

First take

$$\begin{aligned} L(d) &= \log_e \left[\frac{p(d=+1)}{p(d=-1)}\right] \\ &= \log_e \left[\frac{p(d=+1)}{1-p(d=+1)}\right], \end{aligned}$$

or

$$e^{L(d)} = \left[\frac{p(d=+1)}{1-p(d=+1)}\right],$$

or still

$$\begin{aligned} e^{L(d)} - e^{L(d)}p(d=+1) &= p(d=+1), \\ e^{L(d)} &= p(d=+1)(1+e^{L(d)}), \end{aligned}$$

hence

$$p(d=+1) = \frac{e^{L(d)}}{1+e^{L(d)}},$$

and

$$\begin{aligned} p(d=-1) &= 1-p(d=+1) \\ &= 1-\frac{e^{L(d)}}{1+e^{L(d)}} \\ &= \frac{1}{1+e^{L(d)}}. \end{aligned}$$

We can now proceed and define the sum of two log-likelihood ratios $L(d_1)$ and $L(d_2)$ as

$$\begin{aligned} L(d_1) \uplus L(d_2) &= \log_e \left[\frac{p(d_1 \neq d_2)}{p(d_1 = d_2)}\right] \\ &= \log_e \left[\frac{p(d_1=+1)p(d_2=-1)+p(d_1=-1)p(d_2=+1)}{p(d_1=+1)p(d_2=+1)+p(d_1=-1)p(d_2=-1)}\right] \\ &= \log_e \left\{\frac{p(d_1=+1)p(d_2=-1)+[1-p(d_1=+1)][1-p(d_2=-1)]}{p(d_1=+1)p(d_2=+1)+[1-p(d_1=+1)][1-p(d_2=+1)]}\right\} \\ &= \log_e \left[\frac{\frac{e^{L(d_1)}}{1+e^{L(d_1)}}\frac{1}{1+e^{L(d_2)}}+\frac{1}{1+e^{L(d_1)}}\frac{e^{L(d_2)}}{1+e^{L(d_2)}}}{\frac{e^{L(d_1)}}{1+e^{L(d_1)}}\frac{e^{L(d_2)}}{1+e^{L(d_2)}}+\frac{1}{1+e^{L(d_1)}}\frac{1}{1+e^{L(d_2)}}}\right] \\ &= \log_e \left\{\frac{\frac{e^{L(d_1)}+e^{L(d_2)}}{[1+e^{L(d_1)}][1+e^{L(d_2)}]}}{\frac{e^{L(d_1)}e^{L(d_2)}+1}{[1+e^{L(d_1)}][1+e^{L(d_2)}]}}\right\} \\ &= \log_e \left[\frac{e^{L(d_1)}+e^{L(d_2)}}{1+e^{L(d_1)}e^{L(d_2)}}\right]. \end{aligned}$$

From the above it follows that if $L(d_2) = \infty$ then:

$$\begin{aligned} p(d_2=+1) &= 1, \\ p(d_2=-1) &= 0. \end{aligned}$$

Hence,

$$L(d) \uplus \infty = -L(d). \tag{6.63}$$

Similarly, it may be proven that

$$L(d) \uplus 0 = 0. \tag{6.64}$$

The following approximate expressions may be derived

- If $L(d_1), L(d_2) > 0$ and $L(d_1) >> L(d_2)$ then

$$L(d_1) \uplus L(d_2) \approx -L(d_2);$$

- If $L(d_1), L(d_2) > 0$ and $L(d_2) >> L(d_2)$ then

$$L(d_1) \uplus L(d_2) \approx -L(d_1);$$

- If $L(d_1), L(d_2) < 0$ and $|L(d_1)| >> |L(d_2)|$ then

$$L(d_1) \uplus L(d_2) \approx L(d_2);$$

- If $L(d_1), L(d_2) < 0$ and $|L(d_2)| >> |L(d_1)|$ then

$$L(d_1) \uplus L(d_2) \approx L(d_1).$$

It is now possible to write the following expression for the general case:

$$L(d_1) \uplus L(d_2) \approx -\text{sign}[L(d_1)]\text{sign}[L(d_2)] \min[|L(d_1)|, |L(d_2)|]. \tag{6.65}$$

6.11.7 Computing extrinsic log-likelihoods

For the product code the soft decoder output corresponding to data d_i is $L_{sd}(d_i)$ from (6.60)

$$L_{sd}(d_i) = L_c(x_i) + L(d_i) + \{[L_c(x_j) + L(d_j)] \uplus L_c(x_{i,j}\}, \tag{6.66}$$

where:

- $L_c(x_i)$, $L_c(x_j)$, $L_c(x_{i,j})$ are the channel log-likelihood ratios of the received signals corresponding to data bits d_i and d_j and to parity bit $p_{i,j}$;
- $L(d_i)$ and $L(d_j)$ are log-likelihood ratios of the *a priori* probabilities of data bits d_i and d_j;
- $\{[L_c(x_j) + L(d_j)] \uplus L_c(x_{i,j}\}$ is the extrinsic log-likelihood ratio contribution from the code.

Equation (6.66) may be understood as follows: soft decoder output $L_{sd}(d_1)$ is the result of information provided by sample x_1 plus the extrinsic information provided by both data d_2 and parity $p_{1,2}$.

We are now ready to compute the extrinsic log-likelihood ratios, according to the algorithm described in Section 6.11.5. Firstly we recall the input log-likelihood ratios, already computed in Section 6.11.5

$$\mathbf{L}_c(d) = \begin{bmatrix} 1.5 & 0.1 \\ 0.2 & 0.3 \end{bmatrix},$$

$$\mathbf{L}_c(p) = \begin{bmatrix} 2.5 & 2.0 \\ 6.0 & 1.0 \end{bmatrix}.$$

To start with, the external likelihood will be taken as zero:

$$\mathbf{L}(d) = \begin{bmatrix} 0 & 0 \\ 0 & 0 \end{bmatrix}.$$

Next, we compute new values of $\mathbf{L}(d)$ based on row parities:

$$\begin{aligned} L_{er}(d_1) &= [L_c(x_2) + L(d_2)] \uplus L_c(p_{1,2}), \\ L_{er}(d_2) &= [L_c(x_1) + L(d_1)] \uplus L_c(p_{1,2}), \\ L_{er}(d_3) &= [L_c(x_4) + L(d_4)] \uplus L_c(p_{3,4}), \\ L_{er}(d_4) &= [L_c(x_3) + L(d_3)] \uplus L_c(p_{3,4}), \end{aligned}$$

and, using the exact expression (6.62), get

$$\mathbf{L}_{er} = \begin{bmatrix} -0.0848 & -1.2049 \\ -0.2278 & -0.1521 \end{bmatrix}.$$

Then we update the values of $\mathbf{L}(d)$ using $\mathbf{L}_{er}(d)$ and compute new values of $\mathbf{L}(d)$ based on column parities

$$\begin{aligned} L_{ec}(d_1) &= [L_c(x_3) + L(d_3)] \uplus L_c(p_{1,3}), \\ L_{ec}(d_2) &= [L_c(x_4) + L(d_4)] \uplus L_c(p_{2,4}), \\ L_{ec}(d_3) &= [L_c(x_1) + L(d_1)] \uplus L_c(p_{1,3}), \\ L_{ec}(d_4) &= [L_c(x_2) + L(d_2)] \uplus L_c(p_{2,4}). \end{aligned}$$

Using the exact expression (6.62) we get

$$\mathbf{L}_{ec} = \begin{bmatrix} -0.0276 & -0.0682 \\ -1.4056 & 0.4729 \end{bmatrix}.$$

Next we update again $\mathbf{L}(d)$, now using $\mathbf{L}_{ec}(d)$.

Decoder output which started as $\mathbf{L}_c(d)$ after adding external likelihoods based on rows becomes

$$\mathbf{L}_c + \mathbf{L}_{er} = \begin{bmatrix} 1.4152 & -1.1049 \\ -0.0278 & 0.1479 \end{bmatrix}.$$

The decoder has already the correct data bits (+1, −1, −1, +1) but with little confidence on the last two bits. Adding the external likelihoods based on columns

$$\mathbf{L}_c + \mathbf{L}_{er}\mathbf{L}_{ec} = \begin{bmatrix} 1.4428 & -1.1731 \\ -1.4434 & 0.66208 \end{bmatrix}.$$

Repeating the whole process again we finally get

$$\mathbf{L}_c + \mathbf{L}_{er} + \mathbf{L}_{ec} = \begin{bmatrix} 1.8476 & -1.6218 \\ -1.8393 & 1.6526 \end{bmatrix}.$$

The soft decoder output remains the same as before but the confidence builds up and the procedure may be stopped. The soft decoder has improved on the hard decoder which could detect, but not correct, the double error.

The reader is invited to find out that the decoder output would be exactly the same using the approximate formulas. The latter are obviously to be preferred given their much simpler implementation.

6.12 PERFORMANCE LIMITS

6.12.1 Shannon's theorem

The basic theory on coding for bandwidth limited channels derives from the fundamental work of Claude Shannon who, in 1948, published the first of a series of seminal papers on the capacity (maximum transmission rate achievable with negligible error probability) of bandwidth limited channels.

Shannon's theorem [10] [11] states that the capacity c (in binary digits per second or bits/s) of a continuous-input band-limited white Gaussian noise channel, with bandwidth w (in Hz) is given by

$$c \leq w \log_2 \left(1 + \frac{s}{n}\right), \tag{6.67}$$

where s/n is the channel signal-to-noise ratio. Channel capacity may be reached with an arbitrarily low bit error ratio, if suitable coding is employed.

Since all modulation schemes we have dealt with so far are prone to errors, Shannon's theorem, stating that transmission is possible with an arbitrarily low number of errors (provided appropriate coding is used), may come as a surprise. In the following we will try to better our understanding of this fundamental statement.

Denoting bit frequency as f_b, average energy per bit as e_b and noise per unit bandwidth as n_0, Shannon's theorem may be rewritten as

$$c \leq w \log_2 \left(1 + \frac{e_b}{n_0}\frac{f_b}{w}\right). \tag{6.68}$$

6.12.2 Shannon's limit and coding gain

Coding gain for a given bit error ratio will be used to denote the signal-to-noise ratio difference (in dB) between a coded and an uncoded scheme.

For small values of the signal-to-noise ratio we have from (6.67) that

$$c \leq w \frac{\frac{s}{n}}{\log_e(2)}, \tag{6.69}$$

or

$$c \leq \frac{f_b}{\log_e(2)} \frac{e_b}{n_0}. \tag{6.70}$$

Making the bit frequency equal to the channel capacity ($f_b = c$) equation (6.70) yields

$$e_b/n_0 \geq \log_e(2), \tag{6.71}$$

or, in logarithmic units,

$$E_b/N_0 \geq -1.6 \text{ (dB)}. \tag{6.72}$$

Inequalities (6.71) and (6.72) indicate that reliable transmission, that is, error free transmission, imposes a minimum value of e_b/n_0 and provide an upper limit for coding gain. Let us now compare this limit with with the performance of polar transmission for which we had (5.95)

$$ber = \frac{1}{2} erfc\left(\sqrt{\frac{e_b}{n_0}}\right).$$

Figure 6.28 shows the bit error ratio as a function of E_b/N_0 for polar transmission and Shannon's limit $E_b/N_0 = -1.6$ dB. Maximum coding gain, for each value of the bit error ratio, is the difference between the corresponding value of E_b/N_0 and -1.6 dB. For instance, for $ber = 10^{-6}$ maximum coding gain is 11.9 dB.

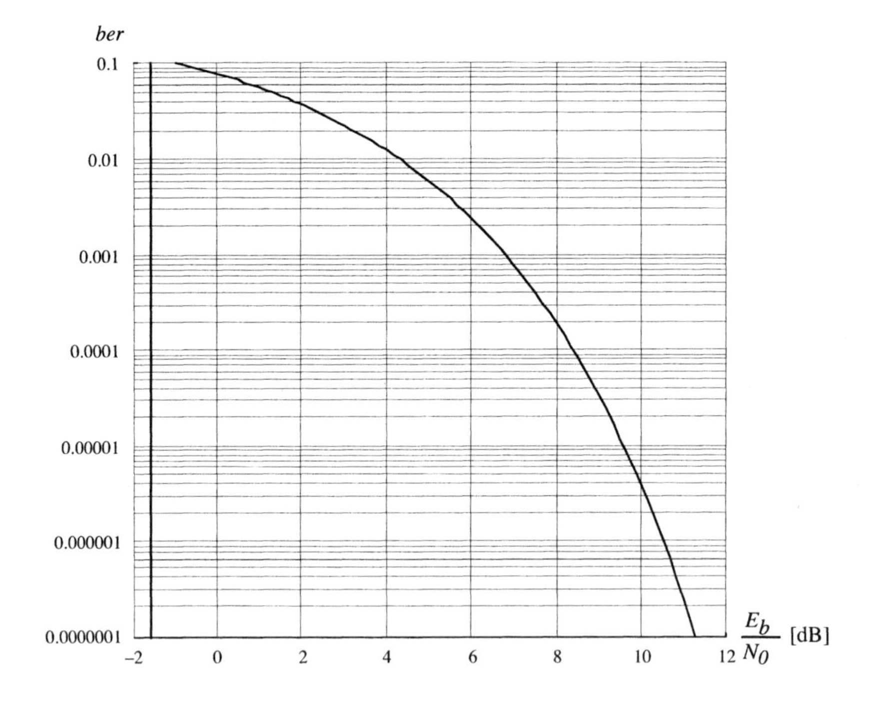

Fig. 6.28 Bit error rate as a function of E_b/N_0 for polar transmission and Shannon's limit.

6.12.3 Shannon's limit and code rate

Another way to look into Shannon's theorem is in terms of code rate. Recalling that for a bandwidth limited channel Nyquist theorem states that the maximum signalling rate (with no inter-symbol interference) is given by twice the bandwidth, the maximum channel code rate r_{max} is

$$r_{max} = \frac{c}{2w}. \tag{6.73}$$

Substituting (6.73) in (6.68) we get

$$r_{max} \leq \frac{1}{2}\log_2\left(1 + 2r_c\frac{e_b}{n_0}\right). \tag{6.74}$$

For $r_c e_b/n_0 << 1$ we get

$$\frac{e_b}{n_0} \geq log_e(2)\frac{r_{max}}{r_c},$$

or, in dB,

$$\frac{E_b}{N_0} \geq -1.6 + 10\log_{10}\left(\frac{r_{max}}{r_c}\right). \tag{6.75}$$

If in (6.74) we substitute the code rate r_c for the maximum code rate r_{max} we obtain an expression for the maximum code rate as a function of the channel e_b/n_0 ratio

$$\frac{e_b}{n_0} \geq \frac{2^{2r_{max}} - 1}{2r_{max}}, \tag{6.76}$$

or

$$\frac{E_b}{N_0} \geq 10\log_{10}\left(\frac{2^{2r_{max}} - 1}{2r_{max}}\right). \tag{6.77}$$

Expression (6.77) is plotted as the lower curve in Figure 6.29, from where we see that the maximum code rate varies from a minimum of 0, for $E_b/N_0 = -1.6$ dB to 1, for $E_b/N_0 = 1.76$ dB.

It is possible [7] to derive a random coding bound by calculating the average sequence error rate *ser* over the ensemble of all possible block codes with a specified code rate and length. Assuming binary block codes, PSK modulation and an continuous-input additive white Gaussian noise channel we have [7]

$$ser \leq 2^{-(n-k)}\left[1 + \exp(-r_c\frac{e_b}{n_0})\right]^n. \tag{6.78}$$

Introducing the exponential bound parameter r_0,

$$r_0 = 1 - \log_2\left[1 + \exp\left(-r_c\frac{e_b}{n_0}\right)\right] \tag{6.79}$$

and recalling than $r_c = k/n$ (6.78) becomes

$$ser \leq 2^{-n(r_0 - r_c)}. \tag{6.80}$$

Equation 6.79 shows that as long as $r_0 > r_c$, that is, as long as the code rate is less than the exponential bound parameter, it is possible to decrease

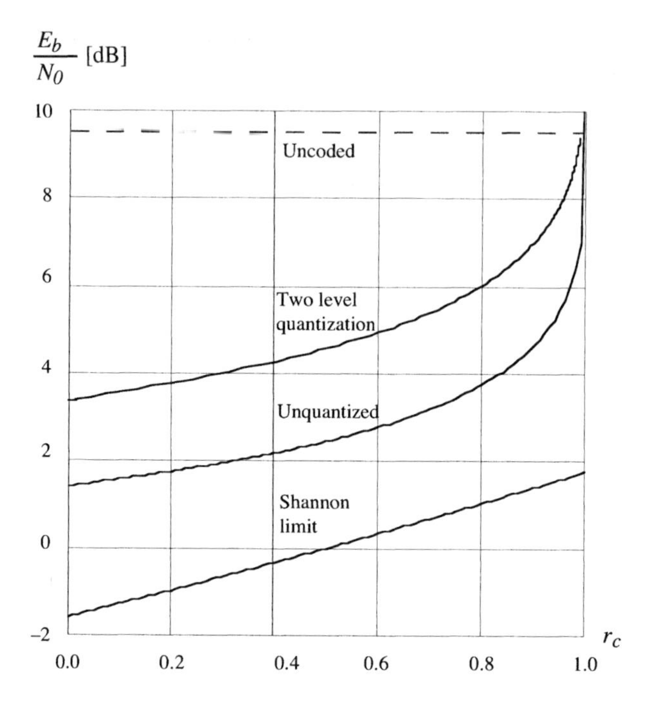

Fig. 6.29 Minimum value of E_b/N_0 in dB as a function of the code rate r_c.

ser by increasing the block length n. Using more sophisticated techniques it is possible to derive tighter bounds than the one given in (6.80), although at the lower code rates (about below 0.4) differences between these bounds are insignificant.

In the limit when r_c tends to r_0 (and the block length tends to infinity) we get

$$r_c = 1 - \log_2 \left[1 + \exp(-r_c \frac{e_b}{n_0}) \right], \tag{6.81}$$

or

$$\frac{e_b}{n_0} \geq -\frac{\log_e(2^{1-r} - 1)}{r}, \tag{6.82}$$

which is yet another limit to the code rate as a function of e_b/n_0. This limit is also plotted in Figure 6.29 and labeled as unquantized.

For binary symmetric channels it is possible to show [14] that

$$r_0 = 1 - \log_2 \left[1 + 2\sqrt{ber(1 - ber)} \right]. \tag{6.83}$$

Assuming polar signalling, for which *ber* is given in (5.95) and repeated here for convenience

$$ber = \frac{1}{2} erfc\left(\sqrt{\frac{e_b}{n_0}}\right),$$

we may derive a performance limit for binary quantization. This is shown as the uppermost curve in Figure 6.29.

Let us now take a closer look to Figure 6.29. From (5.98) or Figure 5.21 we find out that for uncoded polar signalling with $E_b/N_0 = 9.55$ dB we get $ber = 10^{-5}$ (represented as an horizontal dashed line in this figure). Take now a given code rate, for instance, $r_c = 1/2$. We can achieve the same (or better) performance with unquantized decision using polar signalling with $E_b/N_0 = 2.46$ dB, which corresponds to a coding gain of 7.09 dB. Had we used two-level quantization for the same performance we would have required $E_b/N_0 = 4.59$ dB, or a coding gain of 4.96 dB. It is possible to show that the difference between unquantized decision and 8-level quantization amounts to a few tens of a dB. Thus, in practice, the number of quantization levels is usually limited to 8. Shannon's limit for $r_c = 1/2$ is $E_b/N_0 = 0$ dB, which indicates that yet higher coding gains (9.55 dB) are possible, if we do not limit ourselves to binary signals.

Figure 6.29 tells us that:

- coding gains increase with decreasing code rate;
- in spite of Shannon's limit promise of a significant coding gain at (almost) unity code rate, even unquantized decision requires a code rate well below unity to be of interest;
- it is not very rewarding to decrease the coding rate much below about 1/2 since the increase in coding gain is only marginal (about 1.4 dB);
- the difference between two-level quantization and unquantized is almost independent of the code rate.

6.12.4 Modulation efficiency

Let us now compare different modulation schemes in light of Shannon's theorem which, as we have have seen, may be written as (6.76). In the limit, when $r_{max} = r_c$ we have

$$2^{2r_{max}} = 1 + 2r_{max}\frac{e_b}{n_o}, \tag{6.84}$$

or

$$\frac{e_b}{n_0} = \frac{2^{2r_{max}} - 1}{2r_{max}}. \tag{6.85}$$

In the previous chapter, the concept of modulation efficiency $\eta_m = f_b/b_{rf}$ was used. This concept is easily related to the maximum code rate

$$\begin{aligned} r_{max} &= \frac{c}{2w} \\ &= \frac{\eta_m}{2}. \end{aligned} \tag{6.86}$$

Substituting (6.86) in 6.85 we get

$$\frac{e_b}{n_0} = \frac{2^{\eta_m} - 1}{\eta_m}. \tag{6.87}$$

In Figure 6.30 we plot Shannon's limit, η_m, as a function of E_b/N_0, in dB, as defined by (6.87) and the performance of the following modulation schemes, for $ber = 10^{-5}$, computed from equations derived in Chapter 5:

- FSK, with incoherent and coherent detection;
- m-PSK, with $m = 2, 4, 8$ and 16;
- $m - QAM$, with $m = 4, 16, 64, 256$ and 1024.

Analysis of Figure 6.30 shows that:

- incoherent FSK is less efficient than coherent FSK both in terms of power and bandwidth efficiency, although its demodulator is rather simpler;
- 2-PSK is more power efficient than FSK;
- m-PSK and m-QAM improve bandwidth efficiency and decrease power efficiency as m increases;
- m-QAM slowly approaches Shannon's limit as m increases.

6.12.5 Shannon's limit and block size

According to Shannon's theorem, channel capacity, besides assuming a continuous input, sets no limit on block size or, in other words, is only reached for code rates approaching zero. In practice, things may be different. Block size and encoder-decoder complexity have practical limits. Shannon [12] provides exact expressions from which it is possible to derive the channel capacity as a function of the code rate r_c and the number of information bits per block k. The reader is referred to [8] for full details. Here we simply include a plot of Shannon's limit as a function r_c and k.

For simplicity in Figure 6.31 the minimum value of E_b/N_0 as a function of k is ploted as a curve, although only points corresponding to integer values of n and k for each value of $r_c = k/n$ are meaningful.

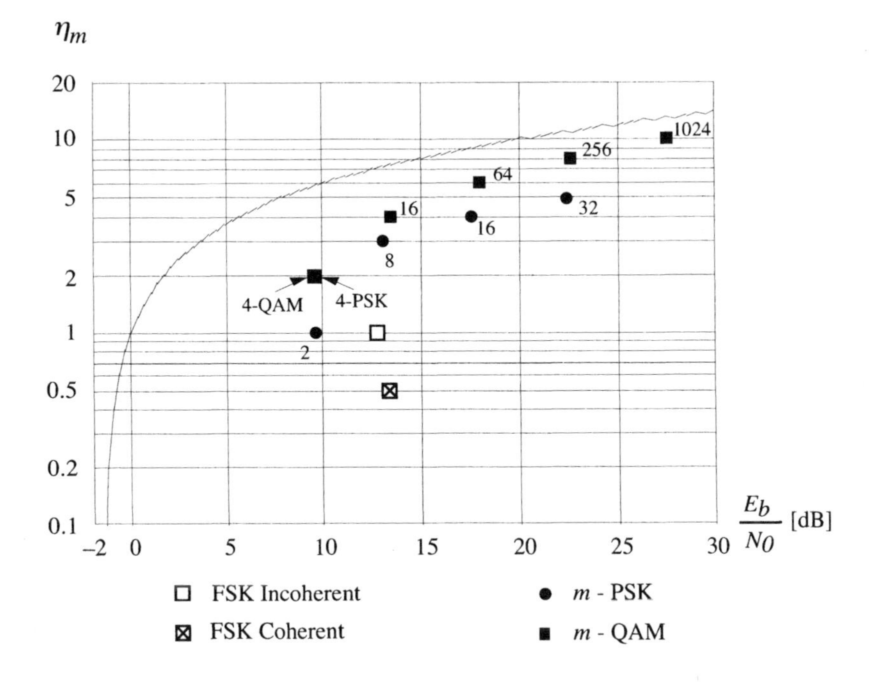

Fig. 6.30 Modulation efficiency η_m as a function of E_b/N_0, in dB, coherent and incoherent FSK, m-PSK, m-QAM and Shannon's limit.

Comparing Figure 6.31 and Figure 6.29 we realize that Shannon's limit, as previously defined, implies very large block sizes ($k > 10000$).

For smaller block sizes Shannon's limits are considerably higher. Take for instance $n = 20$ and $r_c = 1/2$, that is $k = 10$. From Figure 6.31 we get $E_b/N_0 = 5$ dB, rather than 0 dB (for $n \to \infty$), or than -1.59 dB (for $n \to \infty$ and $r_c \to 0$). A possibly even more interesting case is for $n = 2$ and $r_c = 1/2$, for which $E_b/N_0 = 8.4$ dB. Seen under this light, current un-encoded polar transmission is as efficient as can be.

The dependence on the code rate in Figure 6.31 may be practically eliminated, at least for $k \geq 10$ if instead of E_b/N_0 we plot the difference between E_b/N_0 for a given block size and code rate and E_b/N_0 as a function of the code rate given in (6.76). The result is shown in Figure 6.32. For most cases of interest, say $k \geq 10$, the curve is hardly dependent on the code rate and thus can be of universal use. For $k > 10$ the following approximation may be used (error less than about 0.2 dB)

$$\frac{E_b}{N_0} - 10\log_{10}\left(\frac{2^{2r_c}+1}{2r_c}\right) \approx \frac{6.7}{\log_{10}(k)} - 1.33.$$

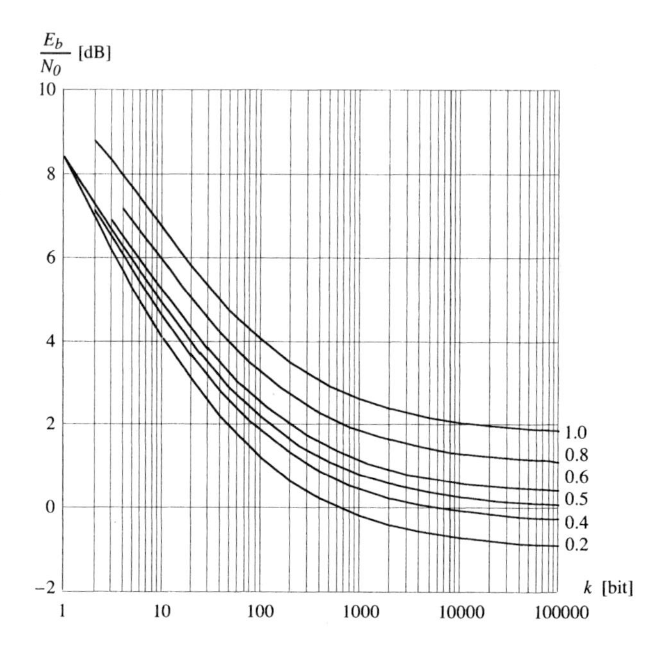

Fig. 6.31 Minimum value E_b/N_0, in dB, as a function of code rate r_c and number of information bits per block k, for a word error probability of 10^{-4}.

6.12.6 Shannon's limits for binary channels

Shannon theorem, as we have stated, assumes a continuous-input channel. In practice, however, many channels have a binary input. What consequence will this have on channel capacity?

The memoryless binary channel capacity c, in bits per symbol, when the bit error probability is p is given by [2]

$$c = 1 + p\log_2(p) + (1-p)\log_2(1-p).$$

Assume now a baseband channel with bandwidth w. According to the Nyquist theorem it is possible to transmit through this channel $2w$ samples per second. Thus the channel capacity c when the input values are constrained to be binary is

$$c = 2w[1 + p\log_2(p) + (1-p)\log_2(1-p)],$$

or, since $r = \frac{c}{2w}$,

$$r \leq 2[1 + p\log_2(p) + (1-p)\log_2(1-p)], \tag{6.88}$$

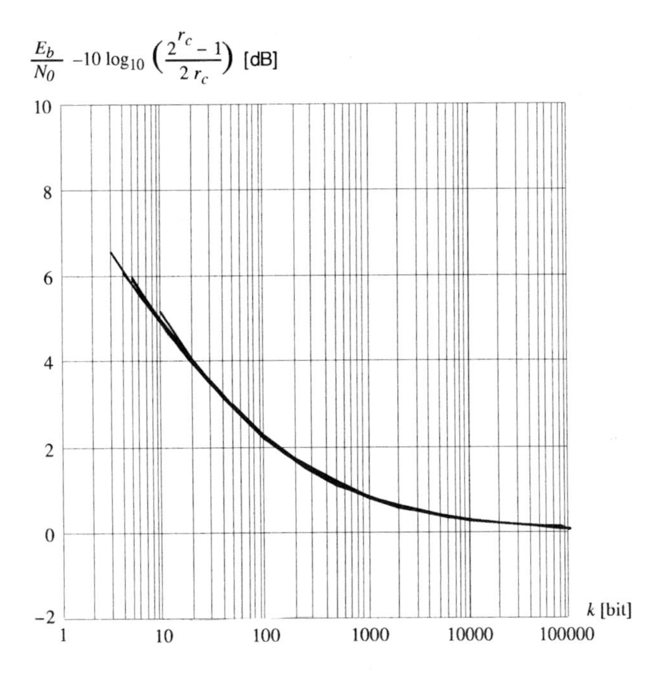

Fig. 6.32 Minimum value E_b/N_0, in dB, above the value given in (6.76) as a function of code rate r_c and the number of information bits per block k, for a word probability of 10^{-4}.

where the bit error probability is given by

$$p = ber = \frac{1}{2} erfc\left(\sqrt{r\frac{e_b}{n_0}}\right).$$

Substituting ber in (6.88) we get an implicit expression for r as a function of e_b/n_0. For $e_b/n_0 << 1$ we may expand the resulting expression in series and take only the first term, obtaining

$$r \leq \frac{2r}{\pi \log_e(2)} \frac{e_b}{n_0}.$$

Thus

$$\frac{e_b}{n_0} \geq \frac{\pi \log_e(2)}{2}, \tag{6.89}$$

or

$$\frac{E_b}{N_0} \geq 0.37 \text{ dB}. \tag{6.90}$$

Shannon's limit for the binary-input channel is thus 1.96 dB higher than for the continuous-input channel.

6.13 COMPARING CODE PERFORMANCES

Different code performances may be compared by reference to Shannon's limit for the same block size as suggested in [8].

Hamming (8,4) and the Golay (24,12) are nearly "perfect codes" its performance differing from Shannon's limit by less than 0.5 dB. Many turbo codes are also quite close to perfection, with differences from Shannon's limit under 1.0 dB. On the other hand, convolutional closed codes are rather good for small block sizes but their performance degrades considerably once block size increases beyond about 10. Interestingly, there is a gap in block sizes, from about $k = 50$ to $k = 500$ for which, apparently, no quite good codes yet exist.

REFERENCES

1. Bahl, L., et *al.*, Optimal Decoding of Linear Codes for Minimizing Symbol Error Rate, *IEEE Transactions on Information Technologies*, Vol. IT-20, March 1974, pp. 248–287

2. Benedetto, S., Biglieri, E. and Castellani, V., *Digital Transmission Theory*, Prentice-Hall International Editions, Englewood Cliffs, NJ, (1987)

3. Berrou, C., Glavieux, A. and Thitimajshima, Near Shannon Limit Error-Correcting Coding and Decoding: Turbo-Codes, *Proceedings of the IEEE ICC'93*, Geneve, May 1993, pp. 1064–1070

4. Biglieri, E., Divsalar, D., McLane, P. and Simon, M., *Introduction to Trellis-Coded Modulation with Applications*, Macmillan Publishing Company, New York, 1991

5. Blahut, R., *Theory and Practice of Error Control Codes*, Addison-Wesley, Reading, Mass., 1983

6. Blahut, R., *Digital Transmission of Information*, Addison-Wesley, Reading, MA., 1990

7. Clark, G. and Cain, J., *Error-Correction Coding for Digital Communications*, Plenum Press, New York, 1982

8. Dolinar, S., Divsalar, D. and Pollara, F., *Code Performance as a Function of Block Size*, TMO Progress Report, May 1998

9. Proakis, J., *Digital Communications*, McGraw-Hill Inc., Tokyo, 1983

10. Shannon, C. E., A Mathematical Theory of Communication: Part I, *Bell System Technical Journal*, Vol. 27, pp. 379–423.

11. Shannon, C. E., A Mathematical Theory of Communication: Part II, *Bell System Technical Journal*, Vol. 27, pp. 623–656.

12. Shannon, C. E., Probability of Error for Optimal Codes in a Gaussian Channel, *Bell System Technical Journal*, Vol. 38, pp. 611–656

13. Sklar, B., *Digital Communications. Fundamentals and applications*, Prentice-Hall International Editions, Englewood Cliffs, NJ, 1988

14. Sklar, B. , A Primer on Turbo Code Concepts, *IEEE Communications Magazine*, Dec. 1997, pp. 94–102

15. Stallings, W., *Data and Computer Communications*, 4th edition, Prentice-Hall International Inc., Englewood Cliffs, NJ, 1996

16. Tanenbaum, A. S. , *Computer Networks*, 3rd edition, Prentice-Hall International Inc., Englewood Cliffs, NJ, 1994

Problems

6:1 Assuming a symbol error ratc of 10^{-2}, calculate the probability that in a block of 100 symbols at least 1 is errored. What happens if the block size and the minimum number of errored blocks in increased tenfold?

6:2 Using the sphere-packing bound, calculate the maximum number of bits that a (7,4) code may correct.

6:3 Calculate the capacity of a continuous input, 3.1 kHz wide, white Gaussian noise channel, with a signal-to-noise ratio of 40 dB.

6:4 Calculate the minimum value of E_b/N_0 (in dB) for code rates r_c equal to 1/2, 2/3, 3/4 and 1 in a continuous input additive white Gaussian noise channel.

6:5 Calculate the exponential bound parameter r_0 for a binary block code with $r = 1/2$, PSK modulation and a continuous input additive white Gaussian noise channel with $E_b/N_0 = 4$ dB and estimate the block length to achieve a symbol error rate less or equal to 10^{-6}.

6:6 Calculate the minimum value of E_b/N_0 for a code rate $r = 1/2$.

6:7 Calculate the minimum value of E_b/N_0 for a code rate $r = 1/2$ with binary quantization.

6:8 Calculate the minimum value of E_b/N_0 for $r = 1/2$ and a block length $n = 200$ using the asymptotic bound.

6:9 Find the smallest block length Hamming code with $d_{min} = 3$ and $r \geq 0.8$.

6:10 Using code word [0111010] as a test, check that Hamming code (7,4) is a cyclic code.

6:11 Compute the syndrome polynomial of the received word [1010001] in the (7, 4) Hamming code. Assuming a single error compute the correct code word?

6:12 Write the generator matrix for the dual code of the (7, 4) Hamming code and find out the code word corresponding to $mathbfu = [101]$.

6:13 Calculate the bit error rate of a BCH (31,16) code for a bit error probability in the uncoded message of 10^{-2}.

6:14 Using CRC-16, a (16, 8) cyclic code, whose error polynomial is $d^{16} + d^{15} + d^2 + 1$, and the test word [11001010], show that can detect a 7-bit error burst.

6:15 Calculate the code word corresponding to [11001] for a convolutional code (3, 2), with constraint $L = 2$, whose generator matrixes are

$$\mathbf{G}_1 = \begin{bmatrix} 1 & 1 \\ 0 & 1 \\ 1 & 0 \end{bmatrix},$$

$$\mathbf{G}_2 = \begin{bmatrix} 0 & 1 \\ 1 & 0 \\ 1 & 1 \end{bmatrix}.$$

6:16 Using a trellis, draw the coding path of sequence [10101] for the (3, 1) convolutional code of Figure 6.7.

6:17 Calculate the bit error rate bound for a t error correcting convolutional $r_c = 1/2$ code with hard and with unquantized soft Viterbi decoding at $E_b/N_0 = 8$ dB.

6:18 Calculate the bit error rate bound for a concatenated code using an outer (7, 3) double error correcting Reed-Solomon code and an inner (7, 4) single error corrrecting Hamming code, when the inner decoder has a symbol error rate $ser = 10^{-3}$.

6:19 Consider the product code described in Section 6.4.8 and assume that sequence [10111111]] is received. Identify the message being transmitted assuming zero or one erorrs may occur.

6:20 Consider the product code described in Section 6.4.8 and assume the received values are $\mathbf{x} = 0.80, 0.02, 0.15, 0.05, 1.25, 1.0, 3.0, 0.2$. Identify the message using the algorithm described in Section 6.11.5 and the approximate expressions to compute the sum of log-likelihood ratios.

7

Link Design

7.1 INTRODUCTION

Link design may be defined as the choice of path and operating frequency plus the specification of the main components required to transmit a given signal, with a predefined quality of service, between two terminals, at the lowest possible cost. It is a difficult or even an impossible task to deal in full detail with the design of a microwave radio link in a book such as this one for quite a few reasons among which:

- the limited number of examples that can be included;
- the large number of factors that must be considered and the varying importance of each factor from case to case;
- the speed of technology changes;
- the constant variation in costs.

Besides technical know-how, the link designer must be familiar with many implementation and exploitation details and make use of lots of common sense, all of which normally require several years of practice.

In spite of the aforementioned shortcomings, this chapter introduces issues, which are vital for a successful link design, such as cost and reliability and thus provides a link between the theoretical and empirical matters of previous chapters and their application.

7.2 COST

In this context cost means total cost, a much broader concept than the simple cost of equipment. Two items are usually identified in cost:

- initial cost,
- exploitation cost.

Initial cost includes not only design and equipment cost, but also infrastructure costs (land rights, buildings, access roads, energy supply), cost of training staff, both for operation and maintenance and, when applicable, cost for the first set of spare parts.

Exploitation cost includes cost of exploitation (staff salaries and social contributions, energy, consumables), maintenance costs (staff and spare parts), financial costs (interest and payback of loans, when applicable) and fiscal costs (at both national and local levels). Exploitation cost excludes the depreciation of equipment and buildings for fiscal reasons.

In a project dues are the counterpart of costs. When costs materialize they become expense. On the other hand dues, upon receipt, become income. The difference between income and expense is usually known as cash flow. Cash flow is taken as positive when income is higher than expense, and negative otherwise.

When we need to compare, add, or subtract income, expenses or cash flows which took place in different dates, first we must refer them to the same date, often the project starting date, using a given annual rate.

Assuming constant prices, total cash flow, also known as net present value npv, at the discount rate j (in percent), corresponding to m expenses d_i and incomes r_i (where i varies from 1 to m), that occur with different time shifts t_i (in years) from the reference moment, is given by

$$npv = \sum_{i=1}^{m} \frac{r_i - d_i}{\left(1 + \frac{j}{100}\right)^{t_i}}. \tag{7.1}$$

When calculating the net present value only expenses and incomes that involve a cash flow should be considered, which excludes, for instance, depreciation of equipment and infrastructures for fiscal reasons.

In telecommunication systems, the usual period of analysis used to be 25 years and the rate of discount, with no inflation, between 4 and 8 percent. Recently, considering the pace of technologic evolution, much shorter periods, as short as 5 to 10 years, and higher rates of discount, inherent to higher risk investments, are used.

When values used to calculate the net present value are not given in constant prices, but rather in current prices, they must be corrected for inflation before calculation of the net present value. To simplify, let us assume that the annual inflation rate inf, expressed in percent, is constant during the period of analysis. Then, the net present value is given by

$$npv = \sum_{i=1}^{m} \frac{\frac{r_i - d_i}{\left(1+\frac{inf}{100}\right)^{t_i}}}{\left(1 + \frac{j}{100}\right)^{t_i}}. \tag{7.2}$$

For low values of the rate of inflation and the rate of discount ($j, inf < 10$) the previous expression may be simplified as

$$npv = \sum_{i=1}^{m} \frac{r_i - d_i}{\left(1 + \frac{j+inf}{100}\right)^{t_i}}, \tag{7.3}$$

which is equivalent to adding the annual inflation rate to the rate of discount (at constants prices).

Different link designs, each with different cash flows at different moments of time, are often compared on the basis of the internal rate of return, or irr, defined as the rate of discount that makes the net present value equal to the remaining value of the project at the end of its life.

It should not be forgotten that taxes are usually paid the year after the exploitation year and thus, for the internal rate of return, taxes due the year after the last year of exploitation (or the end of the project) should be included.

The remaining value of the project at the end of its life is often taken as zero, a quite reasonable assumption for equipment. The same may not be applicable for some infrastructure such as land and buildings.

Two examples will help clarifying the concepts that were introduced in the previous paragraphs. In both cases we will consider constant prices.

Let us assume, to start with, a turn-key project which is payed, all at once, d_0, on the starting day. From then on, during system lifetime, the project has gross results (income less expense) that translate into constant yearly cash-flows cf_a, referred to the year end. We would now like to calculate the internal rate of return as a function of the gross results assuming a zero remaining value and a project lifetime of n years.

In order to normalize the gross results, in relation to the initial investment, we will make use of the concept of payback time t_{pb} defined as

$$t_{pb} = \frac{d_0}{cf_a}. \tag{7.4}$$

Applying equation (7.1) to our first example we get

$$d_0 - \sum_{i=1}^{m} \frac{cf_a}{\left(1 + \frac{irr}{100}\right)^i} = 0,$$

and introducing the payback time we get

$$t_{pb} = \sum_{i=1}^{m} \frac{1}{\left(1 + \frac{irr}{100}\right)^i}. \tag{7.5}$$

In Figure 7.1 we plot the internal rate of return irr as a function of the payback time for $n = 25$ years. As shown, values of the internal rate of return between 4 and 8 percent correspond to payback times between 16 and 10 years.

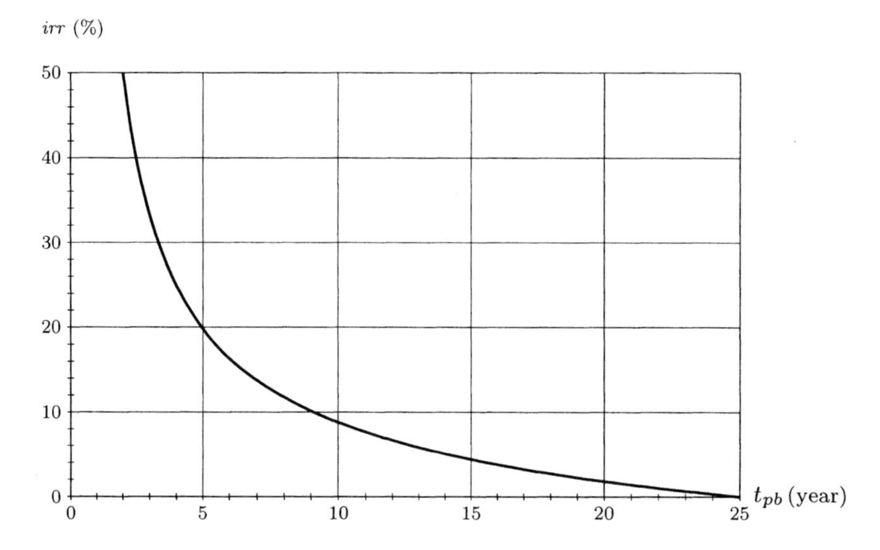

Fig. 7.1 Internal rate of return irr as a function of the payback time t_{pb} for constant gross results throughout project life.

Our second case corresponds to gross results which increase linearly with time, from zero up to a maximum value cf_{max} in the last year of exploitation. Gross results cf_i in year i are

$$cf_i = cf_{max}\frac{i}{n}. \tag{7.6}$$

To simplify the analysis and to allow for payback time to assume non-integer values we will compute payback time t_{pb} as an integral rather than as a sum

$$d_0 = \int_0^{t_{pb}} cf_i di,$$

from where, substituting cf_i by its value in (7.6), we get

$$d_0 = \frac{cf_{max}\, t_{pb}^2}{2n},$$

and, finally

$$t_{pb} = \sqrt{\frac{2d_0 n}{cf_{max}}}. \tag{7.7}$$

Using again (7.1), we have

$$d_0 - \frac{cf_{max}}{m} \sum_{i=1}^{n} \frac{i}{\left(1 + \frac{irr}{100}\right)^i} = 0,$$

and, introducing the payback time given in (7.7)

$$t_{pb} = \sqrt{2 \sum_{i=1}^{n} \frac{i}{\left(1 + \frac{irr}{100}\right)^i}}. \tag{7.8}$$

In Figure 7.2 we plot the internal rate of return irr as a function of the payback time t_{pb} for $n = 25$. Comparing this figure with Figure 7.1 we note that for the same values of the payback time, the internal rate of return is higher when gross results increase linearly with time than when they are constant.

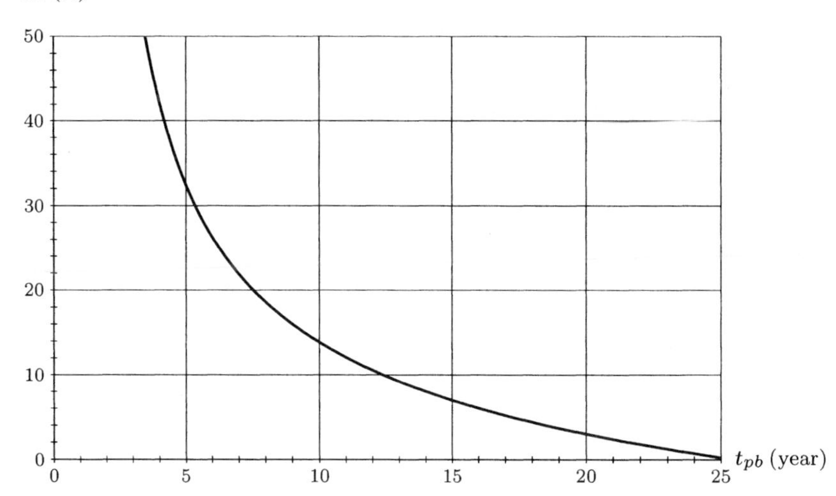

Fig. 7.2 Internal rate of return irr as a function of payback time t_{pb} for gross results increasing linearly with time throughout project lifetime.

In most cases a microwave radio link design cannot be analyzed in isolation and one must take into account its insertion:

- within existing microwave radio links, particularly adjacent ones that may share common infrastructures;
- within the owner, both at the technical level, by sharing common operation and maintenance services, and at the financial level due to budget constraints;

- within the country, the region or the economic community, when the balance of payments or policy may introduce a limit to the import of foreign goods or services.

These constraints, which are obviously rather difficult to quantify without regards to a given project, may often prove to be decisive.

7.3 PATH SELECTION

In this section and in the following one we will describe some criteria pertaining to the choice of path and operating frequency under the assumption that the following items have been previously defined:

- link terminals or, at least, traffic origin and destination;
- link capacity;
- redundancy criteria to overcome equipment faults.

Path selection is the decisive factor in a microwave radio link design and, simultaneously, it is the most difficult factor to present and discuss without an extensive use of examples. Traditionally, path selection has been performed by trial and error, each try being evaluated from the points of view of associated equipment costs and propagation reliability. One usually starts by locating the terminal points on a map such that the whole path is visible. For links with up to 50 or 60 km, the most convenient maps have a 1:25 000 scale with vertical distance between consecutive level lines of 10 m. For longer links, most likely with several hops, other maps, at 1:50 000, 1:100 000 or even 1:200 000, might provide an easier starting point, although they would not usually provide enough data for final path analysis.

Links with more than about 60 km almost always require intermediate relay stations which should be located in such a way that they enable:

- achieving line-of-sight (long) paths;
- sharing of existing infrastructure;
- obtaining the required reliability.

In case relay stations are foreseen, it is possible to start by studying each hop on its own, provided the interdependence between consecutive hops (that share the same relay station) is not forgotten.

Relay station sites may have a pronounced influence in link reliability, both at the level of fault rate (power supply reliability, frequency of thunderstorms, susceptibility to vandalism) and at the level of time to restore interrupted services, which is strongly dependent on access time (for maintenance).

Once terminals for each hop have been identified we must plot the arc of circle that, in a first approach, represents the desired path.

Since manually tracing a terrain profile along the path may take some time, before proceeding it is wise to identify the most remarkable points along the path. Here we define remarkable points as those where the terrain altitude is a relative maximum.

If the path appears to provide line-of-sight between terminals, or if this condition may be achieved using reasonable antenna heights (i.e., up to about 30 m), it probably deserves a more detailed analysis. In many cases simple inspection of remarkable points reveals this condition. In case the path under scrutiny does not fit the description one should start looking for another, more favorable, path:

- by modifying the terminal location, to nearby sites, if that is acceptable;
- by making use of a passive repeater;
- by introducing an extra relay station.

Each of these solutions, including the starting one, has associated costs. Normally obstructed paths are not acceptable, for technical and economical reasons, unless the obstruction is small. On the other hand, modifying the terminal sites may have widely different costs, depending on the existing constraints and the difficulties to transmit the traffic from the initial points. Passive repeaters usually have lower cost than a relay station, both as far as the initial costs go as well as the operation costs. However, small, low-consumption, active repeaters that have recently come into the market may compare favorably in terms of performance and costs with passive repeaters.

Once the likely hop or hops have been identified, we must proceed by plotting the path profile, on a spherical Earth with an equivalent radius $r = k_e r_0$ calculated as referred to in Section 2.5 and using a map with vertical distance between consecutive level lines of 10 m (or less).

Link design will now continue based on this profile and it involves, in the first phase, the next following steps:

- choosing the working frequency;
- defining the antennas and the transmission lines;
- calculating antenna heights;
- computing path attenuation;
- estimating ground reflections;
- specifying the main features of radio equipment;
- checking that the link meets the applicable quality criteria (often ITU-R recommendations).

Procedures to answer such questions as antenna size, tower height, path attenuation and ground reflections were the subject of Chapter 2. Chapter 3 described fading and presented procedures to estimate its effect in the quality of service. Chapters 4 and 5 dealt with analog and digital modulation techniques including quality criteria applicable to international links. Chapter 6 covered error control and correcting codes. In the following we discuss restrictions applicable to the choice of the working frequency, describe a general procedure to compare alternative solutions, and end up with some brief references to commonly available equipment.

7.4 FREQUENCY

The working frequency and the radio electric channel width should be chosen taking into account:

- applicable ITU-R recommendations which consider both the types of link (analog or digital) and their capacities;
- prevailing propagation conditions:
- local conditions, namely, those due to existing or foreseen links, in order to avoid mutual interferences.

Microwave radio links usually require permission from the authority responsible for radio spectrum management (regulator) who specifies the working frequencies.

In Chapter 1 we described the principles on which frequency plans are based. Some frequency plans are detailed in Annex A, with reference to the ITU-R recommendations. Even if, in some cases, the operating frequency (or at least the frequency band) is defined, right from the start, by the licensing authority, we include in the following a general method to choose the best operating frequency band, which may become handy when more than one frequency band is available.

Consider a microwave radio link, for which there are alternative solutions with same total cost that only differ in the operating frequency. For such a link we note that:

- free space attenuation, in dB, varies with $20(m+1)\log_{10}(f)$, where m is the number of passive repeaters between the transmitter and the receiver;
- gain of the transmitter and receiver antennas, in dB, varies with $20\log_{10}(f)$;
- total passive repeater gain, in dB, varies with $40m\log_{10}(f)$
- waveguide attenuation (both rectangular and elliptical) decreases with increasing frequency for the same waveguide, but presents an overall trend to increase with frequency, due to the fact that waveguide cross-section decreases in order to prevent higher order mode propagation;

- attenuation due to atmospheric gas increases (irregularly) with frequency;
- attenuation due to rainfall increases with frequency up to about 100 GHz;
- transmitter power output tends to decrease when frequency increases;
- receiver noise factor tends to increase with frequency;
- uniform fading increases with frequency.

For analog microwave radio links we define margins M_1 and M_2 as the ratios, expressed in dB, between the maximum allowable noise powers in the first and second clause of the ITU-R Recommendation F.395-2 [2] and the link noise powers, under the same conditions. The margin will be taken as positive when the ITU-R recommendations are met and negative otherwise.

Besides quality standards, microwave radio links must also meet availability criteria, defined in ITU-R Recommendation F.557-4 [2], described in detail in the next section.

As we have seen, in Section 2.5.5, an adequate value for k_e guarantees that atmospheric refraction effects do not influence propagation conditions, to a significant degree, and thus most of the unavailability due to propagation will be due to rainfall.

Since for analog links, according to ITU-R Recommendation F.557-4, unavailability corresponds to the situation where noise power in the worst telephone channel is exceeded, it is possible to define another margin, similar to the previous ones, known as rainfall margin, M_{ch}

Although margins are not continuous functions of the frequency (because many frequencies are just not available) it is easier to plot them as such (see Figure 7.3). Experience shows that in many cases, for line-of-sight paths, the margin for the first clause tends to increase with frequency, whereas the opposite is true for the second clause margin. The rainfall margin decreases rather steeply with frequency for frequencies higher than about 11 to 13 GHz.

For each value of frequency, the hardest margin to meet is the smallest one, which will be denoted as the critical margin M_c. The concept of critical margin simplifies the analysis because it reduces all margins to a single one.

For each hop, the optimum frequency will the one that leads to the highest value of the critical margin. If this frequency is not available one should choose, between the available frequencies, the one that yields the highest critical margin.

Identifying, in each case, the critical clause, defined as the clause that at a given frequency is responsible for the critical margin, enables the designer to focus its attention. Thus, for instance, if in a given hop the critical clause is the second one, it may be useful to try to get more accurate data on fast fading. On the other hand, when the critical clause is the rainfall, the designer should confirm local data on rainfall, which may be available from the local meteorologic services.

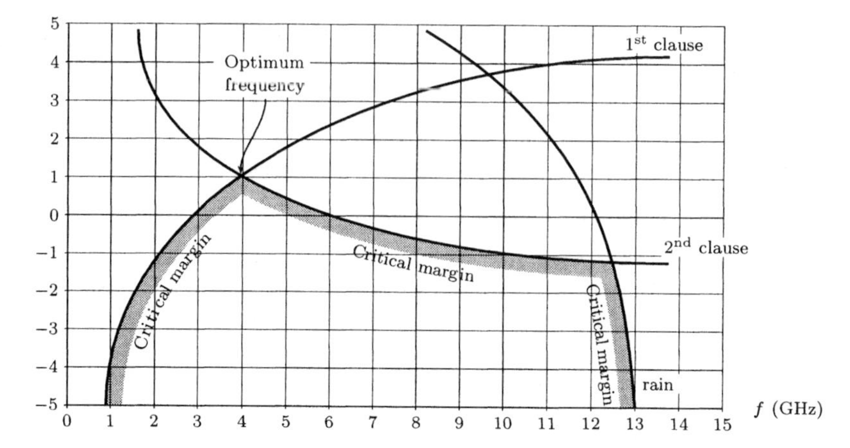

Fig. 7.3 Link margins as a function of frequency, for each ITU-R clause and for rainfall.

When calculating the optimum frequency one should pay attention to the following:

- In many cases, the operating frequency is defined, right at the start, given the type of link, its capacity and siting, by the licensing authority;
- The objective of calculating the optimum frequency is to find the most favorable frequency band (or bands) and not the central frequency of the radio channel:
- In a link, with a number of hops, the frequency band is the same for all hops and it only makes sense to calculate the optimum frequency for the most difficult hop;
- Critical margin differences of less than 1 dB are meaningless, since they are of the same order of magnitude of the changes in the critical margin inside a frequency band.

When the radio channel is not known (or defined) beforehand, the central frequency in the frequency band is often used for calculations because changes in the critical margin are usually small (less than about 1 dB) within the frequency band. When this condition is not met, as in relatively long hops (more than about 20 to 30 km) in the 11 and the 13 GHz bands, the critical margin should be calculated for each radio channel.

Experience shows that:

- in obstructed hops, the optimum frequency is rather low (1.5 or 2 GHz band);
- in hops with a passive repeater (in the far field distance of both terminal antennas) the optimum frequency is rather high (8 and 11 GHz bands);
- in line-of-sight hops, 30 to 50 km in length, the critical margin does not change much with frequency, and the optimum frequency is often in the 4 or 6 GHz bands;
- in shorter line-of-sight hops (length less than about 20 km) the optimum frequency tends to be higher (11 or 13 GHz bands);
- in longer line-of-sight hops (more than about 50 km) the optimum frequency tends to be lower (2 or 4 GHz bands).

For digital links ITU-R recommendations include maximum values for the severely errored seconds ratio, the background block error ratio and the errored seconds ratio, all of which may be converted into a maximum bit error ratio, rather than a maximum noise power as for analog links. We may nevertheless compute, for each modulation type, the carrier-to-noise power (at the demodulator input) that corresponds to the specified bit error ratio, and compare it to the carrier-to-noise power obtained for the system under consideration. We may thus define margins which are similar to those defined for the analog systems and we may compute the optimum frequency in the very same way.

We have seen that under fast fading and for low power levels, the probability $\mathcal{P}$ that the received power p is less than a given value p_0 is given by (3.66) which, for convenience is repeated here

$$\mathcal{P}(p \leq p_0) = \mathcal{F}\,\frac{p_0}{p_n},$$

where the constant $\mathcal{F}$ is dependent on the geoclimatic factor, type of ground, direct ray slope, hop length and operating frequency and p_n is the received power under normal propagation conditions, that is, without fading.

Since the probability $\mathcal{P}$ is linearly dependent on the received power p_0, it is also possible to define margins related to fast fading as ratios, expressed in dB, between the fraction of time (or probability) allowed by the relevant clause and the fraction of time obtained at a given frequency.

The reader may well wonder how is it possible to define the critical margin for a given frequency before this frequency has been defined and before establishing the diameters of the antennas (assumed to be parabolic reflectors) and the area of the passive repeaters, if any. In fact, not all these elements are required beforehand and the critical margin may be used to calculate them.

Antenna size and passive repeater area do not change the way in which the critical margin varies with frequency and, thus, do not change the value of the optimum frequency. The same is true for transmitter power and receiver noise factor, provided we know how they vary with frequency. It is therefore possible to compute the effect of frequency in the critical margin, even if we do not know the sizes of the antennas and the passive repeaters. For simplicity it is usual to start by assuming an antenna diameter of 1 m and a passive repeaters area of 1 m^2.

Available transmitter power as a function of frequency is highly dependent on technology. For all solid-state equipment it is reasonable to assume that the transmitter output power varies, approximately, with the inverse of the frequency. Receiver noise power is also frequency dependent, according to the technology. In the absence of further data we may assume that receiver noise power, in dB, varies linearly with frequency from about 5 dB at 1 GHz to about 8 dB at 10 GHz.

Once the operating frequency has been chosen and the critical margin calculated it is possible to proceed to the final specifications, related to antenna and repeater size and transmitter power, noting that the critical margin should be zero, to meet the ITU-R recommendations exactly. In practice, it is not wise to design for a zero critical margin because:

- unforeseen aspects in the implementation may lead to additional attenuations (waveguide runs longer than expected, additional filters, etc);
- component aging degrades performance (lower output power, decrease in antenna gain);
- imperfections during or after installation (small misalignments in antennas or in passive reflectors, mismatches in radiofrequency circuits due to mechanical shocks);
- calculation errors in obstacle losses, due either to sizable differences between the model and the real obstacle or to model approximations;
- errors in the basic data on fading or on rainfall.

According to designer confidence (and experience) and link difficulties, that may make each additional dB very expensive, the critical margin may be as low as 1 or 2 dB or as high as 5 to 6 dB. In particularly difficult cases, propagation tests may help in reducing the critical margin. In some easy hops the use of standard components (such as transmitter power output and antenna diameter) may lead to much higher values of the critical margin than those previously referred to.

Once the designer has established the minimum acceptable critical margin $M_{c_{min}}$ and compared it with the critical margin M_c for the configuration under study, for analog systems it suffices to distribute the difference $\Delta M_c = M_{c_{min}} - M_c$ between the elements which were assumed to be known (antenna gain, or diameter, passive repeater area, transmitter output power) in the most economical way. This distribution is quite simple for analog links because ΔM_c varies linearly with the antenna and passive repeater gain and with the transmitter power in dB.

For digital links the situation is different. However, when selective fading is not important and the first term in the background block error ratio and in the errored seconds ratio dominates, the previous procedure may still be applied.

When selective fading plays a significant role or when the second term dominates in the background block error ratio or in the errored seconds ratio and any of these two ratios is the critical clause, then the link margin is not linearly dependent on the received power without fading p_n. In this case varying p_n simply changes the uniform margin and, hence, the distribution of $\Delta M_c = M_{c_{min}} - M_c$ does not ensure meeting ITU-R recommendations.

Let us now consider that selective fading is important and that the critical margin is due to the severely errored seconds ratio or that the first term dominates the background block error ratio or the errored seconds ratio in case either of these two parameters becomes the critical one.

Assuming that the critical margin is due fast fading, we may write, from (5.232)

$$\mathcal{P} = \mathcal{P}_u + \mathcal{P}_s, \tag{7.9}$$

where we have from (3.66)

$$\mathcal{P}_u(p \leq p_0) = \mathcal{F}\frac{p_0}{p_n},$$

and, from (5.238),

$$\mathcal{P}_s = k\frac{s}{8000},$$

s representing the area under the receiver signature, in MHz.

Now we may distribute the difference between the minimum acceptable critical margin and the critical margin, obtained with the configuration under study, provided these margins are based on the probability corresponding to the uniform fading $\mathcal{P}_u$ required to meet the applicable clause.

Sometimes the value of the area under the signature s and consequently the value of the probability $\mathcal{P}_s$ are such that the required value of $\mathcal{P}_u$ (needed to meet the value of $\mathcal{P}$) is null or even negative. In this case, it is not possible to meet the specified quality criteria, whatever the increase in antenna gain, in passive repeater area, or transmitter output power, and one must look into ways of reducing the effects of selective fading by one the techniques described in Section 5.9.5.

7.5 LINK AVAILABILITY

In Chapters 4 and 5, while presenting the quality standards for microwave radio links, it was often stated that these standards only apply when the link is available.

For high-quality microwave radio links, with a capacity higher than the second hierarchy, besides the concept of hypothetical reference circuit defined in Section 5.8.2, we make use of the concept of hypothetical reference digital path, or HRDP, defined in ITU-R [2] Recommendation F.556-1, as a bidirectional, digital, link, 2500 km long, at 64 kbit/s that meets the requirements stated in ITU-T Recommendation G.821. The hypothetical reference digital path is made up of nine identical radio electric sections and includes, in each transmission direction, nine sets of multiplexing equipment.

The apportionment of quality objectives and unavailability between the radio electric sections are defined in ITU-R [2] Recommendation F.594-4. According to ITU-T Recommendation G.821, the longest hypothetical reference connection, or HRX, of an integrated services digital link, is 27 500 km long and is made up of three parts. The central part is a high-grade 25 000 km long link, while the edges are each 2 500 km long, medium- and local-grade links.

The high-grade part corresponds to the hypothetical reference circuit which should comply with ITU-R [2] Recommendation F.557-4 which sets an unavailability objective of 0.3 percent for the hypothetical reference circuit (with 2500 km)

$$\mathcal{I}_{fr} = 0.003. \tag{7.10}$$

According to the Annex I of Recommendation F.557-4 the unavailability of microwave links is mainly due to:

- equipment:
  - faults or performance degradation of radio equipment, including modulators and demodulators;
  - faults in auxiliary equipment, such as diversity switches;
  - faults in the power supply system;
  - faults in antennas and feeders;
- propagation:
  - deep fades, sometimes long lasting (from a few minutes to a few hours) associated to abnormal propagation conditions, such as ducts and sub-standard troposphere;
  - very intense rainfall;
  - multipath fast fading;
- interferences;
- building and antennas masts (serious damages caused by acts of God, sabotage or, less frequently, vandalism);
- human activity (usually exploitation or maintenance errors).

Assuming that unavailability causes in one section are uncorrelated with unavailability causes in the remaining sections of the hypothetical reference

Class	Length (km)	sesr	esr
1	280	0.00006	0.00036
2	280	0.000075	0.0016
3	50	0.00002	0.0016
4	50	0.00005	0.004

Table 7.1 Maximum length and quality objective for each class of the medium-grade portions of the hypothetical reference connection, at a bit rate below the first digital hierarchy, according to ITU-R[2] Recommendation F.696-2.

Class	Length (km)	Unavailability (percent)
1	280	0.033
2	280	0.05
3	50	0.05
4	50	0.1

Table 7.2 Maximum length and unavailability objective for each quality class, of the medium-grade portions of the hypothetical reference connection, at a bit rate below the first digital hierarchy, according to ITU-R[2] Recommendation F.696-2.

link and assuming that the unavailability values are very low, it is acceptable to divide the overall unavailability among the various sections according to their length. Under these assumption the maximum unavailability $\mathcal{I}$ of a link d km long should be

$$\mathcal{I} = \mathcal{I}_{fr}\frac{d}{2500}. \tag{7.11}$$

Since the previous expression unduly penalizes short links and since for these links unavailability addition is probably not applicable, ITU-R [2] Recommendation F.695 suggests that the value of d should be limited to a minimum of 280 km.

For the medium-grade part of the hypothetical reference connection (based on digital radio links, at bit rates below the first digital hierarchy) ITU-R[2] Recommendation F.696-2 refers the possibility of using links with up to 4 quality standards. The maximum length and the quality objectives for each of these classes are given in Table 7.1. The corresponding bidirectional unavailability objectives are given in Table 7.2.

For the local-grade part of the hypothetical reference connection, ITU-R Recommendation F697-2 [2] states that the severely errored second ratio

($sesr$) should not exceed 0.000 15, in any month, and the errored second ratio (ESR) should not exceed 0.0012 in any month.

Recommendation F.697-2 does not yet set the unavailability objectives for local-grade links. However it recognizes, in its Annex I, the growing use of microwave radio links, operating at above 17 GHz, and for these links, as an example it presents the followings unavailability values, due to propagation:

- 0.0004 to 0.004 percent, per year, in Japan, for links in the 21 and 26 GHz bands, the lower unavailability limiting the hop length to about 3.5 km;
- 0.001 to 0.01 percent, per year, in the united Kingdom, for links in the 18 GHz band.

It is up to the link designer to distribute the overall unavailability by the different causes. In the absence of other criteria and assuming that in the link design all recommended precautions to avoid interruptions due to long fades have been taken, it usual to assume:

- unavailability due to rainfall $\mathcal{I}_r$

$$0.2\mathcal{I} \geq \mathcal{I}_r \geq 0.1\mathcal{I}; \tag{7.12}$$

- unavailability due to equipment $\mathcal{I}_e$

$$0.4\mathcal{I} \geq \mathcal{I}_e \geq 0.3\mathcal{I}; \tag{7.13}$$

- unavailability due to other reasons (interferences, auxiliaries, human activity) $\mathcal{I}_{or}$

$$\mathcal{I}_{or} = 0.5\mathcal{I}. \tag{7.14}$$

An adequate choice of path and clearance of the direct ray ensures that the unavailability due to propagation is essentially caused by rainfall.

Unavailability due to other reasons $\mathcal{I}_{or}$ is mostly the responsibility of the exploitation team but it is up to the designer to adopt an equipment layout that enables objectives to be met.

Equipment reliability calculations consider that time between faults has a negative exponential distribution, described by the mean time between failures, in short $MTBF$. Let $MTTR$ be the mean time to detect and repair the fault and reestablish service. The unavailability due to the equipment will be

$$\begin{aligned} \mathcal{I}_e &= 1 - \frac{MTBF - MTTR}{MTBF} \\ &= \frac{MTTR}{MTBF}. \end{aligned} \tag{7.15}$$

Unavailability due to equipment depends not only on its own reliability but also on the maintenance team performance which has a directly influence on $MTTR$.

For unmanned microwave relay stations, with easy access, $MTTR$ is of the order of a few hours. Typical values vary between 6 and 12 hours, for well-trained teams, on permanent alert, with their own transport equipment and enough spare parts to repair faults by simple substitution.

For all solid-state microwave transmitters and receivers $MTBF$ may reach a few tens of thousands of hours, leading to an unavailability value around 0.001, per bidirectional link. Although, at first, this value may appear adequate, it is not. In fact, if we take for unavailability due to equipment 0.4 of the total unavailability and for the link length 280 km, as suggested in ITU-R [2] Recommendation F.695, we get $\mathcal{I}_e \leq 0.00013$.

From the point of view of the continuity of service we may consider that a microwave radio link assumes one of two basic configurations:

1. series, when a failure in one element interrupts the link;

2. parallel, when there are two or more independent paths.

At points where the link changes from series to parallel, we must foresee a switch that, at each moment, selects the path in use.

Assume a set of m elements in series with an identical value of $MTTR$. Overall unavailability $\mathcal{I}_s$ is given by

$$\begin{aligned} \mathcal{I}_s &= \sum_{j=1}^{m} \mathcal{I}_j \\ &= MTTR \left(\sum_{j=1}^{m} \frac{1}{MTBF_j} \right). \end{aligned} \tag{7.16}$$

Since:

$$\mathcal{I}_s = \frac{MTTR}{MTBF_s}, \tag{7.17}$$

where $MTBF_s$ is the series mean time between failures, substituting (7.17) into (7.16), we get

$$\frac{1}{MTBF_s} = \sum_{j=1}^{m} \frac{1}{MTBF_j}. \tag{7.18}$$

This expression illustrates a well-known fact: when the complexity of a system increases, by adding elements in series, its availability decreases.

Take now the association, in parallel, of n active elements to which we add r reserve elements in a configuration that we denote as $n+r$. Assuming that:

- switching time, from the active to the reserve elements, is instantaneous,

- all elements have the same unavailability $\mathcal{I}$,

- switching equipment, from active to reserve, has zero unavailability,

- the association only becomes unavailable when at least $r+1$ elements are faulty,

unavailability of the parallel set $\mathcal{I}_p$ is given by

$$\mathcal{I}_p = \sum_{j=r+1}^{n+r} \mathcal{C}_j^{n+r} \mathcal{I}^j (1-\mathcal{I})^{n+r-j}, \tag{7.19}$$

where $\mathcal{C}_j^{n+r}$ represents the number of combinations of $n+r$ elements j to j.

If, as usual, $\mathcal{I} \ll 1$, the previous expression may be simplified as

$$\mathcal{I}_p = \mathcal{C}_{r+1}^{n+r} \mathcal{I}^{r+1}. \tag{7.20}$$

Substituting the definition of $\mathcal{I}_p$ and manipulating we get

$$\begin{aligned} MTBF_p &= \frac{MTTR}{\mathcal{C}_{r+1}^{n+r} \mathcal{I}^{r+1}} \\ &= \frac{MTBF}{\mathcal{C}_{r+1}^{n+r} \mathcal{I}^{r}}. \end{aligned} \tag{7.21}$$

Since $\mathcal{I} \ll 1$ and $\mathcal{I}^r \ll 1/\mathcal{C}_{r+1}^{n+r}$, the unavailability of a set of elements in parallel is less than the unavailability of the same elements in isolation.

To exemplify the need for reserve channels, take a digital, bidirectional, link for which we have:

- transmitter, $MTBF = 120\,000$ hours;
- modulator, $MTBF = 200\,000$ hours;
- receiver, $MTBF = 200\,000$ hours;
- demodulator, $MTBF = 140\,000$ hours;
- switch $2+1$, $MTBF = 78\,000$ hours;
- $MTTR = 6$ hours;.

We will further assume that buildings, towers, antennas and their feeders have a mean time between failures much higher than the remaining elements and that unavailability due to power cuts is not included in unavailability due to equipment.

First, let us assume that there are no reserve channels. Under these conditions, and as far as link interruptions are concerned, a unidirectional link is made up of the following items in series: modulator, transmitter, receiver and demodulator. Applying (7.18) we get for the link$MTBF_u = 39\,252$ hours. Since the bidirectional link is made up of two unidirectional links in series, we get $MTBF_b = 19\,626$ hours and, with $MTTR = 6$ hours, we get $\mathcal{I}_e = 3.1 \times 10^{-4}$.

Although we have not stated link length, the fact that equipment given corresponds to a single hop, places an upper limit to the link length of about 50 to 60 km. Even with the favorable interpretation of ITU-R [2] Recommendation F.695 of a minimum link length of 280 km, if we take for unavailability due to the equipment 0.4 of the overall unavailability, we have $\mathcal{I}_e \leq 1.3 \times 10^{-4}$.

To achieve the recommended unavailability there are two possible solutions:

1. to reduce the $MTTR$ to 2.6 hours;

2. to install a reserve channel.

The reserve channel implies an extra two switches (one at each terminal). Assuming two service channels and one reserve channel, the $MTBF$ will be calculated as a series association of two switches with three bidirectional links, of which one is the reserve.

Using the previous values we get, for the switches

$$\begin{aligned} MTBF_c &= \frac{78000}{2} \\ &= 39\,000 \text{ hours,} \end{aligned}$$

and for the links:

$$\begin{aligned} MTBF_l &= \frac{MTTR}{C_2^3 \left(\frac{MTTR}{MTBF_b}\right)^2} \\ &= \frac{6}{3\left(\frac{6}{19626}\right)^2} \\ &= 21\,398\,882 \text{ hours,} \end{aligned}$$

which yields, for the ensemble, $MTBF \approx 39\,000$ hours and, thus, $\mathcal{I}_e = 1.5 \times 10^{-4}$ which is slightly higher than the ITU-R recommended value, but that may be nevertheless considered as acceptable.

In this example the unavailability is essentially dependent on the switches and is practically not altered by the increase in the number of reserve channels. One should however note that:

- switches are not normally used to protect a single hop but rather a series of hops;

- failure in a switch does not necessarily mean unavailability of the link, but rather that the link is no longer protected against interruption.

Taking, for instance, a link with 6 hops of 50 km with a switch at each end, with the previous data we would get:

$$\begin{aligned} MTBF_{b6} &= \frac{MTBF_b}{6} \\ &= \frac{19626}{6} \\ &= 3271 \text{ hours,} \end{aligned}$$

and for 2 service and 1 reserve link:

$$\begin{aligned} MTBF_l &= \frac{MTTR}{C_2^3(\frac{MTTR}{MTBF_{b6}})^2} \\ &= \frac{6}{3(\frac{6}{3271})^2} \\ &= 594\,413 \text{ hours.} \end{aligned}$$

Now, for the ensemble of the link, including the switches, we get $MTBF \approx$ 39 000 hours and $\mathcal{I}_e = 1.5 \times 10^{-4}$, which is almost exactly the ITU-R recommended value.

If, with a single hop and without switching, the unavailability is about 2.4 times higher than the recommended value, with 6 hops and no switching things get worse and the calculated unavailability

$$\begin{aligned} \mathcal{I}_e &= \frac{6}{3271} \\ &= 1.83 \cdot 10^{-3} \end{aligned}$$

becomes 5.1 times the recommended value.

These examples show that:

- ITU-R recommended unavailability is hard to meet, even using equipment with a high $MTBF$;
- to meet ITU-R recommendations with current equipment, mean time to restore service should reduced to a few hours, 6 or preferably less;
- in the apportionment of the overall unavailability a reasonable fraction (at least about 40 percent) should be attributed to the equipment.

To end this subject, let us consider a link with a single service channel using double diversity (space or frequency). Is a reserve channel required?

As far as unavailability, the existence of two separate channels for transmit and receive enables us to assimilate double diversity to a $1 + 1$ configuration. Since the probability of failures is low, we may neglect the probability of a simultaneous failure of both channels. During failure of one of the channels the link operates without diversity, a fact that increases the probability of errors due to fading.

The probability $\mathcal{P}$ that a given bit error ratio is exceeded is a function of the probabilities that this happens for the link with and without diversity, respectively $\mathcal{P}_d$ and $\mathcal{P}_{wd}$, as

$$\begin{aligned} \mathcal{P} &= \mathcal{P}_d(1 - \mathcal{I}_{e_d}) + \mathcal{I}_{e_d}\mathcal{P}_{wd} \\ &= \mathcal{P}_d + \mathcal{I}_{e_d}(\mathcal{P}_{wd} - \mathcal{P}_d), \end{aligned} \tag{7.22}$$

where $\mathcal{I}_{e_d}$ is the unavailability due to equipment in the link with diversity. If the link with diversity is assimilated to two links without diversity and its unavailability is calculated from the unavailability of the link without diversity $\mathcal{I}_{e_{wd}}$ as

$$\mathcal{I}_{e_d} = 2\mathcal{I}_{e_{wd}}. \tag{7.23}$$

Since $\mathcal{P}_{wd} \gg \mathcal{P}_d$, (7.22) may be simplified

$$\mathcal{P} \approx \mathcal{P}_d + \mathcal{I}_{e_d}\mathcal{P}_{wd}. \tag{7.24}$$

Recalling that

$$\mathcal{P}_{wd} = i\mathcal{P}_d,$$

where i is the diversity gain, (7.24) may be rewritten in terms of $\mathcal{I}_{e_{wd}}$ and $\mathcal{P}_d$

$$\begin{aligned} \mathcal{P} &\approx \mathcal{P}_d + 2\,\mathcal{I}_{e_{wd}}\mathcal{P}_{wd} \\ &\approx \mathcal{P}_d(1 + 2i\,\mathcal{I}_{e_{wd}}). \end{aligned} \tag{7.25}$$

The lack of reserve channel increases the probability of link failure. However, in most cases, this increase does not justify a reserve channel (in links with diversity)

7.6 ANTENNAS

As it was referred in Chapter 1, antennas for microwave radio links above 1 GHz are almost always parabolic reflectors with diameters between 0.5 and 4 m. In special cases, when a high discrimination between orthogonal polarizations and a very low voltage standing wave ratio (vswr) are required, hoghorn antennas may be employed.

For frequencies up to about 2 GHz parabolic reflectors are sometimes made up with discrete metallic elements (tubes or mesh) which does not interfere much with antenna cost but since it reduces antenna weight and wind resistance it may well decrease tower costs.

For frequencies below about 1 GHz, Yagi-Uda or helix antennas are often preferred either isolated or in small arrays, with or without reflector planes, according to the required polarization, bandwidth and gain.

Since transmitter power and receiver noise factor, once the technology is defined, change very little, the designer uses the antennas to adapt the equipment to the link requirements. The following antenna diameters can usually be found on the market:

- 0.6, 1, 1.5, 2, 3 and 4 m;
- 2, 4, 6, 8, 10, 12 and 15 feet.

Besides current series, most antenna makers supply special series antennas that meet stiffer requirements on the radiation pattern such as:

- lower sidelobe level;

- higher front-to-back ratio;
- higher discrimination between orthogonal polarizations.

These results are achieved by one or more of the following solutions:

- more sophisticated primary sources, other than a simple truncated waveguide;
- deeper reflectors, with the focus on, or beneath, the aperture plane;
- use of a secondary reflector in a Cassegrain setup (usually reserved for satellite links);
- side blinds.

Special series antennas may require radomes that, besides offering weather protection, avoid the accumulation of snow and ice and, in some cases, may reduce wind stresses.

Antennas whose radiation pattern meets special requirements have much higher (up to 100 percent) costs than current antennas. On the other hand double polarization antennas cost 40 to 50 percent more than equivalent single polarization antennas.

Interference studies between microwave relay stations and between relay stations and satellite earth stations require knowledge of antenna radiation patterns. In their absence ITU-R [2] Recommendation F.699-5 suggests the following radiation pattern, which is valid in the 1 to 40 GHz range, for circular symmetry antennas with diameter d_a up to 100 λ:

$$G(\phi) = \begin{cases} G_{max} - 2.5 \cdot 10^{-3} \left(\frac{d_a}{\lambda}\phi\right)^2 & \text{for} \quad 0 < \phi < \phi_m, \\ G_l & \text{for} \quad \phi_m < \phi < 100\frac{\lambda}{d_a}, \\ 52 - 10\log_{10}\left(\frac{d_a}{\lambda}\right) - 25\log_{10}\phi & \text{for} \quad 100\frac{\lambda}{d_a} \le \phi < 48^o, \\ 10 - 10\log_{10}\left(\frac{d_a}{\lambda}\right) & \text{for} \quad 48^o \le \phi < 180^o, \end{cases} \tag{7.26}$$

where:

- G_{max} is the maximum gain in the main lobe, in dBi;
- $G(\phi)$ is the gain at the angle ϕ relative to the isotropic lossless antenna, in dBi;
- ϕ is the angle relative to the axis, in degrees;
- G_l is the gain of the first sidelobe, in dB, given by $2 + 15\log_{10}(\frac{d_a}{\lambda})$;
- ϕ_m is given, in degrees, by: $\frac{20\lambda}{d_a}\sqrt{G_{max} - G_l}$.

Figure 7.4 represents the radiation pattern defined in (7.26).

In the market there are flat passive reflectors (mirrors), almost always rectangular in shape, with an aspect ratio (ratio between height and width)

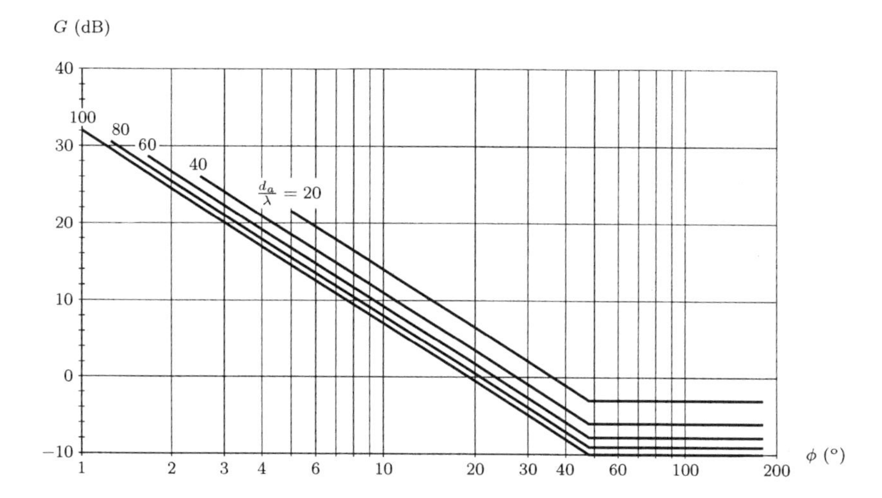

Fig. 7.4 Radiation pattern suggested in ITU-R [2] Recommendation F.699-5.

equal to $\sqrt{2}$:1, and sizes from 2.5 x 3.0 m^2 up to about 12 x 18 m^2. Passive reflectors may be installed close to the ground (clearances between 2.5 to 4.5 m) or on towers. The smaller reflectors (sizes from about 2.5×3.0 m^2 to about 4.2×4.9 m^2) may be installed on fairly simple self-supporting towers, of the type used for antennas. Larger flat reflectors require special towers whose design is closely coordinated with the reflector. Smaller size flat reflectors compete cost-wise with parabolic reflectors in (relatively) low gain passive repeaters. For higher gain flat reflectors are the obvious choice.

Antennas, passive repeaters, towers and masts must be built and installed to withstand severe environmental conditions which include:

- ambient (shadow) temperatures between, at least, −10 and +45°C;
- corrosion, which is particularly serious near the coastline;
- mechanical stresses due to wind and, where applicable, snow and ice.

In many locations, where ice load is not relevant, common requirements include:

- survival, with no permanent damage, for a wind velocity of 180 km/h, or more frequently, 210 km/h;
- guaranteed specifications for a wind velocity of 120 km/h.

As it was stated in Chapter 1, antennas and passive repeaters are installed in towers which, according to the importance of the station and the required height, may be:

- simple, self-standing, metallic structures, for heights up to about 6 m;
- simple, guyed, metallic structures for heights up to about 100 m;
- moderately complex, self-standing metallic structures for heights up to about 100 m;
- complex, metallic, concrete or metallic and concrete, self-standing structures for heights between about 30 and 300 m.

The use of guyed towers, which tend to have the lowest costs, is limited by deformations which occur in normal operation. The top of a well-designed guyed tower may rotate, under normal operating conditions, up to 5 degrees. Additional anti-torsion devices may be used to limit such rotations to about 1 degree. For antennas used in microwave radio links the use of guyed towers in limited to:

- mono channel links in the 400 MHz band;
- in the 1.5 and 2 GHz bands, with anti-torsion devices;
- in the 4 and 6 GHz bands, for small diameter antennas with anti-torsion devices.

Self-standing towers (metallic or concrete) although much more expensive than guyed towers, have some important advantages:

- smaller footprint, an important asset in urban areas;
- small deformations, enabling the use of large antennas, even at the higher frequency bands;
- installation of multiple antennas on the same tower.

Some towers may provide an equipment room (for transmitters and receivers) near the antennas, decreasing installation and maintenance costs and improving performance since feeder length is reduced to a few meters.

Metallic self-standing tower costs changes rather abruptly with height at about 30 m. This is why one tries hard to keep tower height below 30 m.

7.7 CABLES AND WAVEGUIDES

As it was mentioned in Chapter 1, in microwave radio links a few transmitters and receivers may share the same antenna. Hence between the antenna and the transmitters and receivers we have:

- an antenna multiplexer, built with circulators, filters, directional couplers and transmission lines;

- a link between the antenna and the antenna multiplexer, usually known as feeder.

In microwave radio links, transmission lines and feeders are in:

- coaxial cable for frequencies up to about 2 GHz;
- metallic waveguide for frequencies above about 2 GHz.

Waveguide is far more attractive than coaxial cable, from the attenuation point of view, but its use below about 2 GHz is hindered by the large waveguide cross-section (and high costs).

As an alternative to a waveguide it is possible to use (although this is not very common in microwave radio links) a coaxial rigid line. For short connections, between transmitter and receiver modules, or where attenuation is not critical, a special kind of small diameter (about 5 mm) rigid line is often used,

Coaxial cable specifications include the characteristic impedance (50 or 75 Ohm), the dielectric between center and outer conductors and outer conductor diameter. For the latter the most common values are 1/4, 3/8 and 1/2 inch. For the smaller diameters (1/4 and 1/2 inch) it is possible to get extra flexible cables, with a slightly higher attenuation. In microwave radio links minimal attenuation cables are usually preferred.

Attenuation per unit length versus the operating frequency for 50-Ohm coaxial cables at 24°C is plotted in Figure 7.4. For 75-Ohm cable attenuation may be calculated by multiplying the attenuation in Figure 7.4 by 0.95.

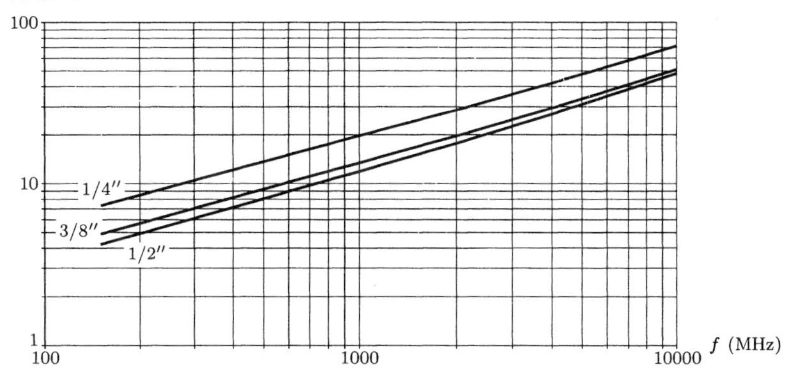

Fig. 7.5 Attenuation per unit length versus frequency for 50-Ohm coaxial cables with outer diameters of 1/4, 3/8, and 1/2 inch [1]. Reproduced by permission of Andrew Corporation.

There are currently available three types of metallic (copper) waveguides which are identified by the cross-section shape:

1. rectangular;
2. circular;
3. elliptic.

Waveguides are usually operated in a rather narrow bandwidth so that a single (fundamental) mode may propagate. The attenuation of a given waveguide decreases with frequency, as shown in Figure 7.6. However, when the frequency increases, enabling propagation of higher order modes, waveguide cross-section has to be reduced and, in this case, attenuation increases with frequency.

Each waveguide has his own characteristics that justify its choice for different purposes:

- Elliptic waveguide is relatively flexible (bending radius of about 1 to 2 m) a feature that greatly simplifies installation; although it exhibits slightly higher values of the voltage standing wave ratio than the rectangular or the circular waveguide it is adopted in most cases;
- Rectangular waveguide offers slightly lower values of voltage standing wave ratio and attenuation than the elliptic waveguide and has an enormous set of accessories that enable complex circuits to be made in compact form; it is often used to build antenna multiplexer units;
- Circular waveguide provides the lowest values of attenuation and standing wave ratio, besides enabling simultaneous transmission of orthogonal polarizations, but it has both higher acquisition and installation costs and may only be used in linear runs; it is restricted to long feeders when attenuation must be kept to a minimum.

At the predesign stage, when exact location of the equipment is yet undecided, the feeder length is usually taken as equal to the antenna height, plus 5 or 10 m. When cables or elliptic waveguides are supplied with jacks or flanges, feeder length must be defined more accurately.

Besides a low attenuation, feeder and antenna multiplexer must ensure a low voltage standing wave ratio, particularly for high-capacity analog links with long feeders. To achieve this objective the following requirements must be met:

- very careful installation, keeping deformations well within the maximum values recommended by the suppliers which, for rectangular waveguides, are much smaller than the usual tolerances for electric circuits;
- keeping the number of flanges and other accessories, particularly flexible pieces and transitions, to a bare minimum;
- maintaining the internal waveguide walls clean and free of oxides;

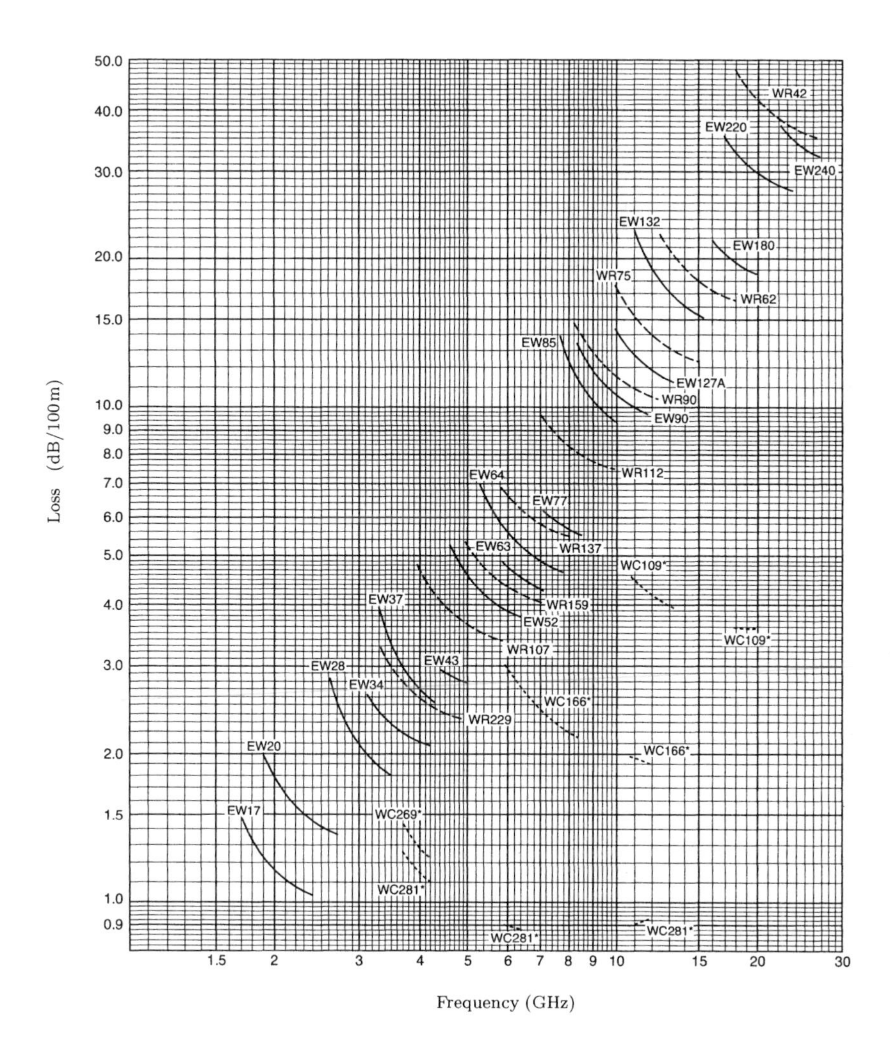

Fig. 7.6 Attenuation per unit length versus frequency for rectangular (WR), circular (WC) and elliptic (WE) metallic waveguides [1]. Reproduced by permission of Andrew Corporation.

- making the whole circuit airtight and pressurizing it with dry air.

Straight pieces of rectangular waveguide, as well as most accessories, provide a voltage standing wave ratio of about 1.05. An important exception are the flexible pieces where the voltage standing wave ratio may be as high as 1.10. Some manufacturers supply rectangular waveguide and accessories with lower voltage standing wave ratios (1.03 to 1.05).

For elliptic waveguide runs up to 90 m, the voltage standing wave ratio varies between 1.07 to 1.09, although in some unfavorable situations it may reach 1.15 to 1.20. With premium waveguide for frequencies above about 3 GHz, it is possible to achieve a voltage standing wave ratio between 1.04 to 1.08. Under normal conditions using elliptic waveguide for the feeder and rectangular waveguide for the antenna multiplexer unit it is hard to get a voltage standing wave ratio below about 1.08 to 1.10.

Feeders made up of circular waveguide achieve very low values of the standing wave ratio (below 1.01). Even including transitions it is possible to obtain an overall voltage standing wave ratio at around 1.05.

The cost of the antenna multiplexer unit is often included in the cost of radio equipment.

7.8 RADIO EQUIPMENT

Radio equipment is usually mounted in racks with 120 mm wide, 225 mm deep and about 2 m high. Typically each rack includes:

- one transmitter and one receiver, including power supply and local oscillator;
- one modulator and one demodulator (analog or digital);
- or a diversity combiner and another receiver.

The transmitter power output in microwave radio links, although dependent on the operating frequency, is at most 10 W. Currently, for all solid-state equipment, the highest power output p_E (in W) at frequency f (em GHz) may be roughly estimated as

$$p_{E_{[W]}} \leq \frac{10}{f_{[GHz]}}. \qquad (7.27)$$

It is possible, although often not desirable for cost, reliability and (or) maintenance reasons, to increase transmitter power output using travelling wave tubes or klystrons. In the former the power output is limited to about some tens of watts, while in the latter hundreds or even thousands of watts are achievable. For frequencies up to 1 GHz output powers up to the kW level are possible, using special vacuum tubes.

The receiver noise factor depends on the technology and varies between 5 dB at 1 GHz to about 10 – 12 dB at 19 GHz.

The use of (uncooled) parametric amplifiers is limited to satellite and troposcatter link. The noise temperature of such amplifiers is often less than 100 K which for a gain of about 40 dB leads to a noise factor about 1.3 dB.

Typical power supply requirements are as follows:

- transmitter-receiver, between 50 and 150 W, depending on transmitter power output;
- analog modulator and demodulator, between 20 and 50 W;
- digital modulator and demodulator, with base band equalizers, between 10 and 100 W, according to capacity.

7.9 POWER SUPPLY

All public telecommunication systems, particularly those included in international links, such as microwave radio links, must ensure very high availability. To this end, must not only the unavailability due to propagation, equipment and human error be very low, but also the electric power supply must provide a much higher availability than usual. In addition many microwave relay stations are remotely located, with no public electric power supply.

In order to quantify the maximum power supply unavailability we recall that the ITU-R [2] Recommendation F.557-4 sets a maximum overall unavailability of 0.3 percent, for the hypothetical reference circuit, with 2500 km. Assuming that unavailability will be apportioned to link length d, with a minimum of 280 km, (ITU-R Recommendation F.695 we have for a typical hop

$$i_{s[\%]} \leq 0.3 \times \frac{280}{2500}.$$

If the power supply contributes with 1/6 of the hop total unavailability, the power supply unavailability for each of the two terminals in the link should be 0.002 8 percent, or less than about 15 minutes per year.

For simplicity reasons we have assumed that interruptions in the power supply lead to identical interruptions in the link, which is obviously not true. In fact, link interruptions last at least about 1 minute, even if the power supply was interrupted for as little as a few tenths of a second. Thus to meet unavailability targets uninterruptible power supplies (in short UPS) are clearly a must.

For most microwave radio link equipment, powered by mains (alternating current) the uninterruptible power supply may be represented by the diagram in Figure 7.7 and includes:

- a battery, with enough capacity to ensure power supply during the longest power cuts;
- a battery charger, to charge the battery from the mains;

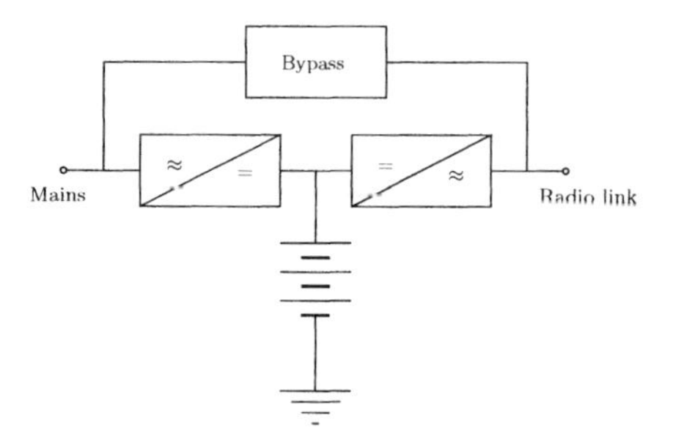

Fig. 7.7 Uninterruptible power supply.

- an inverter, that provides the power required by station from the battery;
- a by-pass switch, to link the station directly to the mains.

The existence of a by-pass switch is debatable, since it contradicts the objective of an uninterruptible power supply. Its existence enables to power the station during short maintenance periods (with no guaranteed power supply) something which would not be possible without replicating the power supply system.

In some cases, microwave radio link equipment may be powered by direct current, thus avoiding need for the inverter and the by-pass switch.

In general, the battery capacity is calculated for an autonomy of 12 to 24 hours. After this period of time, the regular main power supply should be restored or a mobile generator installed. When the station power consumption is too high, or when the period of 12 or 24 hours is insufficient, a motor generator should be foreseen.

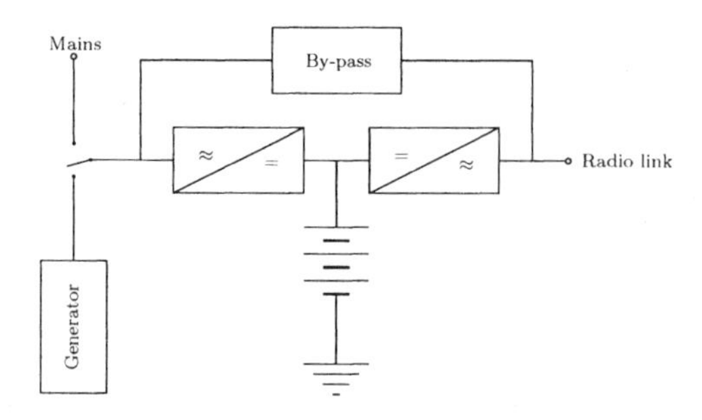

Fig. 7.8 Uninterruptible power supply systems with emergency generator.

When the mains is very irregular and battery autonomy is short (less than about one hour) two emergency generators rather than a single one may be required so as to keep the power supply even during maintenance of one emergency generator. With two emergency generators, battery autonomy may be reduced to a minimum of 15 to 30 minutes.

Emergency generators should start automatically after a mains interruption of about 1 minute (to avoid an excessive number of starts).

When there is no mains, besides the uninterruptible power supply, one (or even two) extra generators must be foreseen, as shown in Figure 7.9. In total we will use three generators: one to power the station, another as a back-up to the first one, while the third is available for periodic maintenance.

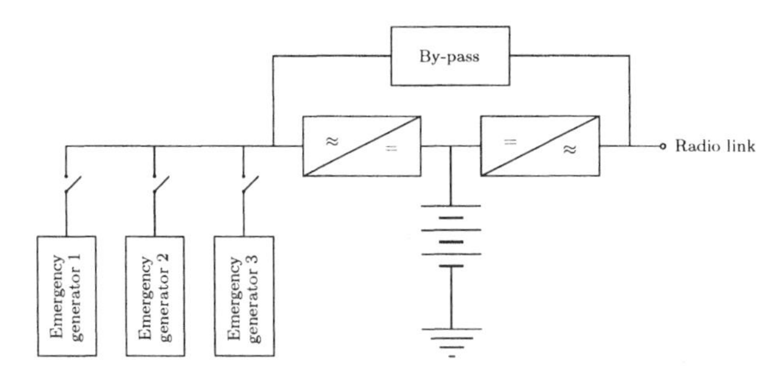

Fig. 7.9 Uninterruptible power supply when there is no mains.

When the station peak power requirements are small, say under about 1000 W, photoelectric cells, wind turbines or even thermoelectric generators may be considered as the primary power source. According to the choice of the primary power source, which is dependent on initial and running costs and on local conditions, the configuration of the uninterruptible power supply may require some adjustments.

7.10 BUILDINGS

Microwave radio link equipment with few exceptions, such as antennas, towers and the feeders, cannot be installed outdoors. Hence the link designer has to look into this subject as well.

Since most relay stations are unmanned the required space houses only the radio equipment and the power supply. In some cases additional storage space for spares may be required.

In most cases radio equipment operates with an ambient temperature in the range −5 to +45 °C, with 90 percent humidity (non-condensing) and, in many cases, may do without air conditioning, provided buildings are adequately ventilated, designed and built.

Recently relay stations have been installed in specially designed prefabricated shelters that may be transported from the factory to the station site, ready to be used. In some cases, station shelters may be buried underground to minimize temperature changes.

REFERENCES

1. Andrew Corporation, *Andrew Catalog 37*, Andrew Co., Orland Park, IL, 1997.
2. ITU, *ITU-R Recommendations on CD-ROM*, UIT, Geneva, UIT, 2000.
3. ITU, *ITU-T Recommendations on CD-ROM*, UIT, Geneva, 1997.

Problems

7:1 Calculate the internal rate of return of a link which cost $1 million, paid as a lump sum on the day it entered commercial exploitation, and which provided a net annual income is $ 100 k for 20 years. Assume fixed prices throughout system life.

7:2 Calculate the internal rate of return of a link which cost $1 million, paid as a lump sum on the day it entered commercial exploitation, and which provided a net annual income increasing linearly from $ 10 k to $ 200 k for 20 years. Assume fixed prices throughout system life.

7:3 Assuming that suitable radio channels are available in the 2, 4, 8 and 13 GHz bands, what would be the optimum frequency band for a 40-km analog line-of-sight link, if fading follows Morita's law and rainfall intensity exceeded for 0.01 percent of the time in the worst month is 40 mm/h? Assume that transmitter power output is $2/f^{1.3}$ W and receiver noise factor is $N_f = 5 + 0.2 * f$ dB, where f is the frequency in GHz. frequency. Neglect excess attenuation caused by oxygen and water vapor, and feeder losses.

7:4 Assuming that suitable radio channels are available in the 2, 4, 8 and 13 GHz bands, what would be the optimum frequency band for a 30-km line-of-sight analog link, employing a passive repeater 2 km from one terminal and 29 km from the other terminal? Consider that fading follows Morita's law and rainfall intensity exceeded for 0.01 percent of the time in the worst month is 40 mm/h. Take transmitter power output as $2/f^{1.3}$ W and receiver noise factor as $N_f = 5 + 0.2 * f$ dB, where f is the frequency in GHz. Neglect excess attenuation caused by oxygen and water vapor, and feeder losses.

7:5 Calculate the required power supply reliability at each terminal of a 6-hop 280-km microwave radio link knowing that power supply failures failures

failures account for 20 percent of link failures and that the link should meet ITU-R maximum unavailability.

7:6 Assuming $MTBF$ 150 000 hours and 200 000 hours for a transmitter and a receiver, respectively, and that other components failures may be neglected, calculate equipment unavailability for a $1 + 0$ and a $1 + 1$ bi-direct ional link with $MTTR = 6$ hours and compare it to ITU-R maximum unavailability.

Appendix A Detailed Frequency Plans

A.1 INTRODUCTION

In this appendix we describe, with some detail, but not exhaustively, the ITU-R recommended frequency plans. The objective is to provide the reader with some insight in the complexity of spectrum management, which is continuously facing the need to adopt systems that make an efficient use of the available bandwidth and the practical difficulties involved in modifying frequency plans in ways that are incompatible with existing equipment, whose useful lifetime may often reach 25 years. The differences between the development stages of ITU-R member states, and hence their priorities, make spectrum management all the more difficult, as recommendations are approved unanimously.

ITU-R [2] recommendation F.746-4 presents criteria to design frequency plans for microwave radio links. This recommendation points out that homogenous plans are to be preferred. These frequency plans are defined by the parameters XS, YS, ZS and DS:

- XS, difference between central frequencies of adjacent channels for the same polarization and transmission direction;

- YS, difference between central frequencies of the nearest go and return channels;
- ZS, difference between central frequencies of the edge channels and the nearest frequency band limit;
- DS, difference between the central frequency of a go channel and the corresponding return channel.

A.2 2 GHZ BAND

According to the ITU-R [2] Recommendation F.283-5 the 2 GHz band is reserved for:

- analog systems for 60, 120 and 300 telephone channels;
- analog systems for 960 telephone channels with 140 kHz r.m.s. frequency deviation, in the subband from 2500 to 2700 MHz;
- digital systems, with bit rates up to 70 Mbit/s, with band widths equivalent to the aforementioned analog systems.

According to this recommendation the 2 GHz band occupies 800 MHz (non-contiguous) and is divided into four subbands, each with it own central frequency f_0:

- 1700 to 1900 MHz (f_0 = 1808 MHz);
- 1900 to 2100 MHz (f_0 = 2000 MHz);
- 2100 to 2300 MHz (f_0 = 2203 MHz);
- 2500 to 2700 MHz (f_0 = 2586 MHz), with the exception of the 2690 to 2700 MHz, where all transmissions are forbidden.

The band between 2300 to 2500 MHz is reserved for industrial, scientific and medical applications of microwaves.

In each subband there are six go and six return radio channels separated by 14 MHz, in a simple frequency plan or, preferably, in a interleaved frequency plan where the central frequencies for radio channels (in MHz) are given by:

- lower half band

$$f_n = f_0 - 108.5 + 14 \cdot n; \tag{A.1}$$

- upper half band

$$f_n^{'} = f_0 + 10.5 + 14 \cdot n; \tag{A.2}$$

where n = 1, 2, 3, 4, 5, 6.

If necessary, additional radio channels may be inserted between the main channels, with a center frequency 7 MHz above the main radio channels center frequency.

According to ITU-R [2] Recommendation F.382-7 the 2 GHz band may also be organized as the 4 GHz band, that is, differently from the previous description.

A.3 4 GHZ BAND

According to the ITU-R [2] Recommendation F.382-7 the 4 GHz band is reserved for:

- analog links with capacity from 600 and 1800 telephone channels, or equivalent;
- 2 × 34 Mbit/s, 2 Mbit/s, 140 Mbit/s or synchronous hierarchy bit rates digital links.

According to the same recommendation the 2 GHz band may be divided into two 400 MHz subbands (upper and lower), for analog links with a capacity between 600 and 1800 telephone channels and 34 MBit/s digital links.

The frequency plan is interleaved and in each 400 MHz subband there are six go and six return radio channels, separated by 29 MHz, whose central frequencies are given by:

- lower half band

$$f_n = f_0 - 208 + 29 \cdot n; \tag{A.3}$$

- upper half band

$$f_n^{'} = f_0 + 5 + 29 \cdot n; \tag{A.4}$$

where $n =$ 1, 2, 3, 4, 5, 6 and the central frequency f_0 is:

- 1903 MHz in the lower 2 GHz subband;
- 2101 MHz in the upper 2 GHz subband;
- 4003.5 MHz, in the 4 GHz band.

If necessary, additional radio channels, interleaved with the main radio channels, may be provided with a central frequency 14.5 MHz below the central frequency of the main channels. When the main radio channels are occupied with 1800 telephone channels links, or 2 × 34 Mbit/s,2 × 45 Mbit/s, 140 Mbit/s or synchronous hierarchy bit rates, additional radio channels may not be possible.

ITU-R [2] Recommendation F.635-5 foresees the use of the 4 GHz band for high-capacity digital links (90, 140, 200 Mbit/s or the first synchronous hierarchy). Two alternative frequency plans are suggested. Onc is simply interleaved and the other superimposed. The channel center frequencies, in MHz, are given by

$$f_n = 4200 - 10 \cdot m, \tag{A.5}$$

where m takes integer values 1, 2, 3, ... , according to the available band width.

A.4 5 GHZ BAND

The 5 GHz band, from 4.4 to 5.0 GHz, is the subject of ITU-R [2] Recommendations F.746-4 (Annex 2) and F.1099-3 . In the first case a simply alternating frequency plan, with 28 MHz between radio channels, is suggested, allowing up to 10 go and 10 return radio channels, each able to take up to 4 times 34 Mbit/s or 140 Mbit/s, with 64-QAM modulation.

In the upper half band the radio channels frequencies, in MHz, are given by

$$f_n = 4700 - 310 + 28n, \tag{A.6}$$

and in the upper half band

$$f_n = 4700 + 2 + 28n, \tag{A.7}$$

with $n = 1, 2, \ldots, 10$.

Recommendation F.1099-3 puts forward a radio channel spacing which is a multiple of 10 MHz with simply alternating or superimposed frequency plans. In the annexes to this recommendation the following examples may be found:

- Doubly alternating frequency plan with 40 MHz channel spacing, allowing up to seven radio channels in each half band, with a capacity up to 2 × 155 Mbit/s (using 256-QAM modulation) with the following center frequencies:

  $$f_n = 4700 - 310 + 40n, \tag{A.8}$$

  in the upper half band, and

  $$f_n = 4700 - 10 + 40n, \tag{A.9}$$

  in the lower half band, with $n = 1, 2, \ldots, 7$.

- Superimposed frequency plan with 60 MHz channel spacing, for multicarrier systems with a capacity of 1 × 155 or 2 × 155 Mbit/s.

- Simply alternating frequency plan, with 40 MHz channel spacing, allowing for four go and four return channels, between 4.540 and 4.900 GHz, each with a capacity of 4 × 45 Mbit/s, 6 × 45 Mbit/s or 2 × 155 Mbit/s (with 512-QAM modulation), with the following center frequencies in MHz:

$$f_n = 4720 - 195 + 40n, \tag{A.10}$$

in the upper half band, and

$$f_n = 4720 - 5 + 40n, \tag{A.11}$$

in the lower half band, with $n = 1, 2, 3, 4$.

- Simply alternating frequency plan, with 20 MHz channel spacing, allowing for eight go and eight return radio channels, between 4.540 and 4.900 GHz, each with a capacity of 2 × 45 Mbit/s, 3 × 45 Mbit/s or 1 × 155 Mbit/s (with 512-QAM modulation), with the following center frequencies in MHz:

$$f_n = 4720 - 185 + 20n, \tag{A.12}$$

in the upper half band, and

$$f_n = 4720 + 5 + 20n, \tag{A.13}$$

in the lower half band, with $n = 1, 2, \ldots, 8$.

A.5 6 GHZ BAND

According to the ITU-R [2] Recommendation F.383-6 the lower 6 GHz band is for analog links with a capacity of 1800 telephone channels (or equivalent) or for 140 or 155 Mbit/s digital links.

The available 500 MHz allow up to eight go and eight return radio channels, in a simply (or preferably doubly) interleaved frequency plan, with the following center frequencies, in MHz

- lower half band:

$$f_n = f_0 - 259.45 + 29.65 \cdot n, \tag{A.14}$$

- upper half band:

$$f_n' = f_0 - 7.41 + 29.65 \cdot n, \tag{A.15}$$

where $n = 1, 2, 3, 4, 5, 6, 7, 8$ and the center frequency f_0 is 6175 MHz.

If necessary, additional radio channels may be foreseen, in between the main radio channels, with a center frequency 14.825 MHz below the center frequency of the main channel. When the main radio channels are occupied with carriers for 1800 telephone channels, the additional radio channels may not be used.

Provided both the main and the additional radio channels are used by up to 600 telephone channels systems, then it is possible to set up to 16 go and 16 return radio channels.

Annex I to Recommendation F.383-6 defines, in alternative, the following homogenous frequency plans:

- a superimposed frequency plan with six go and six return (90 MHz) radio channels, each for 140 Mbit/s, with 4 PSK modulation;
- a superimposed frequency plan with eight go and eight return radio channels, 60 MHz apart, each for 1 × 155 Mbit/s with 16-QAM modulation or 2 × 155 Mbit/s with 256-QAM modulation.

According to the ITU-R [2] Recommendation F.384-7 the upper 6 GHz band may also be used to:

- analog links with a capacity between 1260 and 2700 telephone channels;
- digital links with bit rates of the order of 140 Mbit/s.

The available band width (680 MHz) and the interleaved frequency plan allow for eight go and eight return radio channels, for the higher capacity analog and digital links, or 16 go and 16 return radio channels for the lower capacity analog or digital links.

The center frequencies of the radio channels, in MHz, are as follows:

- analog links for 2700 telephone channels or digital links at 140 or 155 Mbit/s, with $n = 1, 2, 3, 4, 5, 6, 7, 8$ and $f_0 = 6770$ MHz:
  - lower half band
$$f_n = f_0 - 350 + 40 \cdot n, \tag{A.16}$$
  - upper half band:
$$f_n' = f_0 - 10 + 40 \cdot n, \tag{A.17}$$
- analog links up to 1260 telephone channels or medium capacity digital links, with $n = 1, 2, 3, \ldots, 15, 16$ and $f_0 = 6770$ MHz
  - lower half band
$$f_n = f_0 - 350 + 20 \cdot n, \tag{A.18}$$
  - upper half band
$$f_n' = f_0 - 10 + 20 \cdot n. \tag{A.19}$$

A.6 7 GHZ BAND

According to the TU-R [2] Recommendation F.385-6 the 7 GHz band is for analog radio relay links with capacities of 60, 120 or 300 telephone channels but, by agreement of the Administrations involved, it may be used for higher capacity links.

The available 300 MHz allow for 20 go and 20 return radio channels in a simple frequency plan where the center frequency is given in MHz by:

- lower half band

$$f_n = f_0 - 154 + 7 \cdot n, \tag{A.20}$$

- upper half band

$$f_n^{'} = f_0 + 7 + 7 \cdot n, \tag{A.21}$$

where $n = 1, 2, 3, \ldots, 19, 20$ and $f_0 = 7575$ MHz.

Annex I to F.385-6 describes a simple frequency plan with five go and five return radio channels each with capacity for 34 Mbit/s, or 300 telephone channels, with center frequencies given by:

- lower half band

$$f_n = f_0 - 161 + 28 \cdot n, \tag{A.22}$$

- upper half band:

$$f_n^{'} = f_0 - 7 + 28 \cdot n, \tag{A.23}$$

where $n = 1, 2, \ldots, 5$ and $f_0 = 7575$ MHz.

Annex II to the same Recommendation defines a simple frequency plan with 28 go and 28 return radio channels, with 5 MHz spacing and center frequencies given by:

- lower half band

$$f_n = f_0 - 152.5 + 5 \cdot n, \tag{A.24}$$

- upper half band

$$f_n^{'} = f_0 + 7.5 + 5 \cdot n, \tag{A.25}$$

where $n = 1, 2, \ldots, 28$ and $f_0 = 7592.5$ MHz.

The same frequency plan with 10 or 20 MHz spacing offers a capacity per radio channel of 12.6 (8×1.544) or 19 (12×1.544) Mbit/s.

Annex III describes an alternate frequency plan with 10 go and 10 return radio channels spaced 28 MHz, each with a capacity for 140 Mbit/s. In turn, Annex IV defines a simple frequency plan with eight go and eight return radio channels with 28 MHz spacing, double the number of channels with one half the spacing or four times the number of channels with one fourth of the spacing.

A.7 8 GHZ BAND

According to the ITU-R [2] Recommendation F.386-6 the 8 Ghz band is for analog radio relay links for television or up to 960 telephone channels. Additionally, analog links for 300 telephone channels may be foreseen. Alternatively it may be used for low-, medium- or high-capacity digital links with different frequency plans.

The available 300 MHz band width and an interleaved frequency plan allow for six go and six return radio channels, for the higher capacity links, or 12 go and 12 return channels, for the lower capacity links. The center frequencies of the radio channels are given, in MHz, by:

- lower half band

$$f_n = f_0 - 151.614 + 11.662 \cdot n, \tag{A.26}$$

- upper half band

$$f_n' = f_0 + 11.662 \cdot n, \tag{A.27}$$

where $f_0 = 8350$ MHz and n equals:

- 1, 3, 5, 7, 9 or 11, for the links with a capacity of 960 telephone channels;
- 1, 2, 3, ... , 11 or 12, or the links with a capacity of 300 telephone channels.

Annex II to Recommendation F.386-6 describes a superimposed frequency plan with 12 go and 12 return radio channels each with a capacity of 90 Mbit/s. This plan introduces a spacing of 5.56 MHz between center frequencies of radio channels with orthogonal polarizations. The center frequencies are given, in MHz:

- lower half band, with n =1, 3, 5, 7, 9, 11,

$$f_n = 8000 - 275 + 20.37 \cdot n; \tag{A.28}$$

- lower half band, with n = 2, 4, 6, 8, 10, 12,

$$f_n = 8000 - 295.37 + 20.37 \cdot n + 5.56; \tag{A.29}$$

- upper half band, with n = 1, 3, 5, 7, 9, 11,

$$f_n' = 8000 + 30.56 + 20.37 \cdot n; \tag{A.30}$$

- upper half band, with n = 2, 4, 6, 8, 10, 12,

$$f_n' = 8000 + 10.19 + 20.37 \cdot n - 5.56. \tag{A.31}$$

Annex III to Recommendation F.386-6 describes a frequency plan for low capacity (2×8 Mbit/s) or medium capacity (34 Mbit/s) digital links, with 12 (or 6) go and as many return channels spaced 7 (or 14) MHz, whose center frequencies, in MHz, are given by

- 34 Mbit/s per radio channel, with n = 1, 2, ... , 6:
  - lower half band

$$f_n = 8387.5 - 108.5 + 14 \cdot n; \tag{A.32}$$

– upper half band

$$f_n = 8387.5 + 10.5 + 14 \cdot n; \qquad (A.33)$$

- 2×8 Mbit/s per radio channel:
  - lower half band

$$f_n = 8387.5 - 108.5 + 7 \cdot n; \qquad (A.34)$$

  - upper half band:

$$f_n = 8387.5 + 17.5 + 7 \cdot n. \qquad (A.35)$$

Annex IV to ITU-R Recommendation F.386-6 defines a simply interleaved frequency, with eight go and eight return radio channels, each with a capacity up to 1800 telephone channels or high-capacity digital links up to 140 Mbit/s. These 28 MHz channels may be subdivided in 14 MHz or 7 MHz channels. Center frequencies in MHz are given by:

- 28 MHz channels:
  - lower half band, with $n = 1, 2, \ldots, 8$

$$f_n = 8157 - 259 + 28 \cdot n, \qquad (A.36)$$

  - upper half band, with $n = 1, 2, \ldots, 8$

$$f_n = 8157 + 7 + 28 \cdot n; \qquad (A.37)$$

- 14 MHz channels:
  - lower half band, with $n = 1, 2, \ldots, 16$

$$f_n = 8157 - 259 + 14 \cdot n; \qquad (A.38)$$

  - upper half band, with $n = 1, 2, \ldots, 16$

$$f_n = 8157 + 7 + 14 \cdot n; \qquad (A.39)$$

- 7 MHz channels:
  - lower half band, with $n = 1, 2, \ldots, 32$

$$f_n = 8157 - 252 + 7 \cdot n; \qquad (A.40)$$

  - upper half band, with $n = 1, 2, \ldots, 32$

$$f_n = 8157 + 14 + 7 \cdot n. \qquad (A.41)$$

A.8 10 GHZ BAND

The 10 GHz band, which occupies the radio spectrum from 10.5 and 10.68 GHz, is simultaneously used for the mobile and the fixed service, according to ITU-R [2] Recommendation F.747. The recommended frequency plan is simple and homogenous, with a spacing of 1.25 or 3.5 MHz. These radio channels are particularly well suited for the small and medium capacity (up to 8 Mbit/s) digital links.

Radio channel center frequencies f_n are given, in MHz, by:

- 1.25 MHz spacing ($n = 1, 2, \dots , 103$)

$$f_n = 11701 - 1151 + 1.25 \cdot n; \tag{A.42}$$

- 3.5 MHz spacing ($n = 1, 2, \dots , 50$)

$$f_n = 11701 - 1200.5 + 3.5 \cdot n. \tag{A.43}$$

An alternative, described in Annex I to ITU-R Recommendation F.747, is a simple frequency plan with 7 MHz spacing where the center frequencies of the radio channels are given, in MHz, by:

- lower half band

$$f_n = 11701 - 1204 + 7 \cdot n, \tag{A.44}$$

- upper half band:

$$f_n = 11701 - 1113 + 7 \cdot n, \tag{A.45}$$

where $n = 1, 2, \dots , 12$.

A.9 11 GHZ BAND

According to the ITU-R[2] Recommendation F.387-8 the 11 GHz band is for high-capacity analog (up to 1800 telephone channels) and digital (up to 140 and 155 Mbit/s) links. In case of analog links the 1000 MHz band width available allow for 12 go and 12 return radio channels whose center frequencies are given, in MHz, by:

- lower half band

$$f_n = f_0 - 525 + 40 \cdot n, \tag{A.46}$$

- upper half band:

$$f'_n = f_0 + 5 + 40 \cdot n, \tag{A.47}$$

where $n = 1, 2, 3, \dots , 11, 12$ and $f_0 = 11\,200$ MHz.

Although simple frequency plans may be used, interleaved plans are to be preferred.

If needed, 11 secondary channels may be arranged between the main channels, with a center frequency 20 MHz below the adjacent main channel.

For small and medium capacity digital links that require 40 MHz spacing a superimposed frequency plan is recommended. The center frequencies of each pair of radio channels (with orthogonal polarizations) are given, in MHz, by:

- lower half band

$$f_n = f_0 - 545 + 40 \cdot n, \tag{A.48}$$

- upper half band

$$f_n' = f_0 - 15 + 40 \cdot n, \tag{A.49}$$

where n = 2, 3, 4, ..., 11, 12 and f_0 = 11 200 MHz.

For high-capacity digital links any of the frequency plans recommended for the analog links (either simple or interleaved) may be used. It is also possible to use another simply interleaved frequency plan, with 12 go and 12 return radio channels, whose frequencies, in MHz, are given by:

- lower half band

$$f_n = f_0 - 505 + 40 \cdot n, \tag{A.50}$$

- upper half band:

$$f_n' = f_0 - 15 + 40 \cdot n, \tag{A.51}$$

where n = 1, 2, 3, ... , 11, 12 and f_0 = 11 200 MHz.

A.10 13 GHZ BAND

Following the ITU-R [2] Recommendation F.497-6 the 13 GHz band is reserved, among other uses, for analog links, with a capacity up to 960 telephone channels, and for medium and large capacity digital links.

Making use of an interleaved frequency plan, the available 500 MHz band width allows for eight go and eight return radio channels, whose center frequencies, in MHz, are given by:

- lower half band

$$f_n = f_0 - 259 + 28 \cdot n, \tag{A.52}$$

- upper half band

$$f_n' = f_0 + 7 + 28 \cdot n, \tag{A.53}$$

where n = 1, 2, 3, ... , 7, 8 and f_0 is, preferably, 12 996 MHz.

For the 34 Mbit/s digital links the number of radio channels may be doubled with a superimposed frequency plan. For the 70, 140 and 155 Mbit/s digital links only the even (n=2, 4, 6, 8) radio channels of the interleaved plan, or all the channels of the superimposed plan, may be used.

Additional radio channels, with a maximum capacity of 300 telephone channels (analog) or 240 telephone channels (digital), may be interleaved between the main channels with center frequencies 14 MHz above the adjacent main channel.

This band also allows for low capacity (30 telephone channels) digital links in three different ways. The radio channels center frequency are given, in MHz, by:

- Solution I (m = 1, 2, 3, 4 and n = 1, 2)
  - lower half band

$$f_n = f_0 - 276.5 + 28 \cdot n + 7 \cdot m, \tag{A.54}$$

  - upper half band

$$f_n^{'} = f_0 - 10.5 + 28 \cdot n + 7 \cdot m; \tag{A.55}$$

- Solution II (m = 3, 4, 5, 6 and, if required, also m = 1, 2, 7, 8;)
  - lower half band

$$f_n = f_0 - 66.5 + 7 \cdot m, \tag{A.56}$$

  - upper half band:

$$f_n^{'} = f_0 + 3.5 + 7 \cdot m; \tag{A.57}$$

- Solution III (m=1, 2, 3, ... ,7, 8 and n =1, 2 or, by agreement, n = 3, 4, ... , 7, 8)
  - lower half band

$$f_n = f_0 - 273 + 28 \cdot n + 3.5 \cdot m, \tag{A.58}$$

  - upper half band:

$$f_n^{'} = f_0 - 7 + 28 \cdot n + 3.5 \cdot m. \tag{A.59}$$

In all cases the center frequency f_0 is 12 996 MHz.

For some digital links, with a capacity up to 960 telephone channels, it is possible to use a simply interleaved frequency plan, with six go and six return radio channels, whose center frequencies, in MHz, are given by:

- lower half band

$$f_n = f_0 - 259 + 35 \cdot n, \tag{A.60}$$

- upper half band

$$f_n^{'} = f_0 + 21 + 35 \cdot n, \tag{A.61}$$

where n =1, 2, 3, 4, 5, 6 and f_0 = 12 996 MHz.

A.11 15 GHZ BAND

According to the ITU-R [2] Recommendation F.636-3 the 15 GHz band is reserved for small and medium capacity digital links. The available band width may be 950 MHz (14.4 to 15.35 GHz) or 850 MHz (14.5 to 15.35 GHz), according to the region. Frequency plans should preferable be superimposed with 28 MHz spacing, allowing for 32 go and 32 return radio channels in the 950 MHz band width, or 30 go and 30 return radio channels in the 850 MHz band width.

Let N be the number of go (or return) channels. In these conditions radio channel center frequencies are given, in MHz, by:

- lower half band

$$f_n = f_r + a + 28 \cdot n, \tag{A.62}$$

- upper half band

$$f_n^{'} = f_r + 3626 - 28 \cdot (N - n), \tag{A.63}$$

where f_r=11 701 MHz and:

- 14.4 to 15.35 GHz band $a = 2688$ MHz, $N \leq 32$ and $n = 1, 2, ..., 16$,
- 14.5 to 15.35 GHz band: $a = 2786$ MHz, $N \leq 30$ and $n = 1, 2, ..., 15$.

It is also possible to foresee a superimposed frequency plan, with 14 MHz radio channel spacing, where the center frequencies are given, in MHz, by:

- lower half band

$$f_n = f_r + a + 14 \cdot n, \tag{A.64}$$

- upper half band

$$f_n^{'} = f_r + 3640 - 14 \cdot (N - n), \tag{A.65}$$

where f_r=11 701 MHz and:

- 14.4 to 15.35 GHz band: $a = 2702$ MHz, $N \leq 32$ e $n = 1,\ 2,\ ...\ ,\ N$,
- 14.5 to 15.35 GHz band: $a = 2800$ MHz, $N \leq 30$ e $n = 1,\ 2,\ ...\ ,\ N$.

If small capacity channels, with 3.5 or 7 MHz spacing, are required, it is possible to interleave them between the channels of the 14 MHz frequency plan, or the larger spacing channels may be subdivided. In the latter case, the center frequencies of the small capacity channels, in MHz, are:

- 7 MHz spacing:

- lower half band

$$f_m = f_r + a + 28 \cdot n + 7 \cdot m, \tag{A.66}$$

- upper half band

$$f'_m = f_r + 3608.5 - 28 \cdot (N - n) + 7 \cdot m, \tag{A.67}$$

where:

- $m = 1, 2, 3$ or 4;
- n is the channel number of the subdivided base plan;
- N is, as before, the number of radio channels;
- $a = 2670.5$ MHz in the 14.4 – 15.35 GHz band;
- $a = 2768.5$ MHz in the 14.5 – 15.35 GHz band.

• 3.5 MHz spacing:

- lower half band

$$f_m = f_r + a + 28 \cdot n + 3.5 \cdot m, \tag{A.68}$$

- upper half band

$$f'_m = f_r + 3610.25 - 28 \cdot (N - n) + 7 \cdot m, \tag{A.69}$$

where:

- $m =$ 1, 2, 3, 4, 5, 6, 7, 8;
- n is the channel number of the subdivided base plan;
- N is, as before, the number of radio channels;
- $a = 2672.25$ MHz in the 14.4 – 15.35 GHz band;
- $a = 2770.25$ MHz in the 14.5 – 15.35 GHz band.

A.12 18 GHZ BAND

According to ITU-R[2] Recommendation F.595-6 the 18 GHz band, that spans from 17.7 to 19.7 GHz, is reserved for the 34, 140 and 280 Mbit/s digital links. The 2000 MHz available band width may be organized with superimposed or simply interleaved frequency plans.

A.12.0.1 Superimposed frequency plans The radio channels center frequencies are given, in MHz, by:

- Links with a capacity of 280 Mbit/s, eight go and eight return radio channels[1], where $n = 1, 2, 3, 4$:
  - lower half band
$$f_n = f_0 - 1110 + 220 \cdot n, \tag{A.70}$$
  - upper half band
$$f_n' = f_0 + 10 + 220 \cdot n; \tag{A.71}$$
- Links with a capacity of about 140 Mbit/s, 16 go and 16 return radio channels, and $n = 1, 2, 3, \ldots, 7, 8$:
  - lower half band
$$f_n = f_0 - 1000 + 110 \cdot n, \tag{A.72}$$
  - upper half band
$$f_n' = f_0 + 10 + 110 \cdot n; \tag{A.73}$$
- Links with a capacity of about 34 Mbit/s, 70 go and 70 return radio channels, and $n = 1, 2, 3, \ldots, 34, 35$:
  - lower half band
$$f_n = f_0 - 1000 + 27.5 \cdot n, \tag{A.74}$$
  - upper half band:
$$f_n' = f_0 + 10 + 27.5 \cdot n. \tag{A.75}$$

A.12.0.2 Simply interleaved frequency plan

- Links with a capacity about 280 Mbit/s, seven go and seven return radio channels, with $n = 1, 2, 3, \ldots, 6, 7$:
  - lower half band
$$f_n = f_0 - 1000 + 110 \cdot n, \tag{A.76}$$
  - upper half band:
$$f_n' = f_0 + 120 + 110 \cdot n; \tag{A.77}$$
- Links with a capacity about 140 Mbit/s, 15 go and 15 return channels, with $n = 1, 2, 3, \ldots, 15$:
  - lower half band
$$f_n = f_0 - 945 + 55 \cdot n, \tag{A.78}$$

[1] In a superimposed frequency plan, the frequency is shared by two channels with orthogonal polarizations.

– upper half band:

$$f_n' = f_0 + 65 + 55 \cdot n. \qquad (A.79)$$

In all cases the center frequency f_0 is 18 700 MHz.

For the 155 Mbit/s links, using 4-PSK modulation, the preferred frequency plan is superimposed, with 110 MHz spacing, or simply interleaved with 55 MHz spacing. If these systems use 16-QAM modulation the frequency plan should be superimposed with 55 MHz spacing. The center frequency of the radio channels $n = 1, 2, \ldots, 16$ are given in (A.78) and (A.79) for $n = 1, 2, \ldots, 15$. Channels 1 and 17 are located 55 MHz below channel 2 and above channel 16, respectively.

For low-capacity (less than about 10 Mbit/s) links radio channels may be provided inside the higher capacity channels. For medium capacity links the possibility to set up frequency plan similar to those of the higher capacity links is left to the administrations concerned.

A.13 23 GHZ BAND

The 23 GHz band, which spans from 21.2 to 23.6 GHz, is organized according to ITU-R [2] Recommendation F.637-3. The 23 GHz band may be used for analog or digital links, with frequency plans based in 2.5 MHz or 3.5 MHz spacing. The radio channels center frequencies are given by:

- 2.5 MHz spacing, with f_r =21 196 MHz and $n = 1, 2, \ldots, 959$

$$f_n = f_r + 4 + 2.5 \cdot n, \qquad (A.80)$$

- 3.5 MHz spacing, with $f_r = 21\,196$ MHz and $n = 1, 2, \ldots, 685$

$$f_n = f_r + 3.5 + 3.5 \cdot n, \qquad (A.81)$$

where $f_r = 21\,196$ MHz.

A.14 25, 26 AND 28 GHZ BANDS

Frequency plans for the 25 GHz (24.25 to 25.25), the 26 GHz (25.25 to 27.5) and the 28 GHz (27.5 to 29.5) bands are defined in ITU-R [2] Recommendation F.748-3. As for the 23 GHz band, here the adopted frequency plans are based on 2.5 or 3.5 MHz spacing. The radio channels center frequencies are given by:

- 2.5 MHz spacing

$$f_n = f_r + 2 + 2.5 \cdot p, \qquad (A.82)$$

with $f_r = 24\,248$ MHz and:

- $p = 1, 2, \ldots, 399$ in the band from 24.25 to 25.25 GHz,
- $p = 401, 402, \ldots, 1299$ in the band from 25.25 to 27.5 GHz,
- $p = 1301, 1302, \ldots, 2099$ in the band from 27.5 to 29.5 GHz;

- 3.5 MHz spacing

$$f_n = f_r + 3.5 \cdot p, \tag{A.83}$$

with $f_r = 24\,248$ MHz and:

- $p = 1, 2, \ldots, 285$ in the band from 24.25 to 25.25 GHz,
- $p = 287, 288, \ldots, 928$ in the band from 25.25 to 27.5 GHz,
- $p = 930, 931, \ldots, 1500$ in the band from 27.5 to 29.5 GHz.

A.15 38 GHZ BAND

The 38 GHz band is ideal for short links, both analog and digital. The frequency plans, defined in ITU-R [2] Recommendation F.749-1, are simple, based in 2.5 and 3.5 MHz spacing. The radio channels center frequency is given, in MHz, by:

- 2.5 MHz spacing

$$f_n = f_r + 2.5 \cdot p, \tag{A.84}$$

with f_r =36 000 MHz and $p = 1, 2, \ldots, 1799$.

- 3.5 MHz spacing

$$f_n = f_r + 1 + 3.5 \cdot p, \tag{A.85}$$

with f_r =36 000 MHz and $p = 1, 2, \ldots, 1285$.

Some administrations group 2, 4, 8, 16 or 24 channels, with 3.5 MHz spacing to get 160 channels with 7 MHz, 80 channels with 14 MHz, 40 channels with 56 MHz, or 8 channels with 140 MHz, respectively. In all cases the guard bands are 58, 62 and 140 MHz in the lower, the upper band edges and in the center, respectively.

A.16 55 GHZ BAND

The band from 54.25 to 58.20 GHz, also known as the 55 GHz band, is reserved both for the fixed and the mobile services. The specific properties of propagation in the troposphere make the band ideal for very short analog and digital links.

ITU-R [1] Recommendation F.1100 prescribes the use of homogeneous frequency plans, based on 2.5 and 3.5 MHz spacing, as in the 23, 25, 26, 28 and 38 GHz bands. The radio channels center frequencies are given, in MHz, by:

- 2.5 MHz spacing
$$f_n = f_r + 2.5 \cdot p, \tag{A.86}$$
with f_r =54 250 MHz and p = 1, 2, ... , 1579;

- 3.5 MHz spacing
$$f_n = f_r + 3.5 \cdot p, \tag{A.87}$$
with f_r =54 250 MHz and p = 1, 2, ... , 1128.

Grouping 4, 8, 16 or 24 radio channels with 3.5 MHz spacing may be used to obtain 100 channels with 14 MHz, 50 channels with 28 MHz, 25 channels with 56 MHz or 10 channels with 140 MHz. In all cases the guard bands are 42, 38 and 70 MHz, in the lower, the upper band edges and in the center, respectively.

A.17 AUXILIARY RADIO LINKS

Besides main radio links, designed for the intended traffic, sometimes auxiliary links may be used to provide service channels required for command, control and maintenance of main radio links. Frequency plans for such auxiliary links are described in ITU-R [2] Recommendation 389-2 and are repeated here to complete the frequency plans overview.

When auxiliary links share the same frequency band as the main links their radio channels have the following center frequencies in MHz:

A.17.1 2 and 4 GHz bands

- Normal frequencies:
  - Lower half band: $f_0 - 204.5$ and $f_0 - 12$,
  - Upper half band: $f_0 + 8.5$ and $f_0 + 199$ (in the 4 GHz band, only by special agreement between the administrations);
- Interleaved frequencies:
  - Lower half band: $f_0 - 213.5$ and $f_0 - 23$,
  - Upper half band: $f_0 - 2.5$ and $f_0 + 190$,

where the center frequency f_0 is:

- 1903 MHz, for the subband between 1700 and 2100 MHz;
- 2101 MHz, for the subband between 2100 and 2500 MHz;
- 4003.5 MHz, for the 4 GHz band.

Polarization in the auxiliary links should always be orthogonal to the polarization in the nearest (in frequency) main link.

A.17.2 6 GHz band

- Radio links using frequency modulation:
  - Lower half band: $f_0 - 248.9$ and $f_0 - 3.1$,
  - Upper half band: $f_0 + 3.1$ and $f_0 + 248.9$;
- Radio links using amplitude or frequency modulation:
  - Lower half band: $f_0 - 249.5$ and $f_0 - 2.5$,
  - Upper half band: $f_0 + 2.5$ and $f_0 + 249.5$;

where the band center frequency f_0 is:

- 6175 MHz, for the frequency plans with 29.65 MHz radio channels;
- 6770 MHz, or the frequency plans with 40 MHz radio channels.

Polarization in the auxiliary links should always be orthogonal to the polarization in the nearest (in frequency) main link.

A.17.3 11 GHz band

According to ITU-R [2] Recommendation F387-8 the frequency plans for the auxiliary links may should be:

- Simple frequency plans
  - Lower half band: $f_0 - 485$ and $f_0 - 15$,
  - Upper half band: $f_0 + 15$ and $f_0 + 485$;
- Interleaved frequency plans
  - Lower half band: $f_0 - 495$ and $f_0 - 25$,
  - Upper half band: $f_0 + 2.5$ and $f_0 + 465$;

where the band center frequency f_0 is 11 200 MHz.

Auxiliary link polarization shoud be:

- orthogonal to the adjacent main radio link in the band center;
- the same as the adjacent main radio link in the band edges.

REFERENCES

1. ITU, *ITU-R Recommendations on CD-ROM*, UIT, Geneva, 1997.
2. ITU, *ITU-R Recommendations on CD-ROM*, UIT, Geneva, 2000.

Appendix B
Signal-to-Noise Ratio in Angle Modulation

B.1 INTRODUCTION

In this appendix we derive the output signal-to-noise ratio for frequency and phase modulation, as a function of the input carrier to noise ratio and the modulation parameters. Following we compare these two modulation techniques under different viewpoints. In the derivation we follow closely Brown and Glazier [1].

B.2 SIGNAL-TO-NOISE RATIO IN FREQUENCY AND PHASE MODULATION

Take a cosine carrier u_c, with unit amplitude and frequency f_c

$$u_c = \cos(\omega_c t), \tag{B.1}$$

$$\omega_c = 2\pi f_c \tag{B.2}$$

and a modulating signal u_m, also a cosine, with amplitude a_m and frequency f_m

$$u_m = a_m \cos(\omega_m t), \quad \text{(B.3)}$$

$$\omega_m = 2\pi f_m. \quad \text{(B.4)}$$

In phase modulation, carrier phase θ_i varies linearly with the modulating signal. If Δp is the phase change for a modulating signal with unity amplitude we have

$$\theta_i = \omega_c t + \Delta p\, a_m \cos(\omega_m t), \quad \text{(B.5)}$$

thus, the modulated signal becomes

$$u_i = \cos[\omega_c t + \Delta p\, a_m \cos(\omega_m t)]. \quad \text{(B.6)}$$

The phase demodulator outputs a signal u_o whose amplitude is proportional to the instantaneous difference between the modulated signal phase and the carrier phase

$$u_o = k_p(\theta_i - \omega_c t). \quad \text{(B.7)}$$

Substituting (B.5) into (B.7), we get

$$u_o = k_p \Delta p\, a_m \cos(\omega_m t), \quad \text{(B.8)}$$

which, applied to a load resistor R_o, corresponds to a power s_o given by

$$s_o = \frac{1}{2R_o} k_p^2 \Delta p^2 a_m^2. \quad \text{(B.9)}$$

In frequency modulation it is the instantaneous frequency, the derivative of phase with respect to time, that varies linearly with the modulating signal. If $\Delta\omega$ is the change of the angular frequency ω_c for a modulating signal with unity amplitude we get

$$\omega_i = \omega_c + \Delta\omega\, a_m \cos(\omega_m t). \quad \text{(B.10)}$$

Since

$$w_i = \frac{d\theta_i}{dt},$$

we get

$$\theta_i = \int_0^t \omega_i dt, \quad \text{(B.11)}$$

hence, substituting (B.10) into (B.11), yields

$$\theta_i = \omega_c t + \Delta\omega\, a_m \int_0^t \cos(\omega_m t) dt,$$

or

$$\theta_i = \omega_c t + \frac{\Delta\omega}{\omega_m} a_m \sin(\omega_m t), \quad \text{(B.12)}$$

and thus the modulating signal u_i may be written as

$$u_i = \cos\left[\omega_c t + \frac{\Delta\omega}{\omega_m} a_m \sin(\omega_c t)\right]. \tag{B.13}$$

The frequency demodulator, also known as the discriminator, outputs a signal u_o proportional to the difference between the instantaneous input signal frequency and the unmodulated carrier frequency

$$u_o = k_f(\omega_i - \omega_c). \tag{B.14}$$

Substituting (B.10) in(B.14), we get

$$u_o = k_f \Delta\omega \, a_m \cos(\omega_m t), \tag{B.15}$$

which corresponds, to a power s_o, on a a load R_o, given by

$$s_o = \frac{1}{2R_o} k_f^2 \Delta\omega^2 a_m^2. \tag{B.16}$$

Recalling that the radio frequency bandwidth b_{rf} for frequency modulation is approximately given by Carson's formula

$$b_{rf} = 2\left(\frac{\Delta\omega \, a_m}{2} + f_m\right), \tag{B.17}$$

and comparing the modulated signal for phase modulation (B.6) and for frequency modulation (B.13) we may write a similar expression for the radio frequency bandwidth in phase modulation

$$b_{rf} = 2\left(\frac{\Delta p \, a_m \, \omega_m}{2} + f_m\right), \tag{B.18}$$

or

$$b_{rf} = 2(\Delta p \, a_m + 1) f_m. \tag{B.19}$$

For equal bandwidths we get, equating (B.17) and (B.18),

$$\Delta p = \frac{\Delta\omega}{w_m}. \tag{B.20}$$

After having looked into the output of phase and frequency demodulators, we will now look into the response of these devices to noise in the presence of a unmodulated carrier. We will assume that noise at the demodulator input is white, narrow band, Gaussian, centered at angular frequency ω_c. Noise voltage u_n may be written as

$$u_n = x(t)\cos(\omega_c t) + y(t)\sin(\omega_c t), \tag{B.21}$$

where $x(t)$ e $y(t)$ are two independent random functions with zero average and equal standard deviation

$$\overline{x^2(t)} = \sigma^2, \quad (B.22)$$
$$\overline{y^2(t)} = \sigma^2, \quad (B.23)$$

and probability density functions given by

$$f(x) = \frac{1}{\sqrt{2\pi}\sigma} \exp\left(-\frac{x^2}{2\sigma^2}\right), \quad (B.24)$$
$$f(y) = \frac{1}{\sqrt{2\pi}\sigma} \exp\left(-\frac{y^2}{2\sigma^2}\right). \quad (B.25)$$

Considering, at the demodulator input, the simultaneous presence of a unmodulated cosine carrier, with amplitude u_c, and noise

$$u_i = u_c \cos(\omega_c t) + x(t)\cos(\omega_c t) + y(t)\sin(\omega_c t), \quad (B.26)$$

manipulating (B.26) we get

$$u_i = a(t)\cos[\omega_c t + \theta(t)], \quad (B.27)$$

where

$$a(t) = \sqrt{[u_c + x(t)]^2 + y^2(t)}, \quad (B.28)$$
$$\theta(t) = \arctan\left[\frac{y(t)}{u_c + x(t)}\right]. \quad (B.29)$$

If the demodulator is preceded by a limiter (as is usually the case):

$$a(t) = a(\text{constant}).$$

If, on the other hand, most of the time the input signal-to-noise ratio is much higher than unity

$$u_c \gg x(t), \quad (B.30)$$

we may simplify (B.29) into

$$\theta(t) = \frac{y(t)}{u_c}. \quad (B.31)$$

If the passband at the demodulator input is centered about f_c and has a bandwidth b_{rf}, then $y(t)$, responsible for the noise component $y(t)\sin(\omega_c t)$, has a spectrum between 0 and $b_{rf}/2$. We may then consider that $y(t)$ is the result of a sum of m sinewaves

$$y(t) = \sum_{j=1}^{m} a_j \sin(\omega_j t + \phi_j), \quad (B.32)$$

with amplitude, angular frequency and phase given by

$$\frac{\overline{a_j^2}}{2} = \frac{\overline{y^2(t)}}{m}, \tag{B.33}$$

$$\omega_j = 2\pi(j-1)\Delta f + 2\pi\frac{\Delta f}{2}, \tag{B.34}$$

$$m\Delta f = \frac{b_{rf}}{2}, \tag{B.35}$$

$$\phi_j = \text{any between 0 and } 2\pi. \tag{B.36}$$

This approximation is equivalent to considering that the noise $y(t)$ is periodic with period $1/\Delta f$ as large as Δf is small or when m is large.

The voltage at the phase demodulator output will be proportional to the phase difference between the input signal and the phase, at the same instant of time, of the unmodulated carrier. Thus, from (B.7), taking into account (B.27) and (B.31), we get

$$u_o = k_p \frac{y(t)}{u_c}. \tag{B.37}$$

Substituting the value of $y(t)$ given into (B.32) into (B.37), we get

$$u_o = \frac{k_p}{u_c}\sum_{j=1}^{m} a_j \sin(\omega_j t + \phi_j), \tag{B.38}$$

which shows that the output spectrum is constant with frequency.

The output noise power n_o, on a resistor R, within the signal bandwidth b_s may be calculated from (B.38)

$$n_o = \frac{k_p^2}{R u_c^2}\sum_{j=1}^{m_s}\frac{\overline{a_j^2}}{2}, \tag{B.39}$$

where

$$m_s = \frac{b_s}{\Delta f}. \tag{B.40}$$

But, from (B.32), we get

$$\overline{y^2(t)} = \frac{1}{2}\sum_{j=1}^{m}\overline{a_j^2}. \tag{B.41}$$

Considering that the energies (or the average square values of the associated voltages) of the elementary oscillators, that make up the noise voltage $y(t)$, are equal:

$$\begin{aligned} \frac{1}{2}\left(\sum_{j=1}^{m_s}\overline{a_j^2}\right) &= \frac{1}{2}\left(\sum_{j=1}^{m}\overline{a_j^2}\right)\frac{m_s}{m} \\ &= \frac{m_s}{m}\overline{y^2(t)}, \end{aligned}$$

and, substituting in (B.39)

$$n_o = \frac{k_p^2}{Ru_c^2}\,\overline{y^2(t)}\,\frac{m_s}{m},$$

and recalling (B.35) and (B.40)

$$n_o = \frac{k_p^2}{Ru_c^2}\,\overline{y^2(t)}\,\frac{2b_s}{b_{rf}}. \tag{B.42}$$

Since the output signal power of a phase demodulator is given in (B.9), the signal-to-noise ratio s/n_o becomes

$$\frac{s}{n_o} = \frac{u_c^2}{2\,\overline{y^2(t)}}\,\frac{\Delta p^2 a_m^2}{2}\,\frac{b_{rf}}{b_s}. \tag{B.43}$$

The (modulated or unmodulated) carrier power c, with amplitude uc, on a resistor R, is

$$c = \frac{u_c^2}{2R}, \tag{B.44}$$

and the narrow band noise power taken in (B.21) on the same resistor R

$$n = \frac{\overline{x^2(t)} + \overline{y^2(t)}}{2R}, \tag{B.45}$$

then, taking into account (B.23) and (B.44), the carrier to noise ratio c/n is:

$$\frac{c}{n} = \frac{u_c^2}{2\,\overline{y^2(t)}}. \tag{B.46}$$

Substituting (B.46) into (B.43), we arrive at an expression that will enable us to calculate the output signal-to-noise ratio, for a phase demodulator, from the input carrier-to-noise ratio and the modulation parameters

$$\frac{s}{n_o} = \frac{c}{n}\,\frac{\Delta p^2 a_m^2}{2}\,\frac{b_{rf}}{b_s}. \tag{B.47}$$

Take now the frequency demodulator. As we have seen the output voltage is proportional to instantaneous angular frequency difference between the modulated and the unmodulated carrier. Just as we did for the phase demodulator, we will apply at the demodulator input an un modulated cosine carrier with amplitude u_c and frequency f_c and narrow band, white, Gaussian noise. From (B.12) and (B.31) the instantaneous frequency becomes

$$\begin{aligned} \omega_i &= \frac{d[\omega_c t + \theta(t)]}{dt} \\ &= \omega_c + \frac{d\theta}{dt}. \end{aligned} \tag{B.48}$$

Introducing the approximation I (B.29) in (B.46), we get

$$\omega_i = \omega_c + \frac{1}{u_c}\frac{dy}{dt},$$

hence, at the demodulator output we get, from (B.14)

$$u_o = k_f \frac{1}{u_c}\frac{dy}{dt}. \tag{B.49}$$

If the demodulator input has a band width b_{rf} and is centered at f_c then $y(t)$, responsible for the noise component $y(t)\sin(\omega_c t)$, has a frequency spectrum within 0 and $b_{rf}/2$. Assuming again that $y(t)$ is the sum of m sinewaves (B.32), we get

$$u_o = \frac{k_f}{u_c}\sum_{j=1}^{m}[\omega_j a_j \cos(\omega_j t + \phi_j)]. \tag{B.50}$$

For the j elementary oscillator, the power on a resistor R is:

$$n_{o_j} = \frac{1}{2R}\frac{k_f^2}{u_c^2}\,\omega_j^2\,\overline{a_j^2}.$$

Considering (B.33) and (B.35), we get:

$$\begin{aligned} n_{o_j} &= \frac{1}{R}\frac{k_f^2}{u_c^2}\,\omega_j^2\,\frac{\overline{y^2(t)}}{n} \\ &= \frac{2}{R}\frac{(2\pi)^2 k_f^2\,\overline{y^2(t)}}{u_c^2\,b_{rf}} f_j^2\,df. \end{aligned}$$

When df tends to zero, the power n_{o_j} of the j oscillator tends to the noise power density $g_o(f)$

$$g_o(f) = \frac{8\pi^2 k_f^2\,\overline{y^2(t)}}{R u_c^2 b_{rf}} f^2. \tag{B.51}$$

Now the demodulator output noise power is frequency dependent and varies with the square of the frequency. Had we considered the noise voltage instead of the noise power and we would have found that it increases linearly with frequency. That is why the output noise in frequency modulation is said to have a triangular spectrum.

The noise power inside the modulating signal band width, from f_m to f_M, will be given by the integral of the noise power density

$$\begin{aligned} n_o &= \int_{f_m}^{f_M} g_o(f)df \\ &= \frac{8\pi^2 k_f^2\,\overline{y^2(t)}}{R\,u_c^2\,b_{rf}} f^2 \int_{f_m}^{f_M} f^2 df \\ &= \frac{8\pi^2 k_f^2\,\overline{y^2(t)}}{R\,u_c^2\,b_{rf}} f^2 \frac{f_M^3 - f_m^3}{3}. \end{aligned} \tag{B.52}$$

Since the signal power at the demodulator output is given by (B.16), the signal-to-noise ratio at that point is

$$\frac{s}{n_o} = \frac{3\Delta\omega^2 a_m^2 b_{rf} u_c^2}{16\pi^2 \overline{y^2(t)}(f_M^3 - f_m^3)}.$$

Recalling (B.46) and the relation between frequency and angular frequency, the previous expression may be written, in terms of the carrier-to-noise ratio at the frequency demodulator input, as

$$\frac{s}{n_o} = \frac{3\Delta f^2 a_m^2 b_{rf}}{2(f_M^3 - f_m^3)} \frac{c}{n}. \tag{B.53}$$

Since

$$f_M^3 - f_m^3 = (f_M - f_m)(f_M^2 + f_M f_m + f_m^2),$$

or

$$f_M^3 - f_m^3 = b_s(f_M^2 + f_M f_m + f_m^2), \tag{B.54}$$

from (B.51) and (B.52)

$$\frac{s}{n_o} = \frac{3}{2} \frac{\Delta f^2 a_m^2}{f_M^2 + f_M f_m + f_m^2} \frac{b_{rf}}{b_s} \frac{c}{n}. \tag{B.55}$$

When $f_M \gg f_m$ (B.53) may be simplified as follows

$$\frac{s}{n_o} = \frac{3}{2} \frac{c}{n} \frac{\Delta f^2 a_m^2}{f_M^2} \frac{b_{rf}}{b_s}. \tag{B.56}$$

B.3 COMPARING FREQUENCY AND PHASE MODULATION

After deriving the output signal-to-noise ratio as a function of the input carrier-to-noise ratio we may compare frequency and phase modulation.

From (B.47) and (B.56) for equal:

- carrier-to-noise ratio at the demodulator input,
- modulating signal,
- modulating signal bandwidth,
- modulated signal bandwidth,

we find out that frequency modulation compared with phase modulation offers a signal-to-noise improvement m given by

$$m = \frac{3\Delta f^2}{f_M^2 \Delta p^2}.$$

Since to get equal modulated signal band widths, we have from (B.20)

$$\Delta p = \frac{\Delta f}{f_M},$$

hence $m = 3$ (or 4.8 dB).

This improvement justifies the choice of frequency modulation over phase modulation, notwithstanding the unevenness with which the modulating signal frequencies are dealt with.

Another possible way to compare frequency and phase modulation is on the basis of the noise power in the upper channel of the baseband, for equal:

- noise power in the whole of the baseband,
- modulating signal,
- modulating signal bandwidth.

Equalizing the baseband noise power in phase modulation (B.40) and in frequency modulation (B.50), we get

$$\frac{k_p^2}{Ru_c^2}\,\overline{y^2(t)}\,\frac{2b_s}{b_{rf_{PM}}} = \frac{8\pi^2 k_f^2\,\overline{y^2(t)}}{R\,u_c^2\,b_{rf_{FM}}}\,\frac{f_M^3 - f_m^3}{3}. \tag{B.57}$$

Assuming that in the baseband we have $f_M \gg f_m$, from (B.54) we get

$$f_M^3 - f_m^3 = b_s f_M^2. \tag{B.58}$$

Substituting (B.58) in (B.57), we get

$$k_p^2 = \frac{4\pi^2 k_f^2 f_m^2}{3}\,\frac{b_{rf_{PM}}}{b_{rf_{FM}}}. \tag{B.59}$$

Consider now the baseband upper channel with bandwidth b_s and center frequency f_c, which is approximately equal to the highest baseband frequency f_M. In phase modulation the noise power in this channel will be, from (B.40)

$$n_{o_{PM}} = \frac{k_p^2}{Ru_c^2}\,\overline{y^2(t)}\,\frac{2b_s}{b_{rf_{PM}}}.$$

Substituting in the previous expression the value of k_p given in (B.59) we get

$$n_{o_{PM}} = \frac{4\pi^2 k_f^2 f_M^2}{3Ru_c^2}\,\overline{y^2(t)}\,\frac{2b_s}{b_{rf_{FM}}},$$

which, recalling (B.50), may be rewritten as

$$n_{o_{PM}} = \frac{n_{o_{FM}}}{3}.$$

Taking again the concept of improvement m of frequency modulation over phase modulation we get now

$$m = \frac{1}{3},$$

which shows that, under these conditions, and unlike the previous case, phase modulation is to be preferred to frequency modulation. We should note, however, that the bandwidth of the modulated signal b_{rf} is not the same in both modulations.

Yet another possible way of comparing frequency and phase modulation is on the basis of the noise in the least favorable channel (the highest) of a frequency division multiplex modulating signal, for the same radio frequency bandwidth.

In phase modulation, the signal-to-noise ratio in any channel, and obviously also in the least favorable channel, is the same, given by (B.47)

$$\left(\frac{s}{n_o}\right)_{PM} = \frac{c}{n}\frac{\Delta p^2 a_m^2}{2}\frac{b_{rf}}{b_s}.$$

For frequency modulation, we have from (B.55)

$$\left(\frac{s}{n_o}\right)_{FM} = \frac{3}{2}\frac{\Delta f^2 a_m^2}{f_M^2 + f_M f_m + f_m^2}\frac{b_{rf}}{b_s}\frac{c}{n}.$$

Noting that in the highest channel of a frequency division multiplex signal we have $f_M \approx f_m$, we get

$$\left(\frac{s}{n_o}\right)_{FM} = \frac{1}{2}\frac{\Delta f^2 a_m^2}{f_M^2}\frac{b_{rf}}{b_s}\frac{c}{n}.$$

From (B.20) we get for equal radio frequency band width

$$\Delta p = \frac{\Delta f}{f_M}.$$

Hence:

$$\left(\frac{s}{n_o}\right)_{FM} = \frac{\Delta p^2 a_m^2}{2}\frac{b_{rf}}{b_s}\frac{c}{n},$$

which shows that, under these conditions, both types of modulation are equivalent.

This conclusion remains the same even when the amplitudes of the modulating signals are different, provided that phase and frequency deviations caused by the test signal in the upper channel are related through (B.20), that is,

$$(\Delta p a_m)_{PM} = \left(\frac{\Delta f a_m}{f_M}\right)_{FM}. \tag{B.60}$$

In frequency modulation, where the noise in the demodulated signal is frequency dependent, there may be an improvement if we modify the amplitude of the modulating signal – pre-emphasis – and correct it after demodulation – de-emphasis. Obviously both operations should not alter the demodulating signal power. The noise power after demodulation is however modified because noise is affected only by de-emphasis.

Let $a(f)$ be the pre-emphasis curve (in voltage). The de-emphasis obeys:

$$b(f) = \frac{1}{a(f)}.$$

From (B.51) the output noise power in the channel with center frequency f_c and bandwidth b_s becomes:

$$n_o = \frac{8\pi^2 k_f^2\, \overline{y^2(t)}}{R\, u_c^2\, b_{rf}} \int_{f_1}^{f_2} \frac{f^2}{a^2(f)} df, \tag{B.61}$$

where:

$$f_1 = f_c - \frac{b_s}{2}, \tag{B.62}$$

$$f_2 = f_c + \frac{b_s}{2}. \tag{B.63}$$

The improvement m brought about by the ensemble of pre-emphasis and de-emphasis (in short emphasis) will be the ratio of the signal-to-noise ratios with and without emphasis. Since the signal is not affected by emphasis, the improvement may be calculated simply by the ratio of the noise powers without and with emphasis

$$\begin{aligned} m &= \frac{\int_{f_1}^{f_2} f^2 df}{\int_{f_1}^{f_2} \frac{f^2}{a^2(f)} df} \\ &= \frac{f_2^3 - f_1^3}{3 \int_{f_1}^{f_2} \frac{f^2}{a^2(f)} df}. \end{aligned} \tag{B.64}$$

In order not to modify the baseband signal power, the pre-emphasis curve must obey

$$\int_{f_m}^{f_M} a^2(f) df = f_M - f_m. \tag{B.65}$$

Phase modulation may be obtained from frequency modulation using the following pre-emphasis:

$$a(f) = kf, \tag{B.66}$$

where the constant k is chosen taking into consideration the condition (B.65). Substituting (B.66) in (B.65), we get

$$k^2 = \frac{3(f_M - f_m)}{f_M^3 - f_m^3},$$

and recalling (B.66)

$$a(f) = \sqrt{\frac{3}{f_M^2 + f_M f_m + f_m^2}}\, f. \tag{B.67}$$

Substituting $a(f)$ given in (B.67) in (B.64) the improvement factor m becomes

$$m = \frac{f_2^2 + f_1 f_2 + f_1^2}{f_M^2 + f_M f_m + f_m^2}, \tag{B.68}$$

which, recalling (B.62) and (B.63), may be written as

$$m = \frac{3f_c^2 + \frac{b_s^2}{4}}{f_M^2 + f_M f_m + f_m^2}. \tag{B.69}$$

If we accept the following simplifications for the upper channel:

$$f_c \gg b_s,$$
$$f_c \approx f_M,$$

and, noting that for a large number of telephone channels we always have $f_M \gg f_m$, from (B.69) we get the improvement factor m:

$$m = 3(\text{ or } 4.8 \text{ dB}).$$

Note that the improvement factor for the upper channel in the baseband resulting from the ITU-R Recommended emphasis equals 4 dB, very close to the theoretical maximum.

REFERENCES

1. Brown, J. and Glazier E., *Telecommunications*, Chapman and Hall Ltd., Edinburgh, 1969.

Appendix C
Costs

In this appendix we present typical costs for some microwave relay link components, such as parabolic reflector antennas, flat reflectors, guyed masts, towers, waveguides and radio equipment. These costs, which should be taken as merely indicative, are for equipment placed at the customers premises, in the European Union, and do not include taxes (namely, customs and value added tax). They are expressed in Euro (or approximately in American dollars) and, were valid, at end of the 1990s.

Installation and commissioning costs amount to 20 to 30 percent of indicated costs.

- Parabolic reflectors

  Cost c_a of a parabolic reflector with diameter d (in m), normal series, complete with primary feed for a single linear polarization is

$$c_a = 1000 + 75\, d^3; \tag{C.1}$$

- Flat reflectors

  Cost c_e of a flat reflector with area s (in m^2) between about 5 and about 20 m^2 may be estimated as

$$c_e = 2500 + 750(s - 5); \tag{C.2}$$

- Guyed masts

  Cost c_{te} of a guyed mast, with typical anchorages and anti-torsion device (to keep torsion below about ± 0.5 degree), is given as a function of the height h, (in m), by the following approximate formula:

$$c_{te} = 3500 + 300\,h \quad for \quad 10 \leq h \leq 40, \tag{C.3}$$

$$c_{te} = 3500 + 350\,h \quad for \quad 40 < h \leq 60; \tag{C.4}$$

- Self-standing metallic towers

  Cost c_{ta} of a self-standing metallic tower, with typical anchorages, as a function of the height h, in m, changes quite dramatically near $h = 30$ m, as shown in the following approximate formulas:

$$c_{ta} = 4000 + 600\,h \quad for \quad 10 \leq h \leq 30; \tag{C.5}$$

$$c_{ta} = 22500 + 16000\,(h - 30) \quad for \quad 30 \leq h < 80. \tag{C.6}$$

- Coaxial cable

  Cost per meter of normal co-axial cable is 4.5 for 1/2 " cable and about 15 for 7/8 " cable;

- Elliptic waveguide

  Cost, per meter, of elliptic waveguide c_{ge}, for the frequency f, in GHz, may be approximated by:

$$c_{ge} = 15 \times \left(1 + \frac{10}{f}\right); \tag{C.7}$$

- Transmitter and receiver

  Cost of the radio equipment for a digital terminal, in a $1+1$ configuration, including transmitters, receivers, modem for 8 or 34 Mbit/s, and switching gear is in the range of 32 500 and 35 000;

- Shelter and uninterruptible power supply

  Cost of a simple, non-air-conditioned, shelter including uninterruptible power supply, assuming the existence of a local mains supply, is between 50 000 and 60 000. This value does not include the cost of the land nor the cost of access roads.

Appendix D
Link Calculations

D.1 INTRODUCTION

In this appendix we present, as an example, calculations for the design of a high capacity (155 Mbit/s), 50-km-long, digital microwave radio link, which is part of the national portion of the hypothetical reference digital path, obeying the applicable ITU-R Recommendations. Terminals ensure line-of-sight with adequate first Fresnel ellipsoid clearance using antennas on top of 20-m towers, on both terminals. Rain intensity not exceeded for more than 0.01 percent of the time will be taken as 42 mm/h. Vertical polarization is assumed.

D.2 CHOICE OF FREQUENCY

ITU-R frequency plans (Appendix A) show that, in principle, this link may be established in the 4, 6, and 11 GHz bands. We will assume that radio channels are available in all these bands.

Calculations are carried out at the center frequencies of the 4, 6 and 11 GHz, respectively, 4200 MHz (ITU-R [1] Recommendation F.635-2), 6770 MHz (ITU-R [1] Recommendation F.384-5) and 11 200 MHz (ITU-R [1] Recommendation F.387-6), which should be sufficient for the required accuracy.

We will use 64-QAM modulation, with a minimum distance between adjacent channels, using the same polarization, of 40 MHz. Taking the gross bit rate of 159.5 Mbit/s the Nyquist band width is given by (5.218)

$$\begin{aligned} b_{rf} &= \frac{159.5}{\log_2(64)} \text{ (MHz)} \\ &= 26.6 \text{ (MHz)}, \end{aligned}$$

which, according to (5.219), enables an excess band factor such that

$$\begin{aligned} \beta &\leq \frac{40}{26.6} - 1 \\ &\leq 0.5. \end{aligned}$$

According to ITU-R[1] Recommendation F.1189-1, when available this link should obey the following quality criteria:

- $sesr \leq 1.6 \cdot 10^{-4}$;
- $bber \leq 1.6 \cdot 10^{-5}$;
- $esr \leq 0.0128$.

From Table 5.11 we get $ber_{ses} = 2.3 \cdot 10^{-5}$. Following the procedure described in ITU-R [1] Recommendation P.530-8 we adopt $rber = 10^{-13}$.

From Figure 5.33, or solving equations (5.210) and (5.212), we get the corresponding carrier-to-noise ratios C/N:

- $C/N_{rber} = 34.1$ dB,
- $C/N_{ber_{ses}} = 27.4$ dB.

According to Morita's law, the probability that the received signal power p is less than p_0, for the worst month, over average ground, is given by

$$\mathcal{P}(p \leq p_0) = 1.4 \cdot 10^{-8} f d^{3.5} \frac{1}{m}, \tag{D.1}$$

where $m = p_n/p_0$, p_n is the received signal power under normal propagation conditions (i.e., without fading) and f and d have the usual meaning. In a real design, actual path data should be used and fading should be estimated using (3.58) or (3.65).

Excess attenuation due to the presence of oxygen and water vapor in the atmosphere are calculated as described in Section 2.5.2:

- 0.33 dB for 4.2 GHz;
- 0.40 dB for 6.77 GHz;
- 0.64 dB for 11.2 GHz.

We will assume the following equipment data:

- transmitter
  - $P_E = 5$ dBW in the 4 GHz band,
  - $P_E = 3$ dBW in the 6 GHz band,
  - $P_E = 0$ dBW in the 11 GHz band;
- receiver
  - $Nf = 4$ dB in the 4 GHz band,
  - $Nf = 5$ dB in the 6 GHz band,
  - $Nf = 7$ dB in the 11 GHz band;
- (elliptic) waveguide
  - $A_E = A_R = 0.025$ dB/m in the 4 GHz band,
  - $A_E = A_R = 0.051$ dB/m in the 6 GHz band,
  - $A_E = A_R = 0.11$ dB/m in the 11 GHz band.

The receiver will be equipped with transverse adaptive equalizers which reduce the value of K_n in (5.242) (and the receiver signature) by a factor of 10. The calculated selective fading margin is $m_s = 2182$.

Proceeding as in Section 5.12 we calculate the minimum carrier-to-noise ratio at the demodulator input required to meet the required quality criteria, considering only fading, as:

- $C/N_{sesr} = 54.0$ dB;
- $C/N_{bber} = 58.2$ dB;
- $C/N_{esr} = 37.3$ dB.

Intermediate values in the calculations are shown in Table D.1.

Once the equipment has been chosen, the carrier-to-noise ratio at the demodulator input, referred to the receiver input C/N_0 is calculated as:

$$C/N_0 = P_E - A_E - A_R + G_E + G_R - A_0 - A_a + 204 - 10\log_{10}(b_{rf}) - Nf, \quad \text{(D.2)}$$

where we use the same symbols as in Chapters 2 and 4.

We will further assume that:

- parabolic reflectors, 3 m in diameter and a 50 percent aperture efficiency, are used as terminal antennas;
- in each terminal the feeder uses 30 m of waveguide;
- temperature 10 °C, atmospheric pressure 1013 hPa, water vapor content 7.5 gm^3;

Variable	Unit	4200 MHz	6770 MHz	11 200 MHz
m_s		1109	1256	1434
C/N_{sesr}	dB	54.0	56.9	60.8
C/N_{bber}	dB	58.2	60.6	63.2
C/N_{esr}	dB	37.3	39.4	41.6
P_E	dBW	5.0	3.0	0.0
$A_E + A_R$	dB	1.5	3.1	6.6
$G_E = G_R$	dBi	39.4	43.5	47.9
A_0	dB	138.8	143.0	147.4
A_a	dB	0.33	0.40	0.64
$10\log_{10}(b_{rf})$	dB	74.2	74.2	74.2
Nf	dB	4	5	7
C/N_0	dB	69.0	68.3	64.0
M_l	dB	10.8	7.7	0.8

Table D.1 Calculation of the margins with respect to ITU-R clauses for a 155 Mbit/s, 50-km-long, digital microwave radio link.

- area under the receiver signature: 1 MHz, for both minimum and non-minimum phase.

Link calculations are sumarized in Table D.1.

The link margin M_l is the difference between the carrier-to-noise ratio at the demodulator input and the carrier-to-noise ratio required to meet the most stringent clause (in all cases here, the background block error ratio). The link margin is positive in all cases even if it just about meets ITU-R Recommendations in the 11 GHz band.

We may conclude from Table D.1 that the 4 GHz is the preferred frequency band and if that band is adopted we may reduce the transmitter power output or the antenna diameter (and hence gain) to decrease costs. Assuming the choice and aiming at a link margin of about 3 dB, we could reduce antenna diameters (in both terminals) from 3 to 2 m.

The previous analysis must be completed with availability analysis . Rainfall induces unavailability which may prove to be a decisive factor for the higher frequency bands. In this case, where the 4 GHz band is the preferred band, rainfall has a rather small importance, as will be shown in the following.

ITU-R [1] Recommendation F.557-4 implies that the unavailability of a link with less than 280 km should be less than

$$\frac{280}{2500} \times 3 \cdot 10^{-3} \approx 3.36 \cdot 10^{-4}.$$

For vertical polarization and a rainfall intensity of 42 mm/h, not exceeded for more than 0.01 percent (or 10^{-4}) of the time, and for an unavailability due to rainfall of $3.4 \cdot 10^{-5}$, which corresponds to 10 percent of the total unavailability, we get a rain attenuation of .6 dB at 4.2 GHz. ITU-R[1] Recommendation F.557-4 defines unavailability as periods (exceeding 10 s) of severely errored seconds. Hence we must use the corresponding bit error ratio as ber_{ses}, for which we have already calculated the corresponding carrier-to-noise ratio at the demodulator input $C/N_{ber_{ses}} = 27.4$ dB. Adding to this value the atmospheric attenuation and the rain attenuation we obtain 30.3 dB and hence a rainfall margin of 38.7 dB. This confirms that the background block error ratio is the critical one and that the 4 GHz is the most favorable band to implement the link.

Finally, we should look into equipment availability to find out if reserve channels are required to meet unavailability objectives.

REFERENCES

1. ITU, *ITU-R Recommendations on CD-ROM*, UIT, Geneva, 2000.

Index

WILEY SERIES IN TELECOMMUNICATIONS AND SIGNAL PROCESSING

John G. Proakis, Editor
Northeastern University

Introduction to Digital Mobile Communications
Yoshihiko Akaiwa

Digital Telephony, 3rd Edition
John Bellamy

ADSL, VDSL, and Multicarrier Modulation
John A. C. Bingham

Biomedical Signal Processing and Signal Modeling
Eugene N. Bruce

Elements of Information Theory
Thomas M. Cover and Joy A. Thomas

Practical Data Communications, 2nd Edition
Roger L. Freeman

Radio System Design for Telecommunications, 2nd Edition
Roger L. Freeman

Telecommunication System Engineering, 3rd Edition
Roger L. Freeman

Telecommunications Transmission Handbook, 4th Edition
Roger L. Freeman

Introduction to Communications Engineering, 2nd Edition
Robert M. Gagliardi

Optical Communications, 2nd Edition
Robert M. Gagliardi and Sherman Karp

Efficient Algorithms for MPEG Video Compression
Dzung Tien Hoang and Jeffrey Scott Vitter

Active Noise Control Systems: Algorithms and DSP Implementations
Sen M. Kuo and Dennis R. Morgan

Mobile Communications Design Fundamentals, 2nd Edition
William C. Y. Lee

Expert System Applications for Telecommunications
Jay Liebowitz

Polynomial Signal Processing
V. John Mathews and Giovanni L. Sicuranza

Digital Signal Estimation
Robert J. Mammone, Editor

Digital Communication Receivers: Synchronization, Channel Estimation, and Signal Processing
Heinrich Meyr, Marc Moeneclaey, and Stefan A. Fechtel

Synchronization in Digital Communications, Volume I
Heinrich Meyr and Gerd Ascheid

Business Earth Stations for Telecommunications
Walter L. Morgan and Denis Rouffet

Wireless Information Networks
Kaveh Pahlavan and Allen H. Levesque

Satellite Communications: The First Quarter Century of Service
David W. E. Rees

Fundamentals of Telecommunication Networks
Tarek N. Saadawi, Mostafa Ammar, with Ahmed El Hakeem

Microwave Radio Links: From Theory to Design
Carlos Salema

Meteor Burst Communications: Theory and Practice
Donald L. Schilling, Editor

Digital Communication over Fading Channels: A Unified Approach to Performance Analysis
Marvin K. Simon and Mohamed-Slim Alouini

Digital Signal Processing: A Computer Science Perspective
Jonathan (Y) Stein

Vector Space Projections: A Numerical Approach to Signal and Image Processing, Neural Nets, and Optics
Henry Stark and Yongyi Yang

Signaling in Telecommunication Networks
John G. van Bosse

Telecommunication Circuit Design, 2nd Edition
Patrick D. van der Puije

Worldwide Telecommunications Guide for the Business Manager
Walter H. Vignault